T0419831

Evolutionary Biology

Pierre Pontarotti
Editor

Evolutionary Biology

Concept, Modeling, and Application

Editor
Dr. Pierre Pontarotti
UMR 6632
Université d'Aix-Marseille/CNRS
Laboratoire Evolution Biologique et Modélisation, case 19
Place Victor Hugo 3
13331 Marseille Cedex 03
France
Pierre.Pontarotti@univ-provence.fr

ISBN 978-3-642-00951-8 e-ISBN 978-3-642-00952-5
DOI: 10.1007/978-3-642-00952-5
Springer Dordrecht Heidelberg London New York

Library of Congress Control Number: 2009926007

Cover illustration: Canada-France-Hawaii Telescope & Coelum / J.-C. Cuillandre (CFHT) & G. Anselmi (Coelum)

Printed on acid-free paper

Springer is part of Springer Science+Business Media (www.springer.com)

Preface

Since 1997, scientists of different disciplines who share a deep interest in evolutionary biology concepts, knowledge and application have held the Evolutionary Biology Meeting at Marseilles in order to discuss their research, exchange ideas and start collaboration. Lately scientists interested in the application joined the group.

We start the book with the meeting report that gives a general idea of the epistemological posture of the meeting; this is followed by 22 articles.

These articles are, to my mind, a selection of the most representative talks, which have been brought up during the 12th Evolutionary Biology Meeting at Marseilles.

We decide to sort out the articles in several categories: evolutionary biology concepts and knowledge, concepts and knowledge modeling and concepts and knowledge application. The reader will note that evolutionary application transcend biology since two articles in the book show application of evolutionary biology concept to economy and astronomy announcing other field of applications.

I would like to thank all the authors of the book, the meeting participants, the sponsor-CNRS, Université de Provence, Conseil Général 13, Ville de Marseille and GDR BIM.

I also wish to thank the staff of Springer for their cooperation and more especially Andrea Schlitzberger for her considerateness.

Last but not least I wish to thank the "Association pour l'Etude de l'Evolution Biologique" and the meeting coordinator Axelle Pontarotti whose outstanding work enhanced both the quality and friendliness of the meeting.

Marseilles, France
April 2009

Pierre Pontarotti

Meeting Report: 12th Evolutionary Biology Meeting in Marseille

The 12th Evolutionary Biology meeting in Marseille (EBM) was supposed to begin on September 24, 2008, although it actually begun at a dinner table one night before its official opening. People arriving one night before the beginning of the conference got together for informal social activities. Of course it could not be different: a number of passionate researchers seated together around a table must result in a scientific debate. Daniel John Lawson, a postdoc student from the Imperial College of London, posed the central question: "Are there still any basic controversies in evolutionary biology that have not been solved?" The general agreement over the table, during red wine tasting and a delicious "French repas," was a clear and sound "No." Every single evolutionary biologist seems to be satisfied with the modern view of the Darwinian paradigm. They all agreed that it presents the key and the whole convergence point of the enormous research field in biology. Someone remembered Dobzhansky: "Nothing in biology makes sense except in the light of evolution." At that point, it seems that the evolutionary biologists around that restaurant table in the "Vieux Port" at Marseille were underestimating the richness and the deepness of the debates they would have in the following days regarding long-term controversial points in evolutionary biology.

During the very first morning session, a computational model for evolutionary change raised the debate between the strength of natural selection opposed to the application of nearly neutral Kimura-based models. Selectionism and neutralism were under debate. In the afternoon, Guy Hoelzer from the University of Nevada (USA) brought to discussion the relevance of allopatry for speciation and how the genomes of populations got incompatible after geographically specific mutations. Confirming this computational simulated scenario, a group from London showed that the presence of polymorphisms in the Hepatitis B virus is geographically restricted and certain virus strains are enclosed into some delimited regions of the planet. The following day, the relationship between kinship and fertility was also computationally modeled and it was pointed out that the highest fecundity in humans has been observed for couples related at the level of third to fourth cousins (in Iceland??). Stanislaw Cebrat and collaborators produced a model that claims to explain and prove the occurrence of sympatric speciation based on zygotic death associated to effects related a sort of intrapopulation outbreeding depression. Furthermore, the always-under-debate subject about the origin of life was discussed by researchers working in Italy, Russia and Japan. Di Mauro used physical information about the most common organic molecules found in interstellar systems to discuss a putative original living system based on the formamide molecule. Victor Ostrovskii agreed with him that life shall emerge without any great unlikely event, arising from thermodynamically favorable conditions. The point in their thought could be resumed as the following quotation of Victor Stenger in "The comprehensive cosmos": "Something came from nothing because it is more stable than nothing." Tadashi Sugawara addressed the point about the origin of cells showing how giant lipid vesicles could divide and putatively reproduce without any precise teleonomic form of information to guide their duplication. Moreover, the evolution of sex was simplistically explained and theorized using computational models to describe the behavior of preys escaping predators. Robert French from the University of Burgundy presented an interesting model to describe his "Red-Tooth Hypothesis," theorizing on which sort of combined escape strategies should be used for preys to escape being eaten. Using a computational model to describe escape behaviors in sexual and asexual populations, he tried to approach the question about how sexual reproduction could provide an advantage for preys, allowing them to survive under the attack of starving predators at the population level. Finally, the evolution of gene functions through neofunctionalization and subfunctionalization was discussed by Ashley Byun McKay from the Fairfield University (Canada). She showed examples of duplicated genes on which a single nucleotide mutation in the signal peptide would be enough to change the cellular compartment on which a given gene would be expressed and further the gene function. The debates were than shifted into the relevance of saltacionistic theories in evolution. How frequent should these single mutations produce a drastic change on gene function and organisms' behavior?

Apart from these long-term discussions in evolutionary biology, new technologies for evolutionary analysis were presented during this traditional evolutionary biology conference in France. Represented by one talk and three poster presentations, a group from Croatia headed by Tomislav Domazet-Lozo was probably the bigger research group coming to the meeting. Their interesting work was based on a technique they called phylostratigraphy, consisting in (1) the identification of species-specific genes and (2) verifying the expression of these genes in different tissues on these organisms. This way, this group could identify the relevance of new-born genes in tissues and address questions about tissue and organ modification along time in a number of species. Other new technology presented during the EBM was performed by a consortium between France and the United States. They described a conceptual model (ontology) allowing researchers to represent both morphological and molecular characters observed in organisms and genes. They claimed that the description of the most basic evolutionary biology concepts – linked together through semantic relationships – will further allow automatic algorithmic reasoning, helping in the analysis of the enormous amount of comparative and evolutionary biology data available for biology in the petabyte era.

Regarding the evolution of gene families, a group in the Smurfit Institute of Genetics in Ireland showed that immune-related genes are co-localized and putatively co-expressed in the genomes of vertebrates. Selection constraints observed using the number of synonym per non-synonym substitutions were studied by a number of groups around the world: in Japan they were studied in fish galactins; in Ireland, they were related to the three-dimensional structure of proteins. The Ks/Ka was also used in Germany to study the Asr gene family in tomato and a French-American group related the substitution taxes with virulence factors in nematodes.

Changing the subject to the study of fossils, new researches made in USA showed that the development of ^{14}C datation techniques in the last decade have helped to reveal that a great part of molecular changes in the human genome have happened during the Holocene period, encompassing the last 10,000 years of Earth's history. An interesting paleobotanics research made in China has shown evidence of a putative new group of fossil Angiosperms. Another plant research made in Belgium and Germany took profit of both morphological and molecular data to study the evolution of basal perianthless Angiosperms species from the family Piperales. Researchers in France and Italy also reported that dense populations of plants usually produce more viable seeds, a research that could be linked to the previously commented human population-genetics studies highlighting the higher fertility between individuals presenting common ancestors not that far in time. A group in Israel showed evidence for alotetraploydy events happening during the molecular evolution of carps, maybe the single evidence for this sort of event – so common in plants– happening in a vertebrate species.

Brian Kennedy presented an interesting aging study made in the USA. He showed evidences that aging does not seem to be molecular programmed once selective pressures decline with age. Together with his collaborators, they showed that caloric restriction made late in the life of flies, worms and yeasts might help these organisms to extend their average lifespan.

Moreover, a number of open-minded discussions were centered in new-born interdisciplinary fields. The application of cladistic methods to reconstruct the history of galaxies was approached by Didier Fraix-Burnet from the Laboratoire d'Astrophysique de Grenoble. His astrocladistics approach has been used to identify and map a number of putatively evolving characters in galaxies, allowing their classification into phylogenetic trees. As expected, due the tradition of scientific academy, he complained about the difficulties to publish these new and interesting ideas in respected journals. "Galaxies are not living organisms," argued the referees when refusing the papers from his group. Someone working in the origin of life from the audience pointed out: "if galaxies modify along time, if they can be born and die, if they derive from a common ancestor since the Big Bang and if they may interact with others ... why they should not be considered alive?" The discussion about "what is life?" is still a controversy among scientists. By the way, it is noteworthy to remember that the discussion about the main characteristics of a living system probably had one of its highest points when the physicist Schrodinger theorized about a putative "aperiodic cristal" to store genetic information early in 1944. Interdisciplinary approaches were also discussed from the point of view of economics. Michael Turk came from the Fitchburg State College (USA) to speak to the audience about how most of the theories trying to scientifize economy were initially based on models inherited from physics. Nevertheless, a number of other characteristics would make this discipline closer to biology, such as: (1) emphasis on complexity; (2) its dynamic characteristic; (3) the focus on the cumulative change; (4) the competitivity of market; and so on. He introduced a number of philosophers – such as Brian Arthur, Paul David and Paul Krugman – that have been influenced by biologists like Jacques Monod and brought the ideas of "Change and Necessity" to the economics research field. At last, other interdisciplinary topics were discussed during the conference concerning the relationship between evolutionary biology and epistemology. Strict questions coming from the audience generated these off-topic discussions, since some researchers were arguing that a presented given theory had already been "proved false" since a long time ago, making the study presented unfruitful and pointless. The discussions were centered in question whether science actually gives or not some sort of definitive answer to the study of nature. Newtonian physics has been remembered as the most undoubtedly correct of all scientific theories and it has been proven – at least partially – false. Shall scientists dogmatize their enterprise? The general agreement was that open discussions and empirical-based facts are actually what make science progress and dogmatism should not be kept in the science domain; mainly in the field of evolutionary biology that has suffered so much at the hand of religious dogmas.

Last but not least, the variety of different computational, biological and philosophical approaches presented during this meeting highlighted that Darwinian-based thought keeps being at the center of evolutionary biology and encompasses the mainstream linking point of all biological and medical research all over the world.

Francesco Prosdocimi and Pierre Pontarotti

Contents

Contributors

D.A. Afonnikov
Institute of Cytology and Genetics, Siberian Department, Russian Academy of Sciences, Novosibirsk

Juan Ausió
Department of Biochemistry and Microbiology, University of Victoria, V8W 3P6 Victoria, Canada

Pierre U. Blier
Département de Biologie, Université du Québec à Rimouski, 300 Allée des Ursulines, Rimouski, Québec, G5L 3A1, Canada

Branko Borštnik
National Institute of Chemistry, Hajdrihova 19, Ljubljana, Slovenia
branko@hp10.ki.si

Sophie Breton
Department of Biological Sciences, Kent State University, Kent, OH 44242, USA

B. Carroll
Fairfield University, Dept of Biology, 1073 North Benson Road, Fairfield, CT 06824

Brandon Chisham
Department of Computer Science, New Mexico State University, P.O. Box 30001, MSC CS Las Cruces, NM 88003, USA

Fabiana Ciciriello
Dipartimento di Genetica e Biologia Molecolare, Università "La Sapienza"

Joseph D. Coolon
Ecological Genomics Institute, Kansas State University, Manhattan, KS, USA
Division of Biology, Kansas State University, Manhattan, KS, USA
Current address: Department of Ecology and Evolutionary Biology, University of Michigan, Ann Arbor, MI, USA

Giovano Costanzo
Istituto di Biologia e Patologia Molecolari, CNR, P.le Aldo Moro, 5, Rome, 00185, Italy

Etienne G.J. Danchin
INRA, UMR 1301, 400 route des Chappes, F-06903 Sophia-Antipolis, France
CNRS, UMR 6243, 400 route des Chappes, F-06903 Sophia-Antipolis, France
UNSA, UMR 1301, 400 route des Chappes, F-06903 Sophia-Antipolis, France
etienne.danchin@sophia.inra.fr

Deanna Dryhurst
Departamento de Biología Celular y Molecular, Universidade da Coruña, E15071 A Coruña, Spain
jeirin@udc.es

R. Duggan
Fairfield University, Dept of Biology, 1073 North Benson Road, Fairfield CT, 06824

José M. Eirín-López
Departamento de Biología Celular y Molecular, Universidade da Coruña, E15071 A Coruña, Spain
jeirin@udc.es

Jean-Marie Exbrayat
Université de Lyon; Laboratoire de Biologie Générale, Université Catholique de Lyon; Reproduction et Développement Comparés, EPHE, 25 rue du Plat, 69288 Lyon Cedex 02. France
jmexbrayat@univ-catholyon.fr

Mario A. Fares
Evolutionary Genetics and Bioinformatics Laboratory, Department of Genetics, School of Genetics and Microbiology, University of Dublin, Trinity College, Dublin, Ireland
faresm@tcd.ie

Didier Fraix-Burnet
Laboratoire d'Astrophysique de Grenoble, UMR5571 CNRS/Université Joseph Fourier, BP 53, F-38041 Grenoble cedex 9, France
didier.fraix-burnet@obs.ujf-grenoble.fr

R. Geeta
Stony Brook University, Dept. of Ecology and Evolution, 650 Life Sciences Building, Stony Brook NY 11794

Rodrigo González-Romero
Departamento de Biología Celular y Molecular, Universidade da Coruña, E15071 A Coruña, Spain
jeirin@udc.es

Michael A. Herman
Ecological Genomics Institute, Kansas State University, Manhattan, KS, USA
Division of Biology, Kansas State University, Manhattan, KS, USA
mherman@ksu.edu

Walter R. Hoeh
Département de Biologie, Université du Québec à Rimouski, 300 Allée des Ursulines, Rimouski, Québec, G5L 3A1, Canada
Pierre_blier@uqar.qc.ca

Kenneth L. Jones
Ecological Genomics Institute, Kansas State University, Manhattan, KS, USA
Division of Biology, Kansas State University, Manhattan, KS, USA
Current address: Department of Environmental Health Science, University of Georgia, Athens, GA, USA

Magnus Karlsson
Department of Forest Mycology and Pathology, Swedish University of Agricultural Sciences, P.O. 7026, SE-75007, Uppsala, Sweden
Magnus.Karlsson@mykopat.slu.se

Yoshinori Kawabe
College of Life Sciences, University of Dundee, UK
Elisa Alvarez-Curto
College of Life Sciences, University of Dundee, UK

Brian K. Kennedy
Biochemistry, University of Washington, Seattle, WA 98195, USA
bkenn@u.washington.edu

Thomas B.L. Kirkwood
Centre for Integrated Systems Biology of Ageing and Nutrition, Institute for Ageing and Health, Newcastle University, Newcastle upon Tyne, UK

Ernesto Di Mauro
Fondazione "Istituto Pasteur-Fondazione Cenci-Bolognetti" c/o Dipartimento

di Genetica e Biologia Molecolare, Università "La Sapienza" di Roma, P.le Aldo Moro, 5, Rome 00185, Italy.
ernesto.dimauro@uniroma1.it

Takashi Makino
Smurfit Institute of Genetics, University of Dublin, Trinity College, Dublin 2, Ireland
makinot@tcd.ie

S.J. McKay
Cold Spring Harbor Laboratory, 1 Bungtown Rd, Cold Spring Harbor, NY, 11724

S.A. Byun McKay
Fairfield University, Dept of Biology, 1073 North Benson Road, Fairfield CT, 06824
sbyun@mail.fairfield.edu

Aoife McLysaght
Smurfit Institute of Genetics, University of Dublin, Trinity College, Dublin 2, Ireland

Francisco Prosdocimi
Institut de Génétique et de Biologie Moléculaire et Cellulaire (IGBMC), Department of Structural Biology and Genomics, Strasbourg, France. 1 rue Laurent Fries/ BP 10142/67404 Illkirch Cedex, France
fpros@igbmc.fr

Josefina Méndez
Departamento de Biología Celular y Molecular, Universidade da Coruña, E15071 A Coruña, Spain
jeirin@udc.es

Borut Oblak
National Institute of Chemistry, Hajdrihova 19, Ljubljana, Slovenia

M.P. Perepelkina
Institute of Cytology and Genetics, Siberian Department, Russian Academy of Sciences, Novosibirsk

Laetitia Perfus-Barbeoch
INRA, UMR 1301, 400 route des Chappes, F-06903 Sophia-Antipolis, France
CNRS, UMR 6243, 400 route des Chappes, F-06903 Sophia-Antipolis, France
UNSA, UMR 1301, 400 route des Chappes, F-06903 Sophia-Antipolis, France

Samanta Pino
Dipartimento di Genetica e Biologia Molecolare, Università "La Sapienza"
Enrico Pontelli
Department of Computer Science, New Mexico State University, P.O. Box 30001, MSC CS Las Cruces, NM 88003, USA

Danilo Pumpernik
National Institute of Chemistry, Hajdrihova 19, Ljubljana, Slovenia

Michel Raquet
Université de Lyon; Laboratoire de Biologie Générale, Université Catholique de Lyon; Reproduction et Développement Comparés, EPHE, 25 rue du Plat, 69288 Lyon Cedex 02. France
jmexbrayat@univ-catholyon.fr

Armin Rashidi
Centre for Integrated Systems Biology of Ageing and Nutrition, Institute for Ageing and Health, Newcastle University, Newcastle upon Tyne, UK Institute for Ageing and Health, Newcastle University, Campus for Ageing and Vitality, Newcastle upon Tyne NE4 5PL, UK
armin.rashidi@newcastle.ac.uk

Allyson V. Ritchie
College of Life Sciences, University of Dundee, UK

Marc Robinson-Rechavi
Department of Ecology and Evolution, Biophore, Lausanne University, CH-1015 Lausanne, Switzerland. Swiss Institute of Bioinformatics, CH-1015 Lausanne, Switzerland
marc.robinson-rechavi@unil.ch

Pauline Schaap
MSI/WTB/JBC complex, Dow Street, Dundee DD1 5EH, UK Tel.: 44 1382 388078; fax: 44 1382 245286;
p.schaap@dundee.ac.uk

Daryl P. Shanley
Centre for Integrated Systems Biology of Ageing and Nutrition, Institute for Ageing and Health, Newcastle University, Newcastle upon Tyne, UK

T. Sugawara
The University of Tokyo, Japan
tadpole8jp@yahoo.co.jp

Donald T. Stewart
Department of Biology, Acadia University, Wolfville, Nova Scotia, B4P 2R6, Canada
breton.sophie@gmail.com; randy.hoeh@gmail.com

Jan Stenlid
Department of Forest Mycology and Pathology, Swedish University of Agricultural Sciences, P.O. 7026, SE-75007, Uppsala, Sweden

Arlin Stoltzfus
Center for Advanced Research in Biotechnology, University of Maryland Biotechnology Institute, 9600 Gudelsky Drive, Rockville, MD 20850, USA

Romain A. Studer
Department of Ecology and Evolution, Biophore, Lausanne University, CH-1015 Lausanne, Switzerland; Swiss Institute of Bioinformatics, CH-1015 Lausanne, Switzerland

George L. Sutphin
Departments of Pathology, University of Washington, Seattle, WA 98195, USA
The Molecular and Cellular Biology Program, University of Washington, Seattle, WA 98195, USA
lothos@u.washington.edu

Julie D. Thompson
Institut de Génétique et de Biologie Moléculaire et Cellulaire (IGBMC), Department of Structural Biology and Genomics, Strasbourg, France. 1 rue Laurent Fries/BP 10142/67404 Illkirch Cedex, France

Marc Thuillard
La Colline, Creuze 9, CH-2072 St-Blaise, Switzerland
Thuillweb@hotmail.com

Christina Toft
Evolutionary Genetics and Bioinformatics Laboratory, Department of Genetics, School of Genetics and Microbiology, University of Dublin, Trinity College, Dublin, Ireland
faresm@tcd.ie

Timothy Todd
Ecological Genomics Institute, Kansas State University, Manhattan, KS, USA
Department of Plant Pathology, Kansas State University, Manhattan, KS, USA

Michael H. Turk
Professor of Economics, Fitchburg State College, Fitchburg, MA, USA
mturk@fsc.edu

Xin Wang
Nanjing Institute of Geology and Palaeontology, 39 Beijing Dong Road, Nanjing 210008, China, Fairylake Botanical Garden, 160 Xianhu Road, Shenzhen 518004, China
brandonhuijunwang@gmail.com

L.P. Zakharenko
Institute of Cytology and Genetics, Siberian Department, Russian Academy of Sciences, Novosibirsk Novosibirsk State University, Novosibirsk, Russia
zakharlp@bionet.nsc.ru

Part I
Concept and Knowledge

Chapter 1
Spontaneous Generation Revisited at the Molecular Level

Fabiana Ciciriello, Giovanna Costanzo, Samanta Pino, and Ernesto Di Mauro

Abstract A homogeneous chemical frame is described allowing one-pot syntheses from the one-carbon compound formamide NH_2CHO to the whole set of nucleic bases needed as precursors of nucleic acids, as we know them. Formamide also catalyzes the formation of acyclonucleosides and the phosphorylation of nucleosides to nucleotides.

The conditions are described in which the survival of these prebiotically plausible nucleotides is favored. The scenario is simple: the polymeric forms are thermodynamically favored over the monomeric ones. The consequences of this very property are relevant: the formation and the accumulation of (pre)genetic information.

1.1 Introduction

Asking: "where does life come from?" is the wrong way of putting the question. A possibly more correct (and empirically useful) formulation of the problem is: "how did life originate?" Not lessening the relevance of the origin of proteins and of the membrane components with all their cortege of fatty acids, the key components of the living structures as we know them are the genetic information-bearing molecules: RNA and DNA.

F. Ciciriello and S. Pino
Dipartimento di Genetica e Biologia Molecolare Università "La Sapienza" P.le Aldo Moro, 5, Rome 00185, Italy
G. Costanzo
Istituto di Biologia e Patologia Molecolari, CNR, P.le Aldo Moro, 5, Rome 00185, Italy
E. Di Mauro
Fondazione "Istituto Pasteur-Fondazione Cenci-Bolognetti" c/o Dipartimento di Genetica e Biologia Molecolare, Universitá "La Sapienza" di Roma, P.le Aldo Moro, 5, Rome 00185, Italy
e-mail: ernesto.dimauro@uniroma1.it

P. Pontarotti (ed.), *Evolutionary Biology: Concept, Modeling, and Application*,
DOI: 10.1007/978-3-642-00952-5_1,

Of the two categories in which the field is divided: *Metabolism-first* and *Genetics-first*, we are partisans of the latter. The reason is an Occam's Razor logics approach: without a robust system for coding and perpetuating the selection of the energy-and-matter transforming physical–chemical systems involved, the coupling of anabolic and catabolic reactions would only result in a futile series of events. In the absence of a coding apparatus, metabolic activities would not eventually lead to the ur-genetic systems capable of Darwinian selection, commonly considered to be the initial steps of evolution.

This is not meant to say that metabolism and genetics did not come to an early overlap. However, in the absence of a sufficiently firm knowledge of both categories of events, general integrative scenarios cannot at the moment be fruitfully developed further.

We have explored the possibility that a homogeneous and comprehensive chemical frame existed allowing the evolution from the simplest reactive one-carbon compound to complex informational polymers. We started focusing on one defined aspect of HCN chemistry.

1.2 From HCN to Nucleotides

1.2.1 From Formamide to Nucleic Bases

HCN is largely accepted to be the precursor molecule of nucleic bases. The major ground-breaking finding is due to Oró who reported in the 1960s the synthesis of adenosine from HCN (Oró and Kimball 1960, 1961; Oró 1961a, b). In the following decades HCN chemistry was thoroughly explored. For a detailed review see Orgel (2004), for a general overview see Delaye and Lazcano (2005). In spite of the large number of studies reporting the syntheses of nucleic bases from HCN and HCN-derivatives, the progress of the field has been limited. The low pace was essentially due to the fact that a single process (or a small number of coherent reactions) allowing the synthesis of all the basic constituents was not identified. It is all too logic to accept that in order to self-assemble an RNA molecule, all the necessary components must have been present in the same pristine test tube. Limiting for the moment the quest to nucleic bases (and according to known HCN chemistry), in the "warm little pond" first imagined by Darwin (1888) HCN alone could not have provided all the necessary building blocks. Even admitting that some of the present-day bases could have been substituted by their analogs or bioisosters (i.e., isocytosine or hypoxanthine for guanine) the HCN-alone chemistry was clearly not sufficient.

Again, an Occam's Razor logics suggested a possible solution starting from two simple observations: HCN is the most abundant three-atoms organic compound in circum- and interstellar medium and H_2O is the most abundant three-atoms inorganic one (Millar 2004; www.Astrochemistry.net). The combination of HCN and

H_2O affords formamide NH_2COH. Reasoning in terms of probability and chemical potentialities, formamide appeared to be the obvious choice.

We have explored the capacity of formamide as a precursor of nucleic bases. In the course of several years we have analyzed the products obtained by heating formamide at temperatures between 100°C and 160°C in the presence of a comprehensive series of catalysts. The synthetic capacity of formamide was studied in the presence of:

Silica, alumina, kaolin, zeolite, $CaCO_3$ (Saladino et al. 2001)

TiO_2 (Saladino et al. 2003)

Montmorillonite clays (KP-10, K-30, KSF, Al-PiLC) (Saladino et al. 2004)

A set of cosmic dust analogs, from Fayalite to Olivine to Forsterite encompassing compounds with intermediate chemical composition (Saladino et al. 2005)

Phosphate minerals: Augelite $Al_2PO_4(H_2O)_3$, Wavellite $Al_3(OH)_3(PO_4)_2(H_2O)_5$, Hureaulite $Mn^{2+}{}_5(PO_3\{OH\})_2(PO_4)_2(H_2O)_4$, Vivianite $Fe^{2+}{}_3(PO_4)_2(H_2O)_8$, Ludlamite $Fe^{2+}{}_3$ $(PO_4)_2(H_2O)_4$, Libethenite $Cu^{2+}{}_2(PO_4)(OH)$, Lazulite Mg[Al $(PO_4)(OH)]_2$, Vauxite $Fe^{2+}Al_2(PO_4)_2(OH)_2(H_2O)_6$, Childrenite $Mn^{2+}[Al(PO_4)$ $(OH)_2(H_2O)]$, Turquoise $Cu^{2+}Al_6(PO_4)_4(OH)_8(H_2O)_4$, Pyromorphite $Pb_5(PO_4)_3Cl$ (Saladino et al. 2006b).

Iron sulfur and iron copper minerals: Pyrrhotine ($Fe_{(1-x)}S$), Pyrite (FeS_2), Chalcopyrite ($FeCuS_2$), Bornite ($FeCu_5S_4$), Tetrahedrite [(Fe,Cu,Sb)S] and Covellite (CuS) as representative of iron sulfur and iron–copper sulfur minerals differing for elemental composition (Saladino et al. 2008).

The analysis of zirconium and borate minerals is under way.

In each case a large number of nucleic bases was observed, afforded in various yields and in various qualitative and quantitative combinations. The synthesis was observed of purine, 2-aminopurine, adenine, hypoxanthine, N9-formylpurine, N9,N6-diformyl adenine, cytosine, isocytosine, hydroxypirimidine, 2(1*H*)-and 4(3*H*)-pyrimidinone, uracil, thymine and 5-hydroxymethyl-uracil. In numerous instances compounds were also observed that are related to present-day nucleic acids biosynthetic or degradative pathways (as various imidazoles and parabanic acid) or that are known to react with aminoacids and nucleotides, as carbodiimide. The chemical rational for the various syntheses and specific quantitative aspects were reported and discussed in detail (Saladino et al. 2007; Costanzo et al. 2007a, b; and reviews quoted therein).

The general message derived from this comprehensive analysis of the formamide synthetic potentialities is that whatever catalyst is provided, a panel of different nucleic components is obtained in conditions that allow their accumulation.

1.2.2 *From Nucleic Bases to Acyclo-Nucleosides*

Of particular interest is the production of acyclonucleosides, observed as products of the photoreaction of formamide with TiO_2. The formation of nucleosides by linking preformed ribose and nucleic bases is so difficult as to be considered prohibitive in prebiotic terms (Joyce 1989; Zubay and Mui 2001). The question of how ribose and purines were linked together has been only partially solved for inosine (Fuller et al. 1972) but not for the other nucleosides. The syntheses of purine acyclonucleosides potentially alleviate the problem of the abiotic formation of nucleosides. In the presence of TiO_2 formamide condenses to afford the acyclonucleosides N^9-formylpurine and N^9,N^6-diformyladenine (Saladino et al. 2003; see also Saladino et al. 2007 for a detailed description of the chemical aspects of this synthesis). The possibility that RNA might have evolved from acyclonucleosides-containing polymers has been discussed (Tohidi and Orgel 1989; Kozlov et al. 1999).

1.2.3 From Nucleosides to Nucleotides

Polymers will not spontaneously form in an aqueous solution from their monomers because of the standard-state Gibbs free energy change ($\Delta G^{\circ\prime}$), as critically reviewed (van Holde 1980). Thermodynamic considerations impose that the formation of phosphodiester linkages will be spontaneous only under highly dehydrating conditions. In the polymerization process of nucleic acids extant organisms activate the monomers by converting them to phosphorylated derivatives and then utilize the favorable free energy of phosphate hydrolysis to drive the reaction. The prebiotic alternative is that polymerization could have initially occurred in a nonaqueous environment. Thus, life did not arise in water, or (pre)genetic polymers formed from some sort of activated monomers.

Nonenzymatic synthesis of RNA chains was achieved starting from preactivated nucleotides. The use of the phosphoramidate forms (Lohrmann 1977; Ferris and Ertem 1993; Kawamura and Ferris 1994; Kanavarioti et al. 2001; Monnard et al. 2003; Huang and Ferris 2003; Ferris et al. 2004; Mansy et al. 2008), usually phosphorimidazolides (Lohrmann 1977), shows that the accumulation of polymers is possible once suitable activated monomers are available.

Long homo- and hetero-oligomers were obtained with phosphorimidazolides using as catalyst Pb^{++} in eutectic solution (Kanavarioti et al. 2001; Monnard et al. 2003) or Montmorillonite (Ferris and Ertem 1993; Huang and Ferris 2003; Ferris et al. 2004). Earlier studies are reviewed in Orgel (2004). However, these compounds are deemed as non-prebiotic (Orgel 1998, 2004) due to the complex chemistry involved in their formation.

The concept "prebiotic" is somewhat flexible. Nevertheless, in order to be considered as such, a compound or a process should satisfy conditions of simplicity of preparation, abundance, stability and efficiency. In general terms the simpler a process, the larger is its prebiotic interest. Phosphoramidates do not satisfy this condition.

May phosphorylation of nucleosides occur in a robust and simple chemical frame? The answer is positive. In the presence of a source of phosphate and of formamide or, alternatively, of water, nucleosides are actively phosphorylated in every possible position.

1.2.4 Formamide

Adenosine and cytidine were analyzed as model systems of abiotic phosphorylation of purines and pyrimidines. In both instances the sugar moiety is phosphorylated—by simply heating in formamide—in the 2′, or 3′, or 5′ positions or gives rise to the 2′,3′- or 3′,5′-cyclic forms (Costanzo et al. 2007a, b).

Most of the phosphorus in early Earth would have been in the form of water-insoluble minerals like Apatites. Therefore, the origin of the water-soluble (poly) phosphates required for prebiotic evolution has long been a mystery, as discussed by Schwartz (1997). It was shown that volcanic activity produces water-soluble phosphates through partial hydrolysis of P_4O_{10}, providing an at least partial solution to their origin. The phosphorylation of nucleosides to nucleotides in formamide was observed to actively occur using as phosphate donor soluble phosphate salts like KH_2PO_4, or nucleotides (i.e., 5′AMP as donor to cytidine, or 5′CMP as donor to adenosine), or one of a number of phosphate minerals. The minerals tested were: Herderite $Ca[BePO_4F]$, Hureaulite $Mn^{2+}{}_5(PO_3(OH)_2(PO_4)_2(H_2O)_4$, Libethenite $Cu^{2+}{}_2(PO_4)(OH)$, Pyromorphite $Pb_5(PO_4)_3Cl$, Turquoise $Cu^{2+}Al_6(PO_4)_4(OH)_8(H_2O)_4$, Fluorapatite $Ca_5(PO_4)_3F$, Hydroxylapatite $Ca_5(PO_4)_3OH$, Vivianite $Fe^{2+}{}_3(PO_4)_2(H_2O)_8$, Cornetite $Cu^{2+}{}_3(PO_4)(OH)_3$, Pseudomalachite $Cu^{2+}{}_5(PO_4)_2(OH)_4$, Reichenbachite $Cu^{2+}{}_5(PO_4)_2(OH)_4$, Ludjibaite $Cu^{2+}{}_5(PO_4)_2(OH)_4$. The results are detailed in Costanzo et al. (2007a, b). The largely more active minerals were found to be Reichenbachite, Ludjibaite, Libethenite and Cornetite, all of which are copper phosphates. Thus, a simple and robust phosphorylation process could have been possible in the same chemical frame and in the same physical–chemical conditions (formamide and heat) in which nucleic bases could have formed and evolved to their nucleoside forms.

1.2.5 Water

Nucleoside phosphorylation also occurs in water, in conditions similar to those described (Costanzo et al. 2007a, b) for formamide. In water phosphorylation is slower than in formamide (the plateau level being reached in water, in the adenosine model system analyzed, after 100 h relative to the 25 h needed for formamide). The fraction of the phosphorylated nucleoside molecules is also lower (7% relative to 25% in formamide) and only occurs in the presence of high concentrations of phosphate donor molecules (1M KH_2PO_4 was observed to be most the efficient concentration). The degradation of nucleotides back to nucleosides and nucleic bases is rapid, the half lives of the various phosphorylated forms being encompassed between 20 and 50 h. Nevertheless, these results show that phosphorylation occurs in a simple solvent (water), in acceptable temperature conditions (90°C) and at moderate and non-fastidious pH values (between 4.0 and 8.0).

This set of observations (Costanzo et al. 2008) points to the prebiotic possibility of obtaining activated forms of nucleosides by simply treating them in the presence of a concentrate source of a soluble phosphate. A hot "drying lagoon" scenario (for instance: a water-formamide pool formed in an ore rich in phosphate minerals) is not too difficult to imagine. The nucleoside adenosine, analyzed as a model system, is phosphorylated in every position: 2′, 3′, and 5′, and (as observed in the case of formamide) gives rise to the 2′,3′- or 3′,5′-cyclic forms. Both in formamide and in water the cyclic forms are more stable than the open forms.

1.3 Stability

The formation, accumulation and evolution of (pre)genetic information are intimately linked to the problem of the stability of the molecules involved in the process. The consideration is thermodynamically straight: molecules like nucleic bases, nucleosides and their activated derivatives once formed are bound to become part of a futile cycle of syntheses/degradations unless a mechanism is installed allowing their protection and/or increased stability.

A top-down approach tells that extant nucleic macromolecules are protected from external attacks by their interaction with proteins organized in various kinds of chromatins and in capsular structures and by their confinement into a specific and highly evolved environment contained in micellar particles. In order to reach the complex situation that allowed to kick-start the extant forms of life, the convergence was thus necessarily needed of (i) the formation of nucleic polymers, (ii) the evolution of their genetic capacity and (iii) the protein- and membrane-based protective structures.

This endeavor might well have occurred following a typical step-by-step Darwinian evolutionary path. Nevertheless, a property of nucleic acids must exist that *a priori* allowed their very survival as macromolecules. Thus the basic question is: starting from a pool of activated precursors, do intrinsic properties exist that favor the formation of nucleic polymers relative to their nonpolymerized state?

As shown below, the answer is positive. The immediately following question is: in which conditions?

1.3.1 Stability of the Relevant Bonds

In the origin of the informational polymers, their survival as macromolecules is even more problematic than the polymerization process itself. Which physical–chemical conditions provide kinetic and/or thermodynamic advantages over the monomer to the otherwise intrinsically more unstable polymeric forms?

We have measured the stabilities of the bonds that are critical for the half-life of ribonucleotides, namely the β-glycosidic and the 3′- and 5′-phosphoester bonds (Saladino et al. 2006a). Stabilities were measured in a wide range of temperatures and of water/formamide ratios.

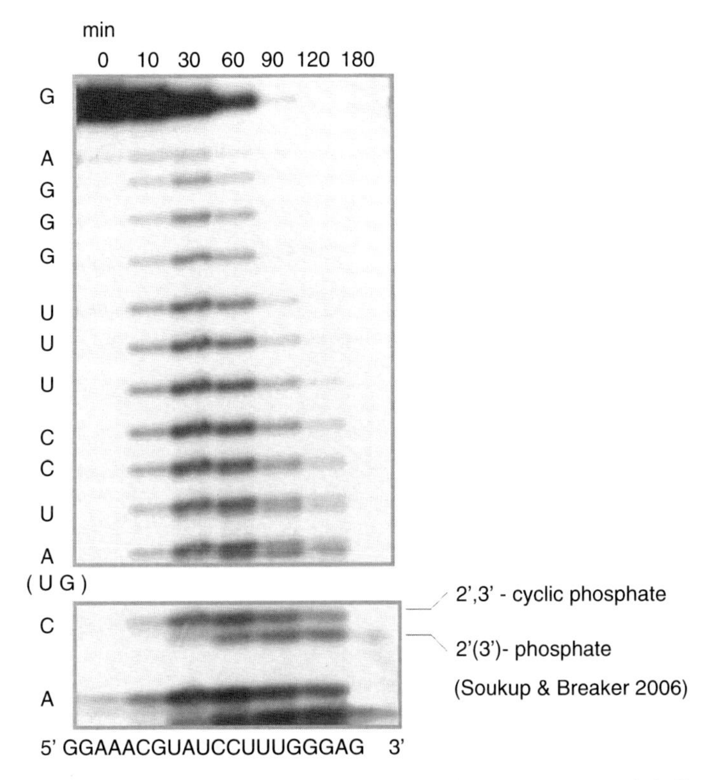

Fig. 1.1 Kinetics of RNA degradation in H_2O. The RNA oligo whose sequence is indicated at the bottom of the figure was P32 labeled at the 5′ extremity, treated in water (90°C, pH 6.2) for the time lapses indicated on top, analyzed by gel electrophoresis. In low molecular-size fragments two forms are evident: the slow migration form is the product of the first step of hydrolysis bearing a 2′,3′-cyclic phosphate, while the faster migration form results from the opening of the ring (as indicated on the right side of the panel and as reviewed in Soukup and Breaker [1999]). For details of the analysis of the determination of the half-life of the 3′ phosphoester bond see Saladino et al. (2006a) and Ciciriello et al. (2008)

The analysis of the stability of the bonds in the various types of precursor monomers was performed following by HPLC the transformation of the nucleoside to its base and sugar components, thus measuring the half-life of the β-glycosidic bond in a large set of physical–chemical conditions. This type of analysis was also performed for the relevant phosphoester bonds in both the ribo and the deoxyribo precursors.

The half-life of the key 3′-phosphoester bond in the RNA polymer was determined in the same environments following the hydrolytic degradation of an RNA oligonucleotide in water, as shown in Fig. 1.1

As an overall result, the comparison of the bond stabilities in the monomeric versus the polymeric forms of the ribo molecules revealed that physical–chemical conditions can be determined in which the polymerized forms are thermodynamically favored. Such direct comparison (Saladino et al. 2006a) was not previously performed and provided information immediately relevant to the monomer versus polymer bias.

When these physical–chemical conditions were compared with those causing a similar behavior in 2′-deoxyribo nucleosides, nucleotides and oligonucleotides (Saladino et al. 2005) they were shown to profoundly differ. From the point of view of the differential monomer versus polymer stability, the comparison between RNA and DNA is actually quite informative. Namely: the stability of the phosphoester bonds in the deoxy system is higher in the polymer relative to the monomer only for the 5′ phosphoester bond and in no condition for the 3′ phosphoester one. Higher stability is observed only at high temperatures (above 80°C) and in solutions containing high formamide concentrations (above 66% formamide in water) (Saladino et al. 2005). Even in these conditions the 5′ phosphoester bond is only a few fold more stable. Thus, polymerization is not the mechanism that provided deoxynucleotides with the means towards stabilization and evolution.

RNA has an opposite behavior. The 3′ phosphoester bond (the key site for RNA molecular fragility) is more stable in the polymer than in the monomer, both in pure water and at high formamide concentrations at temperatures between 60°C and 90°C (Saladino et al. 2006a). Thus, two separate thermodynamic niches (high temperature-water and high temperature-high formamide) exist in which RNA polymerization is favored. Focusing on RNA (and not on DNA) and on water (and not formamide) as a solvent we have explored the conditions in which spontaneous polymerization could have been favored.

1.3.2 pH and Sequence Context

A narrow pH range was identified, centered around pH 5 and 6, in which RNA sequences are markedly more stable than at lower or higher values. It was somehow unexpectedly observed (Ciciriello et al. 2008) that complex sequences resist degradation more than monotonous ones, thus potentially favoring the survival and the evolution of sequence-based genetic information.

Degradation of RNA in water occurs by trans-esterification of phosphodiester bonds, an intensively studied and well described process (Perreault and Anslyn 1997; Soukup and Breaker 1999a, b).

The cleavage of the phosphoester chain normally requires participation of the 2′-OH group as an internal nucleophile (Soukup and Breaker 1999a) in two "nucleophilic cleavage" events: the trans-esterification and hydrolysis reactions. During trans-esterification, the 2′-OH nucleophile attacks the tetrahedral phosphorus to afford a 2′,3′-cyclic monophosphate. This species is then hydrolyzed to a mixture of 3′- and 2′-phosphate monoesters. Both steps are catalyzed by protons, hydroxide, nitrogen derivatives and metal ions. The degradation profile characteristically yields a double-banded profile, due to a first cleavage of the 5′ phosphodiester bond leaving a 2′,3′-cyclic phosphate extremity (Soukup and Breaker

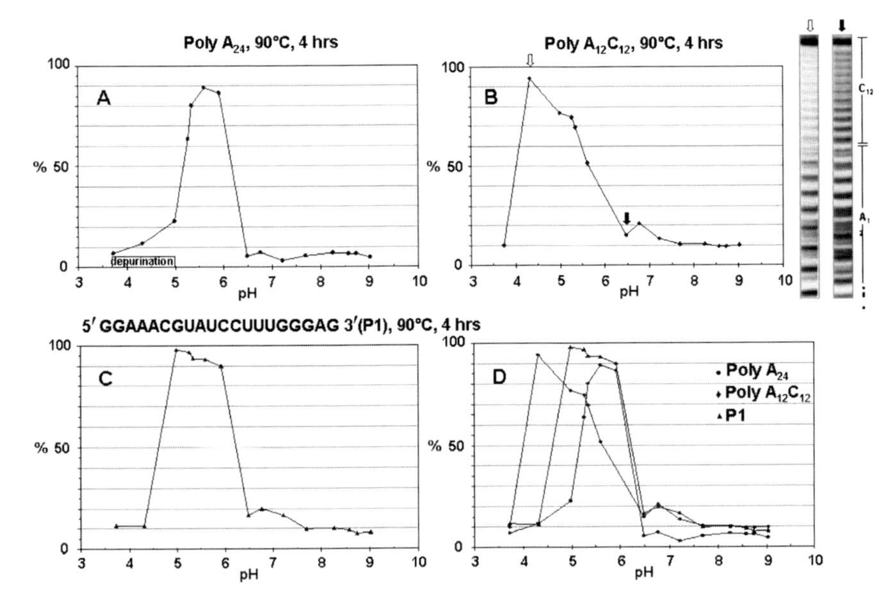

Fig. 1.2 Stability of RNA as a function of pH. Three different RNA oligomers were studied: $PolyA_{24}$ (panel A), $PolyA_{12}C_{12}$ ((Panel B), a mixed sequence 20mer dubbed P1 (panel C), as indicated on top of each panel. The terminally labeled RNAs were treated in buffered water at the indicated pH values (abscissa) for 4 h at 90°C. Each data point shows the percentage of full-length molecules (ordinate) remaining after the treatment. In B, two gel profiles are shown resulting from treatment at the pH values indicated by the corresponding arrows. Panel D shows the comparison of the pH-dependent stabilities of $PolyA_{24}$, Poly $A_{12}C_{12}$, and P1 oligos reported in A–C, respectively (From Ciciriello et al. 2008. With permission)

1999a), which is successively opened resulting in a 2′ or 3′ phosphate extremity (as indicated in the right side of Figs. 1.1 and 1.7).

A detailed analysis of this reaction as a function of the sequence composition and of the pH of the aqueous environment revealed that the half-life of the 3′ phosphoester bond in oligonucleotides was markedly pH- and sequence-dependent (Fig. 1.2). Interestingly, the decrease as a function of temperature was observed not to be linear (Fig. 1.3).

Direct comparison of the half lives of the 3′ phosphoester bonds in the monomers with those of the same bonds when embedded in the polymer revealed that a large area of enhanced resistance to hydrolysis exists favoring the 3′ bonds in the polymer at temperatures higher than 60°C (Fig. 1.3). The increased stability of the polymer lasted at 90°C for 5–6 h. During this period the polymer was not attacked. The onset of rapid hydrolysis followed (Ciciriello et al. 2008).

The enhanced resistance to hydrolysis in the polymer, observed at temperatures between 60°C and 90°C, is markedly sequence-dependent (Fig. 1.2). This strongly suggests the possibility that these properties favor at the same time the polymeric over the monomeric state and the evolution of sequence complexity, as detailed below.

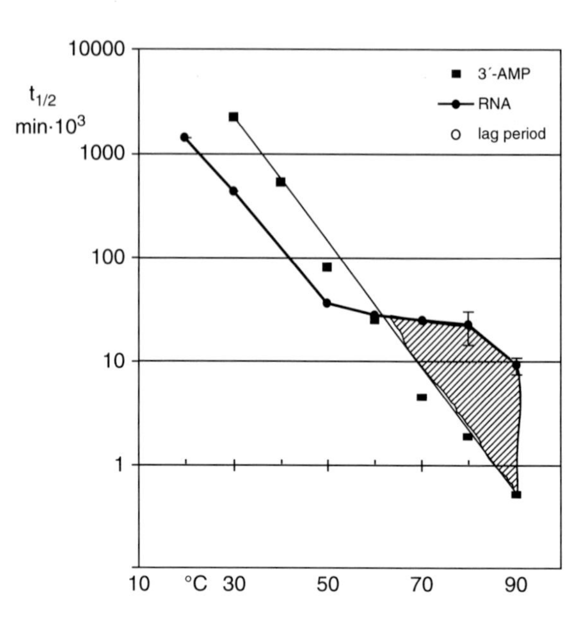

Fig. 1.3 Differential stability of the 3′-phosphoester bond when present in 3′AMP (▪) or in RNA (•). In the oligomer data set, each point represents the half-life (ordinate), as calculated in degradation analyses performed at the indicated temperature (abscissa) (Modified from Ciciriello et al. 2008. With permission)

Factors that enhance trans-esterification have been studied extensively and the effect of base sequence on the reactivity of phosphodiester bonds has been determined (Perreault and Anslyn 1977; Norberg and Nilsson 1995; Kaukinen et al. 2002).

The effect of base composition on the stability of RNA phosphodiester bonds has been frequently attributed to stacking interactions between the adjacent nucleic acid bases (Kierzek 1992; Li and Breaker 1999). A conformational explanation for this effect was indicated (Kierzek 1992) consisting of the fact that strong stacking hinders the cleavage of the intervening phosphodiester bond by preventing the attacking 2′-OH, the phosphorus atom and the departing 5′-oxygen to adopt the colinear conformation necessary for efficient trans-esterification. Stacking farther in the molecule was suggested as a major determinant of the overall structure of the oligomers (Kierzek 1992; Li and Breaker 1999), based on the fact that the reactivity is not particularly sensitive to the nearest neighborhood of the scissile bond. The hydrogen bonding network related to the hydration pattern was also indicated as an effector of this sequence context effect (Bibillo et al. 1999, and references therein).

It was concluded (Kaukinen et al. 2002) that the reactivity of phosphodiester bonds within RNA oligonucleotides having no defined secondary structure is strongly dependent on the base sequence of the substrate, whereas no such differences are observed with dinucleotide monophosphates or tetrameric oligonucleotides.

The oligonucleotides that exhibit rate retardations in their hydrolytic reaction in all likelihood adopt, due to base stacking and hydrogen bonding, a structure that hinders the free rotation of the phosphodiester bond and hence the reaction by an in-line mechanism (Kaukinen et al. 2002).

Transferring these observations into a prebiotic scenario, these analyses show that the pH of the environment in which an RNA happened to be synthesized

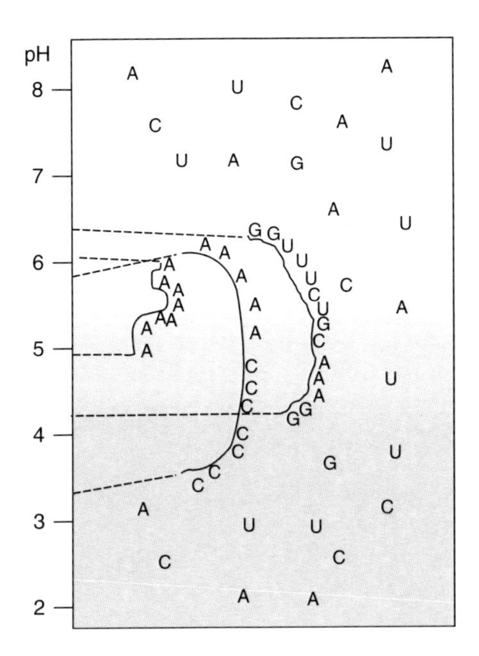

Fig. 1.4 The stability optimum of RNA oligonucleotides changes as a function of the sequence

strongly affected its chances of survival. Figure 1.4 describes in pictorial form this concept. In the same prebiotic frame this property might have conferred to the polymer a sufficient Darwinian edge over its constituent monomers not only to allow its very survival in the polymeric form but also to provide a phenotype for evolution.

1.3.3 Minerals and Protection

The root of the "RNA world" is the origin of the informational properties of RNA. This origin is not really understood. Nevertheless, decades of intense studies do at least allow the formulation of the key questions: was the evolutionary space explored by molecules replicating in aqueous surroundings or was replication/evolution made available only to those molecules encompassed in favorable confined microenvironments? And: if spontaneous synthesis of RNA occurred in an aqueous environment (i.e., in a sort of Darwinian warm little pond) confined by mineral surroundings, what was the influence of the minerals on the stability of RNA?

In order to provide an experimental answer to these questions, we have studied the catalytic properties of a large panel of minerals analyzing in parallel both the synthetic and the protection/degradation reactions. The results were reported (Saladino et al. 2001, 2003, 2004, 2005, 2006a, 2008,) describing the differential effects of the various minerals or nucleic acids. The class of compounds that showed to be protective at the highest degree is the phosphate minerals.

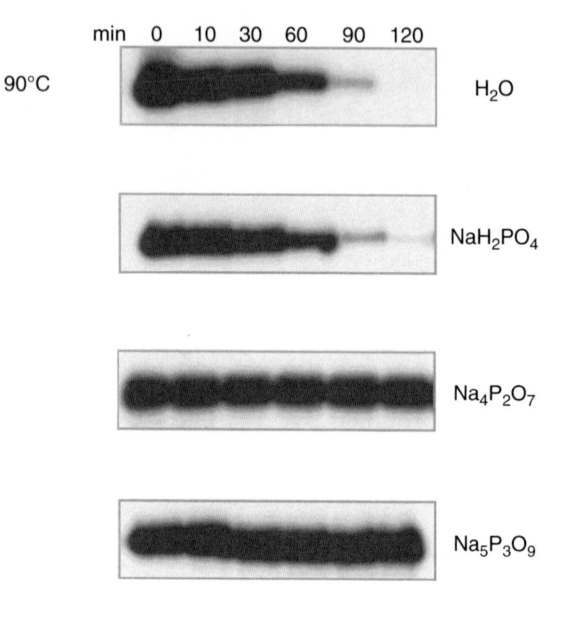

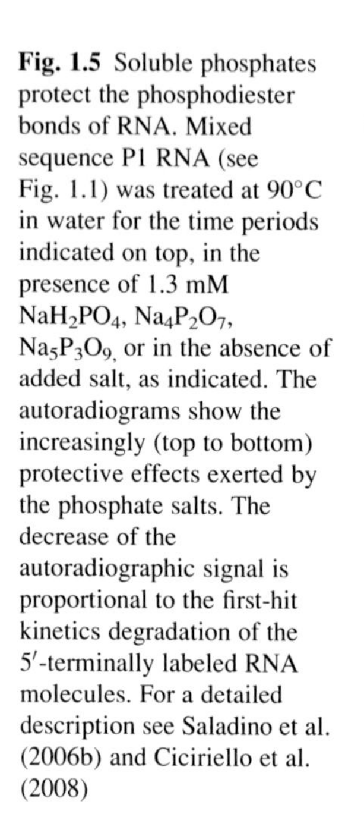

Fig. 1.5 Soluble phosphates protect the phosphodiester bonds of RNA. Mixed sequence P1 RNA (see Fig. 1.1) was treated at 90°C in water for the time periods indicated on top, in the presence of 1.3 mM NaH_2PO_4, $Na_4P_2O_7$, $Na_5P_3O_9$, or in the absence of added salt, as indicated. The autoradiograms show the increasingly (top to bottom) protective effects exerted by the phosphate salts. The decrease of the autoradiographic signal is proportional to the first-hit kinetics degradation of the 5′-terminally labeled RNA molecules. For a detailed description see Saladino et al. (2006b) and Ciciriello et al. (2008)

Figure 1.5 shows the degradation of a 5′ labeled RNA 20mer when treated in H_2O at 90°C for the time periods indicated on top, in the presence of sodium mono-, di-, or tri-phosphate salts. The protection effects are evident. A detailed quantitative analysis (reported in Saladino et al. 2006a; see also Costanzo et al. 2007a, b; Ciciriello et al. 2008) of the protective effects by phosphates on RNA shows that the warm little pond could have been the shrine not only of the synthesis of nucleic bases and of their nucleoside derivatives. If phosphates were present the pond could have been instrumental also for their phosphorylation and, upon polymerization, for their protection and enhanced survival.

1.4 Polymerization

1.4.1 Nonenzymatic Syntheses in Water

Nonenzymatic synthesis of RNA chains was obtained from pre-activated nucleotides, as concisely mentioned above and as reviewed (Orgel 2004). Progress was recently reported (Rajamani et al. 2008), showing that RNA-like polymers can be synthesized nonenzymatically from mononucleotides in lipid environments. In this system, chemical activation of the mononucleotides was not required for polymerization, the synthesis of phosphodiester bonds being presumably driven by the chemical potential of fluctuating anhydrous and hydrated conditions, heat providing the activation energy during dehydration. The RNA polymers were identified by nanopore analysis and by end-labeling followed by gel electrophoresis. The fluctuating hydration of the system provided a solution to the standard-state Gibbs free energy change ($\Delta G^{\circ\prime}$) problem (van Holde 1980), as discussed above. However, technical difficulties intrinsic to the system prevented a detailed analysis of the polymers afforded.

We have observed that 2′,3′- and 3′,5′-cyclic nucleotides actively supported the formation of short oligomers in water, in the absence of catalysts and at moderate temperature. While 5′AMP was much less active and both 2′AMP and 3′AMP did not afford any polymerized product, oligomerization from 2′,3′- and 3′,5′-cyclic nucleotides proceeds up to the esamer from 2′,3′-cAMP and to the octamer from 3′,5′-cAMP. Optimal conditions for polymerization in water were identified at 90°C, requiring time lapses $>$ 500 h for the 3′,5′ -cyclic and circa 50 h for the 2′,3′-cyclic forms (Costanzo et al. 2008).

Analysis of the products with bond-selective RNAses (namely, SVPD phosphodiesterase I and P1 endonuclease) showed that the ribo oligonucletides synthesized from 3′,5′-cAMP are for at least 87% of the 3′-5′ type, while the polymers formed from 2′,3′-cAMP all have 2′-5′ linkages (Costanzo et al. 2008).

Focusing on the products synthesized from the 3′,5′-cyclic form, which are (pre) biologically more relevant, one notes that during the relatively long time periods needed at 90°C to reach a kinetic plateau (500 h), the open forms (i.e., 3′AMP or 5′AMP) are degraded and that also the 3′,5′-cyclic form is on its way to degradation (as detailed in Saladino et al. 2006a). The oligonucleotides produced survive and can be observed simply because of the relatively higher stability of the sensible bonds when present in the polymer than in the monomer, as detailed in the previous section.

The yield is not high but the very presence of material polymerized in water is the proof-of-principle that a spontaneous process is possible. The principle of increased stability as a Darwinian edge applies. If a polymer is more stable than the starting material, the system will inevitably move towards complexity. In the chemical frame described here, however, the increase in size is not expected to reach large values. A limit set by the equilibrium between synthesis and degradation will be rapidly reached, limiting the final size of the polymer.

1.4.2 Ligation of Oligomers as a Way-Out from the Futile Cycle of Syntheses/Degradations

The (presumptive) equilibrium between spontaneous RNA chain elongation and chain degradation favors, in the experimental set-up described, the formation and maintenance of short RNA chains. One can easily envisage that this system lacks the potentialities to reach a (pre)genetically meaningful size. Even not

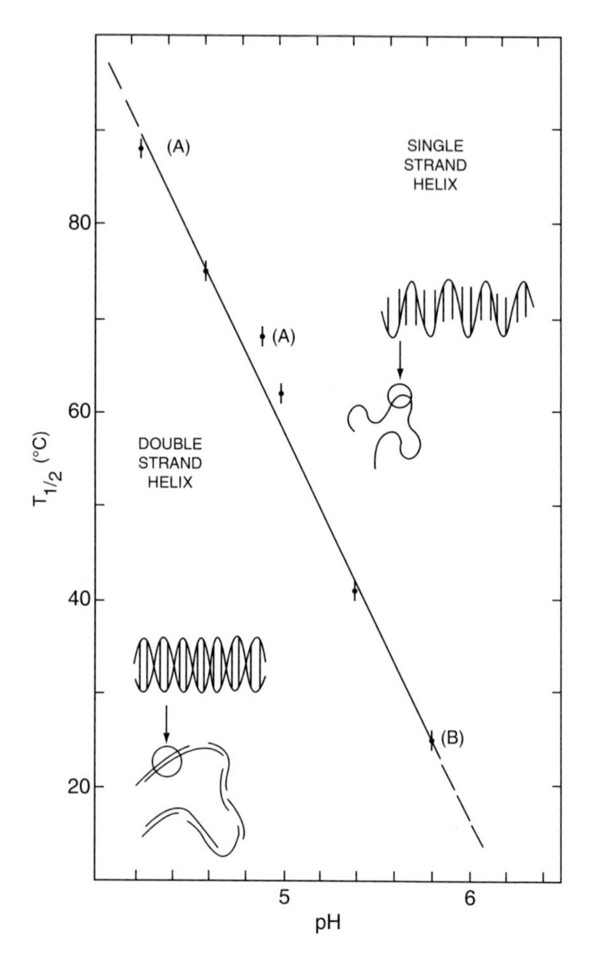

Fig. 1.6 The transition of the PolyA conformation depends on temperature and pH. The conformations of the double-strand and single-strand helices are shown schematically. Both are quite random in terms of long-range conformation but highly ordered in terms of short-range interactions. The double-strand helix is stable to the left of the straight line while the single-strand helix is stable to the right (From Holcomb and Tinoco 1965. With permission)

taking in consideration syntheses from activated precursors, several solutions can be imagined:

- Involvement of lipids, in open systems (as tested in Rajamani et al. 2008) or in micellar structures (Litovchick and Szostak 2008; Mansy and Szostak 2008)
- Physically protected environments, of the kind supporting the Wächtershäuser's scenarios (Wächtershäuser 1988)
- Protective and catalytic interactions with mineral surfaces (i.e., Montmorillonite clays as in Ferris et al. 1996; see also Gallori et al. 2006; Biondi et al. 2007)
- Combinations thereof

We have explored an alternative simpler possibility. Does a property of nucleic polymers exist intrinsically favoring sequence expansion? The answer is positive, as follows:

we have observed that RNA oligomers of various sizes ligate head-to-tail in aqueous solutions (Pino et al. 2008). The reaction does not require metal catalysts,

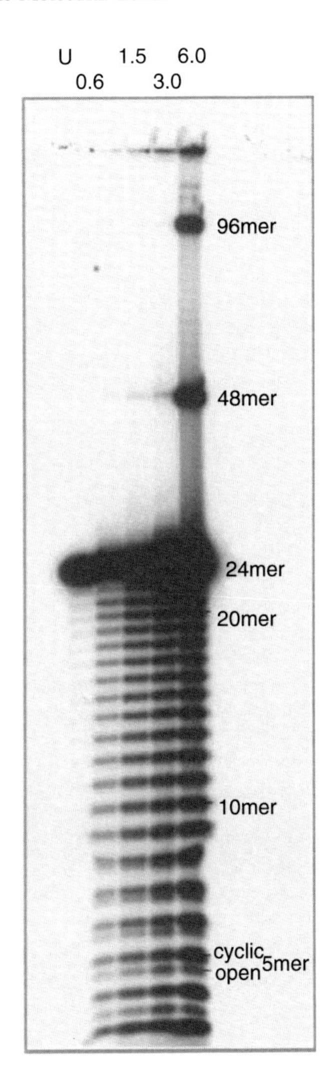

Fig. 1.7 Ligation of RNA oligomers in water. Gel electrophoretic image of the reaction products of a $5'A_{23}U3'$ 24mer RNA, reacted in water at 60°C for 24 h with 6 mM 3′,5′-cAMP. The picomoles of RNA present in the reaction mixture are indicated on top of each lane. U = untreated. During the reaction the 24mer is both hydrolyzed (as shown by the banded ladder below the full-length monomer) and ligated to afford the dimer (48mer) and tetramer (96mer) forms, plus minor amounts of higher molecular weight molecules

occurs at acidic pH values, is more efficient at temperatures encompassed between 40°C and 70°C and may involve up to one fourth of the initial population of molecules. Such a high yield is only obtained in the presence of adenine-based cofactors, 3′,5′-cAMP being the most efficient one.

Two properties of this ligation system are noteworthy: (i) dimers and tetramers are formed but never trimers or other odd-numbered multimers; (ii) higher molecular-size molecules (i.e., esamers and octamers) are also obtained, although at a much lower yield (unpublished observation). Enzymatic analyses (with SVPD phosphodiesterase I and P1 endonuclease) show that at least 66% of the bonds formed in the ligation are of the 3′-5′ type. The analysis of the mechanistic aspects of the reaction (see Pino et al. 2008) show that the ligation starts with the coupling

rationale for polymer evolution

Enhancers of key bonds stability	Ligation enhancers
- molecular context (1,2)	- molecular context (5)
- pH 4-6 (3)	- pH 3-6 (5)
- phosphates (4)	- adenosine phosphates (5)
- minerals (6)	- minerals (6)

(1) R. Saladino et al. (2005) *J. Biol. Chem.* 280: 35658-35669
(2) R. Saladino et al. (2006) *J. Biol. Chem.* 281: 5790-5796
(3) F. Ciciriello et al. (2008) *Biochemistry* 47: 2732-2742
(4) R. Saladino et al. (2006) *Chembiochem* 7: 1707-1714
(5) S. Pino et al. (2008) *J. Biol. Chem.* 283: 36494-36503
(6) G. Costanzo et al. (2007) *BMC Evolutionary Biology, S2*

Fig. 1.8 Both the stability and the sequence expansion by ligation of RNA oligomers are favored by the same effectors

of RNA oligomers two by two. This coupling increases the local concentration of the two reacting groups (the hydroxyl and the phosphate) and allows to consider the formation of double strands as the means through which the reaction-promoting proximity is obtained. Proximity and thermal energy explain the occurrence of the ligation reaction (Fig. 1.7).

The strands to be ligated are brought together by stacking interactions and are presumably further stabilized by adenine cofactors intercalation. In this system the RNA structure has a function similar to that of an enzyme, positioning correctly the reactive species, increasing their local concentration and reducing local water activity. Thus, the thermodynamic foundation of the system is simply that of the polyA interaction.

An evolutionarily appealing corollary of this model is that the size of the oligomers is expanded by a ligation reaction stimulated by the very same monomeric precursors from which they are polymerized, namely 3′, 5′-cAMP.

This is possibly the least demanding molecular sequence amplification reaction yet identified and provides RNA with a simple way-out from the short-size constraint in its possible evolutionary path and is of potential prebiotic interest.

In summary, both polymerization of RNA oligonucleotides and their ligation are possible in water. This provides the indication that RNA has the intrinsic capacity of self-assembly and of sequence expansion required for the spontaneous generation of sequence information (Fig. 1.8).

The described reaction relies on base stacking rather than on base pairing. Given that base stacking is not as sequence-depending as base pairing, in the path of generation of sequence complexity stacking is a priori the favored property. Hence the suggestion that the initial sequence evolution events were based on stacking rather than on the yet-to-be-evolved sequence information

1.5 Conclusion

The spontaneous generation of living things has been a consensus belief for all the thinkers who looked beyond a creationist revealed truth. The major body of discussions and of statements (often highly poetic) remaining from the ancient world is the legacy of the Greek philosophy of the fourth and third centuries BC and of its reflections in the roman culture. An analysis of the logics and of the then accepted beliefs is not within the scope of this discussion. Suffice to remind the verses by Lucretius

Nam si de nihilo fierent, ex omnibus rebus omne genus nasci posset, nil semine egeret.

Taken the things from nothingness, indeed, each of them would be born from any other, and would need no seed.

... from the *Rerum naturae* (V159, 160) and their general acceptation by the Stoic and Epicurean philosophy. From the original formulation by Democritus to popular writers like Virgilius or Censorinus, the spontaneous generation was widely accepted as the solution to the origin of organisms. More than specific examples, what was convincing was the reasoning that organisms were considered capable of transmitting their characters through repeated spontaneous generations, thus indicating a natural repeatable process. This was convincing to the point that such a secular belief was still largely accepted in the European culture in the 17th century, permitting the statement by Descartes that

les mouches, et plusieurs autres animaux sont produites par le soleil, la pluie et la terre.... (Secondes objéctions sur les Méditations, 2°)
... et les sauterelles par les nuages (Météores, Discours 7).

The ensuing experimental demonstrations of the impossibility of a spontaneous process springing living organisms from organic matter caused disillusion but left open the quest for the origins. Present-day molecular biology provides a series of partial truths: half of the amino acids present in extant proteins do form spontaneously in Miller's type of syntheses (Miller 1953, 1955), the formation and duplication of lipid-based micelles is leaving the pioneering period (reviewed in Szostak et al. 2001) and is rapidly becoming a sophisticated discipline (Szostak et al. 2001; Walde 2006; Murtas et al. 2007; Rajamani et al. 2008; Ito et al. 2008; Toyota et al. 2008).

As for the genetic component, the proof-of-principle has been here described that from HCN to cyclic nucleotides to self-polymerizing oligonucleotides, the whole process is possible.

Acknowledgments This work is supported by the Italian Space Agency "MoMa project," by ASI-INAF n. I/015/07/0 "Esplorazione del Sistema Solare," Italy and by "The National Science Foundation," USA (CBC Program).

References

Bibillo A, Figlerowicz M, Kierzek R (1999) The non-enzymatic hydrolysis of oligoribonucleotides. VI. The role of biogenic polyamines. Nucleic Acids Res 27:3931–3937

Biondi E, Branciamore S, Maurel MC, Gallori E (2007) Montmorillonite protection of an UV-irradiated hairpin ribozyme: evolution of the RNA world in a mineral environment. BMC Evol Biol 7 (Suppl 2):S2

Ciciriello F, Costanzo G, Pino S, Crestini C, Saladino R, Di Mauro E (2008) Molecular complexity favors the evolution of ribopolymers. Biochemistry 47:2732–2742

Costanzo G, Saladino R, Crestini C, Ciciriello F, Di Mauro E (2007a) Nucleoside phosphorylation by phosphate minerals. J Biol Chem 282:16729–16735

Costanzo G, Saladino R, Crestini C, Ciciriello F, Di Mauro E (2007b) Formamide as main building block in the origin of life. BMC Evolutionary Biology 7 (Suppl 2):S1

Costanzo G, Pino S, Ciciriello F, Di Mauro E (2008) Nonenzymatic nucleoside phosphorylation and oligomerization in water. (submitted)

Darwin F (1888) The life and letters of Charles Darwin, vol 3. John Murray, London, p 18 (letter to Joseph Hooker)

Delaye L, Lazcano A (2005) Prebiological evolution and the physics of the origin of life. Phys Life Rev 2:47–64

Ferris JP, Ertem G (1993) Montmorillonite catalysis of RNA oligomer formation in aqueous solution. A model for the prebiotic formation of RNA. J Am Chem Soc 115:12270–12275

Ferris JP, Hill AR Jr, Liu R, Orgel LE (1996) Synthesis of long prebiotic oligomers on mineral surfaces. Nature 381:59–61

Ferris JP, Joshi PC, Wang KJ, Miyakawa S, Huang W (2004) Catalysis in prebiotic chemistry: application to the synthesis of RNA oligomers. Adv Space Res 33:100–105

Fuller WD, Sanchez RA, Orgel LE (1972) Studies in prebiotic synthesis: VII. Solid-state synthesis of purine nucleosides. J Mol Evol 1:249–257

Gallori E, Biondi E, Branciamore S (2006) Looking for the primordial genetic honeycomb. Orig Life Evol Biosph 36:493–499

Huang W, Ferris JP (2003) Synthesis of 35–40 mers of RNA oligomers from unblocked monomers. A simple approach to the RNA world. Chem Commun (Camb) 12:1458–1459

Holcomb DN, Tinoco I Jr (1965) Conformation of polyriboadenylic acid: pH and temperature dependence. Biopolymers 3: 121–133

Ito K, Sugawara T, Shiroishi M, Tokuda N, Kurokawa A, Misaka T, Makyio H, Yurugi-Kobayashi T, Shimamura T, Nomura N, Murata T, Abe K, Iwata S, Kobayashi T (2008) Advanced method for high-throughput expression of mutated eukaryotic membrane proteins in Saccharomyces cerevisiae. Biochem Biophys Res Commun 371:841–845

Joyce GF (1989) RNA evolution and the origins of life. Nature 338:217–224

Kanavarioti A, Monnard PA, Deamer DW. (2001) Eutectic phases in ice facilitate nonenzymatic nucleic acid synthesis. Astrobiology 1:271–281

Kaukinen U, Lyytikäinen S, Mikkola S, Lönnberg H. (2002) The reactivity of phosphodiester bonds within linear single-stranded oligoribonucleotides strongly dependent on the base sequence. Nucleic Acids Res 30:468–467

Kawamura K, Ferris JP (1994) Kinetic and mechanistic analysis of dinucleotide and oligonucleotide formation from the 5′-phosphorimidazolide of adenosine on Na^+-montmorillonite. J Am Chem Soc 116:7564–7572

Kierzek R (1992) Nonenzymatic hydrolysis of oligoribonucleotides. Nucleic Acids Res 20:5079–5084

Kozlov IA, Politis PK, Pitsch S, Herdewijn P, Orgel LE (1999) A highly enantio-selective hexitol nucleic acid template for nonenzymatic oligoguanylate synthesis. J Am Chem Soc 121:1108–1109

Li Y, Breaker RR (1999) Kinetics of RNA degradation by specific base catalysis of transesterification involving the 2′-hydroxyl group. J Am Chem Soc 121:5364–5372
Litovchick A, Szostak JW (2008) Selection of cyclic peptide aptamers to HCV IRES RNA using mRNA display. Proc Natl Acad Sci USA 105:15293–15298
Lohrmann R (1977) Formation of nucleoside 5′-phosphoramidates under potentially prebiological conditions. J Mol Evol 10:137–154
Mansy SS, Szostak JW (2008) Thermostability of model protocell membranes. Proc Natl Acad Sci USA 105:13351–13355
Mansy SS, Schrum JP, Krishnamurthy M, Tobé S, Treco DA, Szostak JW (2008) Template-directed synthesis of a genetic polymer in a model protocell. Nature 454:122–125
Millar TJ (2004) Organic molecules in the interstellar medium. In: Ehrenfreud P et al. (eds) Astrobiology: future perspectives. Kluwer, The Netherlands, pp 17–21
Miller SL (1953) A production of amino acids under possible primitive earth conditions. Science 117:528–529
Miller SL (1955) Production of some organic compounds under possible primitive Earth conditions. J Am Chem Soc 77:2351–2361
Monnard PA, Kanavarioti A, Deamer DW (2003) Eutectic phase polymerization of activated ribonucleotide mixtures yields quasi-equimolar incorporation of purine and pyrimidine nucleobases. J Am Chem Soc 125:13734–13740
Murtas G, Kuruma Y, Bianchini P, Diaspro A, Luisi PL (2007) Protein synthesis in liposomes with a minimal set of enzymes. Biochem Biophys Res Commun 363:12–17
Norberg J, Nilsson L (1995) Stacking free energy profiles for all 16 natural ribodinucleoside monophosphates in aqueous solution. J Am Chem Soc 117:10832–10840
Orgel LE (1998) The origin of life—a review of facts and speculations. Trends Biochem Sci 23:491–495
Orgel LE (2004) Prebiotic chemistry and the origin of the RNA world. Crit Rev Biochem Mol Biol 39:99–123
Oró J (1961a) Mechanism of synthesis of adenine from hydrogen cyanide under possible primitive earth conditions. Nature 191:1193–1194
Oró J (1961b) Comets and the formation of biochemical compounds on the primitive Earth. Nature 190:389–390
Oró J, Kimball A (1960) Synthesis of adenine from ammonium cyanide. Biochem Biophys Res Commun 2:407–412
Oró J, Kimball A (1961) Synthesis of purines under possible primitive earth conditions. I. Adenine from hydrogen cyanide. Arch Biochem Biophys 94:217–227
Perreault DM, Anslyn EV (1997) Unifying the current data on the mechanism of cleavage-transesterification of RNA. Angew Chem Int Ed Engl 36:432–450
Pino S, Ciciriello F, Costanzo G, Di Mauro E (2008) Nonenzymatic RNA Ligation in Water. J Biol Chem 283:36494–36503
Rajamani S, Vlassov A, Benner S, Coombs A, Olasagasti F, Deamer D (2008) Lipid-assisted synthesis of RNA-like polymers from mononucleotides. Orig Life Evol Biosph 38:57–74
Saladino R, Crestini C, Costanzo G, Negri R, Di Mauro E (2001) A possible prebiotic synthesis of purine, adenine, cytosine, and 4(3H)-pyrimidone from formamide: implications for the origin of life. Bioorg Med Chem 9:1249–1253
Saladino R, Ciambecchini U, Crestini C, Costanzo G, Negri R, Di Mauro E (2003) One-pot TiO_2-catalyzed synthesis of nucleic bases and acyclonucleosides from formamide: implications for the origin of life. ChemBioChem 4:514–521
Saladino R Crestini C, Ciambecchini U, Ciciriello F, Costanzo G, Di Mauro E (2004) Synthesis and degradation of nucleobases and nucleic acids by formamide in the presence of montmorillonites. ChemBioChem 11:1558–1566
Saladino R, Crestini C, Busiello V, Ciciriello F, Costanzo G, Di Mauro E (2005) Origin of informational polymers. Differential stability of 3′- and 5′-phosphoester bonds in deoxy monomers and oligomers. J Biol Chem 280:35658–35669

Saladino R, Crestini C, Neri V, Brucato J, Colangeli L, Ciciriello F, Di Mauro E, Costanzo G (2005) Synthesis and degradation of nucleic acids components by formamide and cosmic dust analogs. ChemBioChem 6:1368–1374

Saladino R, Crestini C, Ciciriello F, Di Mauro E, Costanzo G (2006a) Origin of informational polymers: differential stability of phosphoester bonds in ribomonomers and ribooligomers. J Biol Chem 281:5790–5796

Saladino R, Crestini C, Neri V, Ciciriello F, Costanzo G, Di Mauro E (2006b) Origin of informational polymers: the concurrent roles of formamide and phosphates. ChemBioChem 7:1707–1714

Saladino R, Crestini C, Ciciriello F, Costanzo G, Di Mauro E (2007) Formamide chemistry and the origin of informational polymers. Chem Biodivers, Helv Chim Acta 4:694–720

Saladino R, Neri V, Crestini C, Costanzo G, Graciotti M, Di Mauro E (2008) Synthesis and degradation of nucleic acid components by formamide and iron sulphur minerals. J Am Chem Soc (Web Release Date: 22 October 2008); doi:10.1021/ja804782e

Schwartz A (1997) Prebiotic phosphorus chemistry reconsidered. Origins Life Evol Biosph 27:505–512

Soukup GA, Breaker RR (1999a) Relationship between internucleotide linkage geometry and the stability of RNA. RNA 5:1308–1325

Soukup GA, Breaker RR (1999b) Nucleic acid molecular switches. Trends Biotechnol 17:469–476

Szostak JW, Bartel DP, Luisi PL (2001) Synthesizing life. Nature 409:387–390

Tohidi M, Orgel LE (1989) Some acyclic analogs of nucleotides and their template-directed reactions. J Mol Evol 28:367–373

Toyota T, Takakura K, Kageyama Y, Kurihara K, Maru N, Ohnuma K, Kaneko K, Sugawara T (2008) Population study of sizes and components of self-reproducing giant multilamellar vesicles. Langmuir 24:3037–3044

van Holde KE (1980) The origin of life: a thermodynamic critique in "The origins of life and evolution." In: Halvorson HO, van Holde KE (eds) The origins of life and evolution. Alan R. Liss, New York, pp 31–46

Wächtershäuser G (1988) Before enzymes and templates, theory of surface metabolism. Microbiol Rev 52:452–484

Walde P (2006) Surfactant assemblies and their various possible roles for the origin(s) of life. Orig Life Evol Biosph 36:109–150

Zubay G, Mui T (2001) Prebiotic synthesis of nucleotides. Origins Life Evol Biosph 31:87–102

Chapter 2
Minimal Cell Model to Understand Origin of Life and Evolution

Tadashi Sugawara

Abstract When we consider the origin of life and its evolution, there is the missing link between inanimate and animate worlds. This chapter aims to connect this missing link by applying a constructive approach that creates an artificial molecular system that expresses similar dynamics as those of a life system using amphiphilic molecules and their self-assemblies. We discuss molecular self-assemblies that express prebiotic dynamics, such as spontaneous movements and self-winding of a helix, or exhibit self-reproducing dynamics of a vesicular cell model and self-replication of informational substances within a vesicular cell. A plausible evolution model of prebiotic cells is also proposed.

2.1 Introduction

When we look back through the history of life, we notice the missing link between the inanimate and animate worlds (Fig. 2.1). The pioneering work of Urey and Miller proved that organic molecules could be generated from simple inorganic molecules, such as CO, CO_2, HCN, and NH_3 (Miller 1953). Oparin further showed that even complicated polymers (coacervates) can be formed from simple organic molecules (Oparin 1952). However, an animate being has not yet been created from a coacervate or self-assembled artificial structure. On the other hand, biologists elucidate sophisticated functions of creatures on the basis of a cell. So that there still remains the missing link between inanimate and animate worlds. To answer profound questions such as "From where has life come?" or "Where is the boundary between animate and inanimate objects," *it may be useful to construct a molecular model system which expresses similar dynamics as those of a life system using*

T. Sugawara
Department of Basic Science, Graduate School of Arts and Sciences, The University of Tokyo, 3-8-1 Komaba, Meguro, Tokyo, Japan
email: tadpole8jp@yahoo.co.jp

P. Pontarotti (ed.), *Evolutionary Biology: Concept, Modeling, and Application*,
DOI: 10.1007/978-3-642-00952-5_2, © Springer-Verlag Berlin Heidelberg 2009

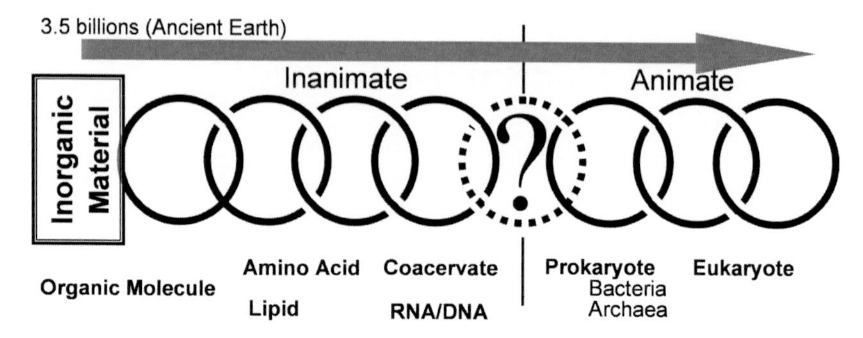

Fig. 2.1 Missing link between inanimate and animate worlds

well-defined organic molecules. This approach is designated as a constructive approach (Kaneko 2006).

Most scientists agree that a useful model life system must be able to (Luisi 2006):

1. Exchange materials with its environment
2. Respond to stimuli
3. Demonstrate spontaneous movement
4. Grow
5. Self-reproduce
6. Evolve as an ensemble

In this chapter, we will focus our attention on the construction of an artificial molecular system that expresses aforementioned dynamics, such as spontaneous movement, self-reproduction, or evolution as an ensemble of prebiotic cells, using amphiphilic molecules and their self-assemblies.

2.2 Spontaneous Movement of Amphiphilic Self-assemblies

Roughly speaking, two thirds of the energy required by animate beings for movement, such as a rotation of a flagellum, is derived through the hydrolysis of adenosine triphosphate (ATP). The remaining third is obtained from the difference in chemical potential of protons across a membrane. Usage of the directional flow of ions or protons is the first candidate to consider for prebiotic movement. This section addresses two primitive models that exhibit spontaneous movement: a self-winding helical self-assembly of amphiphiles and self-propelled oil-in-water (o/w) emulsions. Although these molecular systems convert chemical energy into mechanical energy in a primitive manner, they provide excellent examples to consider the evolution of self-movement from the prebiotic to mature stage.

An amphiphile (Fig. 2.2a), which consists of hydrophilic and hydrophobic parts, constructs a self-assembled structure in water. Among the self-assembled structures

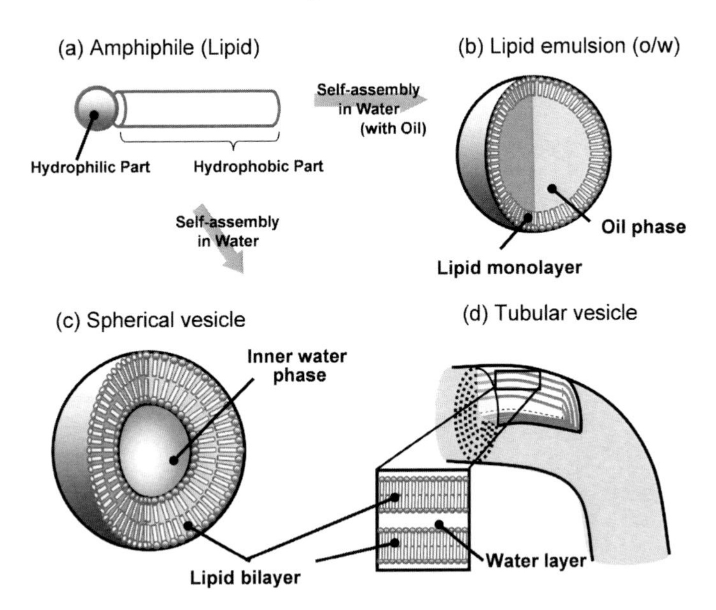

Fig. 2.2 Self-assemblies of amphiphiles (**a**) Amphiphilic molecule with a hydrophilic head and a hydrophobic tail. (**b**) Lipophilic oil droplet covered with surfactants (Oil and Water emulsion). (**c**) Double layered spherical vesicle with a inner water phase. (**d**) Multi-layered (Multilamella) tubular vesicles composed of bimolecular membranes. A water layer exists between bimolecular layers in both vesicles

arising from amphiphiles (Daoud and Williams 1999), a vesicle is often used as a cellular model because its shape is characterized as a hollow and closed structure composed of bimolecular membranes (Luisi and Walde 2000). Vesicles can be spherical or tubular in shape and single, multiple or nested in lamellarity (Fig. 2.2c, d). Vesicles larger than 1 μm (giant vesicles) can be observed by optical microscopy. On the other hand, lipophiles form oil droplets in water but when their surfaces are covered by amphiphiles (surfactants), droplets become miscible with water, forming an oil–water (o/w) emulsion (Fig. 2.2b).

2.2.1 *Self-winding Helix of Oleic Acid*

Figure 2.3 depicts an expanded picture of a bent tubular vesicle. A uni-component tubular vesicle is destabilized unexceptionally when it is bent (Helfrich 1973). However, if the membrane is made of plural amphiphiles with different packing parameters (Fig. 2.3a), the curved structure can be stabilized, provided that the cone-type amphiphiles are allowed to be allocated laterally from the inner surface towards the outer within the same leaflet (unimolecular layer) (Fig. 2.3b) (Tsafrir et al. 2001). Once the helix is stabilized, it is difficult to rewind because the allocation of amphiphiles within the bent membrane is difficult to occur.

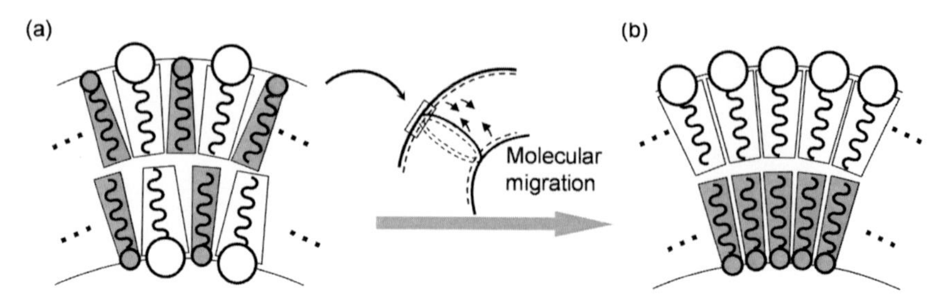

Fig. 2.3 (**a**) Bending of a tubular vesicle consisting of two types of membrane molecules with different packing parameters (Hydrophilic head groups are represented by open and closed circles). (**b**) Bent structure is stabilized by lateral allocation of membrane molecules with different head groups

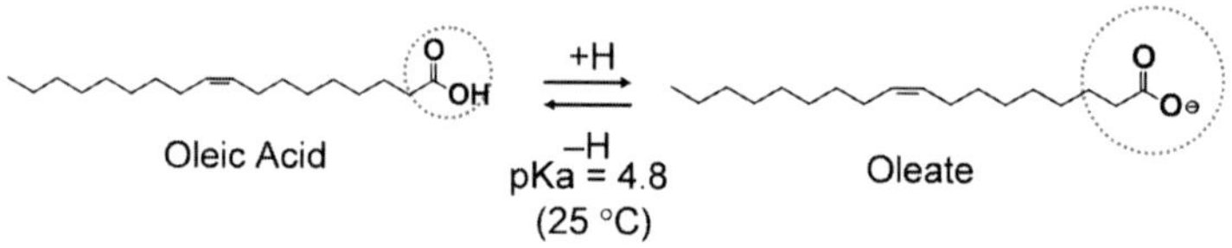

Fig. 2.4 Dissociation of equilibrium of oleic acid

A tubular vesicle composed of fatty acid is promising from the aspect of modeling a multicomponent membrane because even though the membrane is made of a single fatty acid, its protonated and deprotonated (carboxylate) forms are in equilibrium in water (Fig. 2.4), and the two forms have different packing parameters. That is, the effective volume of the ionic head group must be larger than the acid because the anionic head groups repel each other and the carboxylate group is tightly solvated by water molecules. Therefore, the morphology of the fatty acid aggregate can vary depending of the pH of its aqueous solution.

The morphologies of aggregates of oleic acid do, in fact, depend on the pH of the solution. Oleic acid–oleate forms giant vesicles at pH values higher than 7, whereas in acidic conditions (pH < 7), it forms oily droplets. Recently Ishimaru et al. (2005) found that oleic acid–oleate forms a tubular vesicle at pH 8.0 (Fig. 2.5a). According to the chemical shifts of ^{13}C-NMR using ^{13}C-enriched oleic acid, the ratio of oleate to free acid in the membranes was 3:2. Moreover, the oleic acid–oleate tubular vesicle transforms into a helix due to the local strain that accumulates inside the membrane.

Surprisingly, the oleic acid helix formed at pH 8 spontaneously rewound in water (Fig. 2.5a). A series of macroscopic images show the rewinding dynamics of the helix, which has opposite helicity at both ends (Fig. 2.5b). The helix is wound in a counterclockwise manner from the bottom, whereas it is wound in a clockwise manner from the top. The counterclockwise winding predominates over the clockwise, associating the movement of the kink that is a movable boundary between domains.

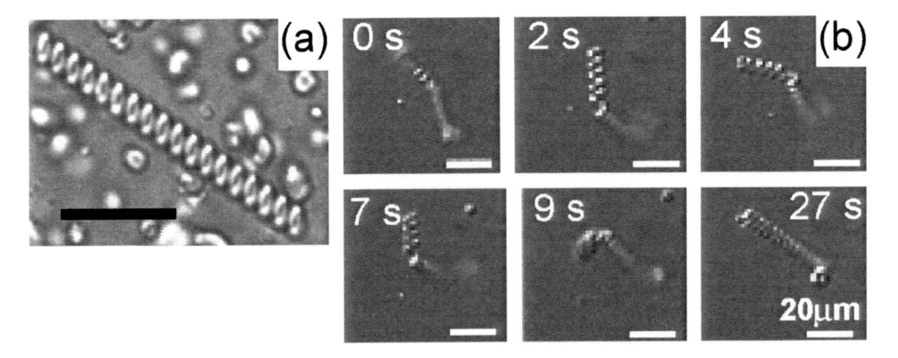

Fig. 2.5 (**a**) Multi-lamellar helical tubes made of oleic acid/oleate at pH = 8. (**b**) Clockwise winding of the upper part of a helical tube is rewound to the counterclockwise from the lower part. A size bar in (**a**) represents 10 μm and size bars in (**b**) represent 20 μm

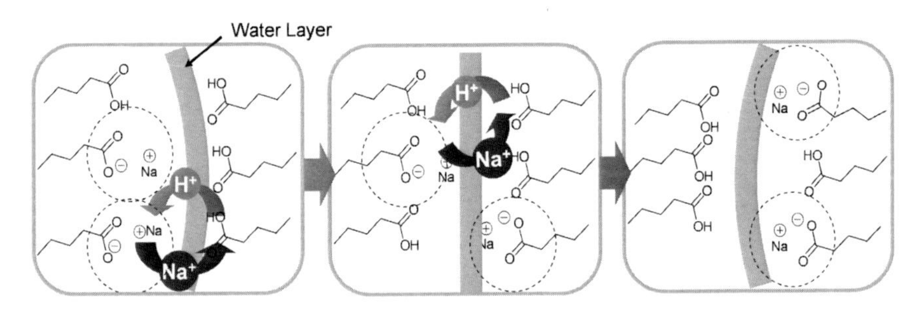

Fig. 2.6 Schematic scheme of microscopic bending fluctuation occurring between facing bimolecular membranes of a multi-lamella tube. (*Left*) Proton and sodium ion exchange in a thin water layer occurring between a facing pair of oleate and oleic acid. (*Middle*) Exchange between protons and sodium ion occurs in the middle pair as well and the bent structure is relaxed to flat. (*Right*) Exchange occurs in the top pair causes the bending to the opposite side

The precise mechanism of these events has not yet been elucidated fully, but the reason for such rewinding dynamics may be explained as follows (Fig. 2.6). The curvature of the bilayer membrane of oleic acid varies, depending on the local ratio of carboxylic acid to carboxylate. Because the effective volume of carboxylate is larger than that of carboxylic acid, the membrane would deform in a convex manner as shown in Fig. 2.6, left. But the local stress can be released just by exchanging a proton with a counter ion through the thin layer of water between them (Fig. 2.5, middle). This dynamic can also proceed the other way around (Fig. 2.6, right).

Here we can see that the morphological dynamics of a whole ensemble, such as macroscopic rewinding dynamics of a multi-lamella tube, is caused by synchronization of the membrane deformation induced by the local stress, which, in turn, is triggered by the exchange of protons and counter ions in a thin water layer between membranes. This is similar to a "wave" performed by an audience in a football stadium, which is created by undulation of a simple stand up-and-sit down motion of an individual person in the seat.

2.2.2 Self-propelled Oil Droplets

The second model of spontaneous movement involves the self-propelled motion of a droplet, which is derived from a chemical reaction that occurs at the interface between the droplet and its surrounding medium. The chemical reaction induces a symmetry-breakage due to the local accumulation and release of the products, and the droplets swim through an aqueous media. In this subsection, two examples of self-propelled oil droplets are discussed. First, an oil droplet is hydrolyzed in an aqueous alkaline solution and this chemical energy is transformed to the mechanical energy to move (Hanczyc et al. 2007). Second, a self-propelled oil droplet gains chemical energy by hydrolyzing a surfactant dissolved in an outer aqueous layer (Toyota et al. 2006). More simplistic physicochemical models of movement have been developed recently. For example, Sumino et al. have shown that oil droplets in water can move laterally by accumulating surfactant molecules adsorbed to a glass substrate (Sumino et al. 2005).

2.2.2.1 Self-propelled Oil Droplets from Lipophilic Precursor of Surfactant (Oleic Acid–Oleate)

When an oil droplet of oleic anhydride (or a mixture of oleic anhydride and nitrobenzene) is added to an alkaline (pH = 12) aqueous solution of oleate, the oil droplet starts to move spontaneously after a brief induction due to hydrolysis of oleic anhydride to oleic acid and oleate at the surface of the droplet (Hanczyc et al. 2007) (Fig. 2.7). Soon after, a pair of convection flows emerges inside the droplet, tubular vesicles develop at the rear of the droplet, and the droplet moves to the opposite direction (Fig. 2.8a). The direction of movement is parallel to that of convection. This process of self-propulsion can be divided into three stages as described in the caption of Fig. 2.8b.

The autonomous and sustained movement of the oil droplet of oleic anhydride is rationalized by Hanczyc et al. as follows. A fluctuation of the rate of hydrolysis of oleic anhydride at the interface between oil surface and aqueous alkaline solution leads to the symmetry-breakage with respect to all the potentially reactive sites on the surface and hydrolyzed products (oleic acid and oleate) are accumulated at the most reactive site. This imbalance in the interfacial tension is indicative of Marangoni instability. As long as this asymmetry is maintained, the oil droplet moves unidirectionally. Since a fresh oleic anhydride is conveyed towards the front edge (reactive site) of the droplet, the convection flow is accelerated, which induces

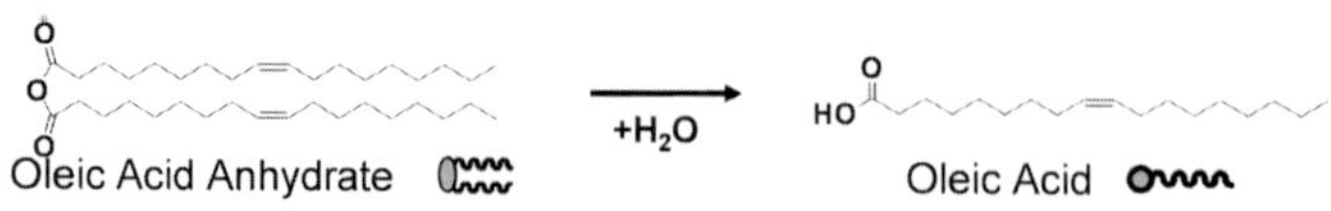

Oleic Acid Anhydrate $\xrightarrow{+H_2O}$ Oleic Acid

Fig. 2.7 Hydrolysis of oleic anhydride

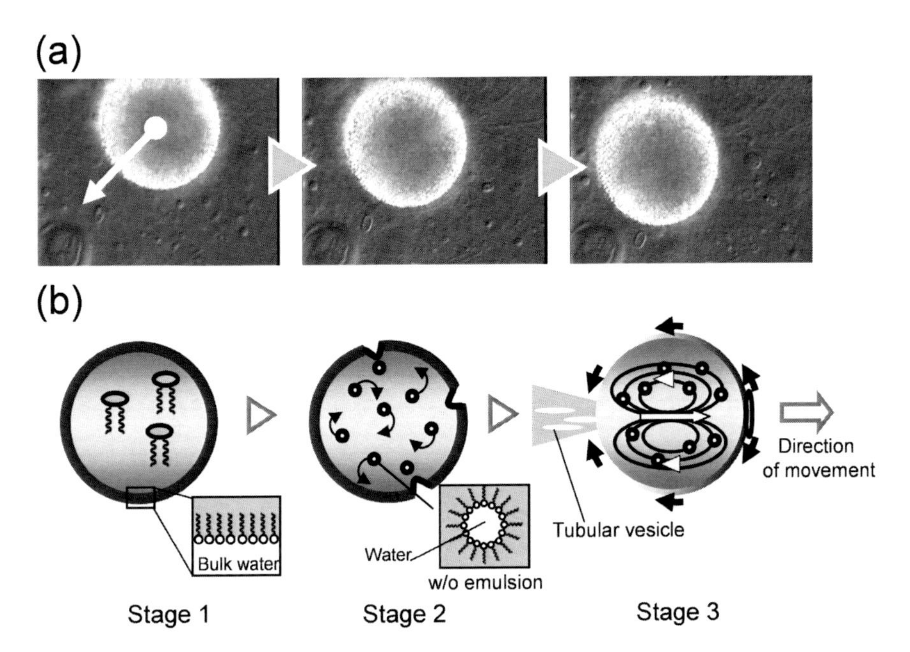

Fig. 2.8 (**a**) Unidirectional spontaneous movement (towards the bottom left corner) of an oil droplet consisting of oleic anhydride covered by surfactants (oleate) in aqueous alkaline solution. (**b**) Schematic drawing of three stages of self-propelled movement of an oil droplet made of oleic anhydride. **Stage 1:** In an alkaline solution the surface of an oil droplet of oleic anhydride is covered with hydrolyzed oleic acid/oleate (Insert: expanded figure). Since oleate is a surfactant, oleate can accommodate transfer of water into the oil droplet. **Stage 2:** Micro-droplets of water (Inset: w/o emulsion) with a microscopically detectable size emerge in the oil droplet, and they start to move in mostly disordered flow patterns. **Stage 3:** A pair of convection flows is generated inside the oil droplet and the hydrolysis of oleic anhydride is activated at the outlet of the convection. The generated oleic acid/oleate is transferred to the rear of the droplet and it is released as tubular vesicles. The oil droplet propels itself towards the opposite direction to the growing tubular vesicles

a positive feedback loop. Simultaneously, the product moves to the posterior surface and are released as tubular vesicles.

The movement of a self-propelled oil droplet is sustained by the transfer and release of surfactants, which is produced at the leading head and conveyed to the posterior surface by the convection flow. It is a surprise that the positive feedback loop between the microscopic chemical reaction at the most active site and the macroscopic propelled movement is realized even in such a primitive reaction system. These dynamics are shaping our understanding of prebiotic movement.

2.2.2.2 Self-propelled Oil Droplets Consuming Surfactant as Fuel

Although this self-propelling oil droplet is an excellent model to examine the prebiotic movement, it is destined to stop when hydrolyzed oleic acid exceeds a critical concentration. Prior to the oil droplet of oleic anhydride, Toyota et al.

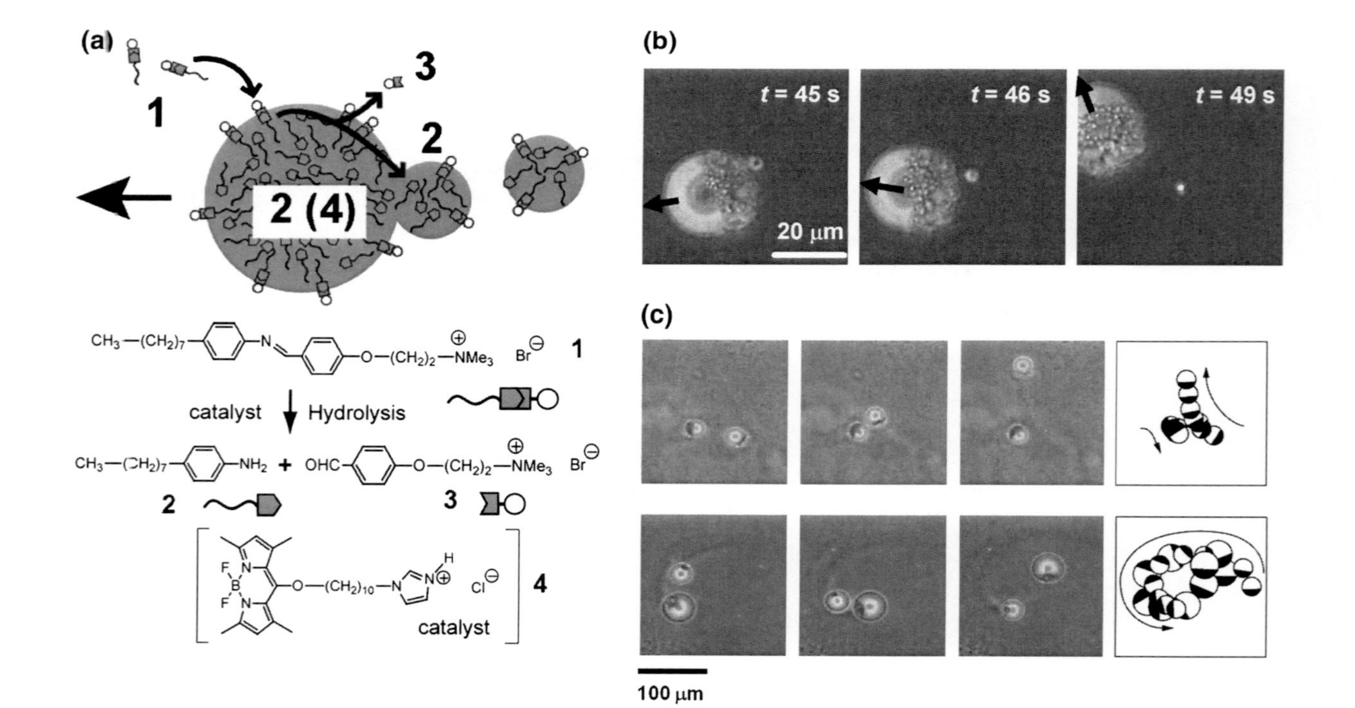

Fig. 2.9 Dynamics and composition of self-propelled oil droplet consuming a fuel surfactant. (**a**) Surfactant (**1**) is hydrolyzed to lipophile (**2**) and electrolyte (**3**) on the surface of an oil droplet composed of lipophile **2** containing catalyst (**4**). Hydrolyzed **2** forms tiny oil droplets and they are conveyed by a convex flow to the posterior surface of the oil droplet and are released as a trail from the bottom. The oil droplet self-propels to the opposite direction to the trail (made of wasted **2** and **3**). (**b**) Microscopic images of self-propelled oil droplet releasing tiny oil droplets made of **2** from the rear side. (**c**) Interactive movements of two self-propelled oil droplets. When two droplets encounter in front, two droplets move apart avoiding the collision (above). When one droplet follows to a precedent, the follower is entrapped by the trail of the precedent and they move together for a while (below). Schematic figures show front (white) and back (black) faces of self-propelled oil droplets in both cases

reported a self-propelled oil droplet made of reactive lipophile containing an amphiphilic acid catalyst (Toyota et al. 2006). It reacts with a surfactant dissolved in an outer water layer to form a membrane molecule. The oil droplet then moves, releasing tubular vesicles from the rear end. The oil droplet is consumed during the propelled movement in this system as well. Hence, Toyota invented a self-propelled oil droplet of a persistent type. Here, an oil droplet exhibits a self-propelled motion by consuming a hydrolyzable surfactant as a "fuel," being supplied from a bulk aqueous dispersion (Toyata et al. 2009).

In this advanced model, an amphiphilic precursor (**1**) of an oil droplet (**2**) is dissolved in the outer aqueous solution. The basic principle is the same as the previous one (Fig. 2.9a). A series of photos in Fig. 2.9b demonstrate the movement of an oil droplet made of **2** and amphiphilic acidic catalyst **4**. The hydrolysis of **1** to give lipophile **2** and electrolyte (**3**) occurs at the surface of the oil droplet containing the catalyst. Similar to the case of an oil droplet of oleic anhydride, a fluctuation of the rates of hydrolysis at the surface of the droplet leads to the accumulation of the products at the most reactive site, and then a pair of convection flows emerges inside the oil droplet. The generated **2** forms micro-droplets on the surface and they assemble and conveyed to the posterior end by the convection flow. Eventually they are released one-by-one from the rear of the mother droplet as a trail, consisting presumably of the dense emulsion of the lipophile **2** and electrolyte **3**. Interestingly, the self-propelling droplet is eventually entrapped by its own trail, since the trail no longer contains the fuel surfactant. If enough time passes for the trail to diffuse, then the droplet starts to move again freely.

An intriguing phenomenon was found when two self-propelled droplets of a fuel-consuming type come closer (Fig. 2.9c). When another droplet approaches a self-propelled droplet, the newcomer follows the precedent oil droplet and the former is going to be entrapped by the trail of the precedent. These two droplets move interactively for a while. It is to be noted that the chemicals released by the droplet influences the movement of another droplet.

This self-propelled droplet exhibits persistent movement by hydrolyzing the fuel dissolved in an outer aqueous phase. Another characteristic of the new system is that a pair of self-propelled droplets exhibits interactive movements. This behavior may be achieved through a delicate balance between positive chemotaxis exhibited by droplets towards the fuel and negative chemotaxis away from the waste released by droplets. The latter event could be regarded as a primitive type of chemical communication.

2.3 Dynamics of Self-reproduction Exhibited by Giant Vesicles

The following sections present the construction of an artificial minimum cell model in terms of the bottom-up method. For this purpose the structure of a cell must be as simple as possible. There are three indispensable elements of a cell: a **compartment**

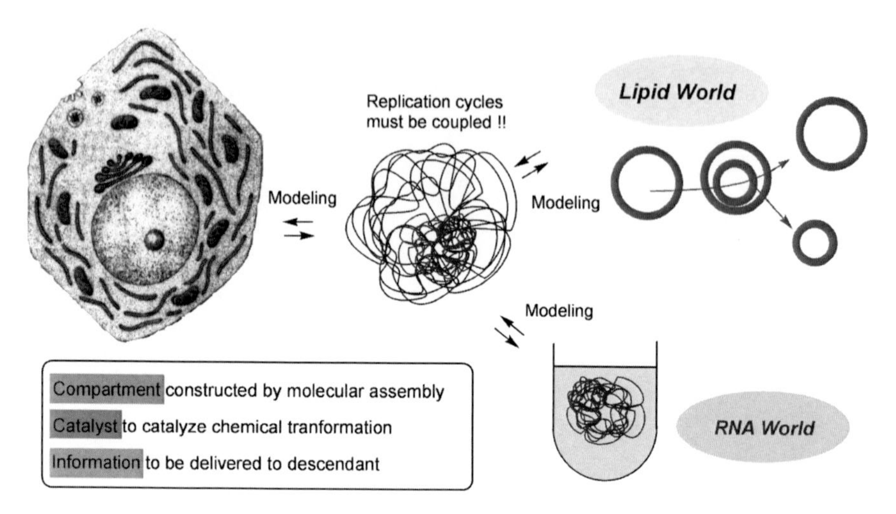

Fig. 2.10 Modeling of a living cell and two indispensable dynamics, self-reproduction of compartments (Lipid world) and self-replication of informational substance (RNA world)

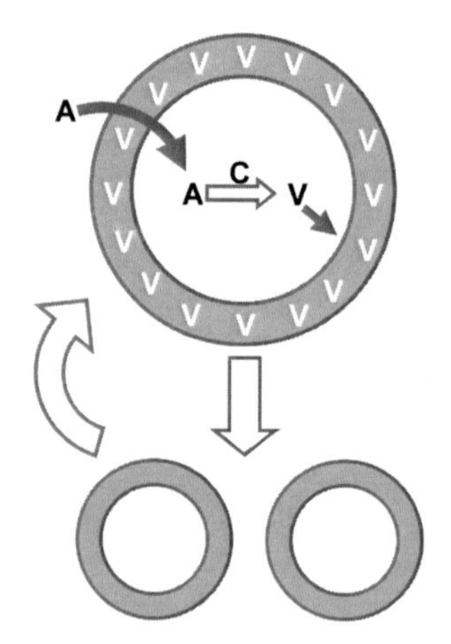

Fig. 2.11 Definition of self-reproduction of compartments (See text)

that separates the inner reaction system from the outer world, a **catalyst** to catalyze important metabolic reactions, and an **informational substance** that delivers information to descendants (Szostak et al. 2001). We notice two important dynamics in this system. The first is the self-production of a cell membrane, which demonstrates the dynamics of a "lipid world." The second is self-replication of the informational substance—that is, an 'RNA world.' When these self-reproducing and self-replicating dynamics are coupled loosely, a minimal cell model emerges. This section focuses on the construction of a self-reproducing giant vesicle in a lipid world (Fig. 2.10).

Before concrete examples are presented, it is necessary to define self-reproduction as a molecular self-assembly in the absence of a template molecule. In the case of vesicular self-reproduction, similar-sized vesicles of the same components as the original are produced from the original vesicle (Luisi 2006). A vesicle consisting of membrane molecules (**V**) is thought to pick up a precursor molecule (**A**) and convert it into membrane molecules (**V**) within the vesicle. After the vesicle becomes corpulent, it self-divides. These dynamics are regarded as self-reproducing dynamics (Fig. 2.11). In contrast, if a vesicle picks up the membrane molecule directly (without first obtaining **A**) and self-divides, this process is not an example of self-reproducing dynamics. This self-reproduction dynamics differs from the self-replication in which a replicate molecule is synthesized with the aid of a template.

2.3.1 Self-reproducing Vesicular System of the Nutrient-Containing Type

Luisi et al. reported that a number of vesicles composed of oleic acid increases in alkaline aqueous solution by adding oleic acid anhydride as a precursor (Walde et al. 1994b). In his model, oleic acid anhydride is transferred to the vesicular membrane by a surfactant (oleate) and it is hydrolyzed to oleic acid, leading to an increase of the number of vesicles. Since hydrolysis is able to occur anywhere in an alkaline solution, the reaction field is not specified to the vesicular membrane. However, the presence of preformed vesicles in the system drastically accelerates

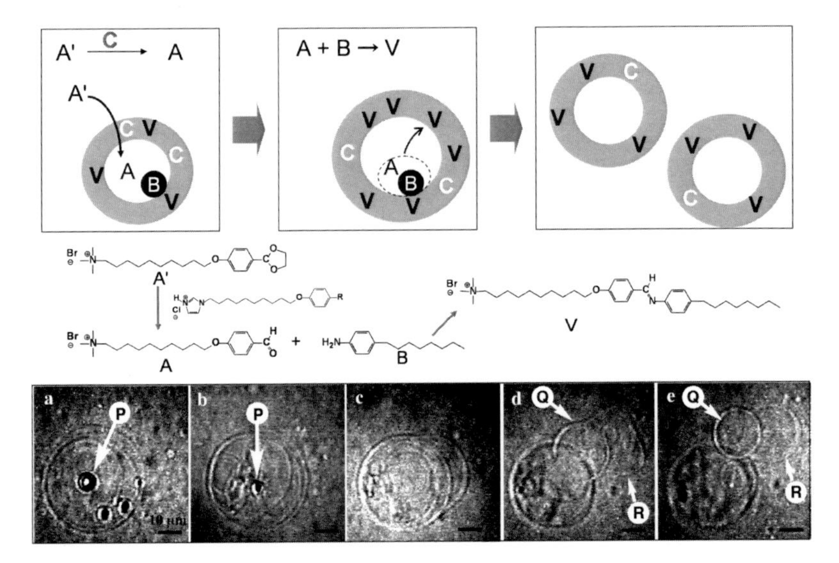

Fig. 2.12 Dynamics and components of a self-reproducing giant vesicle of an ingredient including type. Inactive membrane precursor (**A′**) is converted to an active form (**A**) by the catalyst (**C**) buried in a vesicular membrane and **A** reacts with lipophile (**B**) on the surface of an oil droplet entrapped in a water pool to give membrane molecule **V** (above). A series of photographic images of the morphological changes of a giant vesicle (**GV**) is shown below. When membrane-precursor **A′** is added to the **GV** containing oil droplets of **B**. About 10 min after the addition of **A′**, oil droplets (*P*) in the inner water pool started to decrease. Associated with the disappearance of these oil droplets, new vesicles were generated inside the original vesicle. After 20 min, new vesicles (*Q* and *R*) extrude through the outer membrane of the original vesicle to the bulk water

the production of surfactant, implying some yet unknown interaction of the vesicle membrane with the hydrolysis reaction.

A new self-reproducing giant vesicular system was proposed by Takakura et al., as shown in Fig. 2.12 (Takakura et al. 2003). The original vesicle consists of membrane molecule (**V**), which is formed by a coupling reaction between amphiphile (**A**) and lipophile (**B**). Incidentally, reactive amphiphile **A** is supplied as an inactive form (**A′**) not to react with **B** existing in the outer aqueous phase. The **A′** is converted to the active form (**A**) by catalyst (**C**) dissolving in the vesicular membrane. **A** then reacts with **B** on the surface of an oil droplet entrapped in the inner water pool and they produce membrane molecule **V**. Membrane molecules **V** self-assemble to form daughter vesicles inside the original vesicle and new vesicles come out through the outer membrane to increase the number of vesicles. Accordingly, this whole process can be regarded as the self-reproduction of giant vesicles.

2.3.2 *Robustly Reproducing Giant Vesicular System*

However, there is one problem in the above self-producing system. This system contains lipophilic precursor in the inner water pool, so that the reproduction stops when the oil droplet is consumed. A scenario of an improved self-reproducing giant vesicular system is depicted in Fig. 2.13. The membrane molecule (**V**) forms giant multilamella vesicle (**GMVs**) in water. The precursor of membrane molecule (**V***)

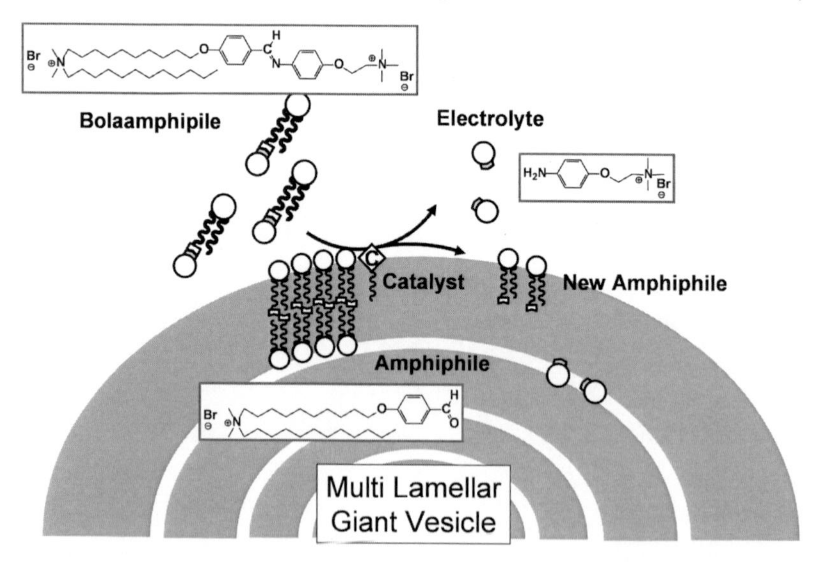

Fig. 2.13 Dynamics and components of a self-reproducing giant multilamellar vesicle of a robust type. Bolaamphiphile, which is a precursor of the membrane molecule is hydrolyzed to a membrane molecule and electrolyte in the vesicular membrane containing a catalyst. The hydrolyzed membrane molecule dissolves into **GMV** and electrolyte is released into a thin water layer between vesicular membranes

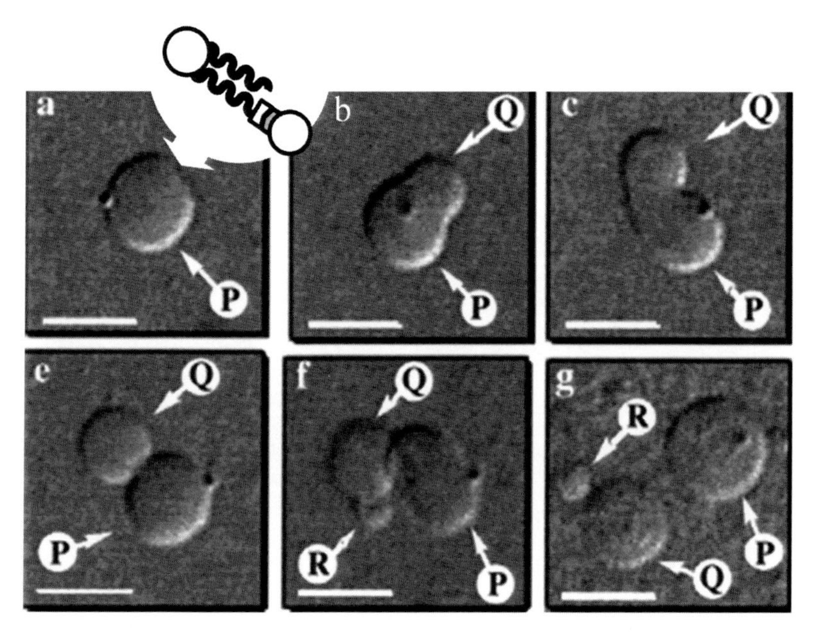

Fig. 2.14 Microscopic images of robust self-reproducing **GMV**. From mother vesicle (*P*), a daughter vesicle (*Q*) is divided, and a granddaughter vesicle (*R*) is divided from *Q*. Size bars represent 10 μm

is a bolaamphiphile (amphiphile carrying a polar head at both ends of a hydrophobic chain), and it does not form vesicles, but small aggregates in water. When **V*** is hydrolyzed on the surface of or inside the vesicular membrane containing catalyst **C**, it gives rise to membrane molecule **V** and electrolyte (**E**). The generated **V** dissolves in the membrane to make **GMV** corpulent. The self-reproducing dynamics are shown in Fig. 2.14 (Takakura and Sugawara 2004).

It is worthwhile mentioning the role of electrolyte **E**, which is produced from **V***. If **E** is solely added to the dispersion of **GMVs**, only the fission of **GMVs** occurs but no increase in its size because membrane molecule **V** is not supplied. This means that **E** induces the division of **GMV.** Since the hydrolysis of **V*** occurs not only on the surface of **GMV** but also within a water layer between the vesicular membranes (Fig. 2.13). The local concentration of **E** in the water layer within **GMV** then becomes higher because released **E** in the outer aqueous phase diffuses away. Due to the osmotic pressure effect, the exchange of **E** and outer water across the membrane takes place, associated with the reorientation of membrane molecules to create a channel for passing **E** and water through. Such reorganization of membranes is transmitted to the inner membranes to induce the division of **GMVs**.

A self-reproduction of cellular membrane is one of the indispensable elements to realize a primitive cell model. A robust self-reproducing vesicular system is achieved by the cooperative dynamics between membrane precursor V, membrane molecule* ***V****, catalyst* ***C****, and, in particular, electrolyte* ***E****, which plays the key role in this spectacular division event.*

2.4 Population Analysis of Self-reproducing Giant Vesicles by Flow Cytometry

As shown by the fluorescence microscopic images of before and after the addition of membrane precursor **V***, the number of vesicles increased to the great extent, since catalyst **C** is tagged with a fluorescent probe (Fig. 2.15a). The presence of shining smaller vesicles after reproduction indicates that the catalyst is transferred to the divided **GVs**. It may be criticized that we are looking at only crowded regions in the dispersion of **GVs**. This robust self-reproducing dynamics was traced by fluorescence activated flow cytometry (FACS) to exclude the possibility that these dynamics are an artifact due to crowding of giant vesicles (Toyota et al. 2008).

2.4.1 Protocol of Flow Cytometric Analysis

The principle of flow cytometry is briefly explained here. A dispersion of vesicles flows through a tube as a rapid stream with a speed of ca. 1,000 vesicles per second. Each vesicle is flashed by an Argon laser apparatus (488 nm), and the forward light scatter (**FS**) of each particle is detected by a photon counting detector. The side light scatter (**SS**) and the fluorescence intensity (**FL**) of the particle are also detected in the perpendicular direction to the light path, through a band-pass filter in the case of FL (Fig. 2.15b).

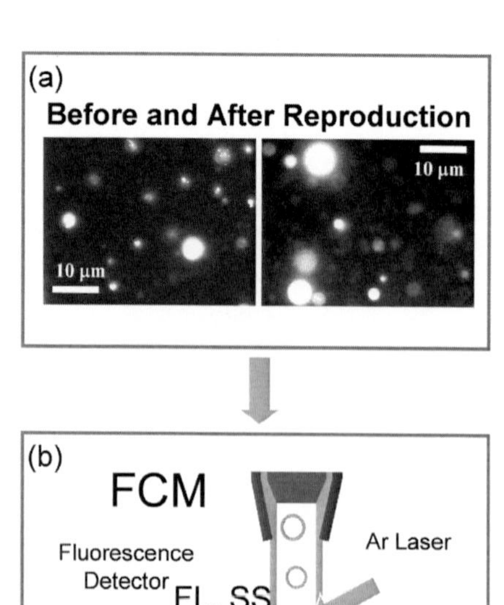

Fig. 2.15 (**a**) Fluorescence microscopic images before and after self-reproduction of **GMV** containing a fluorescence-probe tagged catalyst. Fluorescence emitting **GVs** increase in number after incubation at 25°C. (**b**) Principle of flow cytometric analysis of giant vesicles

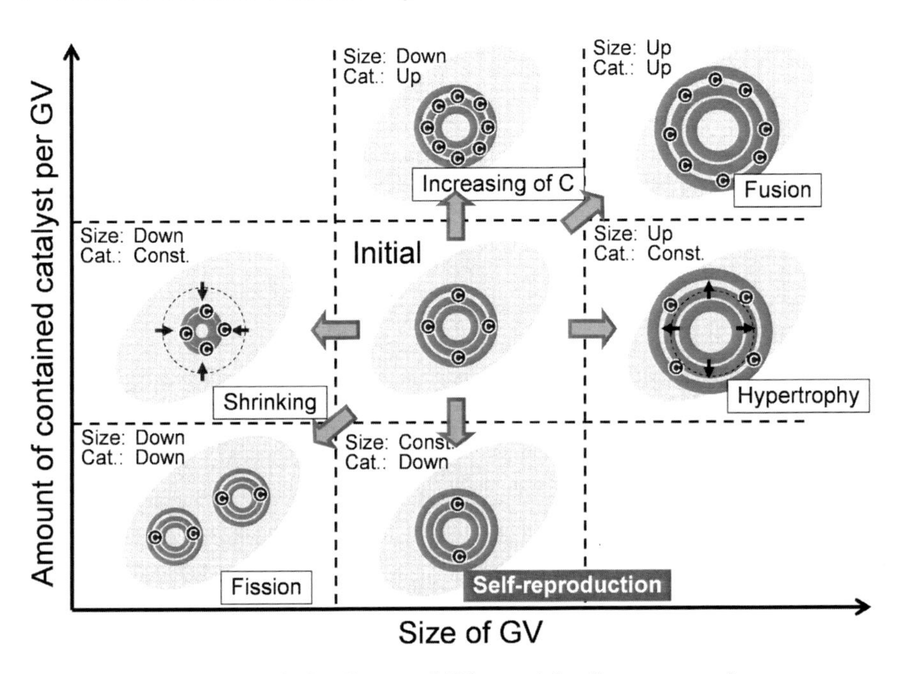

Fig. 2.16 Population change in 2D diagram of **GVs** containing fluorescent catalyst

A schematic two-dimensional (2D) plot of **GVs** containing the fluorescent catalyst is shown in Fig. 2.16. The vertical axis represents the amount of catalyst contained in each vesicle. The horizontal axis is the intensity of the forward scatter and it represents the size of each vesicle. We monitored using flow cytometry the population change of **GVs** during the self-reproduction cycle as shown here. The horizontal shift means that the average size of **GV** shrinks or expands, keeping amount of the catalyst per vesicle the same. Upward diagonal shift means the increase in both the size and catalyst amount presumably due to fusion of vesicles, while downward shift means the decrease in both the size and catalyst amount presumably due to fission of vesicles. Upward vertical shift corresponds to the increase in the catalyst amount keeping the same size due to an uptake of the catalyst from the outer solution, whereas the downward shift means the appearance of almost the same size of **GVs** containing the lesser amount of the catalyst, which means the self-reproduction occurs during this morphological change.

2.4.2 Population Analysis of Self-reproducing Vesicles

Figure 2.17a shows a 2D-plot of **GVs** at the initial stage. The plot shows the positive correlation between the size of **GVs** and the amount of the catalyst. Namely, as the size of the **GV** increases, the amount of the catalyst increases. We also notice a wide distribution in the size of **GVs** in solution, although the average size of **GVs** is

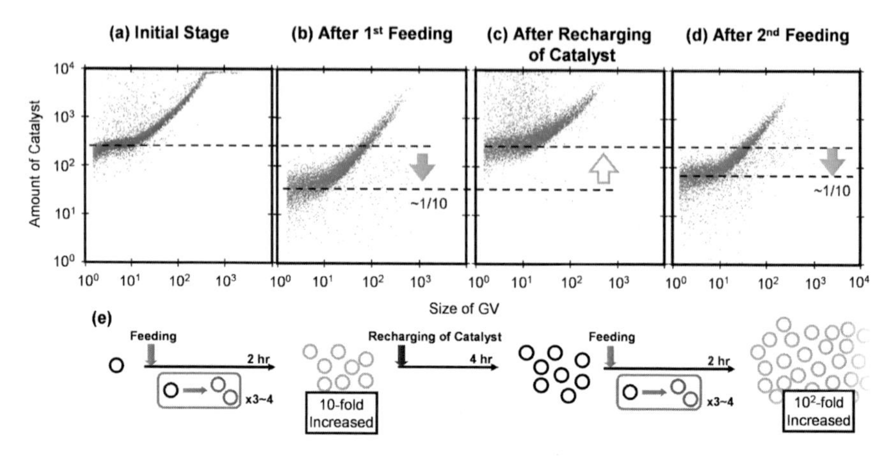

Fig. 2.17 (**a–d**) Population change of self-reproducing **GVs** conducted by feeding-recharging-feeding cycle. (**e**) A black circle represents a vesicle containing enough catalyst, while gray circle represents a vesicle with deficient catalyst. **Preparation of catalyst-containing GVs** Membrane molecule **V**, 10% catalyst **C,** and 1% of fluorescence probe-tagged catalyst $\mathbf{C_f}$ (the fluorescence probe-tagged catalyst was diluted with non-fluorescent catalyst to avoid self-quenching) were dissolved in an organic solvent and dried to form films. Catalyst-containing **GV** was then generated by adding an aqueous solution containing **E** according to a film-swelling method. After incubation of the specimen for 1 h at 24°C to stabilize the population of **GVs** the precursor **V*** of the membrane molecule was added. **Population analysis of GVs after 1st feeding of membrane precursor V*** After addition of **V*** solution to a dispersion of **GV** containing **E**, the mixture was incubated for 2 h at 24°C. The 2D dot plot changed as shown in Fig. 2.17b. As expected, the amount of the catalyst shifted downward along the vertical axis by about 1/10. It means that **GVs** divided three to four times in average as schematically shown in Fig. 2.17e below. However the activity of **GVs** for further division diminished due to the deficiency of the catalyst. **Recharging Catalysts** We can recharge the catalyst in the inactivated **GVs**. In order not to dilute the **GV** dispersion, a small amount of a methanolic solution of **C** and $\mathbf{C_f}$ (0.5 M, 10:1 molar ratio) was added to 5 mL of the vesicular dispersion. As shown in the 2D plot of Fig. 2.17c, the amount of catalysts recovered to the original levels. It means that the system is recharged with catalyst. **Self-production after 2nd feeding of membrane precursor V*** After recharging of the catalysts, **V*** was added again in the same way as the first feeding. Flow cytometric analysis indicated that self-reproduction of the **GVs** restarted and after incubation of 2 h, the amount of catalyst decreased about 1/10 (Fig. 2.17d), keeping the size distribution of the **GVs** almost the same as that of the original

roughly controlled by the packing parameters of the membrane molecules, the surface tension and the concentration of electrolytes, etc. Flow cytometric analysis of the self-reproducing **GVs** was conducted according to the procedures describe in the caption of Fig. 2.17.

As shown in Fig. 2.17a–d, the self-reproduction of **GVs** continues for six or seven cycles on average, associated with sequential additions of the membrane precursor and the catalyst. This means the number of **GVs** increases almost on the order of 10^2. It is to be noticed that the size distribution of **GVs** is almost the same as that of the original after several cycles of the self-reproduction, although only the largest **GVs** disappeared during these processes. Presumably, they are unstable under the current conditions and are divided into smaller vesicles. These results clearly demonstrate that the robust self-reproducing **GV** is accomplished.

2.4.3 Self-reproducing Vesicles as a Molecular Model of Evolution

The population of self-reproducing **GVs** demonstrated in a 2D-plot could be sensitive to an environmental change, such as pH, ionic strength, some sort of chemical. Figure 2.18 shows a couple of possibilities of population drifts of **GVs** (a): The whole group of **GVs** disappears (Extinction), (b) only a certain sub-group survives (Natural selection), (c) a new group appears upon the environmental change (Evolution), induced by the environmental perturbation. The last category, in particular, could be regarded as a molecular model of evolution if **GVs** in the new group self-reproduce themselves robustly. To elucidate the origin of the stability or instability of the population of self-reproducing **GVs**, a sorting technique must be effective. It is able to sort **GVs** composed of mixed populations into specific subgroups. For example, one can sort a new group that appeared after the external perturbation and examine the shape and lamellarity of the sorted **GVs**.

For sorting the vesicular dispersion, an electrolyte solution with the strong ionic strength has to be used as a sheath solution to manipulate the stream of the vesicular dispersion using an electric field. Hence, it is necessary to prepare stable vesicles in a highly ionic dispersion. For this purpose more hydrophobic amphiphiles (a new membrane molecule and a catalyst) are being prepared. These modified amphiphiles and catalysts form **GVs** that can be sorted by a cell sorter (Kurihara et al. to be published).

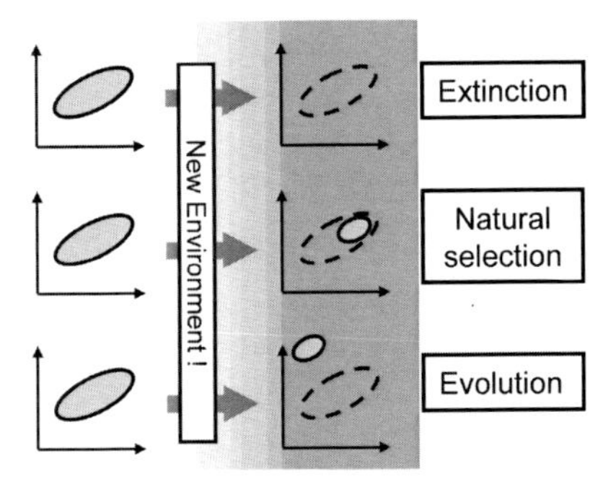

Fig. 2.18 Population change of **GVs** induced by the environmental change leads to evolution of **GV** ensemble. Self-reproducing giant vesicles (**GVs**), which is a prebiotic cell-membrane model, has been created. The population analysis of the self-producing **GVs** reveals that they repeat self-reproduction more than several times to increase their numbers by ca. 10^2 times. It is to be noted that a **GV** ensemble exhibits the recursive distribution in their sizes after reproduction cycles. It would be interesting to see how the distribution of **GVs** is influenced by the change of the environment, which may lead to a molecular model of evolution

2.5 Self-replication of Informational Substances in a Giant Vesicle

The robust self-reproduction model was constructed as described above. A next challenging target is the exploration of the origin of informational substances (RNA, DNA) and its replication. Von Kiedrowski proposed a purely organic template-reaction system without using oligonucleotides (Terfort and von Kiedrowski 1992). We can discuss the origin of self-replication of informational substances on the basis of his excellent model. In this subsection, however, we focus our attention on how to evolve our self-reproducing vesicular model to contain a self-replication system of informational substance, e.g., DNA, RNA. In this respect we are interested in the performance of a polymerase chain reaction (PCR) inside a giant vesicle (**GV**). Another approach has been reported by Hanczyc et al. They observed the association of RNA on clay (montmorillonite) surface encapsulated in fatty acid vesicles (Hanczyc and Fujikawa 2003). Szostak et al. Reported the chemical synthesis of DNA oligonucleotides with ca. 30 bp in **GVs** made of glycerol monoesters of fatty acids (Mansy et al. 2008).

2.5.1 *Enzymatic Reaction in Vesicle*

Recently, giant vesicles have drawn much attention as a micro-reaction capsule made of amphiphiles. One of the preceding investigations of an enzymatic synthesis in vesicles is the polycondensation of adenosine diphosphate (ADP) into poly(A) by encapsulated enzymes (Walde et al. 1994a). Later, gene-expression of green fluorescent protein (GFP) in a cell-sized vesicle was reported (Yu et al. 2001). As for the replication of informational substance in vesicles, Luisi et al. reported PCR in vesicles the diameters of which are less than 500 nm (Oberholzer et al. 1995). They confirmed the PCR products by polyacrylamide gel electrophoresis (PAGE), but the efficiency of DNA amplification was less than 0.1%. This is because the probability for a vesicle ($\phi <$ 500 nm) to encapsulate template DNA in the internal volume is extremely low.

2.5.2 *Performance of Polymerase Chain Reaction in GV*

Recently, Shohda and Tamura et al. conducted a "PCR in giant vesicles (**GVs**)" with a diameter of larger than 1 μm. Since the size of **GVs** is reasonably large, the probability of encapsulation of the template DNA is higher and **GVs** enable direct visualization of the performance of PCR using a fluorescent probe, such as SYBR Green I, under a fluorescence microscope. A protocol of PCR in giant vesicle in bulk water is briefly described in Fig. 2.19a. First, a double strand template DNA is denatured to two single strands. Since two kinds of primers are present in a water pool of **GV**, they hybridize with each 3'-end of the single strand DNA. The DNA

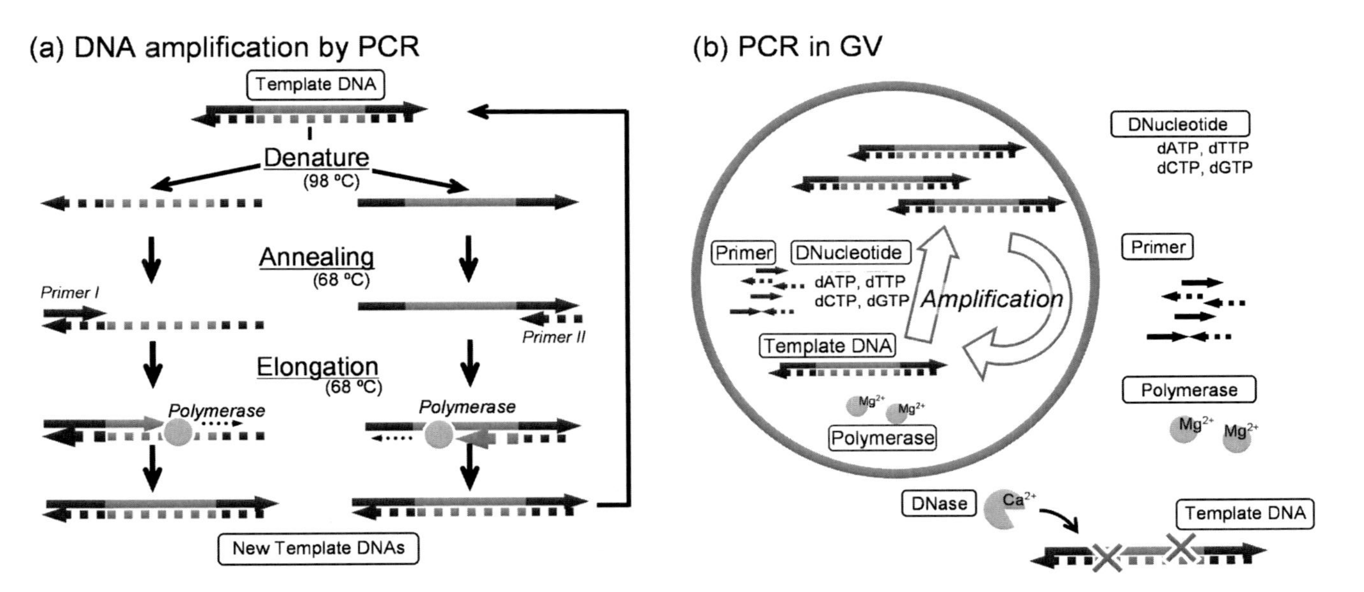

Fig. 2.19 (**a**) Protocol of PCR procedure. (**b**) Ingredients for PCR in **GV** and in an outer water phase. In addition, DNase was added to an exterior solution to prevent the extravesicular reaction. Factors that influences "PCR in **GV**" are listed below. **1. GV** formation in buffered solution PCR proceeds in a buffered solution with high ionic strength for DNA-hybridization. Generally, unilamellar-like vesicles are not formed in a buffered solution. However it turns out that an addition of 10% poly(ethyleneglycol)-grafted phospholipid (DSPE-PEG$_{5000}$) to POPC forms suitable **GVs** even in aqueous ionic solution. **2. Preparation of GV containing PCR reagents** The efficiency of encapsulation of PCR ingredients is desired to be high. In this respect, the lyophilization-and-rehydration method achieves highly efficient encapsulation of macromolecules, such as DNA templates or enzymes [23]. **3. Encapsulation of template DNA** Since the concentration of the template DNA is lowest of the PCR reagents, it is a key substance for the performance of PCR in **GVs** . Therefore we prepared a PCR solution of a relatively high concentration with reference to the template DNA (0.1 nM). The number of template DNAs is 10 for **GV** with a diameter of 10 μm, and only 0.01 for 1 μm **GV**. **4. Optimization of thermal cycle** The probability of nonspecific amplification of template DNA is thought to be higher for PCR in vesicle due to the steric constraints on protein activity. Hence a thermal cycle of PCR in **GV**, usually consisting of three steps [denaturation, annealing, and elongation], should be optimized accordingly. In the current case, the annealing temperature is elevated and set to be equal to the elongation temperature in order to avoid nonspecific amplification. A two step thermal cycle is arranged: [94°C for 15 s and 68°C for 1.5 min] × 20 cycles. Two kinds of primer (24mer and 22mer) are used to hybridize the single stranded template DNAs. **5. Suppression of enzymatic activity in exterior solution** PCR that proceeds in the exterior volume of **GV** must be inhibited to evaluate the efficiency of the reaction exclusively in **GVs**. For this purpose DNase I, which completely digests both single-strand primer and double-strand template DNA to the mononucleotide level, is added to the exterior solution

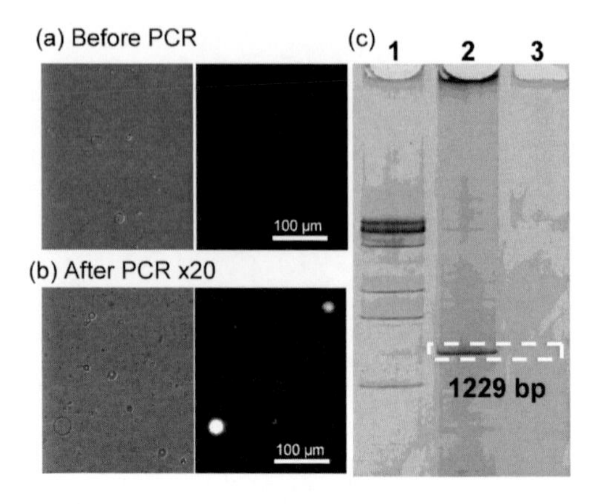

Fig. 2.20 (**a**) Optical and fluorescence images of before and (**b**) after PCR in **GV**. Before the thermal cycling, all the vesicles scarcely emitted fluorescence. After the thermal cycling, most of the vesicles emit intense fluorescence (**c**) Polyacrylamide gel electrophoresis (PAGE) data of the PCR products. Lane 1 represents a marker. Lane 2 represents a chart of PCR products in **GV**, showing a clear band at 1229 bp assignable to the amplified DNA. Lane 3 is of a sample before PCR

polymerase then recognizes the partial duplex and elongates it to the full duplex along the template DNA. This sequence is repeated by thermal cycles to amplify DNA. We can point out crucial problems that influence "PCR in a vesicle" as follows.

1. Buffered solution is necessary for PCR but it is not conducive to **GV** formation
2. Preparation of **GV** containing PCR reagents with high efficiency of encapsulation
3. Encapsulation and distribution of highly diluted template DNA
4. Optimization of thermal cycle to avoid non-specific amplification
5. Suppression of enzymatic activity in exterior solution

These problems and possible answers to them are found in the caption of Fig. 2.19b.

The fluorescence microscopic images of before and after PCR, which was conducted according to the above conditions, are shown in Fig. 2.20a. Whereas almost none of the **GVs** emit fluorescence before PCR (Fig. 2.20a), many **GVs** do emit fluorescence after PCR due to the complexation of SYBR green I with the duplicated DNA (Fig. 2.20b). It means that the performance of PCR in **GVs** is directly visualized by this technique. Since each microscopic area contained a dozen **GVs** emitting strong fluorescence, we could say that "PCR in **GV**" is not a rare but ubiquitous event. PCR products in **GV** were also analyzed by polyacrylamide gel electrophoresis (PAGE). The detection of a band at the position of 1,229 bp is direct proof for the full-length replication of the template DNA (Fig. 2.20c) (Tamura et al. to be published).

2.5.3 Flow Cytometric Analysis of PCR in Vesicles

Population change of **GVs**, accompanied by the progress of PCR, was monitored by flow cytometry. Figure 2.21 shows 2D data of **GV** dispersion containing PCR solution. A horizontal axis corresponds to the size of vesicles (forward scattering intensity) and a vertical axis corresponds the amount of catalyst (fluorescent intensity around 525 nm).

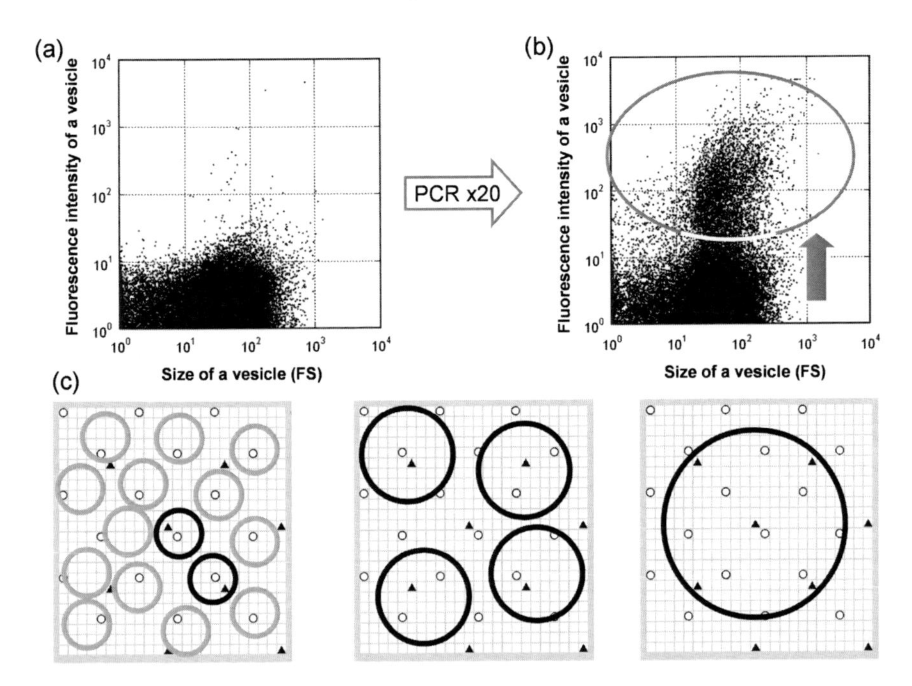

Fig. 2.21 Flow cytometric analysis of PCR in **GV** before PCR (**a**) and after PCR (**b**). Size effect of **GV** upon the entrapment efficiency of PCR reagents (open circle represents DNA polymerase and black triangle represents template DNA) (**c**)

Before the thermal cycling, a ratio of vesicles (total number of ca. 2×10^4) exhibiting the fluorescent intensity larger than a threshold value (FL $= 4 \times 10$; scale arbitrary scale) is only 0.1% (Fig. 2.21a), whereas after 20 cycles the ratio of **GVs** above the FL-threshold becomes 13% (Fig. 2.21b). We also notice that the PCR performance of smaller **GVs** is much lower than that of larger ones and that the performance of larger **GVs** is not necessarily high. This result suggests that the PCR performance depends not only on the size but also on the lamellar structure of **GVs**; if **GVs** are tightly packed or highly nested, those **GVs** are not PCR active.

2.5.4 Meaning of Encapsulation in Enzymatic Reactions

Population analysis on "PCR in **GV**" tells us that there is an intrinsic difference between PCR in bulk solution and PCR in **GV**. The difference is derived from the encapsulation effects of **GVs** on the PCR performance. We can point out following features.

1. Entrapment of precious component

Let us suppose that a substrate and an enzyme are dissolved in a highly diluted solution. If they eventually meet, the substrate is transformed to the product. However, the probability of collision is very low. In contrast, if two compounds at the same diluted concentration are in the same capsule, the chance of collision is extremely high. On the other hand, if they are not entrapped in a same capsule, the reaction never happens. This signifies that local concentrations of components become very important in reactions that are encapsulated, much like natural living cells.

Since the concentration of the template DNA is lowest of the components in the current PCR condition, it becomes the key component for the PCR performance. For example, the concentration of 10^6 DNA in 1 μL water is $10^6 \times 1.7 \times 10^{-24}/10^{-6}$ L = 1.7×10^{-12} mol/L = 1.7 pM, while if one template DNA is included in a vesicle with a diameter of 10 μm, the concentration is 3.4×10^{-9} mol/L = 3.4 nM. Encapsulation in a vesicle exerts a large condensation effect on the reaction (Fig. 2.21c).

2. Barrier effect of membrane

Suppose that an inhibitor and an enzyme are dissolved in solution. Even though the concentrations of the two are very low, the substrate is eventually decomposed by the enzyme. However if a substrate is encapsulated it is never digested, even though the concentration of the enzyme in the outer water phase is high. For example, the PCR performance in **GV** is not quenched even when DNase is added to an exterior solution of **GVs**. The barrier effect protects the encapsulated system from inhibitors and parasites.

3. Electrostatic interaction between membrane and ingredients

Electrostatic interaction between charged membranes and charged ingredients influences a reaction-in-capsule significantly. For example, since cationic vesicular membranes form lipoplex with negatively charged DNA, no **GVs** are formed in solution. On the other hand, anionic vesicular membranes repel negatively charged DNA, so that the efficiency of encapsulation decreases. If the anionic membrane likes to form a complex with DNA polymerase, it may inhibit the enzymatic activity. Therefore charge–charge interactions must be considered when using this technology. Shielding of such direct interactions between the membrane and reaction components may be mitigated by the introduction of counter ions or PEG-lipids.

4. Lamella structure and PCR performance

As revealed by the flow cytometric analysis, PCR in large **GVs** does not necessarily proceed, even though the probability of encapsulation of a template DNA or an enzyme is reasonably high. It means PCR in tightly packed multi-lamella or nested **GVs** does not proceed. Sato et al. drew a contour map of vesicles in a FS (size)-SS (lamella structure) 2D plot and demonstrated microscopic images of sorted vesicles, showing the versatility in the lamella structure of vesicles with a similar size (Fig. 2.22) (Sato et al. 2006). Yomo et al. Evaluated the probability of encapsulation of plasmid DNA based on the fluorescence intensity obtained from the enzymatic cascade reaction in **GV** They claim that the size-dependence of the probability shows an exponential increase (Hosoda et al. 2008).

Additionally, an inner membrane structure may accelerate certain kinds of reaction by holding two components in the right orientation. In such a case, the membrane

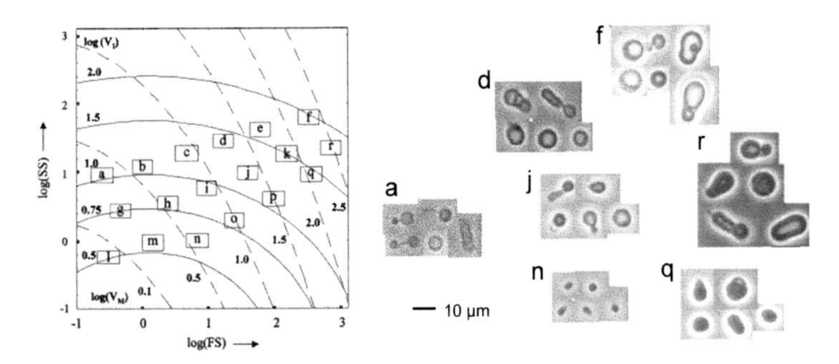

Fig. 2.22 2D-polt (SS vs FS) of giant vesicles consisting of a mixture of phospholipids. SS (side scatter) and FS (forward scatter) are parameters for lamella structure and size of **GVs**, respectively (*left*). Microscopic images of sorted **GVs** (*right*).

serves as a reaction-field for a specific reaction just as a protein does. Therefore, a membrane may serve a function in reactions similar to a protein catalyst.

We find that the enclosed membrane environment offers certain advantages, especially when reaction components are scarce or dilute and that the membrane in some cases may act as an organizing substrate to accelerate some types of reactions. We can take advantage of such particular qualities when constructing a **GV**-based artificial cell model.

2.6 Coupling between Self-reproduction of GV and Self-replication of DNA

Since our vesicular self-reproducing cell model can replicate DNAs inside a capsule, a next challenge is how to distribute amplified DNAs equally to the daughter vesicles during a division process. Otherwise a vesicular division event could produce some daughter vesicles filled with reaction components and others completely empty (Fig. 2.23).

2.6.1 Design and Preparation of DNA-cholesterol Conjugate

To partition the replicated informational substance to dividing vesicles equally, it is worthwhile to prepare a conjugate molecule to couple these two dynamics: self-reproduction of **GV** and self-replication of DNA. Shohda et al. designed and prepared a "conjugate molecule" consisting of a **template** made of DNA 15mer, an **anchor** made of cholesterol, and a **spacer** made of polyethyleneglycol (Fig. 2.24a) (Shohda et al. 2003). Since the both **GV** and DNA are charged, it is necessary to insert a polyethylene glycol spacer between two units to suppress any inhibitory effect on the activity of DNA polymerase. At the same time, the anchor unit fixes the conjugate molecule to the vesicular membrane.

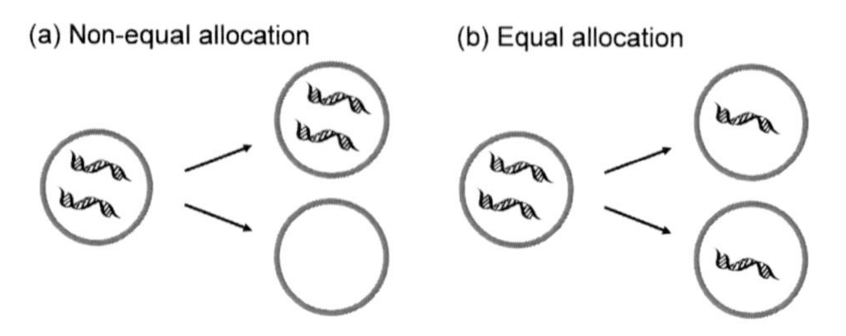

Fig. 2.23 Partition of duplicated DNA to daughter vesicles: (a) Non-equal allocation, (b) equal allocation

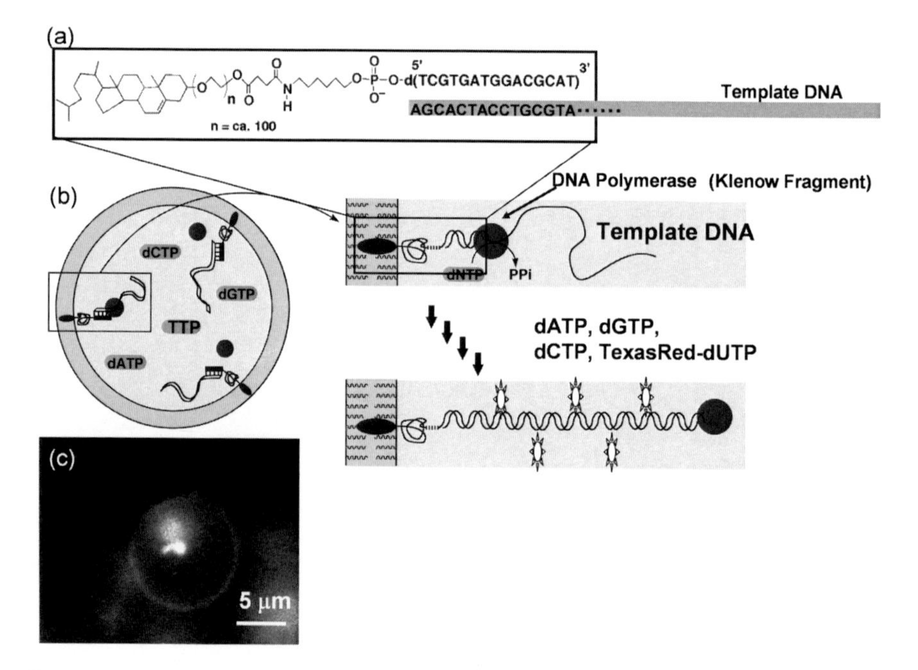

Fig. 2.24 DNA replication using a conjugate primer on the surface of inner membrane of giant vesicle (**a**) Chemical structure of DNA-conjugate. (**b**) Schematic scheme of complimentary DNA-replication occurring on the inner surface of vesicular membrane. DNA-conjugate serves as a primer of complementary replication of single-strand template DNA. The incorporation of the fluorescent signal, TexasRed, is represented by the sun-like symbols along the copied template (**c**) Fluorescence microscopic image of DNA-replication on an inner surface of vesicular membrane

2.6.2 *Replication of DNA Inside GV Carrying DNA–Cholesterol Conjugate*

A vesicle, containing a set of reagents for the DNA replication, is prepared in a similar manner to the specimen for PCR. A vesicular membrane is made of POPC, cholesterol, DSPE-PEG 5000 and contains the conjugate molecule and the inner volume includes a buffered solution of single stranded DNA, DNA polymerase (Klenow fragment), dNTP and magnesium ion (Mg^{2+}). To suppress the activity of DNA polymerase in the outer water layer, EDTA is added to chelate Mg^{2+} which is essential for enzymatic activity.

If the DNA 15 mer part of the conjugate molecule hybridizes with the single strand template DNA, it can serve as a primer of the DNA polymerization. It means that the complementary replication of DNA takes place only on the inner surface of the vesicular membrane (Fig. 2.24b). The full duplication of the template DNA in **GV** was confirmed through gel electrophoresis by detecting the specific fragment of the DNA duplex cleaved by the restriction enzyme (*Hind* III), which cleaves the DNA duplex at the specific site (Shohda and Sugawara 2006).

In order to perform direct visualization of DNA duplication on the inner surface of **GV**, a fluorescence-labeled dUTP* (TexasRed-L12-dUTP) was added to the buffer solution. As shown in Fig. 2.24c, only the membrane of **GV** emitted fluorescence, rather than the internal water volume. This observation strongly suggests that the duplication of DNA did occur only on the inner surface of **GV**.

2.6.3 *Partition Mechanism in Cell Division of Escherichia coli*

A **GV** which grows and divides into two daughter vesicles carrying replicated informational substances would be a sophisticated model of an artificial cell. Here

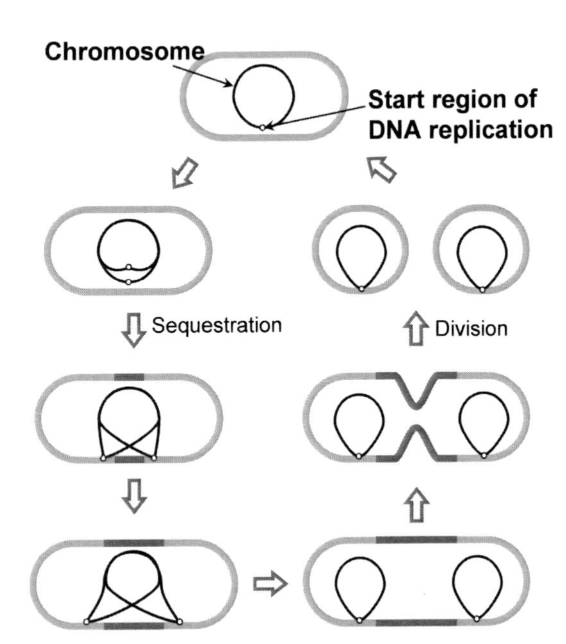

Fig. 2.25 Schematic representation of partition mechanism of chromosome *oriC* in *Escherichia coli*. A prebiotic cell must partition its information substance equally to its offspring. The attachment of the informational molecule to the inner surface of the membrane guarantees the partition of genomic information to the divided daughter cells regardless of whether the division of the membrane is not completely controlled or equal

we consider a cell division of a eucaryotic cell. At the very early stage of replication of the chromosome *oriC* in *E. coli*, two new *oriCs* bind to the cell membrane (Fig. 2.25). Thereafter, the cell membrane between two connected terminals elongates and becomes squeezed. Thus the separated genomic DNAs are equally distributed in two daughter cells. Such dynamics is called a "partition mechanism." The experimental result obtained above by using the DNA-cholesterol conjugate molecule would be interpreted as a simplified model for the partition mechanism of DNA distribution to the daughter vesicles.

2.7 Evolution Towards Artificial Cell

Finally, we will discuss the evolution of a living system starting from a prebiotic molecular system to current life. A simple system to show prebiotic movement appeared from a chemical soup containing a fatty acid (oleic acid and oleate) at a specific pH (pH = 8). Here we see that oleic acid and oleate forms a multi-lamella helical tube that exhibits macroscopic winding dynamics as a result of the **cooperative transmission of local stress** originating from microscopic dynamics. Such a rather complex supramolecular ordering dynamic could be achieved in a prebiotic world consisting of simple organic compounds. The significance of such movements towards the evolution of the first living cells is under investigation.

The other example of this category of simple systems that exhibit movement concerns the self-propelled motion of a droplet made of oleic anhydride. The characteristic point of this dynamic system is that **macroscopic spontaneous movement is triggered by the microscopic chemical reaction** (hydrolysis of oleic anhydride to oleic acid and oleate), which induces **symmetry breakage** and **a positive feedback loop.** We evolved this self-propelled oil droplet in two directions. One is the usage of an externally supplied "fuel" surfactant to induce persistent movement. The other is the emergence of interactive movements between self-propelled droplets through chemicals that dissolved in an outer water layer or in a trail released from the droplets. We see primitive examples of positive and negative **chemotaxis** in such a simple system.

As seen from the above examples, our self-assembled molecular structures can change morphology or behavior responding to stimuli from the environment. Self-reproducing dynamics is another example of this type of dynamic behavior when the membrane molecules are synthesized within the self-assembled structure itself. The **robust self-reproducing dynamics** was, thus, constructed using a catalyst-containing giant vesicle (**GV**) in which a precursor is transformed to the same membrane molecule as the original in the vesicular membrane. The corpulent vesicle was then spontaneously divided into two vesicles of similar size. It was found that this self-reproduction continues repeatedly several times and the number of vesicles increases by 10^2-fold. The size-distribution of **GVs** is also maintained after several self-reproduction cycles determined by the population analysis on the self-reproducing vesicles by flow cytometry. Population analysis on the self-reproducing vesicular model will open a possibility to reveal how the environmental change affects the population of the vesicular ensemble. Such environment-sensitive behavior exhibited by self-reproducing vesicles as an ensemble may be regarded as **a primitive model of the evolution of artificial cells.**

Finally, the informational molecule (DNA) can efficiently be replicated inside a vesicle. We noticed that **natural selection and adaptation** is already present in this simple system from the performance of enzymatic reactions in the capsule. Although giant vesicles (**GV**) are formed under the same conditions, the components of membrane molecules, concentrations of enzymatic reagents, etc., may be distributed differently from vesicle to vesicle. For example, if the content of poly (ethyleneglycol)-grafted phospholipid of a **GV** is too low, then such a **GV** cannot endure under a high electrolyte concentration, which is necessary for DNA amplification. Likewise, if a reasonably large vesicle contains no template DNA, it cannot amplify DNA. As a result, such a vesicle cannot distribute DNA to the self-divided vesicles after self-reproduction.

The last step towards an artificial cell is how to introduce a coupling between two kinds of dynamics, self-reproduction dynamics of a cellular membrane and self-replication of informational molecules. We propose the amphiphilic molecule conjugated to a DNA primer head would be a viable candidate to connect two essential dynamics.

So far the artificial cell system has made a progress along a man-made scenario. If this scenario is an appropriate one and if an artificial cell exceeds a certain critical stage, our artificial cell will start to evolve by itself when energy and matter is supplied. We can then say life is created.

References

Daoud M, Williams CE (1999) Soft matter physics. Springer, Berlin

Fox SW et al. (1994) Experimental retracement of the origins of a protocell: It was also a Proteoneuron. J. Biol. Physics 20:17–36

Hamley IW (2000), Introduction to soft matter. Wiley, New York

Hanczyc MM, Fujikawa SM (2003) Experimental models of primitive cellular compartments: encapsulation, growth, and division. J W Szcostak, Science 302:618

Hanczyc MM, Toyota T, Ikegami T, Packard N, Sugawara T (2007) Chemistry at the oil-water interface: Self-propelled oil droplets. J Am Chem Soc 129:9386–9391

Helfrich W (1973) Elastic properties of lipid bilayers: Theory and possible experiments. Z Naturforsch C 28:693–703

Hosoda K, Sunami T, Kazuta T, Suzuki H, Yomo T (2008) Quantitative Study of the Structure of Multilamellar Giant Liposomes As a Container of Protein Synthesis Reaction. Langmuir 24:13540–13548

Ishimaru M, Toyota T, Takakura K, Sugawara T, Sugawara Y (2005) Helical aggregate of oleic ... dynamics in water at pH 8. Chem Lett 34:46–47

Kaneko K (2006) Life: An introduction to complex systems biology (understanding complex systems). Springer, New York

Kindermann M, Stahl I, Reimold M, Pankau WM, von Kiedrowski G (2005) Systems Chemistry: Kinetic and Computational Analysis of a Nearly Exponential Organic Replicator. Angew Chem Int Ed 44:6750–6755

Kurihara K, Toyota T, Sugawara T, *to be published.*

Luisi PL (2006) The emergence of life: From chemical origins to synthetic biology. Cambridge University Press, New York

Luisi PL, Walde P (2000) Giant vesicles. Wiley, New York

Mansy SM, Schrum JP, Krishnamurthy M, Tobè S, Treco DA, Szostak JW (2008) Template-directed synthesis of a genetic polymer in a model protocell. Nature 454:122–125

Miller SL (1953) A production of amino acids under possible primitive Earth conditions. Science 117:528–529;

Miller SL, Urey HC (1959) Organic compound synthesis on the primitive Earth. Science 130: 245–251

Oberholzer T, Albrizio M, Luisi PL (1995) Chem Biol 2:677–682

Oparin AI (1952) The origin of life, Dover, New York

Oparin AI, Fesenkov V (1961) Life in the universe, Twayne Publishers, New York

Takakura K, Totota T, Sugawara T (2003) A novel system of self-reproducing giant vesicles. J Am Chem Soc 125:8134–8140

Takakura K, Sugawara T (2004) Membrane Dynamics of a Myelin-like Giant Multilamellar Vesicle Applicable to a Self-Reproducing System. Langmuir 20:3832–3834

Terfort A, von Kiedrowski G. (1992) Self-replication by condensation of 3-amino-benzamidines and 2- formylphenoxyacetic acids. Angew Chem Int Ed Engl 31:654–656

Toyota T, Tsuha H, Yamada K, Takakura K, Ikegami T, Sugawara T (2006) Listeria-like motion of oil droplets. Chem Lett 35:708–709

Toyota T, Takakura K, Kageyama Y, Kurihara K, Maru N, Ohnuma K, et al. (2008) Flow cytometric investigation of self-reproducing giant multilamellar vesicles. Langmuir 24:3037–3044

Toyota T, Maru N, Hanczyc MM, Ikegami T, Sugawara T (2009) Self-Propelled Oil Droplets Consuming "Fuel" Surfactant. J Am Chem Soc 131:5012–5013

Tsafrir I, Guedeau-Boudeville MA, Kandel D, Stavans, J (2001) Coiling instability of multi-lamellar membrane tubes with anchored polymers. Phys Rev E 63:031603

Sato K, Obinata K, Sugawara T, Urabe I, Yomo T (2006) Quantification of Structural Properties of Cell-sized Individual Liposomes by Flow Cytometry. J Biosci Bioeng 102:171–178

Shohda K, Sugawara T (2006) DNA Polymerization on the Inner Surface of Giant Liposome for Synthesizing an Artificial Cell Model. Soft Matter 2:402–408

Shohda K, Toyota T, Yomo T, Sugawara T (2003) Direct Visualization or DNA Duplex Formation on the Surface of a Giant Liposome. Chem Bio Chem 4:778–781

Sumino Y., Magome N., Hamada T., Yoshikawa K (2005) Self-running droplet: Emergence of regular motion from nonequilibrium noise. Phys. Rev. Lett. 94:068301

Szostak JW, Bartel DL, Luisi PL (2001) Synthesizing life. Nature 409:387–390

Walde P, Goto A, Monnard P-A, Wessicken M, Luisi PL (1994a) Oparins reactions revisited – Enzymatic – Synthesis of poly(Adenylic acid) in Micelles and Self-reproducing vesicles. J Am Chem Soc 116:7541–7547

Walde P, Wick R, Fresta M, Mangone A, Luisi PL (1994b) Autopoietic self-reproduction of fatty acid vesicles. J Am Chem Soc 116:11649–11654

Yu W, Sato K, Wakabayashi M, Nakaishi T, Ko-Mitamura EP, Shima Y, Urabe I, Yomo T. (2001) J Biosci Bioeng 92:590–593

Nomura SM, Tsumoto K, Hamada T, Akiyoshi K, Nakatani Y, Yoshikawa K. (2003) ChemBio-Chem 4:1172–1175

Tamura M, Shohda K, Kageyama Y, Suzuki K, Suyama A, Sugawara T, *to be published.*

Chapter 3
New Fossils and New Hope for the Origin of Angiosperms*

Xin Wang

**Dedicated to the 200th anniversary of Charles Darwin's birthday*

Abstract Angiosperms are the dominant group in modern vegetation. Despite their importance and over hundred years of research effort since the Darwin age, we still do not know much about the origin of this important group. Recently, *Schmeissneria* has pushed the origin of angiosperms close to the Triassic and at the same time the angiosperms in the early Cretaceous also demonstrate diversity higher than expected, suggesting that angiosperms have occurred long before the currently recognized oldest fossil record, unlike the currently predominant doctrine states. In this chapter, I will briefly summarize the information about several Chinese and German fossil plants ranging from the Early Cretaceous to Early Jurassic in age. These fossils, in addition to others, indicate that much of the angiosperm diversity was extinct before the Cretaceous, many pre-Cretaceous angiosperms may have no direct relationship with living angiosperms, and the key to the abominable mystery may lie in fossil plants that are unknown to scientists yet.

3.1 Introduction

Angiosperms are the most diversified plant taxon in the present world. They dominate in most vegetation types, from desert to tropic rainforest and from equator to poles. They are closely related to the origin, evolution and persistence of the human being. Despite their importance we know very little about the origin of these plants, namely, when, where and how they came into existence. This question is famous but not new at all. In 1879, Charles Darwin described the rapid rise of

X. Wang
Nanjing Institute of Geology and Palaeontology, 39 Beijing Dong Road, Nanjing 210008, China;
Fairylake Botanical Garden, 160 Xianhu Road, Shenzhen 518004, China
e-mail: brandonhuijunwang@gmail.com

P. Pontarotti (ed.), *Evolutionary Biology: Concept, Modeling, and Application*,
DOI: 10.1007/978-3-642-00952-5_3,

angiosperms as an "abominable mystery." Since then, many scientists have been struggling to solve this problem. Historically, it was once thought the problem had been solved in the 1950s–1960s and many "angiosperms" were found in the Triassic and Jurassic but later more careful research (Scott et al. 1960) eliminated those angiosperm-appearing plants from the angiosperms. Information in the 1970s and later, especially that of mesofossils, almost consistently points to the early Cretaceous as the time for the origin of angiosperms (Friis et al. 2005, 2006). Nowadays the so-called Cretaceous-only-angiosperm belief appears well rooted in many palaeobotanists' minds, although it has been more or less undermined by reports of possible angiosperms from the pre-Cretaceous ages now and then (Cornet 1986, 1989a, b, 1993; Cornet and Habib 1992; Duan 1998; Hochuli and Feist-Burkhardt 2004; Ji et al. 2004; Leng and Friis 2003, 2006; Sun et al. 1998, 2001, 2002; Wang et al. 2007a, b; Zavada 2007). There are reports of angiosperms, including *Chaoyangia*, *Archaefructus* and *Sinocarpus*, from the famous Yixian Formation of the Early Cretaceous (Dilcher et al. 2007; Duan 1998; Ji et al. 2004; Leng and Friis 2003, 2006; Sun et al. 1998, 2001, 2002), however, these reports are either ignored, disputed, or even suspected and the claimed Jurassic age has been also highly debated due to various reasons (Dilcher et al. 2007; Duan 1998; Friis et al. 2005, 2006; Ji et al. 2004; Leng and Friis 2003, 2006; Sun et al. 1998, 2001, 2002; Swisher et al. 1998).

It is apparent that there is still a far way to go before this mystery could be deciphered satisfactorily. However, it is clear that the only reliable way to confirm or deny angiosperms in the Jurassic is working on Jurassic fossil plants rather than speculating based on Cretaceous fossil plants. Any reasoning and speculation appear pale in front of nature, since, as Jean Baptiste Lamarck said, "Nature exceeds on all sides the limits we so gratuitously impose on her." As a step to approaching the problem, I would like to introduce some basics about angiosperms, problems and strategies in recognizing them in the fossil record and then I will give some examples of early angiosperms and point out their implications on the origin and evolution of angiosperms.

3.2 Definition of Flower and Angiosperm

Nowadays a primary schooler appears to know what a flower is. But the question becomes extremely difficult when it comes to give a universally applicable scientific definition, especially for fossil plants. In botany, a flower is defined as a reproductive organ of an angiosperm. If you can ascertain that one plant is an angiosperm, you can safely call its reproductive organ a flower. Then the question is translated into another one, what is an angiosperm? This is a question of a science called phytotaxonomy.

Initially, taxonomy was a science putting the same type of organisms together and phytotaxonomy is a branch of taxonomy dealing with plants. Types play a crucial role in taxonomy. According to the International Code of Botanical Nomenclature, a name is connected with a type. Thus even today holotype and various other types are still required and used in taxonomical practice. Later, as science advanced, people found that types alone could not solve all the problems in taxonomy satisfactorily, they extracted certain characters from the entities as features for a taxon and systemized the entities in order based on these characters. These characters are called diagnosis for a taxon. Angiosperm is one of the many taxa phytotaxonomists face everyday. Several characters are used as a diagnosis to unit angiosperms together and distinguish them from gymnosperms. Among them are enclosed seeds/ovules, reticulate leaf venation, tectate-columellate pollen wall, double fertilization, pollen tube, vessel elements and chemical compounds. These features are frequently seen in angiosperms but rarely in gymnosperms. If you have seen all or several of these features in a plant, it is safe for you to declare, “It is an angiosperm!” However, if you are a qualified taxonomist, you should be aware of some exceptions before you go too far. Like Larmarck said above, nature is never so simple as we think. There are always exceptions to rules. The same can said for angiosperms. Actually, none of the above features are unique to angiosperms.

It is almost universally accepted that angiosperms are defined by enclosed seeds, it is what the word “angiosperm” exactly means. Enclosure of seeds/ovules provides seeds of angiosperms extra protection and enhances competition during fertilization and thus gives angiosperms the edge over their gymnospermous rivals. If this advantage was patented by angiosperms, identifying an angiosperm would be much easier. Unfortunately, some of the gymnosperms have also evolved to apply similar strategies to enhance the opportunity for their offspring to survive in the harsh competition. Tomlinson and Takaso (2002) emphasize that some of the conifers have the tendency to enclose and protect their seeds. Actually other fossil and extant gymnospermous groups, such as Caytoniales and Gnetales, have also demonstrated the same tendency (Chamberlain 1957; Harris 1940, 1964; Thomas 1925). Conversely, not all angiosperms have their seeds/ovules physically enclosed, such as those in Amborellaceae, Schisandraceae, Austrobaileyaceae and Trimeniaceae (Endress and Igersheim 2000).

Reticulate leaf venation is rarely seen in gymnosperms but frequently seen in angiosperms, thus it is frequently taken as a sign for angiosperms by a layperson. Actually, reticulate leaf venation has been reported in non-angiosperms, including Dipteridaceae, Gigantopteriales, Caytoniales, Glossopteridales, Bennettitales and Gnetales (Chamberlain 1957; Harris 1940, 1964; Hughes 1994; Kryshtofovich 1923; Li and Taylor 1998; Li et al. 1994; Potonie 1921; Retallack and Dilcher 1981; Shen et al. 1976; Sporne 1971; Sun 1981, 1993; Thomas 1925). Conversely, reticulate leaf venation is absent in monocots, a group of plants on which most people rely to survive in this world.

A tectate-columellate pollen wall structure is usually seen in angiosperms. Pollen grains with such a wall structure have been reported in many pre-Cretaceous plants (Cornet 1989a; Cornet and Habib 1992; Hochuli and Feist-Burkhardt 2004; Pocock and Vasanthy 1988; Zavada 1984) that are regarded as non-angiosperms by others (Friis et al. 2005, 2006). In the meantime, early angiosperms may well have not had such an advanced pollen wall structure (Zavada 1984).

Double fertilization is usually taken as unique to the angiosperms. It makes big differences between angiosperms and gymnosperms in development, reproduction and survival strategies. However, it has been seen in non-angiosperms, as in *Ephedra* and *Abies* (Chamberlain 1957; Friedman 1990, 1991, 1992; Yang et al. 2000). Conversely, not all angiosperms undergo double fertilization: double fertilization does not occur in *Calycanthus* (Stevens 2008).

Pollen tube is a special channel transferring sperm to the egg in plants, especially in angiosperms, during fertilization. However, some structure at least very similar to pollen tube has been seen in Paleozoic seed ferns (Rothwell 1972) and living Coniferales and Gnetales (Crane 1985), although there is some uncertainty about its actual function in the fossils.

Vessel elements are one of the major anatomical features that give angiosperms advantage in water usage. But they have been found in various non-angiosperms, such as *Selaginella*, *Equisetum*, *Pteridium*, Gigantopteriales and Gnetales (Chamberlain 1957; Cronquist 1988; Eames 1961; Li et al. 1996; Sporne 1971). In the meantime, many basal angiosperms, including *Amborella*, do not have vessel elements at all (Doyle 2008; Eames 1961).

The presence and lack of various chemical compounds, including secondary metabolites, DNA, RNA and proteins, can be used to determine the relationships among plants at various levels (Judd et al. 1999). For example, betalains are restricted to Caryophyllaceae while flavonoids are distributed throughout the embryophytes (Judd et al. 1999). However, some chemical compound, such as syringyl lignin, was formerly considered to be restricted in the angiosperms but a recent study indicates it is also present in *Selaginella* (Lycophyta) (Weng et al. 2008). Parallel to this, oleanane that was formerly thought restricted to the angiosperms has been found in Paleozoic and Mesozoic non-angiospermous plant fossils (Taylor et al. 2006).

Facing the above rules and exceptions, many might feel lost and hopeless, "Since all characters appear useless to draw a line between angiosperms and gymnosperms, which character should we use to pin down an angiosperm, especially in the fossil world?" Fortunately, there is a character, physically closed carpel at pollination, namely physically enclosed ovules, which is a sufficient character to identify an angiosperm. Actually, all ovules in angiosperms are at least once exposed and all seeds are more or less protected by enclosure in seed plants. It appears that the closing time matters: angiosperms' ovules are enclosed at the time of the fertilization, while those in gymnosperms are always exposed at the time of fertilization. According to Tomlinson and Takaso (2002), this is the only consistent difference between angiosperms and gymnosperms. This enclosure may be achieved by physical closure of the carpel or by secretion. Physical enclosure, since secretion enclosure is hard to prove in fossil plants, is the character the author will apply to identify angiosperms in the fossil record. By the way, besides this character, more characters, especially those of plants in different developmental stages, are always welcome and helpful for *bona fide* determination.

3.3 Acquisition of the Features

As we have seen above, there are several characters in the diagnosis of angiosperms. With all these features, it is pretty easy for one to identify an angiosperm. This is a very normal and common practice for neobotanists. However, the situation becomes complicated when you go back to the early history of angiosperms. At that time, there was little distinction between angiosperms and gymnosperms, there was no typical angiosperm, either, and the so-called angiospermous characters may well be scattered in several unrelated plant groups. Using all of these characters to identify an angiosperm at that time would only result in no angiosperm. Most living angiosperms are full-fledged angiosperms, which display the feature assemblage typical for angiosperms. Historically, it was a common "logical" practice for palaeobotanists to apply the criteria for full-fledged angiosperms onto early angiosperms. Thus it is non-surprising to find that there were no angiosperms in pre-Cretaceous. This mistake is partially due to the overwhelming influence of phytotaxonomists on palaeobotanists and ignorance of the evolution that angiosperms may have undergone since their earliest age.

Evolutionary, all characters in plants, from cellular to morphological, have their own history and have undergone a process from nothing to something, to full extent. What we see today in angiosperms is the result of long time evolution and it is also the starting point for future evolution. The so-called stable character we see today in plants is just a snapshot of the on-going evolution that started a billion years ago. The origin and development of a character in plants is a process related to time. It is hard to imagine that two, not to mention several, features to come to occur in the same plant at exactly the same time. Therefore it is more plausible that the above-listed characters of angiosperms were acquired one after another (Doyle 2008), as in Fig. 3.1. With this in mind, apparently not all of the characters that were acquired at different times can be used to determine when angiosperms came to existence in geological history, otherwise only chaos and controversy can be expected. To avoid this kind of dilemma and chaos, which character should be chosen as the index feature to determine whether a plant is an angiosperm or not? According to the preceding section, ovule physically enclosed at fertilization is the most favorable choice. This will be the criterion the author will use to identify angiosperms hereafter.

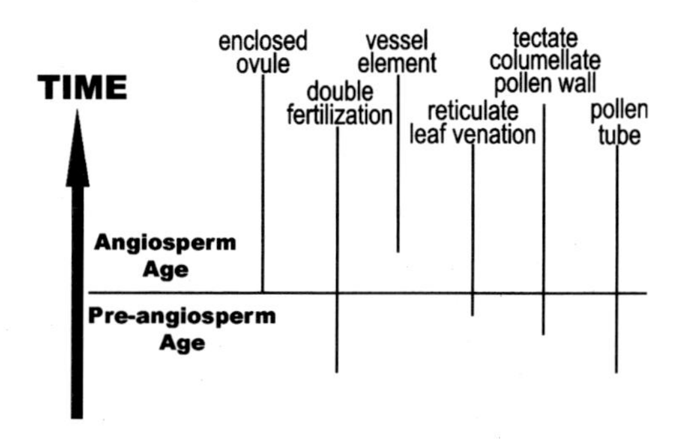

Fig. 3.1 A possible scenario for the acquisition of characters typical for angiosperms during the geological history. Apparently these characters could not come to occur in the same plant at the same time. Picking a different character as the index for angiospermy usually results in debates and controversy on the origin time of angiosperms among various scholars. The author prefers to take physically enclosed ovule as the index character for angiospermy

3.4 Examples of Early Angiosperms

The discoveries of *Archaefructus* and *Sinocarpus* from the Yixian Formation (Early Cretaceous) in western Liaoning, China triggered a great interest in early angiosperms. Before these, mesofossils of the Cretaceous were the major source of information about early angiosperm evolution for a while. Despite controversy around these megafossils of early angiosperms, research effort on early angiosperm never stops. In the following sections, the author will introduce some recent progress on the study of fossil plants that can be put in angiosperms according to the above criterion. Some of these fossil plants are new materials, some are old but restudied materials with new features. Although it is still too early to say how much these fossils can contribute to solving the abominable mystery, they will definitely bring us closer to the target and therefore they deserve our decent attention.

Due to limited space, *Archaefructus* and *Sinocarpus*, which have been well-documented and debated (Dilcher et al. 2007; Doyle 2008; Ji et al. 2004; Friis et al. 2005, 2006; Leng and Friis 2003, 2006; Sun et al. 1998, 2001, 2002), will be excluded from this summary. It is also beyond the author's capability to give a full treatment of each fossil plant, so only a brief history, a discussion on its affinity, a diagnosis and a description accompanied by a few figures are given for each fossil plant. For details, please refer to the original publications or publications in progress.

3.4.1 Chaoyangia

3.4.1.1 Brief History

Chaoyangia was established by Dr. Shuying Duan in 1998 (Duan 1998). At that time, it was put in the angiosperms. Later when *Archaefructus* came into light on Science, *Chaoyangia* was put equivalent to *Gurvanella* and further as a relative of *Welwitschia* (Gnetales) by mistake (Sun et al. 1998). This mistake appeared more like a truth when Krassilov et al. conflated *Chaoyangia* and *Gurvanella* (Krassilov et al. 2004), a genus established by Krassilov (Krassilov 1982). Now it is widely accepted that *Chaoyangia* is a synonym of *Gurvanella*. However, restudy on the holotype and more materials reveals that *Chaoyangia* is a valid genus and is an angiosperm.

3.4.1.2 Discussion

Restudy on the holotype and new materials of *Chaoyangia* indicates (1) it is a monoecious plant; (2) the male flowers are attached below the female flowers along the branch; (3) male flowers have *in situ* monocolpate pollen grains; (4) young female flowers have long straight styles with terminal stigmas; (5) each female flower is spherical and has three carpels that are almost completely enclosed in a hairy receptacle; (6) the fruits are spherical, developed from the three carpels surrounded by the receptacle that has dense hairs, which are on its surface and surrounding the former styles; and (7) each carpel and its counterpart in fruits has an atropous ovule/seed in the spacious locule. All these features together suggest that *Chaoyangia* have their carpels closed before fertilization and their seeds are enclosed and quite different from those in Gnetales and Caytoniales. These conclusions make *Chaoyangia* in place in the angiosperms than anywhere else. Cladistic analysis places *Chaoyangia* close to Hydatellaceae.

3.4.1.3 Diagnosis

Dichasial bisexual flowering branch, with linear leaves. Branch with parallel longitudinal ribs and rare interconnections in-between. Leaf with parallel veins and rare interconnections. Male flower, consisting of two male parts, attached to the branch below female flowers. Each male part including a foliar structure, with numerous pollen sacs sessile on its adaxial surface and upward pricks along its margin. *In situ* pollen grain monocolpate. Female flower terminal, with an urceolate receptacle bearing forked hairs on its surface and enclosing three ovaries. Three carpels inserted on the central bottom of receptacle, each forming an ovary at the bottom and a straight slender style with a terminal stigma at the top. A single atropous ovule inserted to ovary bottom. Fruit indehiscent. Seed single, attached to the base of the fruit.

3.4.1.4 Description

The holotype is about 13 cm long and 11 cm wide, including physically connected male and female flowers of various maturity. Another physically connected specimen including branch and fruits is about 8 cm long and 7 cm wide. All other specimens are isolated female flowers or fruits. The plant includes a monoecious compound dichasium. Two lateral branches are oppositely arranged and subtended by a leaf. The branches are slightly contracted immediately above the joint. Branches of various orders are 0.3–1.6 mm wide, with 4–6 parallel longitudinal ribs on its visible half surface and rare connections between the ribs. Most branches are rigid and straight while some young ones appear fleshy. The leaves are linear, with parallel veins that occasionally branch. Below some young female flowers are male flowers. Each male flower has two male parts oppositely arranged (Fig. 3.2a). A male part is 1.5–2.5 mm long, 1.4–1.7 mm high and 1.3 mm wide, including a foliar structure, pollen sacs and numerous pricks (Fig. 3.2a). The pricks are arranged along the margin of the foliar structure, close to vertical, up to 1.1 mm long. The pollen sacs are about 200 μm wide and 450 μm high, triangular-shaped, sessile on the adaxial surface of the foliar structure. The *in situ* pollen grains are monocolpate, elliptical, 32–51 × 20–36 μm, rough-surfaced in the nonaperturate

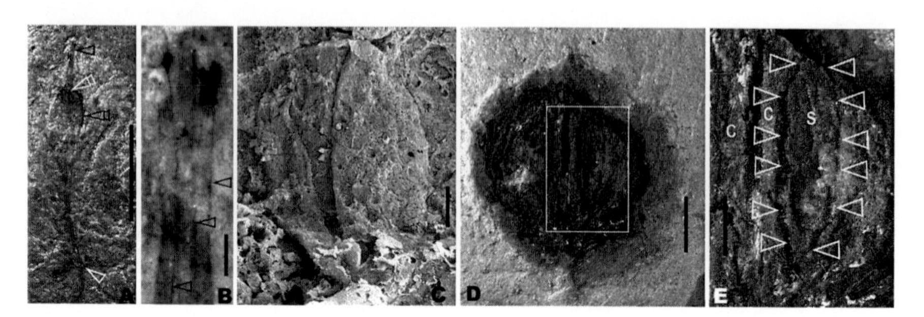

Fig. 3.2 The female and male flowers, *in situ* monocolpate pollen grain, fruit and *in situ* seed of *Chaoyangia*. (**a**) The female flower (*double black arrows*) and male flower (*white arrow*) connected by a branch. Note the connection (*double white arrow*) between the styles (*black arrow*) and the ovaries. Bar = 5 mm. (**b**) The three styles of the female flower shown in (**a**). Note the dark material on the tips of the styles. Bar = 0.2 mm. (**c**) *In situ* monocolpate pollen grain from the male flower in (**a**). Note the oval shape of the pollen grain and longitudinal colpus. Bar = 5 μm. (**d**) A mature fruit with *in situ* seed. Bar = 2 mm. (**e**) *In situ* seed in the fruit, enlarged from the rectangle in (**d**). Note the outline (*white arrows*) of the seed(s) enclosed in the carpel (**c**). Bar = 1 mm

region, relatively smooth in the aperturate region, usually clumped (Fig. 3.2c). The pollen wall is homogeneous, uneven in thickness, with no columellae, thin and nonsolid in the aperturate region. Female flowers are terminal on the branches, elongate to globular in form. Each female flower has a stalk at the bottom, a central unit in the middle and three styles at the top (Fig. 3.2a). The stalk is 1.2–1.8 mm long and 0.2–0.6 mm wide, appear to have three distinct parts when young. The central unit is 1.4–6.3 mm high and 0.6–5.2 mm wide, elongate to globular in form. Each central unit includes an urceolate receptacle covered with forked hairs and three enclosed carpels. The hairs are 40–180 μm wide and up to 3 mm long, forked when mature, tapering to the tip, scattered over the receptacle surface. Hairs are sparse and unforked, not surrounding the style in young flowers but they become dense and forked and form an envelope surrounding the styles in mature flowers and fruits. The receptacles are almost uniformly 0.6 mm thick. Each carpel is base-fixed to the central bottom of the receptacle, forming an ovary at the bottom and a style at the top. An atropous ovule/seed is inserted to the ovary bottom by a funicle. Mature ovary wall is 0.8–1.2 mm thick, with horizontal coarse rumples. The styles are 0.5–3.1 mm long and 67–107 μm wide, straight and slender, corresponding to the three carpels (Fig. 3.2b). The styles are distinct when young but surrounded by hairs when mature. The stigmas are on the style terminal, expanded, lobed, or not, probably secretory, conspicuous in young flowers due to its dark color. There are vascular bundles in the ovary corresponding to the position of the seed. Seeds are 2.8–3.6 mm long, 0.65–1 mm wide, with fine horizontal rumples, much narrower than the ovary cavities, enclosed in but separated from the ovary walls (Figs. 3.2d–e).

Locality: Huangbanjigou and other localities in western Liaoning, China.
Stratigraphic horizon: the Yixian Formation, Lower Cretaceous (>125 Ma).
Holotype: *Chaoyangia liangii* Duan.

3.4.2 *Callianthus*

3.4.2.1 Brief History

The first trace of *Callianthus* appeared in the literature in 1999 when Dr. Shunqing Wu described the Jehol flora from western Liaoning, China (Wu 1999). At that time the specimen just showed a fruit with two persistent styles at the top. Due to the limited material, incomplete preservation and close floristic relationship between the floras in northeastern China and Mongolia, Wu correlated it with *Erenia stenoptera* Krassilov, which was described from the Cretaceous in Mongolia and initially related to Juglandaceae (Krassilov 1982). However, since there were very limited characters for systematic analysis and comparison, Wu's placement appears to be open to question. The current description and diagnosis of the taxon is based on more complete specimens, more specimens and detailed observation on various parts of the reproductive organs, including pedicel, tepal, stamen, carpel and fruit (Wang and Zheng 2009). Character analysis indicates that Wu's *Erenia stenoptera* should be a fruit of *Callianthus dilae*.

3.4.2.2 Discussion

Erenia stenoptera Krassilov is a winged fruit from the Cretaceous in Mongolia (Krassilov 1982). It has a "persistent, funnel-shaped, sessile" stigma, a smooth membraneous wing "avoiding the stalk" and an elliptical endocarp (Krassilov 1982). The holotype of *Callianthus* is distinct from *Erenia* in its two distinct papillate styles, distinct stamens and tepals. It lacks the "stalk-avoiding" wing and "stalk," which are important characters of *Erenia stenoptera* (Krassilov 1982). New material of *Callianthus* has shown enough characters to justify its own identity other than *Erenia*. This fossil, including flower and its mature fruits, demonstrate a floral organization typical for angiosperms. This organization is distinct from those in gymnosperms. The papillate styles imply that the carpels are closed at the time of fertilization, satisfying the criterion for angiosperms. Considering its early age (125 Ma), *Callianthus* is the earliest typical flower, in addition to the contemporary angiosperms (including *Archaefructus* and *Sinocarpus*) that have no typical floral organization. Together with the above *Chaoyangia*, it increases the number of angiospermous genera in the Yixian Formation up to four and suggests that the origin of angiosperms must be much earlier than 125 Ma ago.

3.4.2.3 Diagnosis

Flower small, bisexual, with a perianth, hypogynous, with a slender pedicel. Tepals in two cycles, spatulate, parallel veined, with a long claw and a round tip. Stamen composed of a filament and a globular anther, with numerous bristles at the apex.

In situ pollen grains round-triangular. Fleshy envelope enclosing two separate carpels. Each carpel composed of a hemi-globular ovary and a papillate style, enclosing an ovule in the ovary. Fruit indehiscent, enclosing a hemi-globular seed, with a persistent styles (Wang and Zheng 2009).

3.4.2.4 Description

The flower is small, bisexual, with a perianth, hypogynous, pedicellate, 6.9 mm high, 7.3 mm wide (Fig. 3.3a). The pedicel is up to 1.8 mm long and 0.35 mm wide (Fig. 3.3a). Four tepals and two stamens are visible and attached closely to the pedicel (Fig. 3.3a). The tepals are distinct, in two cycles. Each tepal has two major parallel veins in the distal portion, spatulate, with a long claw and a round tip, up to 6.5 mm long and 0.9 mm wide in the distal portion. A stoma is seen on the tepal, with stomatal aperture 1–2 × 7–8 μm. Stamens are attached to the pedicel just above the inner cycle of the tepals by a thin filament about 1.2 mm long and 0.19 mm wide. Anthers, attached to the terminal of the filaments, are globular, about 0.5 mm wide, with numerous bristles up to 0.8 mm long and 60–65 μm wide at the apex (Fig. 3.3b). Pollen grains *in situ* are compressed into various shapes but two of them appear round-triangular, 28–32 μm in diameter. Similar pollen grains have also been seen three times in the anther region on the transfers. Two stylate carpels are base-fixed and enclosed in a cup-shaped fleshy envelope, which has a rough surface, enclosing the carpels in its center. The fleshy envelope is widest at the middle (about 4.2 mm wide) and about 3.75 mm wide at the top, 0.6–1.6 mm thick, with a 0.4 mm-high raised ring close to the margin. The ovary is hemi-globular, about 3.1 mm high and 1.4 mm thick. Each carpel has a distinct style at the apex and separated from adjacent ones almost to its bottom by a gap about 0.3 mm wide. Style is short, curved, or straight, more than 1 mm long and about 0.2 mm wide, with papillae on its

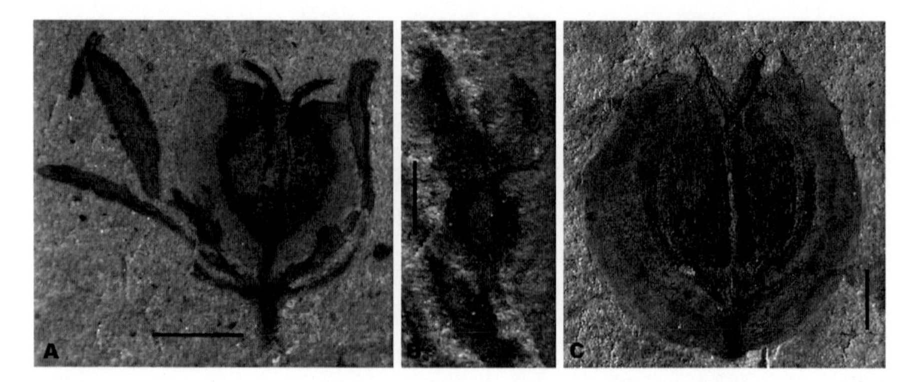

Fig. 3.3 The flower, stamen and fruit of *Callianthus*. (**a**) The monoecious flower showing, from bottom up, the pedicel, tepals, stamens and fleshy envelope surrounding two carpels, which have two styles at the apex. Bar = 2 mm. (**b**) The spherical stamen with a stalk at the bottom and bristles at the top. Bar = 0.5 mm. (**c**) The fruits with two seeds and two persistent styles at the apex. Bar = 1 mm (Wang and Zheng 2009)

surface. The papillae on the style are probably conical-shaped, tapering distally, at least 5 μm long. Two hemi-globular fruits plus the surrounding fleshy envelope are about 4–5.8 mm high and 4–5.5 mm wide, with a raised ring at the top, indehiscent. Each fruit encloses a seed, with a persistent style about 1 mm long. Stamens and tepals fall off when the fruits mature. There appear to be vascular bundles entering the carpels/fruits and the fleshy envelope, respectively, in the proximal. Two seeds are 2.9–3.5 mm high, 1.3–1.7 mm thick, about 3.5 mm wide, taking the position of the former ovaries, separated by a gap in-between, probably with a dorsal ridge (Fig. 3.3c).

Locality: Huangbanjigou, Shangyuan, Beipiao, Liaoning, China.
Stratigraphic horizon: the Yixian Formation, Lower Cretaceous (>125 Ma).
Holotype: *Callianthus dilae* Wang et Zheng.

3.4.3 *Xingxueanthus*

3.4.3.1 Brief History

Xingxueanthus is a new genus established recently (Wang and Wang in press). The study history may be traced back to more than 20 years ago. Dr. Kuang Pan declared that he found angiosperms in the middle Jurassic in western Liaoning, China (Pan 1983). His declaration attracted many eye balls as well as much criticism and suspicion at that time (Xu 1987). One team from the Institute of Botany, Chinese Academy of Sciences was sent to test the truthfulness of his claim. They collected many specimens and found no typical angiosperms in the collection. This collection includes the specimens now called *Xingxueanthus sinensis* and *Schmeissneria sinensis*. Although the specimen of *X. sinensis* caught the author's attention more than 10 years ago, the job was finished just recently (Wang and Wang in press).

3.4.3.2 Discussion

The preservation of *X. sinensis* is coalification, one of the most reliable fossil preservations. The arrangement of ovules around the central column is revealed by the coalified material or its impression on the fine matrix.

The inflorescence of *X. sinensis* is a catkin of female flowers. Its flower is unisexual, composed of an ovule-container with a terminal style, situated in the axil of a bract. The ovary encloses a vertical free central placentation. The ovules are helically attached to a vertical column connecting the base and tip of the ovary. There is a style attached to the top of the ovary. Such an arrangement of floral parts is unique to angiosperms and the ovules are enclosed in the ovary, satisfying the above criterion for angiosperms. The middle Jurassic age of *X. sinensis*, in addition to *Schmeissneria* from the Early-Middle Jurassic in China and Germany, makes its recognition significant in terms of the angiosperm origin and evolution and calls for a reappraisal of the origin of angiosperms (Wang and Wang in press).

Many people may prefer to interpret *Xingxueanthus* from a gymnospermous perspective. This would result in a new class in seed plant, since *Xingxueanthus* apparently does not fit in any existing fossil or extant gymnospermous class. Searching for a position for *X. sinensis* in ferns or other groups has turned out to be futile.

3.4.3.3 Diagnosis

Inflorescence composed of numerous female flowers spirally arranged along inflorescence axis. Each female flower in axil of a bract, composed of an ovule-container and a style at the top. Ovules arranged spirally along a vertical column within the ovule-container (Wang and Wang in press).

3.4.3.4 Description

The inflorescence is spicate, slightly curved, over 23 mm long and 7.5 mm wide at the base, tapering distally, with over 21 female flowers attached, maturing acropetally (Fig. 3.4a). Inflorescence axis is about 1 mm wide at the base, tapering distally, slightly twisted to accommodate the female flowers, with longitudinal striations. Female flowers and their subtending bracts are spirally arranged along the inflorescence axis. Bracts are about 3.5–5 mm long, diverging from the inflorescence

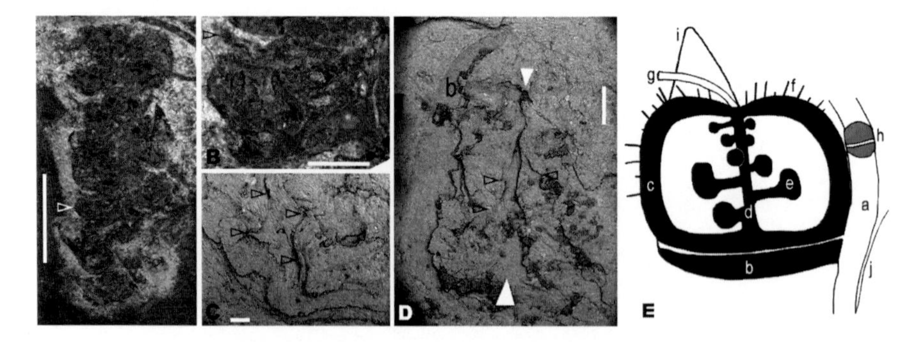

Fig. 3.4 The inflorescence, female flower, ovary with a free central placentation of *Xingxueanthus*. (**a**) A general view of the inflorescence. Bar = 5 mm. (**b**) The flower arrowed in the inflorescence in (**a**). Note the style (*arrow*) at the top of the ovary. Bar = 1 mm. (**c**) The arrangement of the ovules (*arrows*) along the central column, under an SEM. Bar = 0.2 mm. (**d**) Another flower showing the bract (b) protecting the ovary, which has a vertical column (between white triangles) bearing ovules (*arrows*) around. Bar = 0.5 mm. (**e**) A diagram showing inflorescence axis (a), bract (b), ovule-container wall (c), column (d), ovule (e), trichomes (f), style (g), scars of the bract and the female flower (h), bract tip (i) and a portion of another female flower (j) (Wang and Wang inpress)

axis at an angle slightly greater than 90°, upturning at the laterals of the female flowers, with inward-curving tips of about 35°, completely separated from the female flowers in their axils, with their distal ends not extended beyond the bottom extremities of the female flowers. Female flowers are in the axils of the corresponding bracts, up to 3 mm from the adaxial to abaxial side, up to 2 mm from side to side, up to 2.6 mm high (Fig. 3.4b). A female flower is composed of an ovule-container and a style at the top (Fig. 3.4b). The ovule-container has a vertical column within (Figs. 3.4c–e). The column connects the base and the top of the ovule-container, almost parallel to the inflorescence axis, 1.1–2.5 mm long, about 0.5 mm wide at the base and tapering to about 50 μm wide at the top (Figs. 3.4c–e). When the organic material is preserved, the column and its attached ovules are visible; when the organic material falls off, the column and its attached ovules are suggested by their imprints left on the fine sediment (Fig. 3.4c and d). Striations on the column converge where funiculi are attached (Fig. 3.4c and d). More than three ovules are spirally arranged along the column at an angle close to 90° (Fig. 3.4c and d). Funiculi range from 100 to 320 μm in diameter. Ovules are 100–380 μm in diameter. A style 130–190 μm wide and up to 0.9 mm long is inserted on the top of the ovule-container. Trichomes are seen on the ovule-container. Trichomes are about 1–2 cells and 40–50 μm in diameter, up to 328 μm long, single, or in fascicles. Stomatal aperture is about 6–7 μm long and 2–3 μm wide, slightly sunken.

Locality: Sanjiaochengcun, Huludao, Liaoning, China.
Stratigraphic horizon: the Haifanggou Formation, Middle Jurassic (>160 Ma).
Holotype: *Xingxueanthus sinensis* Wang et Wang.

3.4.4 *Schmeissneria*

3.4.4.1 Brief History

Schmeissneria is a genus established in 1994 (Kirchner and Van Konijnenburg-Van Cittert 1994), however, its research history can be traced back to 1838. Presl (1838) reported a few fructifications from Keuper Sandstone of Reundorf near Bamberg, Germany and put them as male flowers of *Pinites microstachys* (Conifers). Later Schenk studied similar fossils from Veitlahm near Kulmbach, regarding them as female flowers of *Stachyopitys preslii* (Schenk 1867). In 1890 Schenk reassigned *S. preslii* as male fructifications of *Baiera münsteriana* (Ginkgoales) in their early developmental stage (Schenk 1890). This association between *Baiera* and *Stachyopitys* was later widely accepted in palaeobotany until 1992. Wcislo-Luraniec cast doubt on the male nature of *S. preslii* and proved its female nature (Wcislo-Luraniec 1992). Meanwhile *S. preslii* was found connected with Weber's *Glossophyllum*? sp. A (Kirchner 1992). More and complete materials prompted the establishment of a new genus, *Schmeissneria* and the new genus was put in Ginkgoales as previously (Kirchner and Van Konijnenburg-Van Cittert 1994).

Wang et al. found the internal structure in the ovary of *Schmeissneria* in a Chinese material and proposed to transfer it to angiosperms (Wang et al. 2007a, b). While questioning Wang et al.'s interpretations, Doyle (2008) recently acknowledged the fact that the ovules in *Schmeissneria* are enclosed. Recently the author studied the syntype and more materials of *Schmeissneria* in the Munich collection. This new effort reveals the details of long-ignored seeds, confirms the septum in the fructifications of *Schmeissneria* and further supports the proposal by Wang et al. (2007a, b).

3.4.4.2 Discussion

Kirchner and Van Konijnenburg-Van Cittert (1994) appeared to focus on the general morphology of *Schmeissneria,* probably due to the fantastic connected parts of the fossils. They somehow underestimated the value of the fossil fructifications, although they mentioned their existence and interpreted the "hole" (actually seed) as a result of desiccation or simply resin body. Their so-called wing on seed is now found to be hairs on the ovary. Study on the materials indicate that (1) there are seeds enclosed in the fruits, satisfying the crude criterion for angiosperms; (2) there is a septum in the fruit, confirming the existence of septum in the ovary, a phenomenon never seen in gymnosperms; and (3) the tip of the ovary is closed before the anthesis and fertilization, satisfying the strict criterion for angiosperms. Thus *Schmeissneria* can be safely regarded as an angiosperm, in spite of its Early Jurassic age (Wang submitted).

The only controversy about the age of *Schmeissneria*, if any, is whether it belongs to the Late Triassic or Early Jurassic in Germany. Such an early age of *Schmeissneria* makes its angiospermous identity especially significant, since recognizing it as an angiosperm prompts a fundamental modification on the currently dominating doctrine on the origin of angiosperms (Wang submitted).

3.4.4.3 Diagnosis

Plants with long and short shoots. Leaves in helical arrangement on short shoots. Short shoots covered with leaf cushions. Leaves slender, slightly cuneiform, apex obtuse. Veins parallel, more than two in the proximal part of the leaf, branching in the lower third of the lamina. Female inflorescence spicate, with a slender axis. Axis of the inflorescence with longitudinal striations. Flower pairs borne on a peduncle, helically arranged along the inflorescence axis. Flower with an ovary surrounded by tepals. Ovary bilocular, subdivided by a vertical septum, with a closed tip, bearing hairs on the surface at anthesis. Hairs long, narrow, straight, scattered on the ovary. Fruit with two locules divided by a septum, enclosing several seeds. Seeds tiny, elongate, or oval in shape, round in cross-section, with a firm seed coat, enclosed in the fruit (Wang submitted).

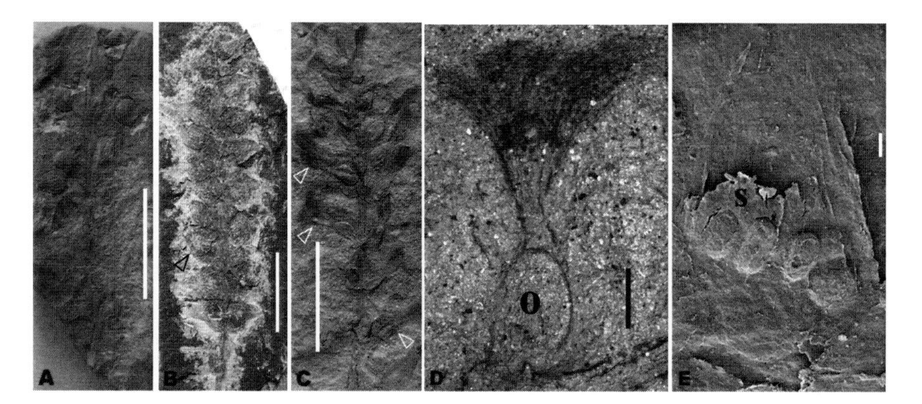

Fig. 3.5 Inflorescences, flower and fruit with *in situ* seeds of *Schmeissneria*. (**a**) Inflorescence of *Schmeissneria microstachys*. Note the paired flowers arranged along the inflorescence axis. Bar = 1 cm. (**b**) Inflorescence of *S. sinensis*. Bar = 1 cm. (**c**) Inflorescence of *S. microstachys*. Note the hairy flowers arranged along the inflorescence axis. Bar = 1 cm. (**d**) One of the flowers in the inflorescence shown in (**c**). Note hair bundle on the apex of the ovary (o). Bar = 1 mm. (**e**) A fruit with *in situ* seeds of *S. microstachys*. Note the seeds are separated by a septum (s) with a broken margin. Bar = 0.2 mm (Wang submitted)

3.4.4.4 Description

Plants have long and short shoots. Leaves are inserted on the apex of the short-shoot, which is covered with leaf cushions. Leaves are slightly cuneiform, with obtuse apex and parallel veins. The female inflorescence is attached to the apex of a short shoot, up to 7.7 cm long and 1.2 cm wide, with flower pairs in dense or loose helical arrangement along the axis, each inflorescence has up to 45 flowers (Figs. 3.5a–c). The inflorescence axis is longitudinally striated, free of flowers proximally, 0.8–1.5 mm wide at the base, tapering distally. Flower peduncles are about 0.5 mm wide, 0.5–2 mm long and bear pairs of flowers. The flower is round-triangular to oval in shape, 1.6–4.6 mm long, 1.2–4 mm in diameter and consists of an ovary surrounded by tepals. Tepals are triangular in shape, 1.6–4.6 mm long, 1–2 mm wide and longitudinally ribbed or not. At least some of the tepals are missing if hairs are present on the ovary. The ovary is round-triangular in shape when young but oval at maturity, 1–3.2 mm in diameter, 1–3.3 mm long, sub-divided into two locules by a vertical septum, hairfree when young but covered with hairs when mature (Fig. 3.5a–d). The septum is thin and extended from the bottom to the tip of the ovary. Hairs are scattered over the ovary, brush-like or arranged in bundles, straight, very fine, up to 0.2 mm wide, up to 5.3 mm long. Fruits are arranged along the infructescence axis, 2.3–3.3 mm long, 1.7–2.1 mm wide and usually enclose more than four seeds in two locules that are divided by the vertical septum (Fig. 3.5e). Seeds are elongate to oval in shape, with a thin seed coat, 0.18–0.46 mm long and 0.11–0.3 mm in diameter (Fig. 3.5e).

Locality: Bamberg, Bayreuth, Kulmbach, Nuremburg in Germany; western Liaoning in China; Odrowaz in Poland.

Stratigraphic horizon: the Lower to Middle Jurassic (160–199 Ma).
Holotype: *S. microstachys* Kirchner and Van Konijnenburg-Van Cittert emend. Wang.

3.5 Conclusion

The origin of angiosperms has been a long-debated topic in palaeobotany. One hundred and thirty years has elapsed since the question started bothering Charles Darwin. More technologies, including molecular biology, are being applied to resolve this problem now. However, little progress has been made on the origin of angiosperms in palaeobotany until recently. Recent progress in molecular biology indicates that the ANITA group and Hydatellaceae are the basal most clades in the angiosperms but the relationship among many basal clades cannot be resolved satisfactorily based on molecular data alone. The perplexing relationship among the basal angiosperms may be due to the missing fossil taxa and their information, lack of which leaves significant gaps in the morphology and genome of the extant basal angiosperms (Zavada 2007). Therefore new fossil taxa of angiosperms from the Early Cretaceous and earlier age are not only important *per se* but also helpful to bridge the gaps and resolve the relationship among living angiosperms.

As for the time of angiosperm origination, the molecular clock frequently points to the Jurassic or Triassic. However, the fossil evidence, up to this time, is not conclusive. On one hand, there have been several reports or claims of Jurassic and even Triassic angiosperms; on the other hand, these evidences do not convincingly prove these claims and are frequently rejected by leading scientists, especially those working on mesofossils. One reason behind this controversy and stalemate is lack of a consensus on the definition of angiosperms that is applicable for the earliest angiosperms; different scholars apply different definitions *ad hoc*. As discussed in the beginning of this chapter, the author thinks that only one character, physically enclosed ovule at the time of fertilization, can be used as the index for angiospermy in the geological history. The above documented fossils taxa are classified as angiosperms based on this index and other characters. Although the controversy and discussion on the placement and other aspects of the above taxa and origin of angiosperms will continue for a while, the author wishes that there would be a consensus on this index character for angiospermy in the geological history. This consensus, if reached, will function as a foundation up on which further research can be built and an important step leading to the final solution of the notoriously abominable mystery.

The fossils briefed above paint a picture of angiosperm origin and evolution different than currently accepted. If these fossils are accepted as angiosperms, then there are two alternative interpretations for the evolution of early angiosperms. (1) If all angiosperms are monophyletic, then all these fossil plants and living angiosperms should be derived from a common ancestor that existed at least as early as the Late Triassic (Fig. 3.6). This is close to some result based on molecular

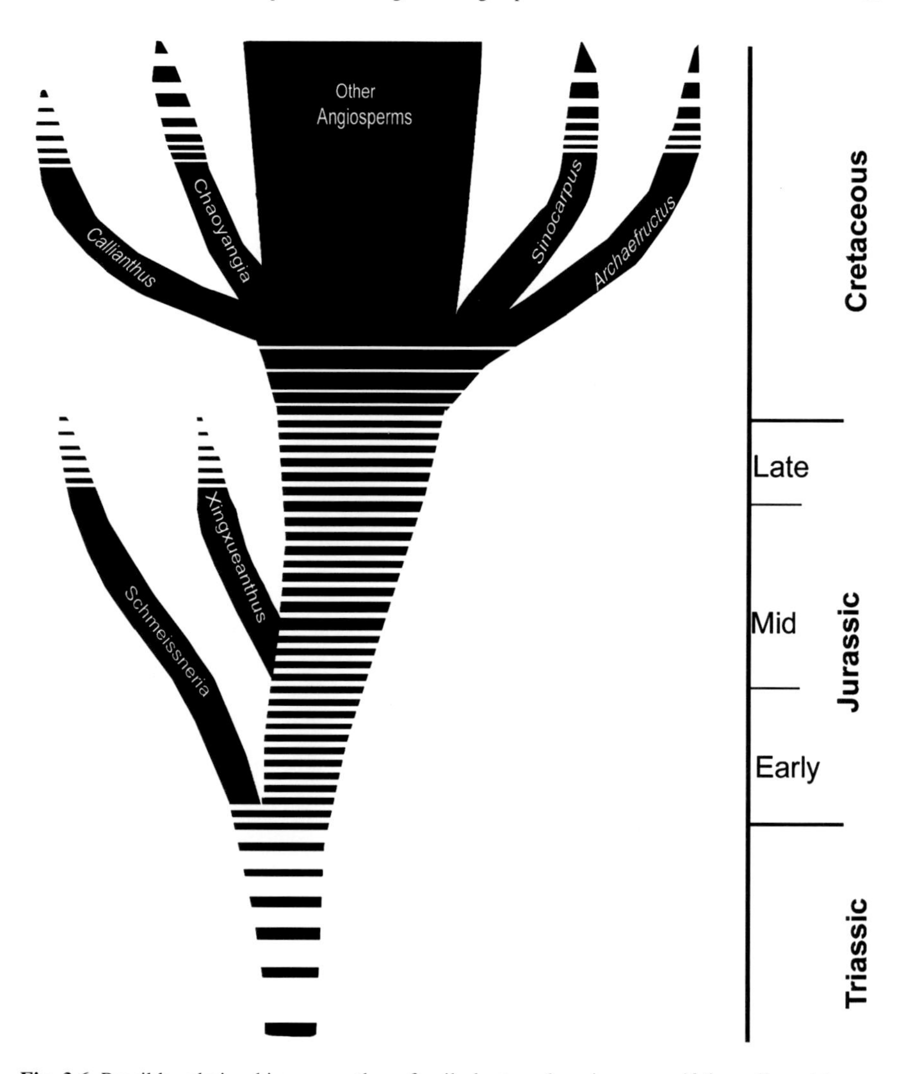

Fig. 3.6 Possible relationship among these fossil plants and angiosperms, if they all are taken as elements of monophyletic angiosperms. See text for details.

data but fundamentally different from the current dominating Cretaceous-only-angiosperm doctrine. (2) If all angiosperms are taken as polyphyletic, then the angiospermy in these fossil plants are just a result of independent parallel evolution in various lineages that have reached the same level of ovule protection. Thus angiospermy becomes a level of evolution rather than a character unique to angiosperms. There must be several groups of seed plants having reached this level before the angiosperms did. Which alternative is more compatible with the reality is something beyond what the currently available evidence can tell. However, as fossil evidence accumulates, nature will release more truth gradually.

Acknowledgments This is a summary of several works that I have been involved in. These works have taken more than a decade to complete. During this period, I have received help and support from colleagues too many to enumerate here. I would like to extend my thanks to them all for their friendship, constructive suggestions and advices. This research is supported by the National Natural Science Foundation of China Programs (No. 40772006, 40372008, 40632010, and J0630967) and SRF for ROCS, SEM. This is a contribution to IGCP 506.

References

Chamberlain CJ (1957) Gymnosperms, structure and evolution. Johnson Reprint Corporation, New York

Cornet B (1986) The leaf venation and reproductive structures of a late Triassic angiosperm, *Sanmiguelia lewisii*. Evol Theory 7:231–308

Cornet B (1989a) Late Triassic angiosperm-like pollen from the Richmond rift basin of Virginia, USA. Paläontogr B 213:37–87

Cornet B (1989b) The reproductive morphology and biology of *Sanmiguelia lewisii*, and its bearing on angiosperm evolution in the late Triassic. Evol Trends Plants 3:25–51

Cornet B (1993) Dicot-like leaf and flowers from the late Triassic tropical Newark Supergroup rift zone, USA. Mod Biol 19:81–99

Cornet B, Habib D (1992) Angiosperm-like pollen from the ammonite-dated Oxfordian (upper Jurassic) of France. Rev Palaeobot Palynol 71:269–294

Crane PR (1985) Phylogenetic analysis of seed plants and the origin of angiosperms. Ann Miss Bot Gard 72:716–793

Cronquist A (1988) The evolution and classification of flowering plants Botanical Garden, Bronx, New York

Dilcher DL, Sun G, Ji Q, Li H (2007) An early infructescence *Hyrcantha decussata* (comb. nov.) from the Yixian Formation in northeastern China. Proc Nat Acad Sci USA 104:9370–9374

Doyle JA (2008) Integrating molecular phylogenetic and paleobotanical evidence on origin of the flower. Intl J Plant Sci 169:816–843

Duan S (1998) The oldest angiosperm—a tricarpous female reproductive fossil from western Liaoning Province, NE China. Sci Chin D 41:14–20

Eames AJ (1961) Morphology of the angiosperms. McGraw-Hill, New York

Endress PK, Igersheim A (2000) Gynoecium structure and evolution in basal angiosperms. Intl J Plant Sci 161:S211–S223

Friedman WE (1990) Sexual reproduction in *Ephedra nevadensis* (Ephedraceae): further evidence of double fertilization in a nonflowering seed plant. Am J Bot 77:1582–1598

Friedman WE (1991) Double fertilization in *Ephedra trifurca*, a non flowering seed plant: the relationship between fertilization events and the cell cycle. Protoplasma 165:106–120

Friedman WE (1992) Double fertilization in nonflowering seed plants. Int Rev Cytol 140:319–355

Friis EM, Pedersen KR, Crane PR (2005) When earth started blooming: insights from the fossil record. Curr Opin Plant Biol 8:5–12

Friis EM, Pedersen KR, Crane PR (2006) Cretaceous angiosperm flowers: innovation and evolution in plant reproduction. Palaeogeo Palaeoclim Palaeoecol 232:251–293

Harris TM (1940) *Caytonia*. Ann Bot 4:713–734

Harris TM (1964) Caytoniales, Cycadales & Pteridosperms. Trustees of the British Museum (Natural History), London

Hochuli PA, Feist-Burkhardt S (2004) A boreal early cradle of angiosperms? Angiosperm-like pollen from the middle Triassic of the Barents Sea (Norway). J Micropalaeont 23:97–104

Hughes N (1994) The enigma of angiosperm origins. Cambridge University Press, Cambridge

Ji Q, Li H, Bowe M, Liu Y, Taylor DW (2004) Early Cretaceous *Archaefructus eoflora* sp. nov. with bisexual flowers from Beipiao, Western Liaoning, China. Acta Geol Sin 78: 883–896

Judd WS, Campbell SC, Kellogg EA, Stevens PF (1999) Plant systematics: a phylogenetic approach. Sinauer, Sunderland, MA

Kirchner M (1992) Untersuchungen an einigen Gymnospermen der Fränkischen Rhät-Lias-Grenzschichten. Paläontogr B 224:17–61

Kirchner M, Van Konijnenburg-Van Cittert JHA (1994) *Schmeissneria microstachys* (Presl, 1833) Kirchner et Van Konijnenburg-Van Cittert, comb. nov. and *Karkenia hauptmannii* Kirchner et Van Konijnenburg-Van Cittert, sp. nov., plants with ginkgoalean affinities from the Liassic of Germany. Rev Palaeobot Palynol 83:199–215

Krassilov V, Lewy Z, Nevo E (2004) Controversial fruit-like remains from the lower Cretaceous of the Middle East. Cret Res 25:697–707

Krassilov VA (1982) Early Cretaceous flora of Mongolia. Paläontogr Abt B 181:1–43

Kryshtofovich A (1982) *Pleuromeia* and *Hausmannia* in eastern Sibiria, with a summary of recent contribution to the palaeobotany of the region. Am J Sci 5:200–208

Leng Q, Friis EM (2003) *Sinocarpus decussatus* gen. et sp. nov., a new angiosperm with basally syncarpous fruits from the Yixian Formation of Northeast China. Plant Syst Evol 241:77–88

Leng Q, Friis EM (2006) Angiosperm leaves associated with *Sinocarpus* Leng et Friis infructescences from the Yixian Formation (mid-Early Cretaceous) of NE China. Plant Syst Evol 262:173–187

Li H, Taylor DW (1998) *Aculeovinea yunguiensis* gen. et sp. nov. (Gigantopteridales), a new taxon of gigantopterid stem from the upper Permian of Guizhou province, China. Int J Plant Sci 159:1023–1033

Li H, Tian B, Taylor EL, Taylor TN (1994) Foliar anatomy of *Gigantonoclea guizhouensis* (Gigantopteridales) from the upper Permian of Guizhou province, China. Am J Bot 81: 678–689

Li H, Taylor EL, Taylor TN (1996) Permian vessel elements. Science 271:188–189

Pan G (1983) The Jurassic precursors of angiosperms from Yanliao region of North China and the origin of angiosperms. Chin Sci Bull 28:15–20

Pocock SAJ, Vasanthy G (1988) *Cornetipollis reticulata*, a new pollen with angiospermid features from Upper Triassic (Carnian) sediments of Arizona (USA), with notes on *Equisetosporites*. Rev Palaeobot Palynol 55:337–356

Potonie H (1921) Lehrbuch der Paläobotanik. Verlag von Gebrüder Borntraeger, Berlin, MA

Presl KB (1838) In: Sternberg KM (ed) Versuch einer geognostisch-botanischen Darstellung der Flora der Vorwelt. Johann Spurny, Prag, pp 81–220

Retallack G, Dilcher DL (1981) Arguments for glossopterid ancestry of angiosperms. Palaeobiol 7:54–67

Rothwell GW (1972) Evidence of pollen tubes in Paleozoic pteridosperms. Science 175:772–724

Schenk A (1867) Die fossile Flora der Grenzschichten des Keupers und Lias Frankens. C.W. Kreidel's Verlag, Wiesbaden

Schenk A (1890) Paläophytologie. Druck und Verlag von R. Oldenbourg, München

Scott RA, Barghoorn ES, Leopold EB (1960) How old are the angiosperms? Am J Sci 258A:284–299

Shen GL, Gu ZG, Li KD (1976) More material of *Hausmannia ussuriensis* from Jingyuan, Gansu. J Lanzhou Univ 3:71–81

Sporne KR (1971) The morphology of gymnosperms, the structure and evolution of primitive seed plants. Hutchinson University Library, London

Stevens PF (2008) Angiosperm Phylogeny Website. http://www.mobot.org/MOBOT/research/APweb/. Cited 12 December 2008

Sun G (1981) Discovery of Dipteridaceae from the upper Triassic of eastern Jilin. Acta Palaeontol Sin 20:459–467

Sun G (1993) Late Triassic flora from Tianqiaoling of Jilin, China. Jilin Science & Technology Press, Changchun

Sun G, Dilcher DL, Zheng S, Zhou ZK (1998) In search of the first flower: a Jurassic angiosperm, *Archaefructus*, from Northeast China. Science 282:1692–1695
Sun G, Zheng S, Dilcher D, Wang Y, Mei S (2001) Early angiosperms and their associated plants from Western Liaoning, China. Shanghai Technology & Education Press, Shanghai
Sun G, Ji Q, Dilcher DL, Zheng S, Nixon KC, Wang X (2002) Archaefructaceae, a new basal angiosperm family. Science 296:899–904
Swisher CC, Wang Y-Q, Wang X-L, Xu X, Wang Y (1998) ^{40}Ar/^{39}Ar dating of the lower Yixian Fm, Liaoning Province, northeastern China. Abstract in the 9th International Conference on Geochronology, Cosmochronology and Isotope Geology (August 20–26, 1998), Beijing, China. Chinese Science Bulletin 43:125
Taylor DW, Li H, Dahl J, Fago FJ, Zinniker D, Moldowan JM (2006) Biogeochemical evidence for the presence of the angiosperm molecular fossil oleanane in Paleozoic and Mesozoic non-angiospermous fossils. Paleobiology 32:179–190
Thomas HH (1925) The Caytoniales, a new group of angiospermous plants from the Jurassic rocks of Yorkshire. Phil Trans Roy Soc Lond 213B:299–363
Tomlinson PB, Takaso T (2002) Seed cone structure in conifers in relation to development and pollination: a biological approach. Can J Bot 80:1250–1273
Zavada MS (2007) The identification of fossil angiosperm pollen and its bearing on the time and place of the origin of angiosperms. Plant Syst Evol 263:117–134
Wang X (submitted) *Schmeissneria*: An early Jurassic angiosperm from Germany. Am J Bot
Wang X, Duan S, Geng B, Cui J, Yang Y (2007a) Is Jurassic *Schmeissneria* an angiosperm? Acta Palaeontol Sin 46:486–490 (in Chinese, with English abstract)
Wang X, Duan S, Geng B, Cui J, Yang Y (2007b) *Schmeissneria*: a missing link to angiosperms? BMC Evol Biol 7:14
Wang X, Wang S (in press) *Xingxueanthus*: An enigmatic Jurassic seed plant and its implications for the origin of angiospermy. Acta Geol Sin
Wang X, Zheng S (2009) The earliest normal flower from Liaoning Province, China. J. Integr. Plant Biol. doi: 10.1111/j.1744-7909.2009.00838.x
Wcislo-Luraniec E (1992) A fructification of *Stachyopitys preslii* Schenk from the lower Jurassic of Poland. Cour Forsch-Institut Senck 147:247–253
Weng J-K, Li X, Stout J, Chapple C (2008) Independent origins of syringyl lignin in vascular plants. Proc Nat Acad Sci USA 105:7887–7892
Wu S-Q (1999) A preliminary study of the Jehol flora from the western Liaoning. Palaeoworld 11:7–37
Xu R (1987) Do fossil angiosperms really occur in Jurassic beds of the Yanshan-Liaoning area, north China. Kexue Tongbao 32:1712–1714
Yang Y, Fu DZ, Wen LH (2000) On double fertilization in *Ephedra*. Adv Plant Sci 3:67–74
Zavada MS (1984) Angiosperm origins and evolution based on dispersed fossil pollen ultrastructure. Ann Miss Bot Gard 71:444–463
Zavada MS (2007) The identification of fossil angiosperm pollen and its bearing on the time and place of the origin of angiosperms. Plant Syst Evol 263:117–134

Chapter 4
Vertebrate Evolution: The Strange Case of Gymnophionan Amphibians

Jean-Marie Exbrayat and Michel Raquet

Abstract Gymnophionan amphibians are lengthened animals with burrowing habits living in tropical countries, with a Gondwanan distribution. Anatomical characteristics are fundamentally amphibian-like with several primitive characters such as teleostean-like scales and evolved ones, such as some morphological features of forebrain, or presence of Hassal's corpuscles in the thymus of some species. Gymnophiona also possess proper characters: general elongation of organs, absence of limbs and girdles (even if blastema have been observed in embryos before metamorphosis), a more or less marked segmentation of several organs (liver, kidneys, fat bodies, gonads). In all gymnophionan species, fertilization is internal. They can be oviparous, sometimes with direct-development, or viviparous, with a placenta-like structure in some species.

The systematic position of Gymnophiona is still discussed, into the order itself, among the other amphibians (Lissamphibia) and within the other vertebrates. When vertebrates began to emerge out of the water it was possible that gymnophionan ancestors acquired a burrowing life. Selective pressure of such a life could favour the development of certain characters and conservation of several old characters.

4.1 Introduction

Amphibians are Anamniota vertebrates. They are characterised by an aquatic larval phase and a terrestrial adult one. Modern amphibians, also called Lissamphibia, are divided into Anura or Salientia, Urodela (or Caudata) and Gymnophiona

---

J.-M. Exbrayat and M. Raquet
Université de Lyon; Laboratoire de Biologie Générale, Université Catholique de Lyon; Reproduction et Développement Comparés, EPHE, 25 rue du Plat, 69288 Lyon Cedex 02, France
e-mail: jmexbrayat@univ-catholyon.fr

P. Pontarotti (ed.), *Evolutionary Biology: Concept, Modeling, and Application*,
DOI: 10.1007/978-3-642-00952-5_4,

(or Caecilians) (Laurent 1986). Gymnophiona have been the subject of several recent synthetic books (Himstedt 1996; Exbrayat 2000, 2006a).

Gymnophionans, still called Caecilians and sometimes Apoda, are limbless amphibians with burrowing habits. They live in tropical areas with a Gondwanan repartition: Central and South America, Africa, Asia. These animals look like snakes or worms. They are elongated, measuring about 10 cm for the shortest (i.e., *Microcaecilia unicolor*) to 1.30 m for the longest (*Caecilia thompsoni*). Generally, their size ranges between 20 and 50 cm (Taylor 1968).

Reproductive modes are variable. Fertilization is always internal. A lot of species are oviparous, sometimes direct-developing. Other species are viviparous. A kind of placenta implicates both the embryonic gills and uterine wall in *Typhlonectes compressicauda*.

In order to appreciate the systematic position of gymnophionans, main morphological tracts, biogeography, molecular data and the position of these animals among the other amphibians and vertebrates will be successively considered.

4.2 Morphological Data

4.2.1 General Anatomy

Gymnophiona are elongated animals (Fig. 4.1). Their body often presents annuli (*Ichthyophis*, *Siphonops*). In several genera such as *Typhlonectes*, there are not annuli but folds on the skin. Some families possess a true tail corresponding to several vertebrae located behind the vent (Ichthyophiidae). In others, there are no posterior vertebra (Typhlonectidae). In several families, a small appendix, called a

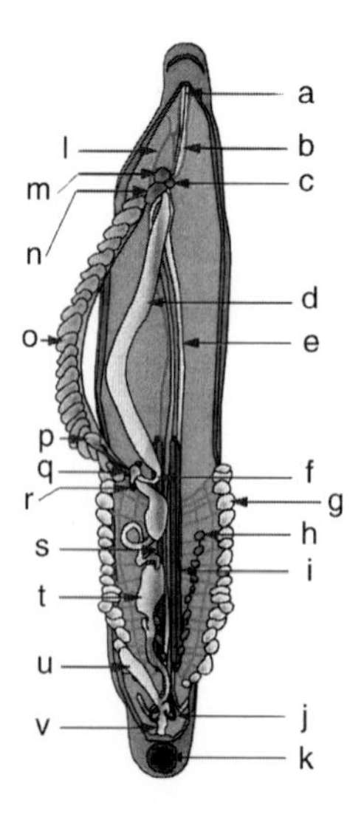

Fig. 4.1 Schematic representation of a male *Typhlonectes compressicauda* dissection modified after Delsol et al. (1980). **a**: trachea; **b**: esophagus; **c**: left auricle; **d**: stomach; **e**: left lung; **f**: duodenum; **g**: left fat body; **h**: left testis; **i**: left kidney; **j**: Mullerian duct; **k**: cloacal opening; **l**: arterial bulb; **m**: right auricle; **n**: ventricle; **o**: liver; **p**: gall bladder; **q**: spleen; **r**: pancreas; s: right kidney; **t**: intestine; **u**: urinary bladder; v: cloaca

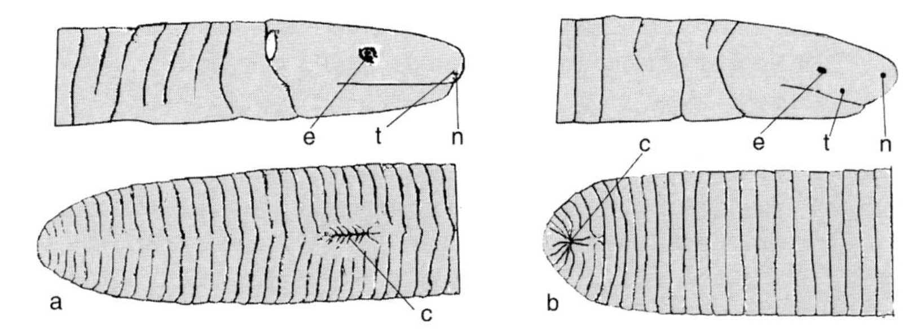

Fig. 4.2 Characteristics of the head and tail in Gymnophiona. (1) *Ichthyophis glutinosus* with a true tail. (2) *Dermophis mexicanus* with a reduced tail (modified after Taylor [1968]). **c**: cloacal opening; **e**: eye; **n**: nostril; **t**: tentacle opening

shield by Taylor (1968), is posterior to the vent (Fig. 4.2). The cloacal aperture is rounded or V-shaped. In *Typhlonectes*, it is different between males and females but in numerous species, it is not possible to determine the sex without dissection.

4.2.2 Integument

The integument is thick and can be covered with scales (Fox 1983, 1987) in the lowest families (Rhinatrematidae, the sister taxon of all other gymnophionans, Datriata that groups Ichthyophiidae, a distinct monophyletic group present on the Indian plate prior to its collision with Laurasia and Uraeotyphlidae, the sister group of Ichthyophiidae (see references in Wilkinson and Nussbaum 2006)). The scales are equivalent to Osteichtyans ones (Cockerell 1912; Casey and Lawson 1979; Zylberberg et al. 1980; Zylberberg and Wake 1990) and cover all the body. In other species, they cover only the posterior part of the body. In this case, the most rostral scales are always smaller than the others and scattered into the skin like in *Geotrypetes seraphinii*. No scale has been found in the most evolved genera, like in *T. compressicauda* in which only small degenerated scales have been described (Wake and Nygren 1987; Riberon and Exbrayat 1996).

4.2.3 Skeleton

The skull is composed of more bones than in other modern amphibians. The lower jaw consists in two halves linked to each other by a connective tissue. A retro-articular process is observed on each lower jaw. On it, powerful muscles are inserted. These processes are involved in both mastication and maintaining closure of the mouth when a prey has been caught (Wake and Hanken 1982; Wake et al. 1985).

The vertebrae are of a primitive type and they only permit horizontal worm-like displacement (Lawson 1963, 1966a; Taylor 1977; Renous and Gasc 1986a, b). The vertebrae are amphicoelic, which is a primitive characteristic found in fishes (plesiomorphy) – the diapophysis and parapophysis are separated, vertebrae are linked to each other by dorsal zygapophysis and basapophysis situated in a dorsal position in relation to the centrum. The structure of the centrum is intermediate between that of Anura and Urodela. During embryonic development, the sclerotoma are metamerised in both Gymnophiona and Amniota but not in other amphibians and their cell density is also high, contrarily to Anura and Urodela (Wake 1986) A collar, corresponding to the larger first three vertebrae, allows a larger mobility and can be compared to the neck of amniota (Lescure and Renous 1992).

There are neither girdles nor limbs. But, in *Ichthyophis glutinosus*, a primitive species, traces of degenerated girdles have been observed and during the embryonic development of *Typhlonectes*, limb buds begin to develop between the third to sixth somites, i.e., at the place where forelimbs develop in other amphibians but they degenerate at metamorphosis (Renous et al. 1997).

4.2.4 Brain

The brain has not been well studied (Kuhlenbeck 1922, 1969; Olivecrona 1964; Senn and Reber-Leutenegger 1986; Schmidt and Wake 1990, 1997, 1998; Martin-Bouyer et al. 1995; Estabel and Exbrayat 1998). In an adult, it is similar to that of other amphibians, intermediary between anurans and urodelans but the thalamus is well differentiated from the telencephalon, a situation that is not found in other amphibians but found in reptiles. During the embryonic development of *I. glutinosus* and *T. compressicauda*, a median septal nucleus appears dorsally to the interventricular foramen, separating the hemispheres. Such a situation is not observed in other amphibians but found in reptiles. Olfactory nerves are particularly massive and two accessory olfactory bulbs associated to vomeronasal organs are also observed on the anterior part of the cerebral hemisphere. A characteristic strong flexion is observed between the diencephalon and mesencephalon.

4.2.5 Sense Organs

The eyes are small, covered with skin that can be more or less thick, transparent and containing gland cells. In *Scolecomorphus*, they are covered with skull bones. The lens is not totally lamellar with nucleated cells, i.e., it is not perfectly transparent. The retina is composed with rods only. Optic nerves are reduced, ocular muscles are combined with tentacles. Cephalic area corresponding to oculomotor muscles is also reduced (Wake 1985; Himstedt and Manteuffel 1985; Brun and Exbrayat 2007). A green pigment characteristic of anuran and urodelan eyes is not found in Gymnophiona (Rage 1985).

A pair of tentacles has been described in several caecilians. Each tentacle is contained in a sac enveloping also the corresponding eye and extruded by means of muscles at the tip of the snout, between the nostril and corresponding eye (Badenhorst 1978; Billo 1986; Fox 1985; Billo and Wake 1987; Exbrayat and Estabel 2006) (Fig. 4.2). Harderian glands empty into the posterior end of the sac. Vomeronasal organs are linked to the tentacles by ductules. Tentacles are continuously extruded and withdrawn. They catch chemical substances from the environment that are dissolved in the Harderian fluid then analysed by vomeronasal organs that are linked to the secondary olfactory bulb with nerve fibres.

The formation of choana is identical in Gymnophiona and Amniota, contrarily to that of other amphibians (Rage 1985).

The ears have been little studied and their anatomy does not present any peculiarity (De Jaeger 1947; Wever 1975; Wever and Gans 1976; Fritzsch and Wake 1988). But, in Anura and Urodela, a bone of the internal ear, the speculum, closes the oral window. This bone is not found in Gymnophiona (Lombard and Bolt 1979; Rage and Janvier 1982).

Taste buds exist in aquatic larvae and adults and their structure resembles that of fishes (Wake and Schwenk 1986).

The lateral line organs are characteristic of aquatic animals and they occur in aquatic gymnophionan larvae and lack in intrauterine ones. They have also been found in adults of some aquatic species. The lateral line organs are composed of neuromasts located on each side of the head and all along the body and ampullary organs that are found on the head, on the upper surface of the snout (Heterington and Wake 1979; Fritzsch and Wake 1986; Dünker et al. 2000). Lateral line organs have been shown as functional organs in *I. kohtaoensis* larva (Himstedt and Fritzsch 1990).

4.2.6 Digestive Tract

The mouth contains the tongue, salivary glands and teeth. The tongue possesses glands with mucous and serous secretions (Zylberberg 1972, 1986; Bemis et al. 1983). Adult teeth are characteristic of amphibians. They consist of a basal pedicel with an external crown (dentine cone), separated from each other with a fibrous connective tissue (Lawson 1965a, b; Wake and Wurst 1979; Casey and Lawson 1981; Greven 1984, 1986). A foetal dentition has been observed during the intrauterine life in all the viviparous species. This is very different from that of an adult one (Parker and Dunn 1964; Wake 1976, 1978, 1980; Hraoui-Bloquet 1995; Hraoui-Bloquet and Exbrayat 1996).

Oesophagus, stomach and intestine do not present any peculiar character, being like those of other amphibians (Hraoui-Bloquet and Exbrayat 1992; Delsol et al. 1995; Exbrayat and Estabel 2006). The liver is elongated, constituted with several units. Like in other lower vertebrates, hepatic blades are constituted with two cell layers. A peripheral haematopoietic layer surrounds the liver in larvae as well as in adults. Layer thickness increases with age. These cells infiltrate the liver to form nodules (Welsch and Storch 1972; Storch et al. 1986; Paillot et al. 1997a, b).

4.2.7 Respiratory System

The respiratory system of adults is composed of an alveolar trachea and one or two, more or less, developed lungs (Baer 1937; Wake 1974; Pattle et al. 1977; Welsch 1981; Kühne and Junqueira 2000). Only one single kind of pneumocyte has been found in *Chthonerpeton indistinctum* and *I. orthoplicata*. Pneumocytes contain lamellar bodies originating the lung surfactant. In *T. compressicauda*, Bastit (2007) has shown two kinds of pneumocytes, like in other amphibians. Larvae are equipped with pair of triradied gills (Taylor 1968; Wake 1967, 1969) that can be transformed in vesiculous organs in viviparous species such as *T. compressicauda* (Exbrayat and Hraoui-Bloquet 2006).

4.2.8 Heart

Like that of other amphibians, the heart of Gymnophiona is composed of an arterial cone, a ventricle, two auricles and a venous sinus. The two auricles are separated by a muscular septum that has been described as complete (Marcus 1935) or fenestrated (Lawson 1966b). The ventricle is unique but a large muscular pillar is found centrally. No particularity has been observed about blood cells.

4.2.9 Immune System

The structure of thymus is classical (Welsch 1982; Paillot et al. 1997a, b). In *T. compressicauda*, when the thymus is differentiated into cortical and medullary regions, it contains Hassal's-like corpuscles that have not been observed in the lowest species *I. kohtaoenesis*. Hassal's corpuscles are observed in Amniota and never in Anamniota, except Gymnophiona (Bleyzac and Exbrayat 2005; Bleyzac et al. 2005). The spleen does not present any peculiar characters: it resembles other amphibians' spleen (Welsch and Storch 1982; Welsch and Starck 1986; Paillot et al. 1997a; Bleyzac and Exbrayat 2005; Bleyzac et al. 2005).

As already signalled, the liver is surrounded with a cortical haematopoietic layer that can be correlated to the lack of bone marrow. Such a cortical layer is observed in embryonic Anura and Urodela and also in adult Urodela.

4.2.10 Excretion System

The excretion system is composed of a pair of elongated kidneys and an urinary bladder (Chatterjee 1936; Garg and Prasad 1962; Wake 1970; Welsch and Storch

1973; Sakai et al. 1986, 1988a; Carvalho and Junqueira 1999). In aquatic species, glomerula of kidneys are more voluminous than that of terrestrial ones. In certain lower species, a pair of pronephros persists in adults and gymnophionan kidneys have often been considered such as the ancestral type of vertebrate kidneys (Brauer 1902). But in other species, pronephros regresses at metamorphosis (Bastit 2007). Like in other amphibians, the urinary bladder is an expansion of cloaca without any direct connexion with Wolffian ducts.

4.2.11 Endocrine Organs

Several works have been devoted to endocrine organs. Epiphysis has been studied in *T. compressicauda*, in adult as well as during development (Leclercq 1995; Leclercq et al. 1995). In this species, the pineal complex is composed of an epiphysis (or pineal organ), a pea-shaped structure formed by the extension of the roof of diencephalon. Like in Amniota, the parapineal organ has never been observed, contrarily to the Anura and Urodela.

The thyroid gland has been little studied and it does not present any original characteristic (Klump and Eggert 1935; Welsch et al. 1974; Raquet and Exbrayat 2007). Parathyroid glands of *Ichthyophis* and *Chthonerpeton* are not very different from that of other amphibians but with only one single type of cell. In Anura and Urodela, this gland is composed of two cell types.

Interrenal glands are paired organs located on the anterior part of kidneys and are like those of other vertebrates (Masood Parveez 1987; Masood Parveez et al. 1992; Dorne and Exbrayat 1996). They contain adrenocortical cells positively reacting with anticortisol serum and adrenal (chromaffin) cells. Cyclic variations linked to season and activity have been observed in both adrenocortical and adrenal cells.

The hypophysis of gymnophionan is always flattened or more or less elongated. Like in other vertebrates, several cell types have been found (Schubert et al. 1977; Zuber-Vogeli and Doerr-Schott 1981; Doerr-Schott and Zuber-Vogeli 1984, 1986; Exbrayat 2006b). Lactotropic cells can be revealed by an anti-prolactine serum, gonadotropic cells react with both anti-FSH and anti LH sera, corticotropic cells react with anti-ACTH serum, thyreotropic cells are revealed by immunocytochemistry using an anti-TSH serum. Like in other amphibians, these cell types are submitted to seasonal variations.

4.2.12 Male Genital Tract

The male genital tract has been well studied in several species (Wake 1968, 1970; Exbrayat and Estabel 2006; Akbarsha et al. 2006; Smita et al. 2006; Scheltinga and Jamieson 2006; Measey et al. 2008). It is composed of a pair of multi-lobed testes associated with segmented fat bodies, a pair of Müllerian glands and an erectile cloaca. Testes are more or less regularly segmented. Segmentation is more regular in the lowest genera (*Ichthyophis*, *Rhinatrema*) than in highest ones (*Typhlonectes*).

Each lobe is composed by several lobules (or locules, after Seshachar 1936) in which spermatogenesis occurs. Germ cells are grouped into isogenic units, moving from the periphery to the centre of lobule, between giant Sertoli cells, giving to the testes a unique structure among the vertebrates.

All the gymnophionan possess a pair of Müllerian ducts that not only degenerate during development, like in other vertebrates but are transformed into true glands with a secretive cycle linked to the sexual cycle (Wake 1981; Exbrayat 1985; Akbarsha et al. 2006). These organs are the equivalent of the prostate of highest vertebrates. All the gymnophionans practice internal fertilization. In all the species, the posterior part of male cloaca is erectile, giving an intromittent phallodeum (Tonutti 1931; Exbrayat 1991, 1997).

4.2.13 Female Genital Tract

Like the male genital tract, the female one has also been well studied in several species (Wake 1968, 1970; Exbrayat and Estabel 2006; Exbrayat 2006c; Raquet et al. 2006). The ovaries are elongated and associated to fat bodies. After ovulation, the staying follicle is transformed into a persistent corpus luteus. The eggs are contained in the oviducts. After ovulation, oocytes are driven to the anterior part of oviduct, in which fertilization occurs. In oviparous species, developing eggs are surrounded by an envelope secreted by the glands of the oviduct. In viviparous species, the posterior part of the oviduct is transformed into the uterus. The uterine wall provides food for embryos that also practice adelphophagy, eating the dead or perhaps live embryos. In *Typhlonectes* a kind of placenta implying both gills of embryo and uterine wall has been studied by Hraoui-Bloquet (1995) and Exbrayat and Hraoui-Bloquet (2006).

4.2.14 Development and Metamorphosis

Embryonic development of caecilians has only been poorly studied in a few species (Sarasin and Sarasin 1887–1890; Sammouri et al. 1990; Dünker et al. 2000; Exbrayat 2006d). The eggs are telolecithic, with a compact mass of yolk. The development is of discoidal type, resembling that of Urodela with a big mass of yolk. Metamorphosis has been observed. It can be either slow in oviparous species with aquatic larvae (10 months in *Ichthyophis*), or quick in viviparous species (estimated to about 2 months in *Typhlonectes*) (Exbrayat and Hraoui-Bloquet 1994).

4.3 Biogeography

The repartition of Gymnophiona has been described by Taylor (1968). Biogeography of these animals has also been studied by Lescure et al. (1986), Wilkinson and Nussbaum (2006). Caecilians are found in Asia with the primitive genus *Ichthyophis*, South America with the other primitive genus *Rhinatrema*, Central America and the genera *Dermophis* and *Gymnopis* and Africa including the Seychelles Islands. Several genera have a wide repartition (*Ichthyophis*, *Rhinatrema*, *Typhlonectes*), others are well localised such as *Schistometopum thomense*, confined in the Sao Tome Islands, or still *Geotrypetes* or *Hypogeophis* from the Seychelles Islands. The highest genera, *Typhlonectes*, *Chthonerpeton* are found in South America. This repartition is fundamentally of a Gondwanan type (Lescure and Renous 1988).

4.4 Fossils

Very few fossils of gymnophionans have been found. First fossils are two vertebrae. One of them, *Apodops pricei* has been found in Brazil (Estes and Wake 1972; Estes 1981) and is dated from highest or lowest Paleocene (−65 to −55 millions years). The second one, discovered in Bolivia, has been described by Rage (1986) and is dated from the Cretacean (−100 to −65 millions years). More recently, a Jurassic Caecilian, *Eocilia micropodia*, has been discovered in Arizona by Jenkins and Walsh (1993). This fossil presents the main characteristics of gymnophionans and, in addition, has a pair of girdles with limbs.

4.4.1 Molecular Data

Few molecular data are available. 12S mitochondrial mRNA has been studied and compared in several species. Immunological data has also been used (Cannatella and Hillis 1993; Hedges and Maxson 1993; Hedges et al. 1993; Wilkinson and Nussbaum 2006).

4.5 Systematic Position

Systematicians have been – and are still – interested by two trends of classification of Gymnophiona: (1) the internal classification of families, genera and species and (2) position of Gymnophiona among Anamniota and especially among other amphibians.

4.5.1 Internal Classification

Several classifications have been given according to the available morphological and molecular data (see Wilkinson and Nussbaum 2006 for references). The first modern classification is that of Taylor (1968) who recognised 157 species, 34 genera and only three families: Ichthyophiidae, Caeciliidae, Typhlonectidae. In 1969, Taylor added a new family: the Scolecomorphidae and divided Caeciliidae into two sub-families: Caeciliinae and Dermophiinae. In 1977, Nussbaum separated the Asiatic Ichthyophiidae from the Southern American Rhinatrematidae. In 1986, Laurent classified Gymnophiona into six families: Ichthyophiidae, Rhinatrematidae, Caecliidae, Scolecomorphidae, Dermophiidae, Typhlonectidae. Several internal modifications were also given by Wake and Campbell (1983) and Duellman and Trueb (1986). In 1986, Lescure et al., using a cladistic method, proposed a new classification that was contested by Nussbaum and Wilkinson (1989).

Today, a new classification using several criteria (morphological, molecular, biogeographical) has been proposed by Wilkinson and Nussbaum (2006). According to this classification, the gymnophionans are grouped into six families: Rhinatrematidae (South America), Ichthyophiidae (Asia), Uraeotyphlidae (a new Asiatic family separated from Caeciliidae), Scolecomorphidae (Africa), Caeciliidae (a heterogenous family with a wide geographical distribution, containing also *Dermophis*, the family Dermophiidae being no longer separated from other Caeciliidae) and Typhlonectidae (South America). These families are narrowly linked to the geographical repartition. Rhinatrematidae present several characters considered as ancestral: they are the lowest gymnophionans. The non-Rhinatrematidae species are also called Neocaecilia by Wilkinson and Nussbaum (2006). In this group, both Ichtyophiidae and Uraeotyphlidae are regrouped in Diatriata, a new super family. Scolecomorphidae, Caeciliidae and Typhlonectidae belong to the super-family Teresomata.

4.5.2 Position of Gymnophiona Among Vertebrates

Rage (1985) explained that Gymnophiona shared a lot of characters with other amphibians. But, in addition, noticeable differences pointed out earlier, have been found between Caecilians and both Anura and Urodela. For Rage (1985), these data let think that amphibian differentiated into two stages. A common ancestor could have given a first branch differentiating in Gymnophiona and a second one being the common ancestor of Anura and Urodela, Anura and Urodela being sister groups (Fig. 4.3a) or Anura being derived from Urodela (Fig. 4.3b).

In 1988 and 1993, Milner analysed the relationships between Amniota and amphibians and proposed several general cladograms depending on these relationships. The first one remained that of Rage (1985). Batrachia (Anura + Urodela considered as sister groups) and Gymnophiona were both Lissamphibians (Fig. 4.4).

Then, he discussed the parental facts between Anura, Urodela and Temnospondyls, fossils suspected to be ancestors of the amphibians, Gymnophiona and Microsaura, a fossil amphibian. Microsaura were considered such as the direct ancestor of Gymnophiona (Fig. 4.5b) or belonging to a different branch. If Gymnophiona

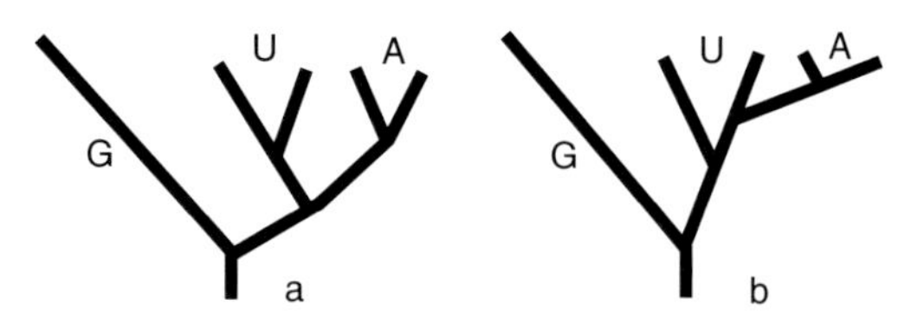

Fig. 4.3 Phylogeny of amphibians modified after Rage (1985), (a) Urodela and Anura are considered such as sister groups. (b) Anura are derived from Urodela. In both two hypotheses, Gymnophiona and Paratoidia (Anura + Urodela) are paraphyletic. Lissamphibia are constituted with Gymnophiona and Paratoidia. A: Anura; G: Gymnophiona; U: Urodela

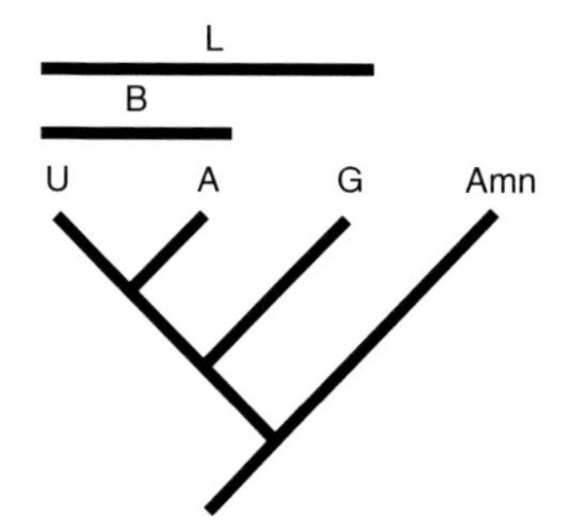

Fig. 4.4 Phylogeny of Amphibia (modified after Milner [1988]), A: Anura (Salientia); Amn: Amniota; B: Batrachia; G: Gymnophiona; L: Lissamphibia; U: Urodela (Caudata)

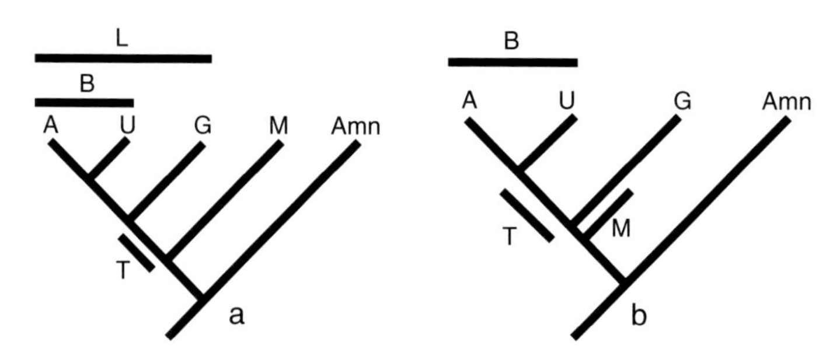

Fig. 4.5 Alternative relationships of Temnospondyls, microsaurs and Lissamphibia modified after Milner (1993), (a) Monophyly of the Lissamphibia: Gymnophiona are included in Lissamphibia; microsaurs are a separated branch; Temnospondyls are the stem-group of Lissamphibia; (b) Monophyly of the Batrachia: Microsaurs are ancestors of Gymnophiona; Temnospondyls are the stem-group of Batrachia. A: Anura (Salientia); Amn: Amniota; B: Batrachia; G: Gymnophiona; L: Lissamphibia; T: Temnospondyls; U: Urodela (Caudata)

were issued from Microsaura, the group Microsaura + Gymnophiona was considered such as the sister group of Amniota (Fig. 4.5a).

Another conclusion has been given after molecular data. Using mRNA 12S in mitochondria, Hedges and Masson (1993) concluded that Gymnophiona and Urodela were sister groups. After these data, Lissamphibia were Anura + (Urodela + Gymnophiona).

4.6 Conclusion

Examination of the main aspects of gymnophionan morphology in adults as well as in larvae permits to classify characters into four categories. (1) Amphibian characters such as the structure of the integument (excepted scales), fat bodies associated to gonads, structure of the heart, haematopoietic layer surrounding liver; gills in larvae and lungs in adults such as respiratory system; development with metamorphosis. (2) Several plesiomorphic characters reminiscent of bony fishes: presence of scale with an Osteichthyan-like structure in lowest genera; structure of the adult vertebrae. (3) Several characters resembling those of reptiles and sometimes other Amniota: structure of brain, presence of Hassal's corpuscles in thymus, presence of a collar equivalent to a neck; a mullerian gland looking like a prostate. In addition, during the embryonic development, the sclerotoma are well metameric, like in Amniota and not at all in amphibian in which metamerisation is not obvious. In addition, the formation of choana looks like that of Amniota and not that of other amphibians. (4) Characters specific to Gymnophiona and their burrowing habits: lengthening of both body and internal organs, absence of limbs and girdles, presence of a retro-articular process on the lower jaw, reduction of ocular system, pair of tentacles, special olfactive organs.

Considering these different characters, it can be assumed that Gymnophiona are fundamentally amphibians. They also present several reminiscent characters of fishes and other characters that, more than in the other amphibians, heralds the reptiles and Amniota. Considering the mode of life of gymnophionans, we can imagine that selection pressure has favoured well specialised characters. Data given by comparative anatomy, biogeography, molecular analysis, show a Gondwanan repartition of these animals. Several questions remain elusive: what are the ancestors of gymnophionans? Microsaurs or others? Are gymnophionans exactly comparable to anurans and urodelans, i.e., other Lissamphibia? To try to give an answer, a scheme of apparition and evolution of Gymnophiona can be proposed – but it is a hypothesis. When vertebrates began to emerge out of the water, becoming amphibians, two possibilities were present: (1) to live **on the ground** and (2) to live **under the ground**. Let us suppose that several animals chose this last possibility and Gymnophiona or more exactly their ancestor, were among them. Selective pressure of burrowing life could favour the development of peculiar characters such as the lengthening of body and organs, the disappearance of limbs and girdles; correlatively several old characters (fish-like) were first conserved. Scales, for instance are still present in the lowest species like in the oldest fossil amphibians as *Ichthyostega*. Scales could be advantageous for burrowing animals (see Riberon and Exbrayat 1996).

During this evolution, several characters announced reptiles and Amniota. It could be a natural trend in Amphibia. In Gymnophiona these trends remained during their own evolution.

During their evolution, Gymnophiona displaced themselves on continents. When continents separated, each population acquired its own specificity, which could explain the occurrence of the current 6 families, each one being linked to a specific biogeographical context (except Caeciliidae). The primitive Asiatic Ichthyophiidae and American Rhinatrematidae, resemble each other; the most highest, the Typhlonectidae acquired the most specialised characters and more especially a true vivipary.

So, Gymnophiona reflect a stage of vertebrate evolution at which animals still show archaic and specialised characteristics, with clues of a future evolution. This mixing of characters is certainly allowed by the burrowing lifestyle.

References

Akbarsha MA, Jancy GM, Smita M, Oommen O (2006) Caecilian male mullerian gland with special reference to *Uroaeotyphlus narayani*. In: Exbrayat JM (ed) Reproductive biology and phylogeny of Gymnophiona (Caecilians), Science Publishers, Enfield/Jersey/Plymouth, pp 157–182

Baer JG (1937) L'appareil respiratoire des Gymnophiones. Revue Suisse de Zoologie 44:353–377

Bastit M (2007) Aspects anatomiques du développement des poumons, des reins et des glandes interrénales chez *Typhlonectes compressicauda* (Duméril et Bibron, 1841), Amphibien Gymnophione. Dipl EPHE, Lyon

Badenhorst A (1978) The development and the phylogeny of the organ of Jacobson and the tentacular apparatus of *Ichthyophis glutinosus* (Linn.). Ann Univ Stellenbosch 1:1–26

Bemis W, Schwenk K, Wake MH (1983) Morphology and function of the feeding apparatus in *Dermophis mexicanus* (Amphibian: Gymnophiona). Zool J Linn Soc 77:75–96

Billo R (1986) Tentacle apparatus of Caecilians. Mém Soc Zool Fr 43:71–75

Billo R, Wake MH (1987) Tentacle development in *Dermophis mexicanus* (Amphibian: Gymnophiona) with an hypothesis of tentacle origin. J Morph 192:101–111

Bleyzac P, Exbrayat JM (2005) Some aspects of the ontogenesis of the immune system organs of *Typhlonectes compressicauda*. In: Ananjeva N, Tsinenko O (eds), Herpetologica petropolitana. Russ J Herp 12(suppl):120–123

Bleyzac P, Cordier G, Exbrayat JM (2005) A morphological des cription of embryonic development of immune organs in *Typhlonectes compressicauda* (Amphibia, Gymnophiona). J Herp 39:57–65

Brauer A (1902). Beitrage zur kenntniss der Entwicklung und Anatomie der Gymnophionen. III. Die Entwicklung der Excretionsorgane. Zoologisches Jahrbuch für Anatomie 16:1–176.

Brun C, Exbrayat JM (2007) Embryonic development of the hypophysis and thyroid gland in *Typhlonectes compressicauda* (Dumeril and Bibron, 1841), Amphibia, Gymnophiona. J Herp 41:702–711

Cannatella DC, Hillis DM (1993) Amphibian relationships: phylogenetic analysis of morphology and molecules. Herp Monographs 7:1–7

Carvalho ETC Junqueira LCU (1999). Histology of the kidney and urinary bladder of *Siphonops annulatus* (Amphibia – Gymnophiona). Arch Hist Cyt 62:39–45
Casey J, Lawson R (1979) Amphibians with scales: the structure of the scale in *Hypogeophis rostratus*. Brit J Herp 5:831–833
Casey J, Lawson R (1981) A histological and scanning electron microscope study of the teeth of Caecilian Amphibians. Arch Or Biol 26:48–58
Charlemagne J (1990) Immunologie des poïkilothermes. In: Pastoret PP, Govaerts A, Bazin H (eds) Immunologie animale, Méd Sci, Flammarion, Paris, pp 447–457
Chatterjee BK (1936) The anatomy of *Uraeotyphlus menoni* Annandale. Part I: digestive, circulatory, respiratory and urogenital systems. Anat Anz 81:393–414
Cockerell TDA (1912) The scales of *Dermophis*. Science 36:681
De Jaeger EFJ (1947) Some points in the development of the staped of *Ichthyophis glutinosus*. Anat Anz 96:203–210
Delsol M, Flatin J, Exbrayat JM (1995) Le tube digestif des Amphibiens adultes. In: Grassé PP, Delsol M (eds) Traité de Zoologie, tome XIV, fasc.I A. Masson, Paris, pp 497–508
Doerr-Schott J, Zuber-Vogeli M (1984) Immunohistochemical study of the adenohypophysis of *Typhlonectes compressicaudus* (Amphibia, Gymnophiona). Cell Tissue Res 235:211–214
Doerr-Schott J, Zuber-Vogeli M (1986) Cytologie et immunoicytologie de l'hypophyse de *Typhlonectes compressicaudus*. Mém Soc Zool Fr 43:77–79
Dorne JL, Exbrayat JM (1996) Quelques aspects de l'anatomie et de l'histologiue des glandes interrénales chez trois Gymnophiones. Bull Soc Zool Fr 121:146–147
Duellman WE, Trueb L (1986) Biology of amphibians. McGraw-Hill, New York
Dünker N, Wake MH, Olson WM (2000) Embryonic and larval development in the Caecilian *Ichthyophis kohtaoensis*. J Morph 243:3–34
Estabel J, Exbrayat JM (1998) Brain development of *Typhlonectes compressicaudus* (Dumeril and Bibron, 1841). J Herp 32:1–10
Estes R (1981) Encyclopaedia of paleoherpetology. II : Gmnophiona, caudata. Gustav Fisher, Stuttgart
Estes R, Wake MH (1972) The first fossil record of Caecilian amphibians. Nature 239:228–231
Exbrayat JM (1985) Cycle des canaux de Müller chez le mâle adulte de *Typhlonectes compressicaudus* (Duméril et Bibron, 1841), Amphibien Apode. C r séanc Acad Sci Paris 301:507–512
Exbrayat JM (1991) Anatomie du cloaque chez quelques Gymnophiones. Bull Soc Herp Fr 58:31–43
Exbrayat JM (1996) Croissance et cycle du cloaque chez *Typhlonectes compressicaudus* (Duméril et Bibron, 1841), Amphibien Gymnophione. Bull Soc Zool Fr 121:93–98
Exbrayat JM (2000) Les Gymnophiones, ces curieux amphibians. Société Nouvelle des éditions Boubée, Paris
Exbrayat JM (ed) (2006a) Reproductive biology and phylogeny of Gymnophiona (Caecilians). Vol 5 of series. In: Jamieson GMB (ed) Science Publishers, Enfield/Jersey/Plymouth
Exbrayat JM (2006b) Endocrinology of reproduction in Gymnophiona. In: Exbrayat JM (ed) Reproductive biology and phylogeny of Gymnophiona (Caecilians), Science Publishers, Enfield/Jersey/Plymouth, pp 183–229
Exbrayat JM (2006c) Oogenesis and folliculogenesis. In: Exbrayat JM (ed) Reproductive biology and phylogeny of Gymnophiona (Caecilians), Science Publishers, Enfield/Jersey/Plymouth, pp 275–290
Exbrayat JM (2006d) Fertilization and embryonic development. In: Exbrayat JM (ed) Reproductive biology and phylogeny of Gymnophiona (Caecilians), Science Publishers, Enfield/Jersey/Plymouth, pp 359–386
Exbrayat JM, Estabel J (2006) Anatomy with particular references to the reproductive system. In: Exbrayat JM (ed) Reproductive biology and phylogeny of Gymnophiona (Caecilians), Science Publishers, Enfield/Jersey/Plymouth, pp 79–155
Exbrayat JM, Hraoui-Bloquet S (1994) Un exemple d'hétérochronie: la metamorphose chez les Gymnophiones. Bull Soc Zool Fr 110:117–126

Exbrayat JM, Hraoui-Bloquet S (2006) Viviparity in *Typhlonectes compressicauda*. In: Exbrayat JM (ed) Reproductive biology and phylogeny of Gymnophiona (Caecilians), Science Publishers, Enfield/Jersey/Plymouth, pp 325–357

Fox H (1983) The skin of *Ichthyophis* (Amphibia: Caecilia): an ultrastructural study. J Zool 199:223–248

Fox H (1985) The tentacles of *Ichthyophis* (Amphibia: Caecilia) with special reference to the skin. J Zool 205:223–234

Fox H (1987) On the fine structure of the skin of larval juvenile and adult *Ichthyophis* (Amphibia: Caecilia). Zoomorph 107:67–76

Fritzsch B, Wake MH (1986) The distribution of ampullary organs in Gymnophiona. J Herp 20:90–93

Fritzsch B, Wake MH (1988) The inner ear of gymnophione amphibians and its nerve supply: a comparative study of regressive events in a complex sensory system (Amphibia, Gymnophiona). Zoomorph 108:201–217

Garg BL, Prasad J (1962) Observations of the female urogenital organs of limbless amphibian *Uraeotyphlus oxyurus*. J An Morph Phys 9:154–156

Greven H (1984) The dentition of *Gegenophis ramaswamii* Taylor 1964 (Amphibia, Gymnophiona), with comments on monocuspid teeth in the Amphibia. Z Zool System Evol Forsch 22:342–348

Greven H (1986) On the diversity of tooth crowns in Gymnophiona. Mém Soc Zool Fr 43:85–86

Hedges SB, Maxson LR (1993) A molecular perspective on lissamphibian phylogeny. Herp Monographs 7:27–42

Hedges SB, Nussbaum RA, Maxson LR (1993) Caecilian phylogeny and biogeography inferred from mitochondrial DNA sequences of the 12S rRNA and 16S genes (Amphibia: Gymnophiona). Herp Monographs 7:64–76

Heterington TE, Wake MH (1979) The lateral line system in larval *Ichthyophis* (Amphibian: Gymnophiona). Zoomorph 93:209–225

Himstedt W (1996) Die Blindwühlen. Westarp-Wissenschaft, Magdeburg

Himstedt W, Fritzch B (1990) Behavioral evidence for electroception in larvae of the Caecilian *Ichthyophis kohtaoensis* (Amphibia, Gymnophiona). Zool Jahrb Physiol 94:484–492

Himstedt W, Manteuffel G (1985) Retinal projections in the caecilian *Ichthyophis kohtaoensis* (Amphibia, Gymnophiona). Cell Tissue Res 239:689–692

Hraoui-Bloquet S (1995) Nutrition embryonnaire et relations materno-foetales chez *Typhlonectes compressicaudus* (Duméril et Bibron, 1841), Amphibien Gymnophione vivipare. Thèse de Doctorat E.P.H.E., Lyon, France.

Hraoui-Bloquet S, Exbrayat JM (1992) Développement embryonnaire du tube digestif chez *Typhlonectes compressicaudus* (Duméril et Bibron, 1841), Amphibien Gymnophione vivipare. Ann Sci Nat, Zool., 13éme sér 13:11–23

Hraoui-Bloquet S, Exbrayat JM (1996) Les dents de *Typhlonectes compressicaudus*, Amphibia, Gymnophiona) au cours du développement. Ann Sci Nat, Zool, 13éme série 17:11–23

Jenkins, Walsh (1993) An Early Jurassic Caecilian with limbs. Nature 365:246–250

Klumpp W, Eggert B (1935) Die Schilddrüse und die branchiogenen Organe in *Ichthyophis glutinosus*. L. Zeit wissen Zool 146:329–381

Kuhlenbeck H (1922) Zur Morphologie des Gymnophionengehirns. Jena Zeit Naturwiss 58:453–484

Kuhlenbeck H (1969) Observations on the rhombencephalon in the Gymnophion *Siphonops annulatus*. Anat Res 163:311

Kühne B, Junqueira LCU (2000) Histology of the trachea and lung of *Siphonops annulatus* (Amphibia, Gymnophiona). Rev Braz Biol 60:167–172

Laurent RF (1986) Sous-classe des Lissamphibiens (Lissamphibia). Systématique. In: Grassé PP, Delsol M (eds) Traité de zoologie, tome XIV, fasc 1B, Masson, Paris, pp 595–797

Lawson R (1963) The anatomy of *Hypogeophis rostratus* Cuvier (Amphibia: Apoda or Gymnophiona). I. The skin and skeleton. Proc Univ Durham Phil Soc Series A, 13:254–273

Lawson R (1965a) The development and replacement of teeth in *Hypogeophis rostratus* (Amphibia, Apoda). J Zool 147:352–362
Lawson R (1965b) The teeth of *Hypogeophis rostratus* (Amphibia, Apoda) and tooth structure in the amphibia. Proc Zool Soc London 145:321–325
Lawson R (1966a) The development of the centrum of *Hypogeophis rostratus* (Amphibia, Apoda) with special reference to the notochordal (intra-vertebral) cartilage. J Morph 118:137–148
Lawson R (1966b) The anatomy of the heart of *Hypogeophis rostratus* (Amphibia, Apoda) and its possible mode of action. J Zool 149:320–336
Leclercq B (1995) Contribution à l'étude du complexe pinéal des Amphibiens actuels. Dipl E.P.H. E., Lille, France.
Leclercq B, Martin-Bouyer L, Exbrayat JM (1995) Embryonic development of pineal organ in *Typhlonectes compressicaudus* (Dumeril and Bibron, 1841), a viviparous Gymnophionan Amphibia. In: Llorente GA, Montori A, Santos X, Carretero MA (eds) Scientia Herpetologica, Proceedings of the 7th General Meeting of the Societas Europaea Herpetologica, Barcelona, Spain, pp 107–111
Lescure J, Renous S (1988) Biogéographie des Amphibiens Gymnophiones et histoire du Gondwana. C r Soc Biogeogr 64: 19–40
Lescure J, Renous S (1992) Signification du collier chez les Amphibiens Gymnophiones. Bull Soc Herp Fr 61:45–51
Lescure J, Renous S, Gasc JP (1986) Proposition d'une nouvelle classification des amphibiens Gymnophiones. Mém Soc Zool Fr 43 :144–147
Lombard RE, Bolt JR (1979) Evolution oft he tetrapod ear: an analysis and reinterpretation. Biol J Linn Soc Lond 11:19–76
Marcus H (1935) Zur stammensgeschichte der Herzens. Morph Jahrb 76:92–103
Martin-Bouyer L, Godard JP, Mairie JM (1995) 3-D reconstitution of the brain of *Typhlonectes compressicaudus* (Dumeril and Bibron, 1841) (Amphibia, Gymnophiona). In: Llorente GA, Montori Ai, Santos X, Carretero MA (eds) Scientia Herpetologica, Proceedings of the 7th General Meeting of the Societas Europaea Herpetologica, Barcelona, Spain, pp 98–100
Masood Parveez U (1987) Some aspects of reproduction in the female Apodan Amphibian *Ichthyophis*. Ph.D., Karnataka University, Dharwad, India
Masood Parveez U, Bhatta GK, Nadkarni VB (1992) Interrenal of a female gymnophione amphibian, *Ichthyophis beddomei*, during the annual reproductive cycle. J Morph 211:296–382
Measey J, Smita M, Beyo RS, Oommen OV (2008) Year-round spermatogenic activity in an oviparous subterranean caecilian, *Boulengerula taitanus* Loveridge 1935 (Amphibia Gymnophiona Caceciliidae). Trop Zool 21:109–122
Milner AR (1988) The relationships and origin of living amphibians. In: Benton J (ed) The phylogeny and classification of the Tetrapods, vol. 1. Clarendon Press, Oxford, pp 59–102
Milner A (1993) The paleozoic relatives of lissamphibians. Herp Monographs 7:8–27
Nussbaum R (1977) Rhinatrematidae: a new family of Caecilian (Amphibia: Gymnophiona). Occ pap mus zool Michigan 682:1–30
Nussbaum RA, Wilkinson M (1989) On the classification and phylogeny of Caecilians (Amphibia: Gymnophiona), a critical review. Herp Monographs 3:1–42
Olivecrona H (1964) Notes on forebrain morphology in the Gymnophion (*Ichthyophis glutinosus*). Acta Morph 6:45–53
Paillot R, Estabel J, Exbrayat JM (1997a) Organes hématopoiétiques et cellules sanguines chez *Typhlonectes compressicaudus* et *Typhlonectes natans*. Bull Mens Soc Linn Lyon 66:124–134
Paillot R, Estabel J, Exbrayat JM (1997b) Liver, bone marrow substitut in adult Gymnophionan. 3rd World Congress Herp., Prague, Czech Republic, 157
Parker HW, Dunn ER (1964) Dentitional metamorphosis in the Amphibia. Copeia 1964:75–86
Pattle RE, Schock C, Creasey JM, Hughes GM (1977) Surpelling film, lung surfactant, and their origin in newt, caecilian, and frog. J Zool 182: 125–136
Rage JC (1985) Origine et phylogénie des Amphibiens. Bull Soc Herp Fr 34:1–19

Rage JC (1986) Le plus ancien amphibien apode (Gymnophiona) fossile. Remarque sur la répartition et l'histoire paléobiogéographique des Gymnophiones. C r séanc Acad Sci Paris 302:1033–1036.

Rage JC, Janvier P (1982) Le problème de la monophylie des Amphibiens actuels à la lumière des nouvelles données sur les affinités des tétrapodes. Geobios Mem sp 6:65–83

Raquet M, Exbrayat JM (2007) Embryonic development of the hypophysis and thyroid gland in *Typhlonectes compressicauda* (Dumeril and Bibron, 1841), Amphibia, Gymnophiona. J Herp 41:702–711

Raquet M, Measey J, Exbrayat JM (2006) Premières observations histologiques de l'ovaire de *Boulengerula taitanus* Loveridge, 1935, amphibien gymnophione. Rev Fr Histotechnol 19:9–15

Renous S, Gasc JP (1986a) Le fouissage des Gymnophiones (Amphibia). Hypothèse morphofonctionnelle fondée sur la comparaison avec d'autres Vertébrés tétrapodes. Zool Jahrb, Anat 114:95–130

Renous S, Gasc JP (1986b) Hypothèse d'étude concernant la locomotion des Gymnophiones. Mém Soc Zool Fr 43:133–143

Renous S, Exbrayat JM, Estabel J (1997) Recherche d'indices de membres chez les Amphibiens Gymnophiones. Ann Sci Nat 18:11–26

Riberon A, Exbrayat JM (1996) Quelques aspects de la structure du tégument des Amphibiens Gymnophiones. Bull Soc Herp Fr 79:43–56

Rogers DC (1965) An electon microscope study of the parathyroid gland of the frog (*Rana clamitans*). J Ultrastr Res 13:478–499

Sakai T, Billo R, Kriz W (1986) The structural organization of the kidney of *Typhlonectes compressicaudus* (Amphibia, Gymnophiona). Anat Embr 174:243–252

Sakai T, Billo R, Kriz W (1988a) Ultrastructure of the kidney of a South American Caecilian, *Typhlonectes compressicaudus* (Amphibia, Gymnophiona). II: distal tubule, connecting tubule, collecting duct and Wolffian duct. Cell Tissue Res 252:601–610

Sakai T, Billo R, Nobiling R, Gorgas K, Kriz W (1988b) Ultrastructure of the kidney of a south american Caecilian, *Typhlonectes compressicaudus* (Amphibia, Gymnophiona) I: renal corpuscule, neck segment, proximal tubule and intermediate segment. Cell Tissue Res 252: 589–600

Sammouri R, Renous S, Exbrayat JM, Lescure J (1990) Développement embryonnaire de *Typhlonectes compressicaudus* (Amphibia Gymnophiona). Ann Sci Nat, Zool, 13ème sér 11:135–163

Sarasin P, Sarasin F. (1887–1890). Ergebnisse Naturwissenschaftlicher Forschungen auf Ceylon. Zur Entwicklungsgeschichte und Anatomie der Ceylonischen Blindwuhle Ichthyophis glutinosus. C.W. Kreidel's Verlag, Wiesbaden.

Schmidt A, Wake MH (1990) Olfactory and vomeronasal system of caecilians (Amphibia: Gymnophiona). J Morph 205:255–268

Schmidt A, Wake MH (1997) Cellular migration and morphological complexity in the caecilian brain. J Morph 231:11–27

Schmidt A, Wake MH (1998) Development of the tectum in Gymnophiones, with comparison to other Amphibians. J Morph 236:233–245

Schubert C, Welsch U, Goos H (1977) Histological, immuno and enzyme-histochemical investigations on the adenohypophysis of the Urodeles, *Mertensiella caucasica* and *Trituris cristatus* and the Caecilian, *Chthonerpeton indistinctum*. Cell Tissue Res 185:339–349

Senn DG, Reber-Leutenegger S (1986) Notes on the brain of Gymnophiona. Mém Soc Zool. Fr 43:65–66

Seshachar BR (1936) The spermatogenesis of *Ichthyophis glutinosus* (Linn.) I. The spermatogonia and their division. Z Zellforsch mikr Anat 24:662–706

Scheltinga DM, Jamieson BGM (2006) Ultrastructure and phylogeny of Caecilian spermatozoa. In: Exbrayat JM (ed) Reproductive biology and phylogeny of Gymnophiona (Caecilians), Science Publishers, Enfield/Jersey/Plymouth, pp 245–274

Smita M, Jancy GM, Akbarsha MA, Oommen OV, Exbrayat JL (2006) Caecilian spermatogenesis. In: Exbrayat JM (ed) Reproductive biology and phylogeny of Gymnophiona (Caecilians), Science Publishers, Enfield/Jersey/Plymouth, pp 231–246

Storch V, Prosi F, Gorgas K, Hacker HJ, Rafael J, Vsiansky P (1986) The liver of *Ichthyophis glutinosus* Linn? 1758 (Gymnophiona). Mém Soc Zool Fr 43:91–106

Taylor EH (1968) The Caecilians of the world. A taxonomic review. Univ Kansas Press. Lawrence, Kansas, USA

Taylor EH (1969) A new family of African Gymnophiona. Univ Kansas Sci Bull 48:297–305

Taylor EH (1977) Comparative anatomy of Caecilian anterior vertebrae. Kansas Univ Bull 51:219–231

Tonutti E (1931) Beitrag zur Kenntnis der Gymnophionen. XV Das genital System. Morph Jahrb 70:101–130

Wake MH, Campbell JA (1983) A new genus and species of Caecilian from the Sierra de Las Minas of Guatemala. Copeia 1983:857–863

Wake MH (1967) Gill structure in the Caecilian genus *Gymnopis*. Bull So Calif Acad Sci 66:109–116

Wake MH (1968) Evolutionary morphology of the Caecilian urogenital system. Part I: the gonads and fat bodies. J Morph 126:291–332

Wake MH (1969) Gill ontogeny in embryos of *Gymnopis* (Amphibia: Gymnophiona). Copeia 1969:183–184

Wake MH (1970) Evolutionary morphology of the caecilian urogenital system. Part II: the kidneys and urogenital ducts. Acta Anat 75:321–358

Wake MH (1974) The comparative morphology of the caecilian lung. Anat Rec 178:483

Wake MH (1975) Another scaled caecilian (Gymnophiona, Typhlonectidae). Herpetologica 31:134–136

Wake MH (1976) The development and replacement of teeth in viviparous Caecilians. J Morph 148:33–63

Wake MH (1978) Comments on the ontogeny of *Typhlonectes obesus* particularly its dentition and feeding. Pap av Zool 32:1–13

Wake MH (1980) Fetal tooth development and adult replacement in *Dermophis mexicanus* (Amphibia: Gymnophiona): Fields versus clones. J Morph 166:203–216

Wake MH (1981) Structure and function of the male Mullerian gland in Caecilians (Amphibia: Gymnophiona) with comments on its evolutionary significance. J Herp 15:17–22

Wake MH (1985) The comparative morphology and evolution of the eyes of Caecilians (Amphibia, Gymnophiona). Zoomorph 105:277–295

Wake MH (1986) A perspective on the systematics and morphology of the Gymnophiona (Amphibia). Mém Soc Zool Fr 43:21–38

Wake MH, Campbell JA (1983) A new general species of Caecilian from the Sierra Leone de Las Minos of Guatemala. Copeia 1983:857–863

Wake MH, Hanken J (1982) Development of the skull of *Dermophis mexicanus* (Amphibia: Gymnophiona) with comments on skull kinesis and amphibian relationships. J Morph 173:203–223

Wake MH, Nygren KM (1987) Variation in scales in *Dermophis mexicanus*: are Gymnophione scales of systematic utility? Fieldiana 36:1–8

Wake MH, Schwenk K (1986) A preliminary report on the morphology and distribution of taste buds in Gymnophiones, with comparison to other Amphibians. J Herp 20:254–256

Wake MH, Wurst GZ (1979) Tooth crown morphology in Caecilians (Amphibia: Gymnophiona). J Morph 159:331–341

Wake MH, Exbrayat JM, Delsol M (1985) The development of the chondrocranium of *Typhlonectes compressicaudus* (Gymnophiona), with comparison to other species. J Herp 19:568–577

Welsch U (1981) Fine structural and enzyme histochemical observations on the respiratory epithelium on the Caecilian lungs and gills. A contribution to the understanding of the evolution of the Vertebrate respiratory epithelium. Arch Histol Jap 14:117–133

Welsch U (1982) Morphologische Beobachtungen am thymus larvaler und adulter Gymnophionen. Zool Jahrb Anat 107 288–305

Welsch U, Starck M (1986) Morphological observations on blood cells and blood cells forming tissue of Gymnophione. Mém Soc Zool Fr 43:107–115

Welsch U, Storch V (1972) Electronmikroskopische Untersuchungen an der Leber von *Ichthyophis kohtaoensis* (Gymnophiona). Zool Jahrb Anat 89:621–635

Welsch U, Storch V (1973) Elektronenmikroskopische Beobachtungen am Nephron adulter Gymnophionen (*Ichthyophis kohtaoensis* Taylor). Zool JahrbAnat 90:311–322

Welsch U, Storch V (1982) Light and electron microscopical observations on the Caecilian spleen. A contribution to the evolution of lymphatic organs. Dev Comp Imm 6:293–302

Welsch U, Schubert C, Storch V (1974) Investigation on the thyroid gland of embryonic, Larval and adult *Ichthyophis glutinosus* and *Ichthyophis kohtaoensis* (Gymnophiona, Amphibia). Histology, fine structure and studies with radioactive iodide (^{131}I). Cell Tissue Res 155: 245–268

Wever EG (1975) The Caecilian ear. J Exp Zool 191:63–71

Wever EG, Gans C (1976) The Caecilian ear. Further observations. Proc Natl Acad Sci, USA 73:3744–3746

Wilkinson M, Nussbaum RA (2006) Caecilian phylogeny and classification. In: Exbrayat JM (ed) Reproductive biology and phylogeny of Gymnophiona (Caecilians), Science Publishers, Enfield/Jersey/Plymouth, pp 39–78

Zuber-Vogeli M, Doerr-Schott J (1981) Description morphologique et cytologique de l'hypophyse de *Typhlonectes compressicaudus* (Duméril et Bibron) (Amphibien Gymnophione de Guyane fançaise). C r Séanc Acad Sci Paris sér D 292:503–506

Zylberberg L (1972) Données histologiques sur les glandes linguales d'*Ichthyophis glutinosus* (L.) Batracien gymnophione. Arch Anat Micr Morphol exper 61:227–242

Zylberberg L (1986) L'épithélium lingual de deux Amphibiens Gymnophiones: *Typhlonectes compressicaudus* et *Ichthyophis kohtaoensis*. Mém Soc Zool F 43:83–84

Zylberberg L, Wake MH (1990) Structure of the scales of *Dermophis* and *Microcaecilia* (Amphibia: Gymnophiona), and a comparison to dermal ossifications to other Vertebrates. J Morph 206:25–43

Zylberberg L, Castanet J, de Ricqles A (1980) Structure of the dermal scales in Gymnophiona (Amphibia). J Morph 165:41–54

Chapter 5
The Evolution of Morphogenetic Signalling in Social Amoebae

Yoshinori Kawabe, Elisa Alvarez-Curto, Allyson V. Ritchie, and Pauline Schaap

Abstract Multicellular organisms have evolved several times from unicellular protists giving rise to the familiar forms of animals, plants and fungi. An important question in biology is how such transitions occurred. Multicellular life is typically dependent on complex communication between cells, whereas unicellular organisms respond mainly to environmental signals. Social amoebae are eminently suited to study the evolution of multicellularity, since they still combine a unicellular feeding stage with a stage where thousands of cells aggregate to form motile slugs and fruiting structures. In this chapter we summarize the signalling mechanisms that coordinate multicellular development in social amoebae and we discuss how these signalling mechanisms evolved from a response to environmental stress in solitary amoebae.

5.1 Introduction

This chapter introduces the social amoebae as a new group of model organisms that offer unique opportunities for understanding how complex multicellular organisms evolved from unicellular ancestors. The dictyostelid social amoebae are members of the amoebozoans, a supergroup of genetically highly diverse organisms that are

---

Yoshinori Kawabe
College of Life Sciences, University of Dundee, MSI/WTB/JBC complex, Dow Street, Dundee, DD1 5EH, UK

Elisa Alvarez-Curto
Neuroscience and Molecular Pharmacology, Wolfson Building, University of Glasgow, G12 8QQ, UK

Allyson V. Ritchie
Department of Haematology, Cambridge Institute for Medical Research Hills Road, Cambridge, CB2 0XY, UK

P. Schaap
MSI/WTB/JBC complex, Dow Street Dundee, DD1 5EH, UK
e-mail: p.schaap@dundee.ac.uk

P. Pontarotti (ed.), *Evolutionary Biology: Concept, Modeling, and Application*,
DOI: 10.1007/978-3-642-00952-5_5,

most closely related to the opisthokonts, the supergroup that contains animals and fungi (Baldauf et al. 2000). The majority of amoebozoans are unicellular amoebae or amoeboflagellates that feed on bacteria or other microbes that are small enough to be engulfed. When experiencing nutrient depletion or other forms of stress, cells typically encapsulate to form dormant cysts. Two groups of amoebozoans additionally show alternative life cycles. Myxomycete amoebae fuse to form a sexual zygote, which grows into a large syncytium by undergoing nuclear division without cytokinesis. When food runs out the syncytium partitions to form multiple spore-bearing fruiting structures (Stephenson and Stempen 1994). Dictyostelid social amoebae collect into aggregates when starved, and then proceed to form a fruiting structure by a combination of coordinated cell migration and cell differentiation. Similar to metazoan embryogenesis, this developmental programme is largely regulated by intercellular communication.

Dictyostelids combine the ease of culture and genetic manipulation of microbes with a sophisticated programme of multicellular development. The model organism *Dictyostelium discoideum* is widely used to understand cellular and developmental processes, such as cell signalling, cell differentiation, directional cell movement, cytokinesis, vesicle trafficking and phagocytosis (Kessin 2001). Many developmental signals, including their mechanisms of production and detection, have been characterized. The deeply conserved second messenger cyclic AMP (cAMP) is the most dominant signal in *D. discoideum* development. It is secreted as a chemoattractant to bring cells together in aggregates (Konijn et al. 1967) and as a morphogen to initiate the differentiation of spores (Wang et al. 1988). cAMP also acts intracellularly to mediate the effect of various peptide- and low-molecular-weight signals that control initiation of development, maturation of spore and stalk cells and the germination of spores (Anjard et al. 1998b; Van Es et al. 1996).

Dictyostelids are found in all types of soils, including those in arctic regions and above the tree line (Swanson et al. 2002). At present over a 100 species are known and this number is rapidly increasing due to recent interest in mapping the biodiversity of these organisms. The different species show considerable variation in the morphology of their fruiting bodies and the proceeding developmental stages (Raper 1984). They also differ in the use of chemoattractants, with a subset of species using cAMP but the large majority using other compounds.

With their combined single- and multi-celled life style the social amoebae are well suited to explore how multicellularity evolved. Multicellular organisms are typically dependent on highly orchestrated signalling between cells, whereas unicellular organisms respond mainly to environmental signals. A major aspect of the evolution of multicellularity is therefore how the sensory systems of unicellular organisms were adapted to allow for more sophisticated signalling between cells.

In this chapter we first present an overview of the signalling mechanisms that control development of the model organism *D. discoideum*. Secondly, we will summarize recent efforts to create a framework for studying the evolution of developmental signalling in the Dictyostelia. Thirdly, we will show how the developmental roles of cAMP in the social amoebae evolved from a response to water stress in their unicellular ancestors.

5.2 The Life Cycle of *D. discoideum*

5.2.1 Cell Differentiation and Morphogenesis

The developmental programme of *D. discoideum* is initiated by starvation and peptide signals that are secreted by cells to estimate cell density relative to food availability (Clarke and Gomer 1995). Together, these factors initiate expression of early genes and of the cAMP signalling genes that are required for aggregation. Once basal levels of these genes are expressed, some cells start to spontaneously secrete pulses of cAMP. In surrounding cells this evokes two responses: (i) extension of a pseudopod and movement towards the cAMP source; and (ii) synthesis and secretion of a cAMP pulse. The latter response causes the original pulse to travel outward through the population and the former causes the cells to move inwards and collect at the oscillating centre (Fig. 5.1). In addition to mediating aggregation, the cAMP pulses also serve another function. They trigger rapid upregulation of all proteins that are involved in the aggregation process and induce competence for the subsequent stages of development (Gerisch et al. 1975; Schaap et al. 1986).

The formation of well-proportioned fruiting bodies from loose heaps of cells depends on an interplay between cell differentiation and cell movement. Cell movement remains under control of cAMP signalling. The top of the aggregate takes over the function of the aggregation centre and emits waves of cAMP (Dormann et al. 1998). Cells become more adhesive and secrete matrix proteins and cellulose, which form an elastic sheath around the aggregate (Wilkins and Williams 1995). The cells, which are now confined, can only go upwards in response to cAMP waves and form the slug-shaped sorogen. Gravity causes the slugs to topple over and they start migrating towards light and warmth (Bonner et al. 1950). In nature, this serves to reach the top layer of the soil, the most propitious spot for fruiting body formation and spore dispersal (Bonner and Lamont 2005).

Meanwhile, cells differentiate into spore and stalk precursors. In the mound, these cell types are at first intermixed (Fig. 5.1). The prespore cells largely loose the properties of aggregating cells but the prestalk cells retain the ability to sense and propagate cAMP waves (Matsukuma and Durston 1979; Verkerke-van Wijk et al. 2001). They therefore move with greatest efficiency towards the oscillating tip, thus creating an anterior–posterior prestalk-prespore pattern (Sternfeld and David 1981). At the posterior, cell-type regulation occurs to maintain a fixed proportion of prespore and anterior-like cells. The prestalk cells at the front provide most of the motive force during slug migration and are essentially used up. They are replenished by the forward movement of the anterior-like cells (Sternfeld and David 1982).

During fruiting body formation the slug contracts and points its tip upwards (Fig. 5.1). Prestalk cells start to synthesize a central cellulose tube (Raper and Fennell 1952). They continue their movement towards the cAMP waves emitted by the tip but now move into the upper open end of the stalk tube and differentiate into stalk cells. This involves synthesis of a cellulose cell wall, uptake of water, formation of one large central vacuole and eventually cell death. The prespore

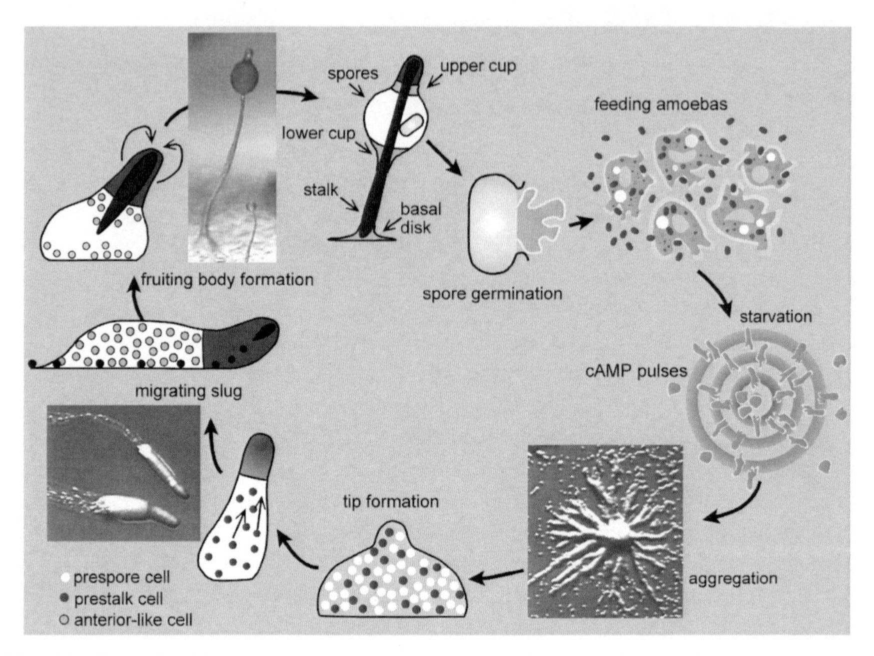

Fig. 5.1 Life cycle of *Dictyostelium discoideum*. *D. discoideum* amoebae emerge from spores and start feeding on bacteria. When food runs out, amoebae secrete cAMP pulses, which trigger chemotactic movement and aggregation of cells. Once aggregated, the amoebae differentiate into prestalk and prespore cells in a regulated ratio. The organizing tip continues to emit cAMP pulses, which shape the cell mass by coordinating cell movement. The cAMP pulses also cause the prestalk cells, which are chemotactically most responsive, to move towards the front. At the posterior, a subset of prespore cells dedifferentiates into anterior-like cells, which replenish the prestalk cells when needed. At the onset of culmination the cells synthesize a cellulose tube, the apical prestalk cells move into the tube and mature into stalk cells, the remaining prestalk cells form support structures, such as the upper- and lower cup and the basal disk. The prespore cells move up the stalk and mature into spores

cells move up the newly formed stalk and mature into spores, starting from the prestalk boundary downwards. Spore maturation is a rapid process that involves fusion of prespore vesicles with the plasma membrane. This brings the first layer of the spore coat *in situ* at the surface, while the remaining two layers are rapidly being produced by ejected enzymes and prefabricated materials (West 2003).

5.2.2 Signals that Control D. discoideum *Development*

The most remarkable aspect of *D. discoideum* development is the very dominant role of cAMP. As a secreted signal it controls cell movement and differentiation throughout the developmental programme but in its more common role as intracellular messenger, it mediates the effect of many other developmental signals.

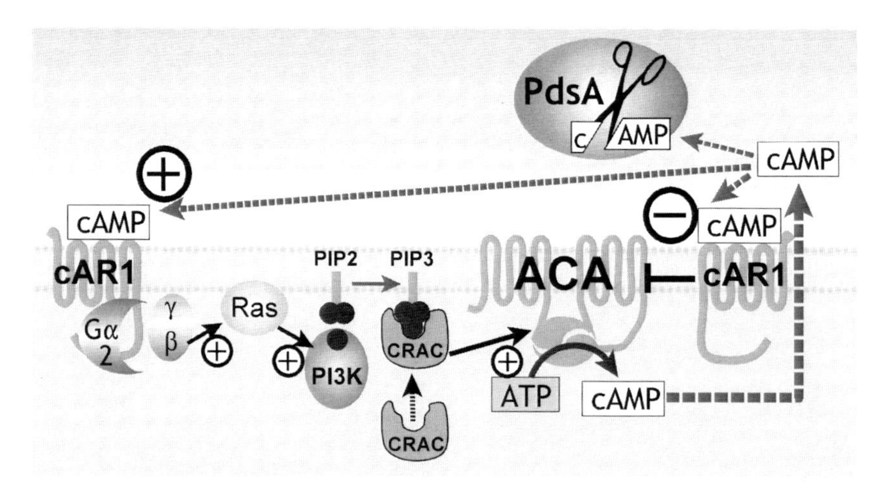

Fig. 5.2 Production of cAMP pulses. Pulsatile cAMP signalling results from positive and negative feedback loops acting on cAMP production. Cells secrete a low basal level of cAMP, which binds to cAR1 to activate the G-protein G2, which splits up into its α and β-γ subunits. The latter activate PI3-kinase, which phosphorylates the membrane phospholipid PIP2 to form PIP3. PIP3 brings the Cytosolic Regulator of Adenylate Cyclase (CRAC) to the membrane, where it activates ACA. This positive feedback loop terminates because cAMP also causes desensitization of cAR1, which blocks further cAMP accumulation. Removal of extracellular cAMP by PdsA resets the system for the next pulse

Extracellular cAMP is detected by four homologous cAMP receptors (cARs1–4), members of the very large family of G-protein coupled receptors, which detect a broad variety of hormones, neurotransmitters and odorants in metazoa (Devreotes 1994; Ginsburg et al. 1995). The main intracellular target for cAMP is cAMP-dependent protein kinase or PKA, which like the fungal PKA consists of a single catalytic (PKA-C) and a single regulatory subunit (PKA-R) (Mutzel et al. 1987). Active PKA is essential for expression of early genes and for basal expression of the aggregation genes, cAR1, cAMP phosphodiesterase A (PdsA) and adenylate cyclase A (ACA), which are necessary for the production of cAMP pulses (Schulkes and Schaap 1995).

The biochemical network that enables pulsatile cAMP signalling involves the following sequence of events: (i) ACA produces a low basal level of cAMP, which is rapidly secreted. (ii) Secreted cAMP binds to cAR1 and upregulates ACA activity. This involves activation of the G-protein, G2 and a series of intermediate steps that are outlined in Fig. 5.2. (iii) The cAMP pulse reaches its peak when persistent stimulation with cAMP brings cAR1 in an inactive desensitized state, blocking further activation of ACA. (iv) Hydrolysis of extracellular cAMP by PdsA frees up cAR1 and resets the system for the next pulse (Devreotes 1994; Kriebel and Parent 2004).

Spontaneous cAMP oscillations are initiated by a few cells and propagate as waves through the cell population, causing rapid cell aggregation. The tip of the aggregate takes over the function of the aggregation centre and continues to emit cAMP waves autonomously. These waves shape the organism during slug and fruiting body formation by coordinating the movement of its component cells. Extracellular cAMP also gains other roles in cell differentiation and new layers of complexity in cAMP signalling become apparent after aggregation.

Once collected in mounds, overt phenotypic differences between cells become apparent. Firstly, the proteins that are involved in oscillatory cAMP signalling, such as ACA and PdsA are lost from the prespore cells. Three new cARs appear: cAR3 in all cells and cAR2 and cAR4 in prestalk cells (Ginsburg et al. 1995). Two new adenylate cyclases appear, ACG and ACB. ACG is translationally upregulated in the posterior region of the early slug (Alvarez-Curto et al. 2007). The differentiation of prespore cells requires both extracellular cAMP acting on cAR1 and intracellular cAMP acting on PKA (Hopper et al. 1993; Verkerke-VanWijk et al. 1998; Wang et al. 1988). ACG produces cAMP intracellularly, where it can activate PKA. However, cAMP is also secreted to activate cAR1 (Alvarez-Curto et al. 2007). A third adenylate cyclase, ACB, is predominantly expressed in the prestalk region. Its activity increases strongly at early fruiting body formation, where it is required for both spore and stalk cell maturation (Meima and Schaap 1999; Soderbom et al. 1999). The prespore cells produce polyketide-based compounds that cause redifferentiation of a fixed proportion of prespore cells into prestalk-like cells. One of these compounds is the differentiation-inducing-factor, DIF, which triggers differentiation of a subpopulation of prestalk-like cells that will sort downward during fruiting body formation to form the basal disk and lower cup (Saito et al. 2008).

The final stages of development depend on interplay between secreted catabolites, environmental stimuli and cell–cell communication. They ensure that the immobile stalk and spores cells mature in the right place and at the right time. The starving cells depend on autophagy for survival. Protein degradation produces large amounts of ammonia, which accumulates in the fluid surrounding the migrating slugs and act as an inhibitor of stalk cell maturation (Wang and Schaap 1989). Fruiting body formation is initiated by incident light, which causes migrating slugs to point their tips upwards (Fig. 5.3a). Ammonia is now lost by gaseous diffusion, allowing stalk cell maturation. The maturing stalk cells produce a peptide, SDF2, which in turn triggers the maturation of spores.

Ammonia, SDF2 and several other signals that regulate terminal differentiation are detected by sensor-linked histidine kinases/phosphatases, common sensors for external stimuli in bacteria, fungi and plants (Thomason and Kay 2000; Thomason et al. 1998). Binding of a signal to the sensor triggers either its kinase or phosphatase activity, which initiates forward or reverse phosphotransfer from histidine to aspartate. In *Dictyostelium*, this leads to phosphorylation or dephosphorylation of an aspartate in the response regulator domain of the intracellular cAMP phosphodiesterase RegA (Shaulsky et al. 1996). This enzyme requires the phosphorylated aspartate to be active (Fig. 5.3b). Ammonia activates the histidine kinase DhkC and consequently RegA (Singleton et al. 1998), resulting in cAMP degradation and inhibition of PKA. PKA activation is essential for spore and stalk cell maturation, and ammonia activation of RegA therefore effectively blocks both processes. (Harwood et al. 1992; Hopper et al. 1993). SDF2 binds to the sensor histidine

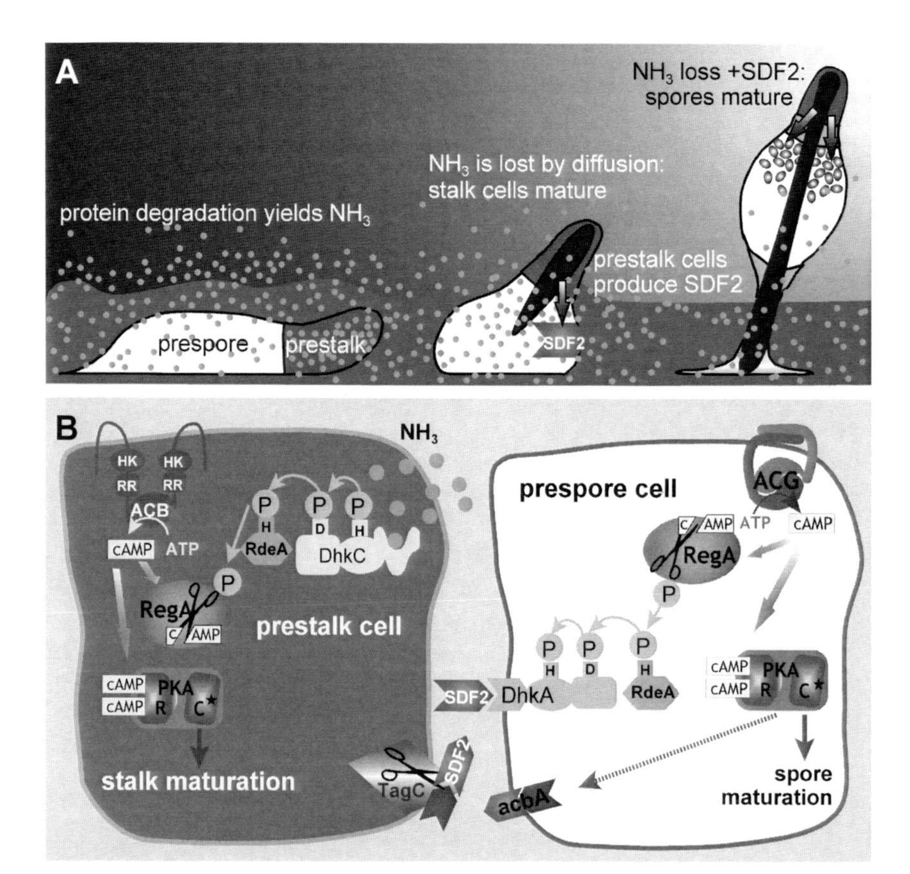

Fig. 5.3 Regulation of terminal differentiation by ammonia and SDF2. (**a**) *The process of culmination*. Protein degradation in starving cells yields large amounts of ammonia (NH_3), which accumulates in soil moisture. Once slugs reach the soil surface, light from above redirects the tip upwards across the air–water interface. Ammonia is now lost by gaseous diffusion and ammonia depletion allows stalk cells to mature. Maturing prespore cells secrete a protein acbA, which is processed by prestalk cells to produce the spore differentiation factor SDF2, which together with ammonia depletion triggers spore encapsulation (**b**) *Ammonia and SDF2 regulate PKA activity*. Terminal spore and stalk maturation require high levels of PKA activation by cAMP. In prestalk cells cAMP is produced by ACB and in prespore cells by ACG. The cAMP phosphodiesterase RegA requires phosphorylation to be active. Its phosphorylation state is controlled by sensor histidine kinases/phosphatases that initiate forward or reverse phosphorelay, respectively, when a signal molecule binds to their sensor domain. The phosphate is carried over from histidine to aspartate to histidine and uses a small protein, RdeA, as an intermediate. Ammonia activates the histidine kinase activity of DhkC, thereby causing activation of RegA and preventing cAMP accumulation. The protease TagC that is expressed on maturing stalk cells cleaves acbA to yield SDF2, which activates the phosphatase activity of DhkA. This results in dephosphorylation of RegA and consequent cAMP accumulation and PKA activation in prespore cells, resulting in spore maturation

phosphatase DhkA on prespore cells. This results in reverse phosphorelay and inactivation of cAMP hydrolysis by RegA (Fig. 5.3b). cAMP now accumulates and activates PKA, thus inducing spore encapsulation (Anjard et al. 1998a; Wang et al. 1999).

Dictyostelid survival critically depends on spore dispersal to novel feeding grounds. In the fruiting body the germination of spores is inhibited by high ambient osmolality (Cotter 1977). High osmolality acts on the intrinsic osmosensor of ACG to activate cAMP synthesis. cAMP in turn activates PKA, which blocks the transition of dormancy to germination (Saran and Schaap 2004; Van Es et al. 1996). High intracellular cAMP levels and PKA activation maintain the dormant state (Fig. 5.6c).

In summary, the transduction mechanisms of the different signals that control the final stages of development converge at regulating intracellular cAMP levels and the activation of PKA. cAMP is the most deeply conserved signalling intermediate in both pro- and eukaryotes, with a broad repertoire of functions in environmental sensing, hormone action, neurotransmission and developmental control. While intracellular use of cAMP is common to most organisms, only a subset of social amoebae use cAMP as an extracellular chemoattractant. To understand how the different roles of cAMP in the social amoebae evolved and how they contributed to the evolution of multicellular complexity in these organisms, we initiated a programme to firstly determine the directionality of phenotypic evolution in social amoebae and to secondly investigate how changes in cAMP signalling genes contributed to the appearance of novel phenotypes.

5.3 Phenotypic Evolution in the Social Amoebae

To retrace in what order the different morphological and behavioural features of the social amoebae evolved, we constructed a molecular phylogeny of all known social amoebae, based on small subunit ribosomal RNA (SSU RNA) and α-tubulin sequences (Schaap et al. 2006). The phylogeny shows a subdivision of all known taxa into four major groups with the model *D. discoideum* in the most recently diverged group 4 (Fig. 5.4).

Social amoebae differ in shape and size of fruiting bodies, which can be solitary or clustered, unbranched or carrying branches with specific architectures. Cellular structures such as basal disks or triangular supporters may buttress the stalk, while upper and lower cups may support the spore head. A migrating slug stage can be present and often a stalk is laid down during migration. Spores vary in size and shape and may carry granules at their poles. Species can go through alternative life cycles, such as encystation to form microcysts and sexual fusion to form macrocysts. The chemoattractant used for aggregation may be cAMP, folic acid, glorin or pterin but is unknown for most species. Twenty consistently described traits were plotted to the molecular phylogeny (Schaap et al. 2006). Many traits showed no group-specific distribution. The traits that showed clear evolutionary trends are summarized in Fig. 5.4.

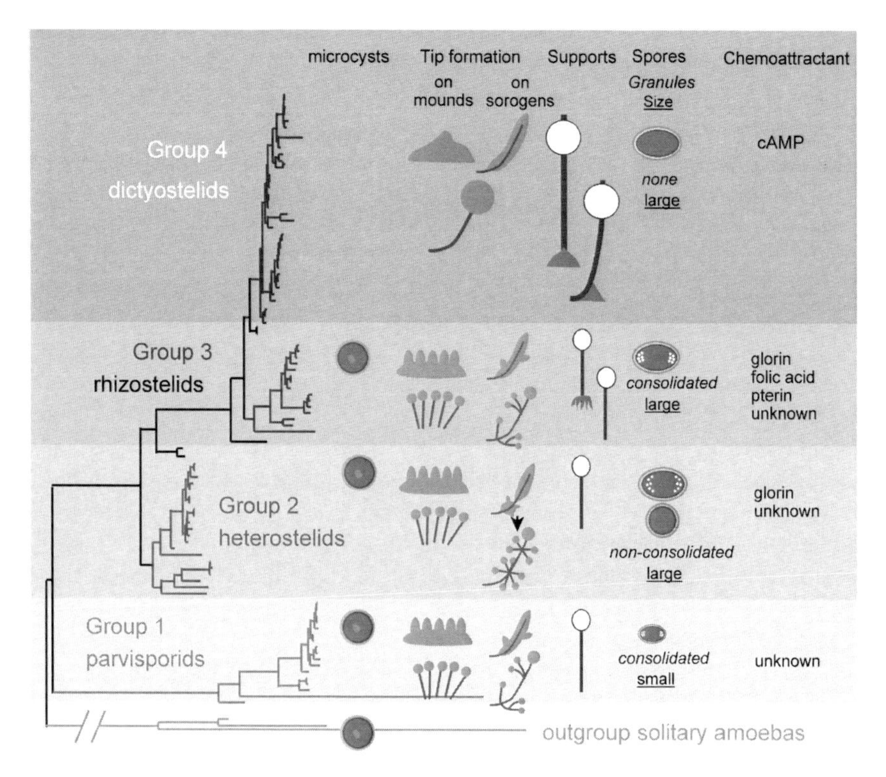

Fig. 5.4 Character evolution in the social amoebae. The phylogenetic tree of the social amoebae shows a subdivision of all species into four major groups. The tree is based on small subunit ribosomal RNA sequences and is rooted on sequences of solitary amoebozoa. A total of 20 described morphological characters were mapped to the phylogeny. Many of those are more or less randomly scattered over the four taxon groups but a subset displays clear evolutionary trends. The ancestral solitary amoebozoa encyst individually in response to stress. The ability to form such microcysts is retained in many species in groups 1–3 but is lost from the most recently diverged group 4. Taxa in groups 1–3 generally form many tips on aggregates and sorogens, which give rise to clusters of small branched fruiting bodies. Group 4 taxa usually form only one or a few tips per aggregate and fruiting bodies are consequently solitary, unbranched and large. Virtually all group 4 taxa form cellular supporters or basal disks to hold up the stalk. A cluster of 4 taxa in group 3 forms unique crampon-like supports. Spores are relatively small in group 1 and larger in the other groups. Groups 1–3 all carry granules at the poles of their spores, which are tightly consolidated in groups 1 and 3. Group 4 spores carry no pronounced granules. None of the taxa in groups 1–3 use cAMP as an chemoattractant but all investigated species in group 4 do

Sexual macrocysts are found in all groups but microcysts appear to be lost from the most derived group 4. Spores are smaller in group 1 than in any of the other groups and they have lost their polar granules in group 4. A single aggregate usually gives rise to a cluster of fruiting bodies in groups 1–3 and these fruiting bodies are also often branched. However, group 4 aggregates generally give rise to a solitary unbranched fruiting body. Consequently, group 4 fruiting bodies are significantly larger than those of other groups and usually have basal disks or supporters.

Outside group 4, there is a single species with a basal disc in group 1 and a cluster of species in group 3 with a crampon-shaped support. Remarkably, for all tested species in group 4, the chemoattractant is cAMP, while the tested species in the other groups all use other chemoattractants. There is apparently a correlation between the use of cAMP during aggregation, large fruiting body size with additional cell-type specialization, loss of polar granules in spores and loss of microcysts.

5.4 The Evolution of cAMP Signalling

5.4.1 Extracellular cAMP

Extracellular cAMP pulses control cell movement and cell differentiation during aggregation and fruiting body formation in the model *D. discoideum*. However, until recently, it was unclear to what extent the roles of extracellular cAMP were conserved. All effects of extracellular cAMP are mediated by serpentine cAMP receptors, which are present as a family of four homologous proteins (cARs1–4) in *D. discoideum* (Ginsburg et al. 1995).

cARs are most diagnostic for the presence of extracellular cAMP signalling and to establish to what extent this process is conserved in other taxon groups, we used a polymerase chain reaction (PCR) approach to amplify cAR genes from group-representative taxa (Fig. 5.5a). It appeared that orthologs of the *cAR1* gene were present in taxa from all four groups, while cAR2 was detected in another group 4 taxon (Alvarez-Curto et al. 2005). *D. discoideum* null mutants in *cAR1* loose the ability to aggregate and form fruiting bodies (Sun and Devreotes 1991). Complementation of the *car1* null mutant with a cAR from the group 3 taxon *D. minutum* restored aggregation and development, including oscillatory cAMP signalling; thus demonstrating that the basal cARs are fully functional cAMP receptors.

The *cAR* genes in groups 1 and 2 were transcribed into a single mRNA species after aggregation. *D. discoideum* and another group 4 taxon, *D. rosarium,* transcribed a second mRNA from a more distal promoter before and during aggregation, while the group 3 taxon *D. minutum* expressed the second mRNA during growth (Fig. 5.5b). The differences in stage-specific expression of the cARs mirror the stages at which the disruption of cAR function blocks the developmental programme in the four groups. Group 4 taxa cannot aggregate without functional cARs, whereas group 1, 2 and 3 taxa show normal aggregation but blocked or abnormal fruiting body formation (Alvarez-Curto et al. 2005).

These results demonstrated that the ancestral role of extracellular cAMP signalling is to coordinate fruiting body formation. Its role in the aggregation of group 4 taxa is evolutionary derived. The expression of a cAR during growth of a group 3 taxon is somewhat enigmatic, but suggests that here aggregation to cAMP has not yet been fully gained. Growth-specific expression suggests that the *D. minutum* cAR may serve a function in food-seeking, since the bacterial food source is known to

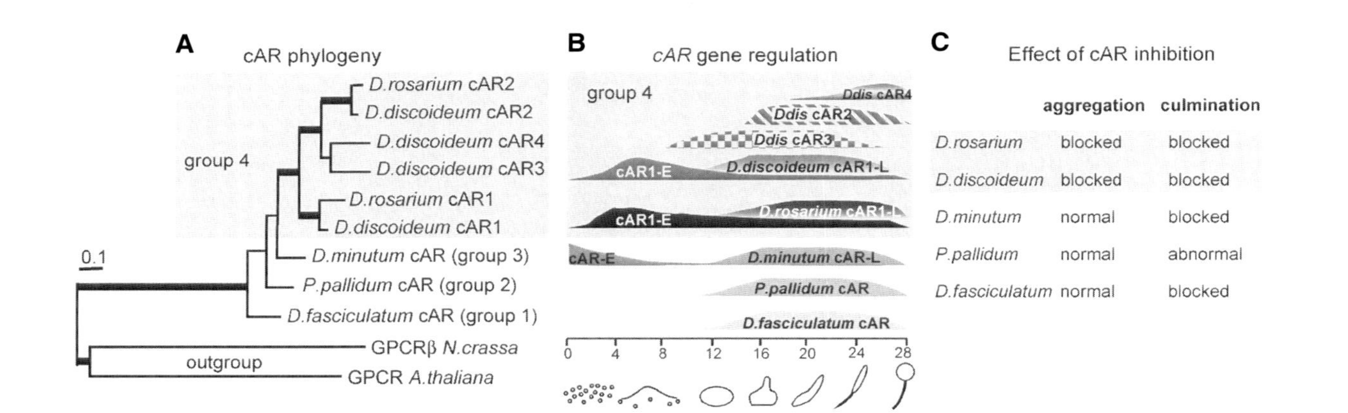

Fig. 5.5 cAMP receptor evolution in social amoebae. (**a**) *cAR phylogeny*. Conserved regions of cAR genes were amplified by PCR from group-representative taxa. After alignment of their derived protein sequences with the known four *D. discoideum* cARs and the most closely related G-protein coupled receptors in other organisms, phylogenetic relationships were determined by Bayesian inference (**b**) *cAR expression*. Schematic representation of cAR mRNA levels during development of group-representative taxa (**c**) *cAR function*. Taxa were developed on agar containing the non-degradable cAMP analogue SpcAMPS, which causes desensitization of cAMP pulse-induced responses, such as chemotaxis and cAMP relay. In groups 1–3 taxa this only interferes with fruiting body formation, while in group 4 taxa it also blocks aggregation (Alvarez-Curto et al. 2005)

secrete cAMP (Konijn et al. 1969). As a transitory phase in the sequence of events that led to co-option of cAMP signalling for aggregation, food-seeking has the advantage of requiring only cAMP detection and not oscillatory cAMP production.

In addition to using cAMP for aggregation, group 4 taxa stand out by having large solitary unbranched fruiting structures, as opposed to the clustered and branched small fruiting bodies that are found in groups 1–3. Are cAMP signalling and size related? The segmentation of aggregates into clusters of fruiting bodies as well as the formation of side branches all represent the formation of multiple body axes. These axes are typically initiated by newly emerging tips. *D. discoideum* tips are pacemakers for cAMP waves. Although all cells can in theory become pacemakers, existing pacemakers suppress emergence of new ones, thereby keeping a mass of cells under control. Two aspects of cAMP signalling are specific to group 4: (i) oscillatory cAMP signalling occurs much earlier in development than in groups 1–3; and (ii) the cAMP receptor gene was duplicated three times and both expression and affinity of the daughter cARs were altered. It is conceivable that either of these novelties acts on autonomous cAMP oscillations in a manner that allows it to dominate a larger number of cells.

5.4.2 *Intracellular cAMP*

The use of cAMP as an extracellular signal for cell–cell communication is thus far unique to social amoebae. However, its use as an intracellular second messenger for external stimuli is common to almost all cellular organisms. In prokaryotes, cAMP directly binds to transcription factors such as the catabolite activator protein (CAP). In eukaryotes, cAMP-dependent protein kinase (PKA) is its most common target. Both PKA and CAP contain a deeply conserved cAMP binding domain. In the group 4 taxon *D. discoideum,* PKA activity is essential for initiation of development, induction of prespore differentiation, maturation of stalk and spore cells and control of spore dormancy (Saran et al. 2002). Some and perhaps all of these functions are conserved in the group 2 taxon *Polysphondylium pallidum*, where inhibition of PKA blocks proper development. The few aberrant fruiting structures that are formed, carry atypical spores with poor viability (Funamoto et al. 2003).

To gain deeper insight in the origins of the different roles of intracellular cAMP, we used a PCR approach to identify genes encoding the two adenylate cyclases, ACG and ACB. ACG produces cAMP for induction of prespore differentiation and control of spore dormancy (Alvarez-Curto et al. 2007; Van Es et al. 1996), while ACB is essential for spore maturation and for proper stalk formation (Soderbom et al. 1999). We were unable to amplify *ACB* from taxa in groups 1–3 but have recently detected *ACB* genes in ongoing *P. pallidum* (group 2) and *D. fasciculatum* (group 1) genome sequencing projects. We have only just started to analyse ACB function in early diverging species.

We did amplify the *ACG* catalytic region from taxa representing the four major groups (Fig. 5.6a) and cloned full length ACG from the group 3 taxon *D. minutum*

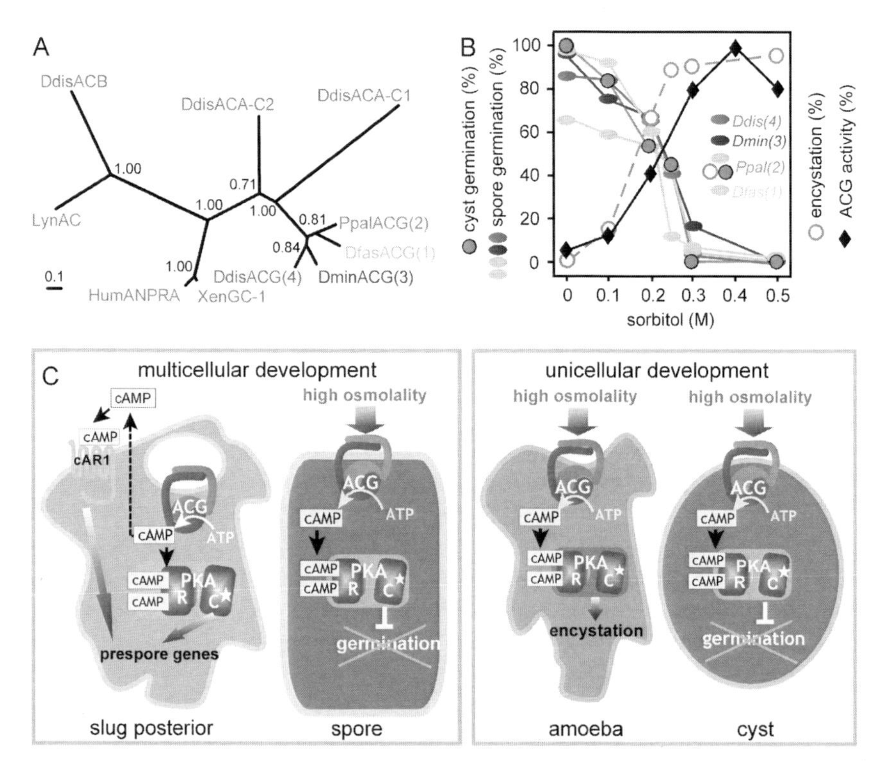

Fig. 5.6 Origins of intracellular cAMP signalling. (**a**) *Adenylyl cyclase G conservation.* The *ACG* catalytic domain was amplified from the species *Dfas* (group 1), *Ppal* (group 2) and *Dmin* (group 3) and aligned with all *Ddis* (group 4) adenylate cyclases (ACA, ACB, ACG) and their most closely related non-dictyostelid relatives. Phylogenetic reconstruction by Bayesian inference shows that the ACG's are most closely related to each other (**b**) *Effect of osmolyte on ACG activity, germination and encystation.* ACG is activated by the osmolyte sorbitol with halfmaximal activation occurring around 0.2 M. This sorbitol concentration also causes halfmaximal inhibition of spore germination in group-representative taxa and halfmaximal induction of encystation and inhibition of excystation in the group 2 taxon *Ppal,* suggesting that ACG mediates the osmolyte effects on encystation and germination (**c**) *ACG has homologous roles in encystation and sporulation.* During multicellular development, ACG triggers prespore differentiation in slugs and inhibits the germination of spores by ambient high osmolality in the fruiting body (Van Es et al. 1996; Alvarez-Curto et al. 2007). In unicellular soil amoebae high osmolality, caused by drought, acts on ACG to induce encystation and to inhibit excystation (Ritchie et al. 2008)

(Ritchie et al. 2008). The catalytic domains of the newly identified ACG genes were closely related to *D. discoideum* ACG but not to ACA or ACB. The *D. minutum* ACG showed the same domain architecture as the *D. discoideum* ACG with two transmembrane regions that flank an extracellular sensor domain. Similar to *D. discoideum* ACG, *D. minutum* ACG was also activated by high osmolality. In *D. discoideum*, ACG mediates the inhibition of spore germination by high ambient osmolality in the fruiting body. We found that spore germination in all group-representative taxa was inhibited by high osmolality (Fig. 5.5b), suggesting that this function of ACG is deeply conserved in the social amoebae.

The solitary ancestors of the social amoebae form dormant cysts when exposed to starvation. Many of the early diverging social amoeba taxa have retained encystation as an alternative survival strategy (Schaap et al. 2006). We found that similar to spore germination, cyst germination is also inhibited by high osmolality. However, unlike spore formation, encystation is actually triggered by high osmolality, even when sufficient food is still available (Fig. 5.5c). A pronounced increase in intracellular cAMP accompanied osmolyte induced encystation, indicating that ACG mediates this response. Direct activation of PKA with membrane-permeant cAMP analogues induced encystation without the need for high osmolality, while inhibition of PKA prevented osmolyte induced encystation. This indicated that similar to osmolyte-inhibited spore germination, osmolyte-induced encystation was mediated by ACG and PKA (Ritchie et al. 2008).

What is the significance of osmolyte-induced encystation? Increasing osmolality is most likely a signal for approaching drought, when soil mineral concentrations start to increase. By triggering timely encystation, death of soil amoebae by drying out is prevented. The roles of ACG in encystation and spore formation show a striking resemblance (Fig. 5.5c). ACG blocks excystation and triggers encystation in individual amoebae. Similarly, ACG blocks spore germination in fruiting bodies and triggers prespore differentiation in the slug stage (Alvarez-Curto et al. 2007). This most likely signifies that the roles of ACG in cell-type specification and spore germination during multicellular *Dictyostelium* development are evolutionary derived from a role as drought sensor, triggering encystation in individual amoebae. Osmolyte-induced encystation was also observed in the solitary amoebozoans *Acanthamoeba castellani* and *Hartmannella rhysodes* (Band 1963; Cordingley et al., 1996), while PKA agonists were reported to trigger encystation of *Hartmanella culbertsoni* and *Entamoeba invadens* (Raizada and Murti 1972; Coppi et al., 2002). Evidently, also in these distantly related amoebozoans high osmolality and PKA activation are associated with encystation.

In contrast to intracellular cAMP signalling, which is common to both eukaryotes and prokaryotes, extracellular cAMP signalling is thus far unique to the social amoebae. It is therefore most likely that the extracellular roles of cAMP are derived from its intracellular roles, probably by the invention of cAMP secretion and detection in an early stage of social amoeba evolution.

5.5 Conclusions

Dictyostelid social amoebae have evolved a unique form of multicellularity that is based on aggregation and collective formation of fruiting structures. To study the evolution of multicellular development, a molecular phylogeny of all known species was constructed. Mapping of 20 morphological characters onto the phylogeny shows a trend towards the formation of large unbranched fruiting structures with up to five differentiated cell types in the more derived species, as opposed to small, clustered and branched structures with one or two cell types in groups that diverged earlier. The evolution of larger fruiting structures is accompanied by the use of secreted cyclic AMP (cAMP) as a chemoattractant to bring cells together in aggregates and the loss of the ancestral survival strategy of encystation.

Orthologues of developmental signalling genes were identified and functionally analysed in species that span the Dictyostelid phylogeny. Analysis of cAMP receptor (cAR) genes showed that the role of extracellular cAMP in cell aggregation is evolutionary derived from a deeply conserved role in the organization of fruiting body morphogenesis and the differentiation of prespore cells. In its more common intracellular role, cAMP induces the maturation of spore- and stalk cells and mediates inhibition of spore germination by ambient high osmolality in the spore head. The osmolyte-activated adenylate cyclase ACG produces cAMP for induction of prespore differentiation and control of spore germination. Functional analysis of ACG throughout the phylogeny showed that the roles of cAMP in spore differentiation and germination are evolutionary derived from drought-induced encystation of solitary amoebae.

Combined, the data suggest that a cAMP-mediated stress response of unicellular protists was at the origin of developmental cAMP signalling in the social amoebae.

References

Alvarez-Curto E, Rozen DE, Ritchie AV, Fouquet C, Baldauf SL, Schaap P (2005) Evolutionary origin of cAMP-based chemoattraction in the social amoebae. Proc Natl Acad Sci USA 102:6385–6390

Alvarez-Curto E, Saran S, Meima M, Zobel J, Scott C, Schaap P (2007) cAMP production by adenylyl cyclase G induces prespore differentiation in *Dictyostelium* slugs. Development 134:959–966

Anjard C, Chang WT, Gross J, Nellen W (1998a) Production and activity of spore differentiation factors (SDFs) in *Dictyostelium*. Development 125:4067–4075

Anjard C, Zeng CJ, Loomis WF, Nellen W (1998b) Signal transduction pathways leading to spore differentiation in *Dictyostelium discoideum*. Dev Biol 193:146–155

Baldauf SL, Roger AJ, Wenk-Siefert I, Doolittle WF (2000) A kingdom-level phylogeny of eukaryotes based on combined protein data. Science 290:972–977

Band RN (1963) Extrinsic requirements for encystation by soil amoeba, *Hartmannella rhysodes*. J Protozool 10:101–106

Bonner JT, Lamont DS (2005) Behavior of cellular slime molds in the soil. Mycologia 97:178–184

Bonner JT, Clarke Jr WW, Neely Jr CL, Slifkin MK (1950) The orientation to light and the extremely sensitive orientation to temperature gradients in the slime mold *Dictyostelium* discoideum. J Cell Comp Physiol 36:149–158

Clarke M, Gomer RH (1995) PSF and CMF, autocrine factors that regulate gene expression during growth and early development of *Dictyostelium*. Experientia 51:1124–1134

Coppi A, Merali S, Eichinger D (2002) The enteric parasite *Entamoeba* uses an autocrine catecholamine system during differentiation into the infectious cyst stage. J Biol Chem 277:8083–8090

Cordingley JS, Wills RA, Villemez CL (1996) Osmolarity is an independent trigger of *Acanthamoeba castellanii* differentiation. J Cell Biochem 61:167–171

Cotter DA (1977) The effects of osmotic pressure changes on the germination of *Dictyostelium discoideum* spores. Can J Microbiol 23:1170–1177

Devreotes PN (1994) G protein-linked signaling pathways control the developmental program of *Dictyostelium*. Neuron 12:235–241

Dormann D, Vasiev B, Weijer CJ (1998) Propagating waves control *Dictyostelium discoideum* morphogenesis. Biophys Chem 72:21–35

Funamoto S, Anjard C, Nellen W, Ochiai H (2003) cAMP-dependent protein kinase regulates *Polysphondylium pallidum* development. Differentiation 71:51–61

Gerisch G, Fromm H, Huesgen A, Wick U (1975) Control of cell-contact sites by cyclic AMP pulses in differentiating *Dictyostelium* cells. Nature 255:547–549

Ginsburg GT, Gollop R, Yu Y, Louis JM, Saxe CL, Kimmel AR (1995) The regulation of *Dictyostelium* development by transmembrane signalling. J Euk Microbiol 42:200–205

Harwood AJ, Hopper NA, Simon M-N, Driscoll DM, Veron M, Williams JG (1992) Culmination in *Dictyostelium* is regulated by the cAMP-dependent protein kinase. Cell 69:615–624

Hopper NA, Harwood AJ, Bouzid S, Véron M, Williams JG (1993) Activation of the prespore and spore cell pathway of *Dictyostelium* differentiation by cAMP-dependent protein kinase and evidence for its upstream regulation by ammonia. EMBO J 12:2459–2466

Kessin RH (2001) *Dictyostelium*: evolution, cell biology and the development of multicellularity. Cambridge University Press, Cambridge

Konijn TM, Van De Meene JG, Bonner JT, Barkley DS (1967) The acrasin activity of adenosine-3′, 5′-cyclic phosphate. Proc Natl Acad Sci USA 58:1152–1154

Konijn TM, van de Meene JGC, Chang YY, Barkley DS, Bonner JT (1969) Identification of adenosine-3′, 5′-monophosphate as the bacterial attractant for myxoamoebae of *Dictyostelium discoideum*. J Bacteriol 99:510–512

Kriebel PW, Parent CA (2004) Adenylyl cyclase expression and regulation during the differentiation of *Dictyostelium discoideum*. IUBMB Life 56:541–546

Matsukuma S, Durston AJ (1979) Chemotactic cell sorting in *Dictyostelium discoideum*. J Embryol Exp Morph 50:243–251

Meima ME, Schaap P (1999) Fingerprinting of adenylyl cyclase activities during *Dictyostelium* development indicates a dominant role for adenylyl cyclase B in terminal differentiation. Dev Biol 212:182–190

Mutzel R, Lacombe M-L, Simon M-N, De Gunzburg J, Veron M (1987) Cloning and cDNA sequence of the regulatory subunit of cAMP- dependent protein kinase from *Dictyostelium discoideum*. Proc Natl Acad Sci USA 84:6–10

Raizada MK, Murti CRK (1972) Transformation of trophic *Hartmannella culbertsoni* into viable cysts by cyclic 3′5′-adenosine monophosphate. J Cell Biol 52:743–748

Raper KB (1984) The Dictyostelids. Princeton University Press, Princeton, NJ

Raper KB, Fennell DI (1952) Stalk formation in *Dictyostelium*. Bull Torrey Club 79:25–51

Ritchie AV, van Es S, Fouquet C, Schaap P (2008) From drought sensing to developmental control: evolution of cyclic AMP signaling in social amoebas. Mol Biol Evol 25:2109–2118

Saito T, Kato A, Kay R (2008) DIF-1 induces the basal disc of the *Dictyostelium* fruiting body. Dev Biol 317:444–453

Saran S, Schaap P. (2004) Adenylyl cyclase G is activated by an intramolecular osmosensor. Mol Biol Cell 15:1479–1486

Saran S, Meima ME, Alvarez-Curto E, Weening KE, Rozen DE, Schaap P (2002) cAMP signaling in *Dictyostelium* – complexity of cAMP synthesis, degradation and detection. J Muscle Res Cell Motil 23:793–802

Schaap P, Van Lookeren Campagne MM, Van Driel R, Spek W, Van Haastert PJ, Pinas J (1986) Postaggregative differentiation induction by cyclic AMP in *Dictyostelium*: intracellular transduction pathway and requirement for additional stimuli. Dev Biol 118:52–63

Schaap P, Winckler T, Nelson M, Alvarez-Curto E, Elgie B, Hagiwara H, Cavender J, Milano-Curto A, Rozen DE, Dingermann T, Mutzel R, Baldauf SL (2006) Molecular phylogeny and evolution of morphology in the social amoebas. Science 314:661–663

Schulkes C, Schaap P (1995) cAMP-dependent protein kinase activity is essential for preaggegative gene expression in *Dictyostelium*. FEBS Lett 368:381–384

Shaulsky G, Escalante R, Loomis WF (1996) Developmental signal transduction pathways uncovered by genetic suppressors. Proc Natl Acad Sci USA 93:15260–15265

Singleton CK, Zinda MJ, Mykytka B, Yang P (1998) The histidine kinase dhkC regulates the choice between migrating slugs and terminal differentiation in *Dictyostelium discoideum*. Dev Biol 203:345–357

Soderbom F, Anjard C, Iranfar N, Fuller D, Loomis WF (1999) An adenylyl cyclase that functions during late development of *Dictyostelium*. Development 126:5463–5471

Stephenson SL, Stempen H (1994) Myxomycetes: a handbook of slime molds. Timber Press, Portland, OR

Sternfeld J, David CN (1981) Cell sorting during pattern formation in *Dictyostelium*. Differentiation 20:10–21

Sternfeld J, David CN (1982) Fate and regulation of anterior-like cells in *Dictyostelium* slugs. Dev Biol 93:111–118

Sun TJ, Devreotes PN (1991) Gene targeting of the aggregation stage cAMP receptor cAR1 in *Dictyostelium*. Genes Dev 5:572–582

Swanson AR, Spiegel FW, Cavender JC (2002) Taxonomy, slime molds, and the questions we ask. Mycologia 94:968–979

Thomason P, Kay R (2000) Eukaryotic signal transduction via histidine-aspartate phosphorelay. J Cell Sci 113:3141–3150

Thomason PA, Traynor D, Cavet G, Chang W-T, Harwood AJ, Kay RR (1998) An intersection of the cAMP/PKA and two-component signal transduction systems in *Dictyostelium*. EMBO J 17:2838–2845

Van Es S, Virdy KJ, Pitt GS, Meima M, Sands TW, Devreotes PN, Cotter DA, Schaap P (1996) Adenylyl cyclase G, an osmosensor controlling germination of *Dictyostelium* spores. J Biol Chem 271:23623–23625

Verkerke-VanWijk I, Kim JY, Brandt R, Devreotes PN, Schaap P (1998) Functional promiscuity of gene regulation by serpentine receptors in *Dictyostelium discoideum*. Mol Cell Biol 18:5744–5749

Verkerke-van Wijk I, Fukuzawa M, Devreotes PN, Schaap P (2001) Adenylyl cyclase A expression is tip-specific in *Dictyostelium* slugs and directs StatA nuclear translocation and CudA gene expression. Dev Biol 234:151–160

Wang M, Schaap P (1989) Ammonia depletion and DIF trigger stalk cell differentiation in intact *Dictyostelium discoideum* slugs. Development 105:569–574

Wang M, Van Driel R, Schaap P (1988) Cyclic AMP-phosphodiesterase induces dedifferentiation of prespore cells in *Dictyostelium discoideum* slugs: evidence that cyclic AMP is the morphogenetic signal for prespore differentiation. Development 103:611–618

Wang N, Soderbom F, Anjard C, Shaulsky G, Loomis WF (1999) SDF-2 induction of terminal differentiation in *Dictyostelium discoideum* is mediated by the membrane-spanning sensor kinase DhkA. Mol Cell Biol 19:4750–4756

West CM (2003) Comparative analysis of spore coat formation, structure, and function in *Dictyostelium*. Int Rev Cytol 222:237–293

Wilkins MR, Williams KL (1995) The extracellular matrix of the *Dictyostelium discoideum* slug. Experientia 51:1189–1196

Chapter 6
On the Surprising Weakness of Pancreatic Beta-Cell Antioxidant Defences: An Evolutionary Perspective

Armin Rashidi, Thomas B.L. Kirkwood, and Daryl P. Shanley

Abstract Pancreatic beta-cells are unique among other cells in possessing an extremely weak antioxidant defence system. Furthermore, beta-cell defences against oxidative stress in both mice and humans are weaker in females than in males. These observations need an evolutionary explanation given that oxidative stress in beta-cells has an important contribution to the pathogenesis of type 2 diabetes. It turns out that a satisfactory explanation is feasible only by a systems-level approach to the glucose homeostatic system and beta-cell physiology. The connection between physiology and ecology is particularly important in major transitions in the evolutionary history of beta-cells. We have proposed that beta-cells evolved weak antioxidant protection to allow for reactive oxygen species (ROS) to exert a regulatory function directly linked with reproductive fitness of the organism. By down-regulating insulin synthesis and secretion, ROS in beta-cells cooperate with the stress-induced whole-body insulin resistance to divert glucose from insulin-dependent organs to an important insulin-independent structure: the evolving brain. Mammalian evolution provoked further reduction of beta-cell antioxidant defences in females in order to provide the fetus (through the insulin-independent placenta) with a reliable energy supply in pregnancy. The stress response has driven, within physiological constraints, the co-evolution of beta-cells, the brain and placenta and shaped their current status. We review the hypothesis and explore its implications.

T.B.L. Kirkwood and D.P. Shanley
Centre for Integrated Systems Biology of Ageing and Nutrition, Institute for Ageing and Health, Newcastle University, Newcastle upon Tyne, UK
A. Rashidi
Institute for Ageing and Health, Newcastle University, Campus for Ageing and Vitality, Newcastle upon Tyne NE4 5PL, UK; Centre for Integrated Systems Biology of Ageing and Nutrition, Institute for Ageing and Health, Newcastle University, Newcastle upon Tyne, UK
e-mail: armin.rashidi@newcastle.ac.uk

P. Pontarotti (ed.), *Evolutionary Biology: Concept, Modeling, and Application*,
DOI: 10.1007/978-3-642-00952-5_6,

6.1 Introduction

From an evolutionary perspective, pancreatic beta-cells and their weak protection against oxidative stress represent an intriguing puzzle. We here describe the hypothesis we recently proposed, that beta-cells evolved their weak antioxidant protection to allow for reactive oxygen species (ROS) to exert a regulatory function directly linked with reproductive fitness of the organism and we discuss further the implications of this idea. We believe the hypothesis throws new light on the trade-offs that are involved in the evolutionary optimisation of metabolism. The use of this kind of evolutionary physiology approach has already provided insights into the underpinning of important features of human aging and age-related diseases. For example, there are interesting inferences to be made about the phenomenon of the human menopause and whether or not it is functionally distinct from the generalised reproductive senescence that occurs in other species (Rashidi and Shanley 2009). If our hypothesis concerning beta-cells is correct, it may yield new ways to think about the pathogenesis of a common age-related disorder, type 2 diabetes.

6.1.1 Beta-Cells and Glucose Homeostasis

Beta-cells are major constituents of the endocrine pancreas and are localised within specialised structures (the islets of Langerhans) throughout the pancreas (Fig. 6.1). Beta-cells are believed to have first evolved as scattered insulin-producing cells in the intestinal tissue of primitive proto-chordates (>500 million years ago). During the evolution of the hagfish and lampreys (~500 million years ago), our most primitive vertebrate species of today, they formed the endocrine pancreas within the abdominal cavity (Madsen 2007). Embryonic development of the endocrine

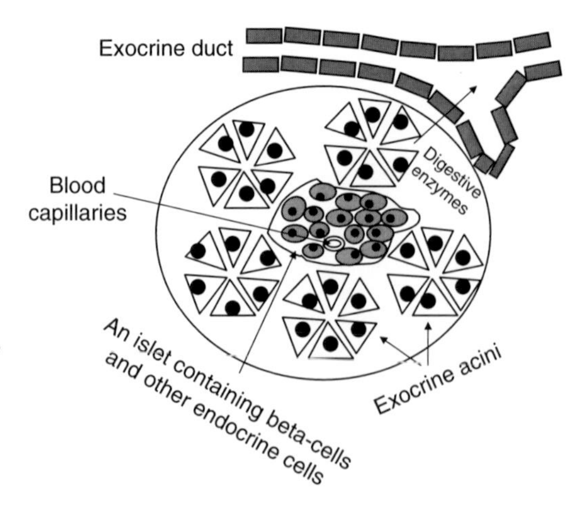

Fig. 6.1 Schematic histology of the pancreas. Pancreatic islets are collections of endocrine cells (including beta-cells) scattered between the exocrine acini

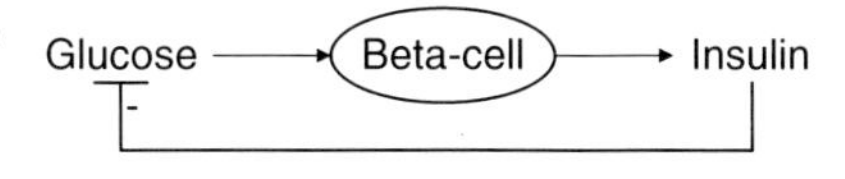

Fig. 6.2 Main elements of the glucose homeostatic system. Beta-cells sense the blood glucose and secrete insulin, which in turn reduces blood glucose and maintains homeostasis

pancreas involves sequential, cell-specific expression of transcriptional factors that determine the fate of individual cell types within the fetal pancreas (Habener et al. 2005). For example, expression of pancreatic duodenal homeobox-1 (PDX-1) and MafA (activators of insulin gene transcription) during gestation (8–21 weeks in humans) is associated with beta-cell differentiation and insulin production (Kaneto et al. 2007; Lyttle et al. 2008).

Beta-cells play a central role in whole-body glucose homeostasis. The enormously complicated organism-level process that results in glucose homeostasis can be simplified, for the sake of the present discussion, as a feedback loop between plasma glucose and insulin (Fig. 6.2). A rise in blood glucose, e.g., after a meal, is converted to a stimulatory signal inside beta-cells, leading to increased insulin synthesis and secretion. Following exocytosis and entry into the bloodstream, insulin promotes, via insulin receptors on the cell membrane, glucose disposal in various tissues such as the liver and skeletal muscles (for exceptions see below). Consequently, plasma glucose is robustly maintained within a narrow range (~70–100 mg/dl in humans), thus providing exclusively glucose-dependent organs such as the brain with a reliable energy supply. Uncorrected deviations from homeostasis can have detrimental outcomes, most notably seizure (with hypoglycemia) and diabetes (with hyperglycemia). Maintenance of glucose homeostasis is contingent upon proper beta-cell function, well-regulated beta-cell mass (i.e., number) and appropriate peripheral response to insulin.

Figure 6.3 provides a simplified overview of glucose-induced insulin secretion in beta-cells and insulin-induced glucose disposal in other cell types in the body. With a few exceptions including beta-cells, the brain and the placenta, glucose disposal depends on insulin availability in the vicinity of the cell (Watve and Yajnik 2007). Binding of insulin to insulin receptors on the surface of the cell activates an intracellular signalling pathway among the effects of which is the transport of glucose receptors to the cell membrane allowing for glucose entry (Fröjdö et al. 2008). Impairment in this signalling pathway is known as insulin resistance and is a central feature of an increasingly common constellation of pathologies collectively known as metabolic syndrome. In beta-cells, the ATP-controlled potassium channels on the cell surface close following the rise in cellular ATP concentrations that are produced by glucose metabolism. The consequent depolarisation of the cell membrane promotes calcium entry to the cell via calcium channels. Increased amounts of calcium in the cell cause release of previously synthesized insulin. Glucose also regulates insulin synthesis at both transcriptional and post-transcriptional levels (Andrali et al. 2008).

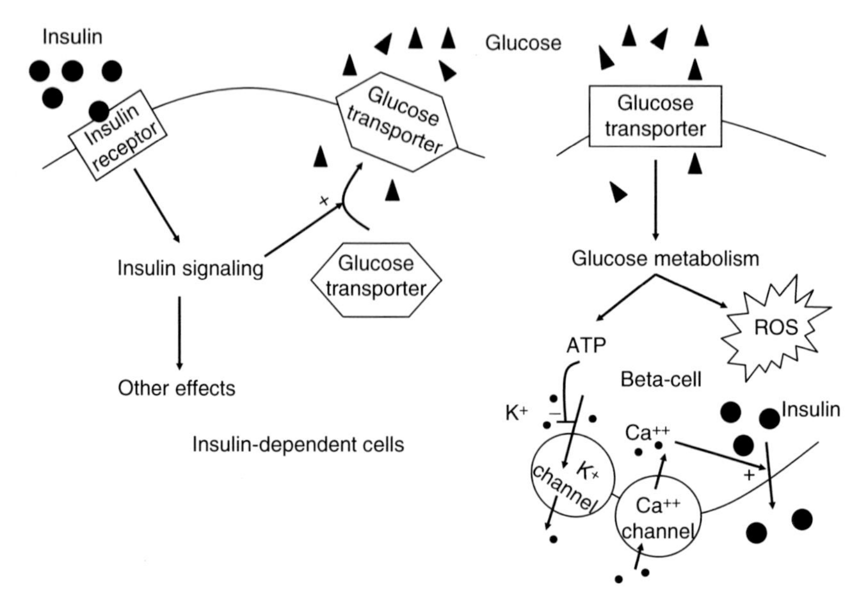

Fig. 6.3 Glucose-induced insulin secretion in beta-cells and peripheral insulin-induced glucose disposal. The ATP generated following glucose metabolism results in calcium entry into the beta-cell and insulin secretion. In peripheral tissues, one effect of insulin signalling is the translocation of glucose transporters to the cell membrane and glucose disposal

6.1.2 Reactive Oxygen Species: Effects on Beta-Cells

Glucose metabolism in the cell leads to the generation of reactive oxygen species (ROS), which negatively affect the insulin synthesis/secretion machinery via several mechanisms (Wiederkehr and Wollheim 2006). Insulin gene transcription is regulated by a number of transcription factors, most notably PDX-1 and RIPE3b1a/MafA. ROS negatively affect both of these transcription factors (Harmon et al. 2005; Kaneto et al. 2005; Poitout et al. 1996; Read et al. 1997; Robertson et al. 2003). Fusion of insulin vesicles with the cell membrane depends on intracellular concentrations of calcium, which rise as a result of calcium flow into the cell through calcium channels. Calcium channels open following the closure of ATP-dependent potassium channels and membrane depolarisation. Any event that recues cellular ATP levels may result in the closure of potassium channels. In particular, superoxide ions activate uncoupling protein-2 (UCP-2)-mediated proton leak and lower ATP levels, impairing glucose-mediated insulin secretion (Krauss et al. 2003).

The effects of transiently elevated glucose concentrations on ROS generation are opposite to those of prolonged glycemia. Glucose acutely suppressed ROS formation in rat beta-cells by increasing cellular NADH levels (Martens et al. 2005). However, this effect was limited. ROS production quickly surpassed its rate of elimination, leading to intracellular ROS accumulation (Gurgul et al. 2004; Pi et al. 2007; Tiedge et al. 1997). That beta-cells show different behaviors under conditions of acute versus prolonged stress is probably a more universal finding applicable to other systems and can be evolutionarily explained by considering what is 'good' for the organism. Homeostatic systems evolve to allow the organism to resist transient environmental noise; the organism should not change its strategies when the same environment is likely to be encountered shortly after a transient perturbation. However, when the environment seems to be deviating, for a long period, from its current status, adherence to a non-flexible homeostatic system can be detrimental. Physiological systems are neither too fragile nor too strictly homeostatic. It should be noted that our hypothesis (see below) only concerns prolonged stress conditions.

6.1.3 Antioxidant Defences in Beta-Cells Versus Other Cell Types

Given their slow turnover (Teta et al. 2005) and their central role in glucose homeostasis, beta-cells are expected to possess an efficient defence system that protects them against various insults, including ROS. The three main antioxidant enzymes are superoxide dismutase (SOD), catalase (CAT) and glutathione peroxidase (GPx) (Chance et al. 1979). SOD converts superoxide (O_2^-) molecules to less toxic molecules of H_2O_2, which may then be converted to oxygen and water by the enzymes catalase (CAT) and glutathione peroxidase (GPx) (Pi et al. 2007; Rhee et al. 2005). In human beta-cells, the expression pattern of the three main antioxidants is SOD > CAT > GPx (Tonooka et al. 2007). The same situation holds true in human liver cells (Asikainen et al. 1998) but antioxidants in human lung cells seem to have a different expression pattern: CAT > GPx > SOD (Asikainen et al. 1998). Antioxidant expression is cell-specific and different cell types may differ in their antioxidant responsiveness to intracellular redox changes (Asikainen et al. 1998; Jornot and Junod 1993).

Comparative studies have demonstrated that antioxidant defences are remarkably weaker and less responsive to oxidative challenge in beta-cells than in other tissues such as liver, kidney, brain, lung, skeletal muscle, heart muscle, adrenal gland and pituitary gland (Grankvist et al. 1981; Lenzen et al. 1996; Malaisse et al. 1982; Tiedge et al. 1997; Zhang et al. 1995). For example, GPx is expressed at very low levels and has almost undetectable activity in both human (Robertson and Harmon 2007; Tonooka et al. 2007) and rodent beta-cells (Grankvist et al. 1981; Tiedge et al. 1997). Catalase activity in human beta-cells is at the same level as, if not even lower than, in rat beta-cells (Sigfrid et al. 2003). Catalase gene expression is nearly undetectable in mouse beta-cells (Lenzen et al. 1996) and in rats is approximately 5% of the liver (Tiedge et al. 1997).

The weakness of antioxidant defences in beta-cells is clinically relevant. Several lines of evidence point to oxidative stress as a causative factor in the pathogenesis of type 2 diabetes. Compared to healthy controls, islets of type 2 diabetic patients contain significantly higher levels of oxidative stress markers (Chang-Chen et al. 2008), possibly because beta-cell function and turnover are highly susceptible to the adverse effects of prolonged glycemia. The insufficiently countered oxygen radicals generated over time as a result of chronic exposure to glycemia are believed to damage beta-cell function and turnover mechanisms (Tanaka et al. 2002). Taken together, the lack of a strong (antioxidant) defence system has been an increasingly likely suspect for beta-cell sensitivity to ROS (Kajimoto and Kaneto 2004; Lenzen 2008; Robertson et al. 2003).

6.1.4 Beta-Cell Antioxidant Defences: Gender Differences

Beta-cell antioxidants have lower levels of activity in female mice than in males (Cornelius et al. 1993). Humans are no different. Tonooka et al. (2007) performed one of the few studies on human beta-cell antioxidants. They observed that GPx protein expression was remarkably low and GPx activity was almost undetectable in the islets. Furthermore, they demonstrated under varying glucose concentrations that GPx protein expression was significantly lower in islets from females. There are no studies, to our knowledge, that have looked at gender-specific patterns in beta-cell antioxidant expression in species other than mice and humans.

Beta-cells seem to be rather unique among other cells in this regard. Indeed, females are generally better protected against oxidative stress (Proteggente et al. 2002). For example, mitochondria from female rats had higher antioxidant gene expression than those from males in one study. Ovariectomy removed this difference, which re-appeared following estrogen replacement therapy (Borras et al. 2003). Among tissues where ovariectomy causes a significant reduction in antioxidant expression and activity are bones (Muthusami et al. 2005) and the liver (Oztekin et al. 2006).

The proximate mechanisms underlying this gender difference are not clear. Estrogen has a phenol ring containing a hydroxyl group that may be donated. This structure offers estrogen a possible chain-breaking antioxidant mechanism of action, like that of vitamin E (Subbiah et al. 1993; Sugioka et al. 1987). However, physiological concentrations of estrogen are not high enough to exert significant effects on the cellular redox state. Indirect effects such as membrane stabilisation are a more likely mechanism for the antioxidant properties of estrogen at physiological concentrations (Paroo et al. 2002). Beta-cells do not seem to follow the same response pattern to sex hormones as other cell types. It is androgens, rather than estrogen, that provide some protection for beta-cells, by reducing mega-islet formation (Rosmalen et al. 2001) and apoptosis (Morimoto et al. 2005).

6.2 Concepts from an Evolutionary Hypothesis

We have recently developed an evolutionary hypothesis in response to the following two questions (Rashidi et al. 2009):

1. Is there any organism-level fitness advantage associated with having weakly protected beta-cells against oxidative stress?
2. Why are females different from males in their beta-cell antioxidant status?

The hypothesis is built upon both physiological and evolutionary considerations. Physiological aspects concern changes that occur in the glucose homeostatic system during stress in general and pregnancy in particular. Evolutionary aspects include co-evolution of the brain, beta-cells and stress response. As explained below, a reduced antioxidant status in beta-cells may be associated with a fitness advantage for the organism that outweighs the high price that is paid in terms of cellular vulnerability to ROS.

Physiological considerations: With a few exceptions, namely the brain and placenta, glucose disposal in cells depends on insulin availability in the micro-environment (Peters et al. 2002; Watve and Yajnik 2007). Therefore, decreased insulin sensitivity does not affect insulin-independent cell types that become preferential destinations for the net extra glucose available in the blood in insulin resistant states. Corticosteroids (released during the stress response) and tumour necrosis factor-α (released during pregnancy) cause impairments in the insulin signalling pathway that leads to insulin-induced glucose disposal, thus making stress conditions and pregnancy two important insulin resistant states (Kirwan et al. 2004; Qi and Rodrigues 2007). Both acute and chronic stress activate the hypothalamic-pituitary-adrenal (HPA) axis and result in glucocorticoid release into the bloodstream. In spite of differences in upstream events, the response elicited by both types of stress is qualitatively similar (Chowdrey et al. 1995; Harbuz et al. 1992; Lightman 2008).

Table 6.1 provides a mechanistic overview of the important changes that occur in the glucose homeostatic system during pregnancy. For the purpose of our hypothesis, it is sufficient to remember that plasma glucose rises within the normal range during pregnancy, particularly in the third trimester (Parretti et al. 2001). Beta-cells respond by secreting more insulin to compensate.

Microevolutionary considerations: Given the importance of the evolving brain (in non-pregnant states) and the growing fetus (in mammals), it might have been worth investing less in somatic maintenance (e.g., antioxidant protection) and accept higher risks of oxidative stress-related problems (e.g., reduced beta-cell proliferation, survival and function), yet provide the brain and fetus with a reliable energy supply and avoid an otherwise inevitable compromise in times of stress. This can be achieved by insulin resistance-induced re-allocation of energy provided that the compensatory hyperinsulinemia is unable to return glucose back to the set-point. Given that ROS have negative effects on insulin synthesis/secretion, we believe antioxidants in beta-cells are expressed at a sufficiently low level to allow for ROS to perform their important regulatory function during stress (including pregnancy). One would expect the evolutionarily most important tissues in the body to be insulin-independent.

For the following reasons, pregnancy has been an excellent opportunity for evolution to aid the mother in producing high-quality offspring. First, the relationship between the mother and offspring is never closer than it is in pregnancy. Second, pregnancy is a relatively long period of metabolic relationships between two genetically related individuals from successive generations. The maternal disadvantages associated with poorly protected beta-cells are offset, according to

Table 6.1 Summary of the changes that occur in glucose homeostasis in pregnancy

Beta-cell mass regulation	–Late gestational increase in glucocorticoids increase beta-cell apoptosis and decrease beta-cell proliferation.	–Freemark 2006; Ranta et al. 2006; Weinhaus et al. 2000
	–Lactogens decrease beta-cell apoptosis and increase beta-cell proliferation.	–Arumugam et al. 2008; Fujinaka et al. 2007
	–Net result: increased beta-cell mass during gestation	–Sorenson and Brelje 1997
Glucocorticoid metabolism	–Glucocorticoids are important for decidualisation, implantation and placental development but have undesired effects if not counteracted.	–Arcuri et al. 1996; Malassine and Cronier 2002
	–The placental 11ß-hydroxysteroid dehydrogenase type 2 converts cortisol to inactive cortisone and protects the fetus.	–Shams et al. 1998
	–Attenuated HPA axis stress responsiveness is another protective mechanism.	–Weinstock 1997
Insulin sensitivity	–Post-receptor defects in insulin signalling mainly in skeletal muscles and adipose tissues	–Catalano et al. 2002; Gonzalez et al. 2002; Kirwan et al. 2004; Saad et al. 1997; Shao et al. 2000
	–Blunting of insulin sensitivity more than 50% in the third trimester	–Butte 2000; Catalano et al. 1991; Catalano et al. 1993; Yamashita et al. 2000
	–TNF-α plays the role of stress hormones	–Kirwan et al. 2004
Glucose[a]	–Prolactin lowers the glucose set-point by up-regulating the expression of glucose transporters and glucokinase (glucose sensor) activity in beta-cells (first trimester)[b]	–Butte 2000; Mills et al. 1998; Weinhaus et al. 1996
	–Increased fasting hepatic glucose production, increased plasma volume in the first trimester and increased fetoplacental glucose utilisation in late gestation	–Catalano et al. 1992
	–Net result: lower plasma glucose levels than in non-pregnant women, with a gradual rise in the third trimester	–Parretti et al. 2001
Insulin	–Increased metabolic clearance of insulin, due partly to placental insulinase[c]	–Catalano et al. 1998; Posner 1973
	–Compensatory rise of fasting plasma insulin	–Assel et al. 1993; Catalano et al. 1993

[a]The maternal plasma glucose level is highly correlated to utero-placental glucose consumption, glucose transfer to the fetus and fetal plasma glucose (Hay 1995; Hay and Meznarich 1989).
[b]This mechanism protects the mother against glycemia as insulin sensitivity declines in the course of pregnancy.
[c]This mechanism may aid ROS in reducing plasma insulin (see text).

the hypothesis, by the resulting increments in offspring fitness. Adapting to pregnancy is indeed a prominent feature of mammalian evolution. Evolution of beta-cell antioxidant defences is an example of the link between antioxidant protection and reproductive success. This link, however, is not without precedent. Negative correlations have been described in birds between reproductive success and protection against ROS (Alonso-Alvarez et al. 2006; Wiersma et al. 2004). The disposable soma theory is built on such trade-offs and concerns the evolutionarily optimised balance between cellular investment in reproduction and survival (Kirkwood and Holliday 1979).

Oxygen radicals act in many systems as second messengers in signal transduction (Finkel 2003). Relevant to protein misfolding and protein aggregation diseases (e.g., Alzheimer disease), it has been shown that ROS are signals generated by misfolded proteins in the endoplasmic reticulum that cause cell death (Malhotra et al. 2008). In addition, protein misfolding in the endoplasmic reticulum has been suggested to be a causative upstream event in insulin resistance (Ozcan et al. 2006). Mitochondrial oxidants act as facilitators of communication between the mitochondria and the cytosol compatible with an evolutionary adaptation following the ancient incorporation of mitochondria into eukaryotic cells (Nemoto et al. 2000). Why natural selection has so frequently chosen potentially dangerous molecules such as ROS for intracellular signalling may be explained by the fact that these molecules are inevitable by-products of cellular metabolism. By using ROS as signalling molecules, the system did not need to invent and pay additional costs for a distinct regulatory system.

Macroevolutionary considerations: We have proposed two major transitions in beta-cell evolutionary history (Table 6.2). Primitive intestinal beta-cells sensed gut glucose and were in a more direct contact with the external environment surrounding the organism than pancreatic beta-cells, which sensed blood glucose and were exposed only to the internally regulated microenvironment. Concurrent

Table 6.2 The proposed evolutionary history of beta-cells

Time	Species	Evolutionary features
>500 million years ago	Proto-chordates	No brain No cortisol/corticosteroid receptors Beta-cells in the intestinal tissue Beta-cells sensed gut glucose Beta-cell antioxidants normal Beta-cells in males equal to females
450–500 million years ago	Vertebrates	Evolution of the brain Evolution of cortisol/corticosteroid receptors Evolution of the endocrine pancreas Beta-cells sensed blood glucose Reduction in beta-cell antioxidants Beta-cells in males equal to females
~100 million years ago	Placental mammals	Evolution of the placenta Further loss of antioxidants in beta-cells

with this increase in complexity was the evolution of the most primitive brain and stress response (Bury and Sturm 2007; Ortlund et al. 2007). Simultaneous evolution of beta-cells, the brain and stress response represents an interesting example of within-organism co-evolution. This involved an ecological component (stress) and an organismal component (the need for a brain), which together produced a sufficiently high selection pressure to partially sacrifice beta-cell protection against oxidative stress. The few million years that might have elapsed between the initiation of evolution of beta-cells, stress response and the brain is a short period on evolutionary scales.

Evolution of placental mammals around 100 million years ago demarcates the second major transition. It required beta-cells in females to lose even more of their antioxidants in order to provide the fetus with a reliable source of energy at times of maternal stress (Kitazoe et al. 2007).

6.3 Robustness, Homeostasis and Allostasis

Robustness is an inherent, fundamental feature of complex biological systems and a milestone in the course of their evolution. Complete elimination of environmental perturbations, as occurs in homeostasis, is not always advantageous. Rather, maintenance of function in unstable environments frequently requires temporary or permanent departure from homeostasis. Under such circumstances, the fitness advantages of changing may outweigh those of remaining the same (Kitano 2004). The concept of robustness is being incorporated, using firm mathematical definitions, into our current understanding of fitness in biological systems (Kitano 2007). Natural selection appears to have used ROS, which tend to impair homeostasis in many systems, as regulatory molecules in beta-cells that help the organism appropriately re-allocate the available energy resources under stress conditions. Corticosteroid response to stress and ROS-regulated beta-cell reaction to the resulting insulin resistance act in harmony to protect the brain and fetus against hypoglycemia.

Homeostasis means remaining stable by staying the same, whereas allostasis refers to the process of remaining stable through change. Both homeostasis and allostasis are important for maintaining the stability of an organism (Sterling and Eyer 1988). Allostasis makes the organism capable of actively adjusting in the long run to both predictable and unpredictable events. The costs of this long-term adaptation to stress, referred to as 'allostatic load', is manifested as pathology (McEwen and Wingfield 2003). The term 'glucose allostasis' has recently emerged in the literature to explain the process by which prolonged environmental stress and the adaptations they provoke lead ultimately to severe insulin resistance and type 2 diabetes (Stumvoll et al. 2004). Stumvoll et al. (2003) suggest that glycemia is a signal to inform beta-cells of the presence of insulin resistance, that is to say, the organism is likely to be facing prolonged stress. There is strong observational evidence for the involvement of allostasis in the pathogenesis of type 2 diabetes

(Stumvoll et al. 2003). Our hypothesis demonstrates how beta-cell defence mechanisms might have evolved to confer the organism the ability to respond to stress in an optimal manner. The role of pregnancy as a common stress condition in the past 100 million years in shaping the evolution of the glucose homeostatic system should not be underestimated.

6.4 Clinical Implications and Future Directions

The weakness of antioxidant defences in beta-cells along with the definitive contribution of oxidative stress to the pathogenesis of type 2 diabetes has provoked the idea of helping beta-cells better protect themselves against ROS to prevent development and progression of diabetes. This idea has been successfully tested in the lab (Yamamoto et al. 2008). Over-expression of thioredoxin, an antioxidant protein, in mouse beta-cells prevented progression of type 2 diabetes. This was associated with suppression of the ROS-induced reduction of PDX-1 and MafA expression in beta-cells. However, ignoring the evolutionary history of beta-cells and the adaptive mechanisms that drove their evolution may have undesired outcomes. In a system that has evolved under the influence of several trade-offs, trying to modulate one feature might leave the organism unfit in other functions. Without considering the evolutionary history of beta-cells, any attempt aimed at improving beta-cell sensitivity to ROS may result in compromised stress responsiveness and reduced reproductive success.

Figure 6.4 shows one pathway to pathology. If not countered, stress chronicity might cause irreversible and undesired changes in the glucose homeostatic system. There are, however, a number of mechanisms by which the system tries to resist chronic stress. Glucocorticoid secretion has a circadian rhythm. After each pulse of secretion, there is a refractory period during which the HPA axis is not activated by

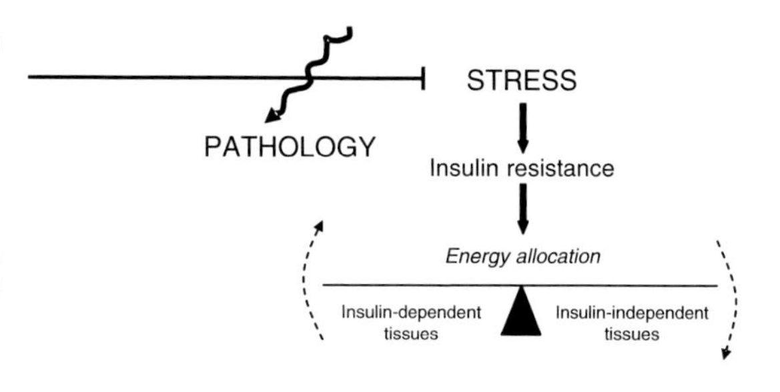

Fig. 6.4 The relation between stress and energy allocation. Insulin resistance mediates the link between stress and response. Glycemia-induced ROS signal beta-cells of the presence of insulin resistance, beta-cells partially correct the glycemia and energy is re-allocated to different tissues as appropriate. The system fails to return to the original steady state if the stress persists for a long time (Adapted from Rashidi et al. 2009)

mild stressors (Windleet al. 1998). Therefore, with increased frequency of pulsatility due to chronic stress, the animal will spend more time in stress hyporesponsiveness (Windle et al. 2001). Stress responsiveness is under both genetic (Windle et al. 1998) and epigenetic control. It is known that the HPA axis can be programmed by early life events (Levine 1967). Perinatal exposure to stress makes animals more stress responsive in adulthood. The underlying mechanism seems to be related to increased frequency of HPA pulses and glucocorticoid pulse amplitude (Shanks et al. 1995, 2000).

We have discussed how pregnancy might have contributed to the evolution of a weaker beta-cell antioxidant system in female mammals. The above observations suggest the possibility of the HPA axis neonatal programming having some impact on beta-cell antioxidant responsiveness. Experimental evidence is to date lacking on this possibility. Although neonatal programming may have a fine-tuning effect on beta-cell antioxidants, we believe this mechanism has not significantly influenced the evolution of beta-cell antioxidants. The proximate mechanism that down-regulates beta-cell antioxidants in females is not known at the moment. The time point during development (e.g., fetal period, perinatal period) when this modification occurs is also unclear. Sex hormones and changes in the intrauterine environment are good candidates for future investigations. The connection between stress, pregnancy and beta-cell antioxidants becomes even more complicated when considering the reduction that occurs in HPA responsiveness to emotional and physical stressors in pregnancy (Neumann et al. 1998). It has been suggested that this mechanism might have evolved to protect the mother and her fetus from excess corticosteroids (Weinstock 1997).

Why did natural selection select for organisms that had beta-cells with permanently reduced antioxidant defences, rather than those with beta-cells that were capable of tuning down their defence mechanisms only when encountering stress? In doing the latter, the organism could probably enjoy a lower life-time risk for development of type 2 diabetes because beta-cells would be more resistant to oxidative stress. It certainly requires more work before one can conclusively answer the question. Nevertheless, there are at least two reasons why organisms with permanently weakly protected beta-cells might have had the advantage. First, such organisms were able to avoid the possible costs of a facultative antioxidant system, which could become significant if stress (e.g., pregnancy, infection) was a sufficiently frequent problem at the time of the evolution of the system. Second, prolonged and severe exposure to an environment with high insulin demand were significantly less common in our evolutionary past than in today's world with abundant food, obesity, physical inactivity and increasing life expectancy (Kitano et al. 2004). Termination of pregnancy helped the organism avoid one of the few potential sources of prolonged stress and disease susceptibility. It is therefore likely that beta-cell susceptibility to oxidative stress was not considered by natural selection as a major problem that could reduce reproductive fitness. One can speculate that natural selection might be on its way to make further modifications to beta-cell physiology in response to the rather sudden changes that have recently occurred in our lifestyles.

Acknowledgments AR is supported by a Dorothy Hodgkin Postgraduate Award. TBLK and DPS are supported by the BBSRC Centre for Integrated Systems Biology of Ageing and Nutrition (CISBAN). Part of this work was supported by and carried out within the EU-funded Network of Excellence LifeSpan (FP6 036894).

References

Alonso-Alvarez C, Bertrand S, Devevey G, Prost J, Faivre B, Chastel O, Sorci G (2006) An experimental manipulation of life-history trajectories and resistance to oxidative stress. Evolution Int J Org Evolution 60:1913–24

Andrali SS, Sampley ML, Vanderford NL, Ozcan S (2008) Glucose regulation of insulin gene expression in pancreatic beta-cells. Biochem J 415:1–10

Arcuri F, Monder C, Battistini S, Hausknecht V, Lockwood CJ, Schatz F (1996) Potential role of 11 beta-hydroxysteroid dehydrogenase in human trophoblast-endometrial interactions. Ann N Y Acad Sci 784:433–438

Arumugam R, Horowitz E, Lu D, Collier JJ, Ronnebaum S, Fleenor D, Freemark M (2008) The interplay of prolactin and the glucocorticoids in the regulation of beta cell gene expression, fatty acid oxidation, and glucose-stimulated insulin secretion: implications for carbohydrate metabolism in pregnancy. Endocrinology 149:5401–5414

Asikainen TM, Raivio KO, Saksela M, Vuokko L (1998) Kinnula expression and developmental profile of antioxidant enzymes in human lung and liver. Am J Respir Cell Mol Biol 19:942–949

Assel B, Rossi K, Kalhan S (1993) Glucose metabolism during fasting through human pregnancy: comparison of tracer method with respiratory calorimetry. Am J Physiol 265:351–356

Borrás C, Sastre J, García-Sala D, Lloret A, Pallardó FV, Viña José (2003) Mitochondria from females exhibit higher antioxidant gene expression and lower oxidative damage than males. Free Rad Biol Med 34:546–552

Brownlee M (2003) A radical explanation for glucose-induced beta cell dysfunction. J Clin Invest 112:1788–1790

Butte NF (2000) Carbohydrate and lipid metabolism in pregnancy: normal compared with gestational diabetes mellitus. Am J Clin Nutr 71:1256–1261

Catalano PM, Drago NM, Amini SB (1998) Longitudinal changes in pancreatic beta-cell function and metabolic clearance rate of insulin in pregnant women with normal and abnormal glucose tolerance. Diabetes Care 21:403–408

Catalano PM, Nizielski SE, Shao J, Preston L, Qiao L, Friedman JE (2002) Downregulated IRS-1 and PPARgamma in obese women with gestational diabetes: relationship to FFA during pregnancy. Am J Physiol Endocrinol Metab 282:522–533

Catalano PM, Tyzbir ED, Roman NM, Amini SB, Sims EA (1991) Longitudinal changes in insulin release and insulin resistance in nonobese pregnant women. Am J Obstet Gynecol 165:1667–1672

Catalano PM, Tyzbir ED, Wolfe RR, Calles J, Roman NM, Amini SB, Sims EA (1993) Carbohydrate metabolism during pregnancy in control subjects and women with gestational diabetes. Am J Physiol 264:60–67

Catalano PM, Tyzbir ED, Wolfe RR, Roman NM, Amini SB, Sims EA (1992) Longitudinal changes in basal hepatic glucose production and suppression during insulin infusion in normal pregnant women. Am J Obstet Gynecol 167:913–919

Chance B, Sies H, Boveris A (1979) Hydroperoxide metabolism in mammalian organs. Physiol Rev 59:527–605

Chang-Chen KJ, Mullur R, Bernal-Mizrachi E (2008) beta-cell failure as a complication of diabetes. Rev Endocr Metab Disord 9:329–343

Chowdrey HS, Larsen PJ, Harbuz MS, Jessop DS, Aguilera G, Eckland DJ, Lightman SL (1995) Evidence for arginine vasopressin as the primary activator of the HPA axis during adjuvant-induced arthritis. Br J Pharmacol 116:2417–2424

Finkel T (2003) Oxidant signals and oxidative stress. Curr Opin Cell Biol 15:247–254

Freemark M (2006) Regulation of maternal metabolism by pituitary and placental hormones: roles in fetal development and metabolic programming. Horm Res 65:41–49

Fröjdö S, Vidal H, Pirola L (2008) Alterations of insulin signaling in type 2 diabetes: A review of the current evidence from humans. Biochim Biophys Acta [Epub ahead of print]

Fujinaka Y, Takane K, Yamashita H, Vasavada RC (2007) Lactogens promote beta cell survival through JAK2/STAT5 activation and Bcl-XL upregulation. J Biol Chem 282:30707–30717

Gonzalez CG, Alonso A, Balbin M, Diaz F, Fernandez S, Patterson AM (2002) Effects of pregnancy on insulin receptor in liver, skeletal muscle and adipose tissue of rats. Gynecol Endocrinol 16:193–205

Grankvist K, Marklund SL, Taljedal IB (1981) CuZn-superoxide dismutase, Mn-superoxide dismutase, catalase and glutathione peroxidase in pancreatic islets and other tissues in the mouse. Biochem J 199:393–398

Gurgul E, Lortz S, Tiedge M, Jorns A, Lenzen S (2004) Mitochondrial catalase overexpression protects insulin-producing cells against toxicity of reactive oxygen species and proinflammatory cytokines. Diabetes 53:2271–2280

Harbuz MS, Rees RG, Eckland D, Jessop DS, Brewerton D, Lightman SL (1992) Paradoxical responses of hypothalamic corticotropin-releasing factor (CRF) messenger ribonucleic acid (mRNA) and CRF-41 peptide and adenohypophysial proopiomelanocortin mRNA during chronic inflammatory stress. Endocrinology 130:1394–1400

Hay WW Jr (1995) Regulation of placental metabolism by glucose supply. Reprod Fertil Dev 7:365–375

Hay WW Jr, Meznarich HK (1989) Effect of maternal glucose concentration on uteroplacental glucose consumption and transfer in pregnant sheep. Proc Soc Exp Biol Med 190:63–69

Habener JF, Kemp DM, Thomas MK (2005) Minireview: transcriptional regulation in pancreatic development. Endocrinology 146:1025–1034.

Harmon JS, Stein R, Robertson RP (2005) Oxidative stress-mediated, post-translational loss of MafA protein as a contributing mechanism to loss of insulin gene expression in glucotoxic beta cells. J Biol Chem 280:11107–11113

Jornot L, Junod AF (1993) Variable glutathione levels and expression of antioxidant enzymes in human endothelial cells. Am J Physiol Lung Cell Mol Physiol 264:482–489

Kajimoto Y, Kaneto H (2004) Role of oxidative stress in pancreatic beta-cell dysfunction. Ann N Y Acad Sci 1011:168–176

Kaneto H, Kawamori D, Matsuoka TA, Kajimoto Y, Yamasaki Y (2005) Oxidative stress and pancreatic beta-cell dysfunction. Am J Ther 12:529–533

Kaneto H, Miyatsuka T, Shiraiwa T, Yamamoto K, Kato K, Fujitani Y, Matsuoka TA (2007) Crucial role of PDX-1 in pancreas development, beta-cell differentiation, and induction of surrogate beta-cells. Curr Med Chem 14:1745–1752

Kirkwood TB, Holliday R (1979) The evolution of ageing and longevity. Proc R Soc Lond B Biol Sci 205:531–546

Kirwan JP, Varastehpour A, Jing M, Presley L, Shao J, Friedman JE, Catalano PM (2004) Reversal of insulin resistance postpartum is linked to enhanced skeletal muscle insulin signaling. J Clin Endocrinol Metab 89:4678–4684

Kitano H (2004) Biological robustness. Nature 5:826–837

Kitano H (2007) Towards a theory of biological robustness. Mol Syst Biol 3:137

Kitano H, Oda K, Kimura T, Matsuoka Y, Csete M, Doyle J, Muramatsu M (2004) Metabolic syndrome and robustness trade-offs. Diabetes 53:1–10

Kitazoe Y, Kishino H, Waddell PJ, Nakajima N, Okabayashi T, Watabe T, Okuhara Y (2007) Robust time estimation reconciles views of the antiquity of placental mammals. PLoS ONE 2:384

Krauss S, Zhang CY, Scorrano L, Dalgaard LT, St-Pierre J, Grey ST, Lowell BB (2003) Superoxide-mediated activation of uncoupling protein 2 causes pancreatic beta cell dysfunction. J Clin Invest 112:1831–1842

Lenzen S (2008) Oxidative stress: the vulnerable beta-cell. Biochem Soc Trans 36:343–347

Lenzen S, Drinkgern J, Tiedge M (1996) Low antioxidant enzyme gene expression in pancreatic islets compared with various other mouse tissues. Free Radic Biol Med 20:463–466

Levine S (1967) Maternal and environmental influences on the adrenocortical response to stress in weanling rats. Science 156:258–260

Lightman SL (2008) The neuroendocrinology of stress: a never ending story. J Neuroendocrinol 20:880–884

Lyttle BM, Li J, Krishnamurthy M, Fellows F, Wheeler MB, Goodyer CG, Wang R (2008) Transcription factor expression in the developing human fetal endocrine pancreas. Diabetologia 51:1169–1180

Madsen OD (2007) Pancreas phylogeny and ontogeny in relation to a 'pancreatic stem cell'. C R Biol 330:534–537

Malaisse WJ, Malaisse-Lagae F, Sener A, Pipeleers DG (1982) Determinants of the selective toxicity of alloxan to the pancreatic B cell. Proc Natl Acad Sci USA 79:927–930

Malassine A, Cronier L (2002) Hormones and human trophoblast differentiation: a review. Endocrine 19:3–11

Malhotra JD, Miao H, Zhang K, Wolfson A, Pennathur S, Pipe SW, Kaufman RJ (2008) Antioxidants reduce endoplasmic reticulum stress and improve protein secretion. Proc Natl Acad Sci USA 105:18525–18530

Martens GA, Cai Y, Hinke S, Stange G, Van de Casteele M, Pipeleers D (2005) Glucose suppresses superoxide generation in metabolically responsive pancreatic beta cells. J Biol Chem 280:20389–20396

McEwen BS, Wingfield JC (2003) The concept of allostasis in biology and biomedicine. Horm Behav 43:2–15

Mills JL, Jovanovic L, Knopp R, Aarons J, Conley M, Park E, Lee YJ, Holmes L, Simpson JL, Metzger B (1998) Physiological reduction in fasting plasma glucose concentration in the first trimester of normal pregnancy: the diabetes in early pregnancy study. Metabolism 47:1140–1144

Morimoto S, Mendoza-Rodriguez CA, Hiriart M, Larrieta ME, Vital P, Cerbon MA (2005) Protective effect of testosterone on early apoptotic damage induced by streptozotocin in rat pancreas. J Endocrinol 187:217–224

Muthusami S, Ramachandran I, Muthusamy B, Vasudevan G, Prabhu V, Subramaniam V, Jagadeesan A, Narasimhan S (2005) Ovariectomy induces oxidative stress and impairs bone antioxidant system in adult rats. 360:81–86

Nemoto S, Takeda K, Yu ZX, Ferrans VJ, Finkel T (2000) Role for mitochondrial oxidants as regulators of cellular metabolism. Mol Cell Biol 20:7311–7318

Neumann ID, Johnstone HA, Hatzinger M, Liebsch G, Shipston M, Russell JA, Landgraf R, Douglas AJ (1998) Attenuated neuroendocrine responses to emotional and physical stressors in pregnant rats involve adenohypophysial changes. J Physiol 508:289–300

Ozcan U, Yilmaz E, Özcan L, Furuhashi M, Vaillancourt E, Smith RO, Görgün CZ, Hotamisligil GS (2006) Chemical chaperones reduce ER stress and restore glucose homeostasis in a mouse model of type 2 diabetes. Science 313:1137–1140

Oztekin E, Tiftik AM, Baltaci AK, Mogulkoc R (2006) Lipid peroxidation in liver tissue of ovariectomized and pinealectomized rats: effect of estradiol and progesterone supplementation. Cell Biochem Funct 25:401–405

Paroo Z, Dipchand ES, Noble EG (2002) Estrogen attenuates postexercise HSP70 expression in skeletal muscle. Am J Physiol Cell Physiol 282:245–251

Parretti E, Mecacci F, Papini M, Cioni R, Carignani L, Mignosa M, La Torre P, Mello G (2001) Third-trimester maternal glucose levels from diurnal profiles in nondiabetic pregnancies: correlation with sonographic parameters of fetal growth. Diabetes Care 24:1319–1323

Peters A, Schweiger U, Fruhwald-Schultes B, Born J, Fehm HL (2002) The neuroendocrine control of glucose allocation. Exp Clin Endocrinol Diabetes 110:199–211

Pi J, Bai Y, Zhang Q, Wong V, Floering LM, Daniel K, Reece JM, Deeney JT, Andersen ME, Corkey BE, Collins S (2007) Reactive oxygen species as a signal in glucose-stimulated insulin secretion. Diabetes 56:1783–1791

Poitout V, Olson LK, Robertson RP (1996) Chronic exposure of betaTC-6 cells to supraphysiologic concentrations of glucose decreases binding of the RIPE3b1 insulin gene transcription activator. J Clin Invest 97:1041–1046

Posner BI (1973) Insulin metabolizing enzyme activities in human placental tissue. Diabetes 22:552–563

Proteggente AR, England TG, Rehman A, Rice-Evans CA, Halliwell B (2002) Gender differences in steady-state levels of oxidative damage to DNA in healthy individuals. Free Radic Res 36:157–162

Qi D, Rodrigues B (2007) Glucocorticoids produce whole body insulin resistance with changes in cardiac metabolism. Am J Physiol Endocrinol Metab 292:654–667

Ranta F, Avram D, Berchtold S, Dufer M, Drews G, Lang F, Ullrich S (2006) Dexamethasone induces cell death in insulin-secreting cells, an effect reversed by exendin-4. Diabetes 55:1380–1390

Rashidi A, Shanley DP (2009) Evolution of the menopause: life histories and mechanisms. Menopause Int 15:26–30

Rashidi A, Kirkwood TBL, Shanley DP (2009) Metabolic evolution suggests an explanation for the weakness of antioxidant defences in beta-cells. Mech Ageing Dev 130:216–221

Read ML, Masson MR, Docherty K (1997) A RIPE3b1-like factor binds to a novel site in the human insulin promoter in a redox-dependent manner. FEBS Lett 418:68–72

Rhee SG, Yang KS, Kang SW, Woo HA, Chang TS (2005) Controlled elimination of intracellular H(2)O(2): regulation of peroxiredoxin, catalase, and glutathione peroxidase via post-translational modification. Antioxid Redox Signal 7:619–626

Robertson RP, Harmon JS (2007) Pancreatic islet beta-cell and oxidative stress: the importance of glutathione peroxidase. FEBS Lett 581:3743–3748

Robertson RP, Harmon J, Tran PO, Tanaka Y, Takahashi H (2003) Glucose toxicity in beta-cells: type 2 diabetes, good radicals gone bad, and the glutathione connection. Diabetes 52:581–587

Rosmalen JG, Pigmans MJ, Kersseboom R, Drexhage HA, Leenen PJ, Homo-Delarche F (2001) Sex steroids influence pancreatic islet hypertrophy and subsequent autoimmune infiltration in nonobese diabetic (NOD) and NODscid mice. Lab Invest 81:231–239

Saad MJ, Maeda L, Brenelli SL, Carvalho CR, Paiva RS, Velloso LA (1997) Defects in insulin signal transduction in liver and muscle of pregnant rats. Diabetologia 40:179–186

Shams M, Kilby MD, Somerset DA, Howie AJ, Gupta A, Wood PJ, Afnan M, Stewart PM (1998) 11Beta-hydroxysteroid dehydrogenase type 2 in human pregnancy and reduced expression in intrauterine growth restriction. Hum Reprod 13:799–804

Shanks N, Larocque S, Meaney MJ (1995) Neonatal endotoxin exposure alters the development of the hypothalamic-pituitary-adrenal axis: early illness and later responsivity to stress. J Neurosci 15:376–384

Shanks N, Windle RJ, Perks PA, Harbuz MS, Jessop DS, Ingram CD, Lightman SL (2000) Early-life exposure to endotoxin alters hypothalamic-pituitaryadrenal function and predisposition to inflammation. Proc Natl Acad Sci USA 97:5645–5650

Shao J, Catalano PM, Yamashita H, Ruyter I, Smith S, Youngren J, Friedman JE (2000) Decreased insulin receptor tyrosine kinase activity and plasma cell membrane glycoprotein-1 overexpression in skeletal muscle from obese women with gestational diabetes mellitus (GDM): evidence for increased serine/threonine phosphorylation in pregnancy and GDM. Diabetes 49:603–610

Sigfrid LA, Cunningham JM, Beeharry N, Lortz S, Tiedge M, Lenzen S, Carlsson C, Green IC (2003) Cytokines and nitric oxide inhibit the enzyme activity of catalase but not its protein or mRNA expression in insulin-producing cells. J Mol Endocrinol 31:509–518

Sorenson RL, Brelje TC (1997) Adaptation of islets of Langerhans to pregnancy: beta-cell growth, enhanced insulin secretion and the role of lactogenic hormones. Horm Metab Res 29:301–307

Sterling P, Eyer J (1988) Allostasis: A new paradigm to explain arousal pathology. In: Fisher S, Reason J (eds) Handbook of life stress, cognition and health. Wiley, New York

Stumvoll M, Tataranni PA, Bogardus C (2004) The role of glucose allostasis in type 2 diabetes. Rev Endocr Metab Disord 5:99–103

Stumvoll M, Tataranni PA, Stefan N, Vozarova B, Bogardus C (2003) Glucose allostasis. Diabetes 52:903–909

Subbiah MT, Kessel B, Agrawal M, Rajan R, Abplanalp W, Rymaszewski Z (1993) Antioxidant potential of specific estrogens on lipid peroxidation. J Clin Endocrinol Metab 77:1095–1097

Sugioka K, Shimosegawa Y, Nakano M (1987) Estrogens as natural antioxidants of membrane phospholipid peroxidation. FEBS Lett 210:37–39

Tanaka Y, Tran PO, Harmon J, Robertson RP (2002) A role for glutathione peroxidase in protecting pancreatic beta cells against oxidative stress in a model of glucose toxicity. Proc Natl Acad Sci USA 99:12363–12368.

Teta M, Long SY, Wartschow LM, Rankin MM, Kushner JA (2005) Very slow turnover of beta-cells in aged adult mice. Diabetes 54:2557–2567

Tiedge M, Lortz S, Drinkgern J, Lenzen S (1997) Relation between antioxidant enzyme gene expression and antioxidative defense status of insulin-producing cells. Diabetes 46:1733–1742

Tonooka N, Oseid E, Zhou H, Harmon JS, Robertson RP (2007) Glutathione peroxidase protein expression and activity in human islets isolated for transplantation. Clin Transplant 21:767–772

Watve MG, Yajnik CS (2007) Evolutionary origins of insulin resistance: a behavioral switch hypothesis. BMC Evol Biol 7:61

Weinhaus AJ, Stout LE, Sorenson RL (1996) Glucokinase, hexokinase, glucose transporter 2, and glucose metabolism in islets during pregnancy and prolactin-treated islets in vitro: mechanisms for long term up-regulation of islets. Endocrinology 137:1640–1649

Weinhaus AJ, Bhagroo NV, Brelje TC, Sorenson RL (2000) Dexamethasone counteracts the effect of prolactin on islet function: implications for islet regulation in late pregnancy. Endocrinology 141:1384–1393

Weinstock M (1997) Does prenatal stress impair coping and regulation of hypothalamic-pituitary-adrenal axis? Neuroscience and Biobehavioral Reviews 21:1–10

Wiederkehr A, Wollheim CB (2006) Minireview: implication of mitochondria in insulin secretion and action. Endocrinology 147:2643–2649

Wiersma P, Selman C, Speakman JR, Verhulst S (2004) Birds sacrifice oxidative protection for reproduction. Proc Biol Sci 271:360–363

Windle RJ, Wood SA, Kershaw YM, Lightman SL, Ingram CD, Harbuz MS (2001) Increased corticosterone pulse frequency during adjuvant-induced arthritis and its relationship to alterations in stress responsiveness. J Neuroendocrinol 13:905–911

Windle RJ, Wood SA, Lightman SL, Ingram CD (1998a) The pulsatile characteristics of hypothalamo-pituitary-adrenal activity in female Lewis and Fischer 344 rats and its relationship to differential stress responses. Endocrinology 139:4044–4052

Windle RJ, Wood SA, Shanks N, Lightman SL, Ingram CD (1998b) Ultradian rhythm of basal corticosterone release in the female rat: dynamic interaction with the response to acute stress. Endocrinology 139:443–450

Yamashita H, Shao J, Friedman JE (2000) Physiologic and molecular alterations in carbohydrate metabolism during pregnancy and gestational diabetes mellitus. Clin Obstet Gynecol 43:87–98

Yamamoto M, Yamato E, Toyoda S, Tashiro F, Ikegami H, Yodoi J, Miyazaki J (2008) Transgenic expression of antioxidant protein thioredoxin in pancreatic beta cells prevents progression of type 2 diabetes mellitus. Antiox Redox Signal 10:43–49

Zhang H, Ollinger K, Brunk U (1995) Insulinoma cells in culture show pronounced sensitivity to alloxan-induced oxidative stress. Diabetologia 38:635–641

Chapter 7
The Importance of Transpositions and Recombination to Genome Instability According *hobo*-Element Distribution Pattern in Completely Sequenced Genome of *Drosophila melanogaster*

L.P. Zakharenko, M.P. Perepelkina, and D.A. Afonnikov

Abstract Transposable elements (TEs) constitute a considerable part of many eukaryotic genomes; therefore, the interest in the patterns of TE distribution in genomes is natural. We have analyzed the distribution of *hobo* transposon in the *Drosophila melanogaster* strain *y cn bw sp*, whose genome is almost completely sequenced. Based on the experimental and *in silico* data, we have grouped all *hobo* into the "old" and "new" sites and demonstrated that the old *hobo* sites are short variable regions predominantly located near the chromocenter, whereas the new sites are spread over the genome with the preference of chromosome central regions. As a rule, the defective *hobo* copies have lost the terminal repeats and the central part of DNA sequence, containing potential TATA boxes. Except for the complete *hobo*, only one defective *hobo* variant is able to transpose in this genome. The active (capable of transposing) *hobo* variants are excised, whereas the passive variants accumulate in the pericentromeric regions. The inactive deficient *hobo* sequences are accumulated in the chromosome regions with a decreased recombination frequency, whereas the active copies, moved mainly by transposases, do not follow this rule. Analysis of the direct repeat formed at the sites of *hobo* insertion and belonging to *hobo* sequences has demonstrated that some number of *hobo* copies can increase due to an unequal crossing-over. Recombinations are more frequently observed between the neighboring than between remote *hobo* elements, as defective *hobo* from the neighboring chromosome regions display a higher homology to each other than to the *hobo* elements located on different chromosomes. Such recombination can produce locus-specific instability as was shown before. Computer analysis of the number of hypothetical direct repeats remaining after the excision of *hobo* confirms the earlier published hypothesis that the last *hobo* invasion into *D. melanogaster* genome could occur nearly 100 years ago.

L.P. Zakharenko, M.P. Perepelkina and D.A Afonnikov
Institute of Cytology and Genetics of the Siberian Division of the Russian Academy of Sciences, pr. Lavrent'eva, 10, Novosibirsk-90, Russia, 630090
e-mail: zakharlp@bionet.nsc.ru

P. Pontarotti (ed.), *Evolutionary Biology: Concept, Modeling, and Application*,
DOI: 10.1007/978-3-642-00952-5_7,

Nevertheless, in spite of high *hobo* transposition and recombination frequency there is no one detected *hobo* dependent genetic event in the *D. melanogaster y cn bw sp* genome.

7.1 Introduction

Transposable elements (TEs) occupy a considerable part of many eukaryotic genomes; this is why they cannot but influence the genome functioning and evolution. Recent years have brought about new possibilities for analyzing the distribution patterns of mobile elements in the genome, as many complete or almost complete genomic sequences are now available. This allowed both the general patterns and interspecific distinctions to be detected; however, the made conclusions are sometimes contradictory.

The TE insertion into vitally important genes frequently interferes with their function; therefore, it is not surprising that the study of TE distribution in various eukaryotic genomes has demonstrated that the rate of mobile elements in the regions with a low gene density is higher, except for the human SINEs (Bergman et al. 2006; Duret et al. 2000; Medstrand et al. 2002; Wright et al. 2003). Nonetheless, a TE inserted into the gene coding region in certain cases can be fixed during the evolution and provide the appearance of a new function; this gave rise to the term exonization. TEs are rarer met in exons as compared with other genomic regions either because they are less frequently inserted into exons due to the specific conformational features of this gene region or because such inserts are subject to negative selection and not preserved. The TEs are inserted into genomic DNA by transposases, which are encoded by TEs themselves. The insertion sites of some TEs are conserved in either the nucleotide composition or, at least, the conformation of the region adjacent to insertion; however, such a pattern was undetectable for other TEs (Oshchepkov et al. 2002a). Usually, the total TE distribution pattern is analyzed; this pattern is the sum of the two oppositely directed processes—TE insertion and subsequent excision. The rate of TE detection in *Drosophila melanogaster* introns is equal to the rate expected considering a random TE distribution over the genome and is higher than the expected rate in the intergenic regions (Lipatov et al. 2005). Some papers report preferred insertion sites for some TEs; for example, *P*-element is most frequently found in the 5′-regulatory regions of *Drosophila* genes and *piggyback*, in introns (Thibault et al. 2004). According to Fontanillas et al. (2007), the rate of TE detection is higher in genes with a more active transcription, i.e., which are less compactly packaged.

The ability of TEs to recombine with one another via an unequal crossing-over at the sites of ectopic contacts also influences the number of TEs in the genome. The transposons of *C. elegans* display a negative correlation between the recombination frequency in various chromosome regions and the density of mobile elements, whereas the retrotransposons are evenly distributed over the genome (Duret et al. 2000). According to Rizzon et al. (2002), the density of mobile elements is higher in the *D. melanogaster* chromosome regions with decreased recombination frequency. However, this pattern was undetectable in some *D. melanogaster* strains (Hoogland and Biemont 1996). Thus, there is no unanimous opinion on the patterns of TE distribution.

We have analyzed the distribution of *hobo* in the *D. melanogaster* strain *y cn bw sp*, whose genomic sequence is almost completely known. Based on experimental (Zakharenko et al. 2007) and *in silico* (Fly Base) data, we grouped the *hobo* sites into "old" and "new" sites. The old *hobo* sites are short variable regions that lost their terminal repeats and the ability to transpose and are predominantly located near the chromocenter, whereas the new sites, which had changed their location, are distributed over the genome with a certain preference of the central chromosome regions. Thus, the old *hobo*, unable to actively transpose and destroyed possibly due to ectopic recombinations, accumulate in the chromosome regions with decreased recombination frequency, whereas the active copies, transposed mainly by transposases, are more frequently met in the central chromosome regions, where the recombination rate is the highest.

There is no one described *hobo* dependent genetic event in spite of its high transposition and recombination frequency in the *D. melanogaster y cn bw sp* genome.

7.2 Material and Methods

FlyBase data were used; the sequences were compared with the help of Blast Two Sequences; the sequences were searched for according to the specified coordinates using NCBI Mapview (Download Sequence Region *D. melanogaster* (fruit fly) (Build 5.1)). The samples of insertion site sequences aligned with respect to the target site were analyzed by the SITECON technique (Oshchepkov et al. 2002b).

7.3 Results and Discussion

7.3.1 The Hobo Sequences with Preserved Activity Are More Conserved than the Inactive Sequences

The distribution of *hobo* sequences was analyzed in the complete genome of the *D. melanogaster* strain *y cn bw sp*. Overall, 59 *hobo* sequences have been annotated in this strain (flybase.bio.indiana.edu). One of the copies is complete and able to encode an active transposase, whereas the rest of the copies are defective. According to the FlyBase data, these sequences were annotated in two stages: the main part of the *hobo* sequences annotated in 2003 (24 sequences) retained their terminal inverted repeats, had a long size and a larger degree of homology to the complete *hobo* copy as compared with the *hobo* sequences that were added to the database in 2005. Characteristic of the sequences annotated in 2005 is a high heterogeneity; they are represented by short fragments from different parts of the complete *hobo* sequence located at different distances from one another. For example, H{ }84, a complete *hobo* copy with a length of 2957 bp and H{ }236, with a length of 6626 bp, display only a partial similarity at the sequence ends with a total length of slightly longer that 1 kbp, as if a sequence was inserted into the central *hobo* part and part of the *hobo* sequence was lost.

Defective *hobo* elements are able to transpose, if they preserve the terminal inverted repeats. The genome of strain *y cn bw sp* contains 20 such sequences; all these sequences are 1,406 bp long and highly homologous to one another and the complete *hobo* variant. This fits the concept that the more active TEs are more conserved than the passive TEs. There is an analogy with the linguistic concepts on a considerable similarity between the evolutions of genes and language, namely, that the frequently used words evolve slower than the rarely used words (Whitfield 2008).

The complete TE variants are usually present in the genome in a smaller copy number as compared with their defective variants, presumably, because the genomes with actively transcribing TEs are more unstable. The central *hobo* part (1,643 nucleotides) differs from the flanks (1,435 nucleotides) preserved in the defective sequence with a length of 1,406 bp by a higher content of TATA (11 versus 4) and especially ATAT (18 versus 1) sequences, although polyA or polyT ($N > 2$) sequences are homogeneously spread over *hobo* DNA (Fig. 7.1). On the one hand, it is possible that a larger content of TATA and ATAT sequences in the central *hobo* region, part of which can function as a full-fledged TATA box, keeps the genes with an inserted *hobo* sequence (both in a direct and inverse orientation) in a switched-on state. For example, a *hobo* insertion in the regulatory region of *D. melanogaster* gene *yellow* can be indifferent for its function (Zakharenko et al. 2004) as well as a *hobo* insertion in the exon of *decapentaplegic* gene (Blackman et al. 1987) or in intron of *Notch* gene (Eggleston et al. 1996). On the other hand,

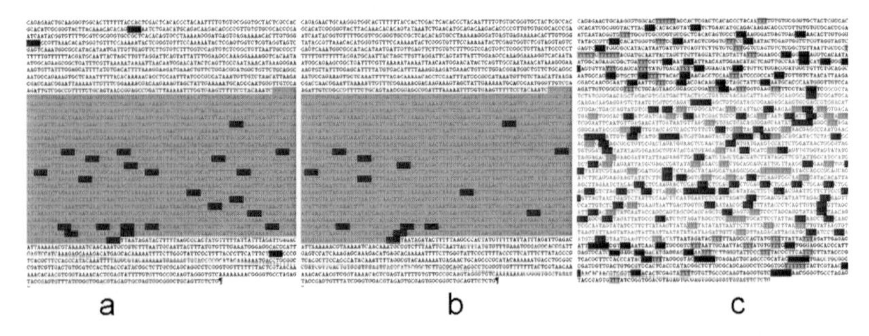

Fig. 7.1 The distribution of TATA (**a**), ATAT (**b**), polyA and polyT (**c**) sequences in the full-sized *hobo*. The deleted central part of the most prevalent defective *hobo*-derivative (**a**, **b**) and polyT (**c**) are marked by light gray. TATA, ATAT and polyA are marked by dark gray

the presence of TATA boxes can influence the activity regulation of the genes neighboring *hobo*. It has been demonstrated for the noncoding human and mouse SINE sequences (*Alu* and *B1*, respectively) that their transcripts are involved in the repression of host genes during heat shock, although this role was traditionally ascribed to proteins (Mariner et al. 2008). Considering the short RNAs as regulators of gene activity, it cannot be excluded that the transcripts of defective *hobo* sequences can also have a certain effect on the function of other genes.

7.3.2 The Hobo Elements from Different Chromosomes Display Less Similarity than the Neighboring Hobos

It seems that *hobo* elements are more frequently transposed or duplicated within the same chromosome rather than between different chromosomes. The distribution pattern of similar defective *hobo* sequences over chromosomes favors the assumption that *hobo*s more frequently transpose along the chromosome. As a rule, long sequences alternate with short sequences; however, several clustered identical defective *hobo* sequences were detected in the 2L and 2R chromosomes (Fig. 7.2). On 2L, these are defective sequences with a length of 1,406 bp with preserved ability to transpose; on 2R, they are shorter (800–1,000 bp) and have lost the ability to transpose.

The short defective *hobo* sequences with a length of 800–1,000 bp from different chromosomes are less similar than the sequences from the same chromosome. Similar *hobo* sequences are frequently located in the neighboring bands. For example, the neighboring *hobo*s from the regions 41B1 H{}3284 and 41B2 H{} 3285 as well as 42A14 H{}1788 and 42A15 H{}1790 of the 2R chromosome are similar. As these *hobo* sequences lack terminal repeats, necessary for transposase to recognize a *hobo* element, these copies are likely to result from duplication of *hobo* sequences rather than their transposition. The more so as not only the *hobo*

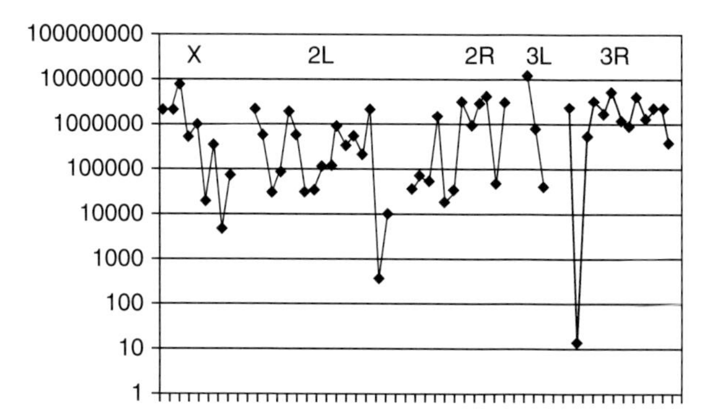

Fig. 7.2 The distribution of *hobo* according to its length (*in silico* data) on the *y cn bw sp* chromosomes

fragment but also the adjacent regions are duplicated. At least 1,000 nucleotides adjacent to the defective *hobo* in the region 41B1 display a 98% homology to the sequences neighboring the *hobo* from the region 41B2. The 500 nucleotides upstream and 100 nucleotides downstream of the *hobo* from 42A14 are also identical to the corresponding sequences adjacent to the *hobo* from 42A15.

The direct repeats at the insertion sites of the *hobo* elements that preserved their activity have been detected in 17 of 20 cases. 35D3 H{ }380 contains the chimeric insertion site composed of the insertion sites of the neighboring *hobo*s. The *hobo* elements 35D3 H{ }377 and 35D4 H{ }382 contain the direct repeats with a length of 8 bp (GTATACAC and ATCCATTT, respectively) at the insertion sites. The element 35D3 H{ }380, located between them, contains to the left the sequence ATCCATTT, which coincides with the direct repeat flanking H{ }382 and to the right, GTATACAC, coinciding with the repeat of H{ }377. The chimeric insertion site of 35D3 H{ }380 could result from an unequal crossing-over between two neighboring *hobo*s (Fig. 7.3), especially because the sequences between 35D3 H{ }377 and 35D3 H{ }380 with a length of about 32 kbp are highly homologous to the sequences of the same length located between 35D3 H{ }380 and 35D4 H{ } 382. The recombination could take place at the sites of ectopic contacts between neighboring *hobo*s. Our observations coincide with the data of Innan (2008), which demonstrate that the recombination is more frequently observed between the repeats located in the neighborhood than between the distant repeats. The

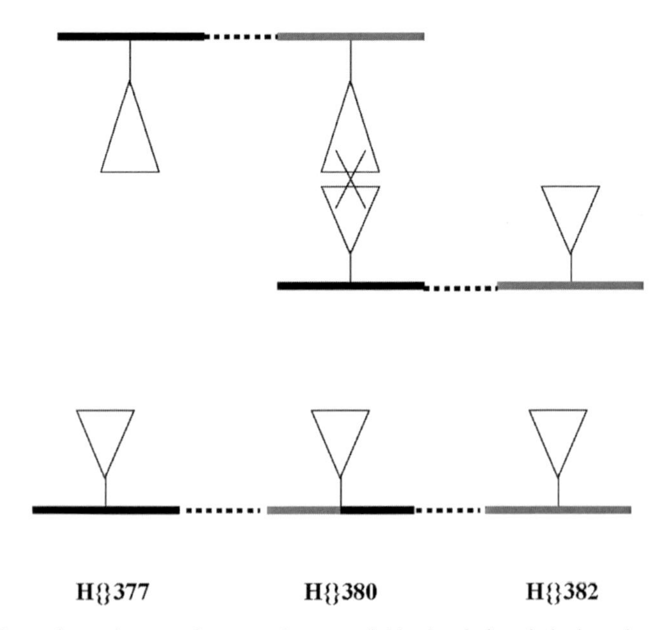

Fig. 7.3 Unequal crossing-over between the two neighboring *hobos*. *hobo* insertions – triangles; insertion sites (direct repeats) – solid lines, sequences between *hobos* – broken lines. The similar sequences are colored similarly

complete *hobo* variant does not contain a direct repeat in the insertion site, although the pattern nTnnnnAn at both sides of the insertion site is preserved (CTATAAAC to the left of *hobo* and ATTGCCAT to the right). Such chimeric variant at the insertion site can appear as a result of the recombination between two unidirectional *hobo* sequences with a subsequent deletion of the DNA sequence located between them.

Thus, at least in three cases, the number of *hobo* elements in the genome of *D. melanogaster* strain *y cn bw sp* increased due to *hobo* duplications. In one case, it is possible with a certain probability to assume deletion of an active *hobo*. The remaining insertions of active *hobo*s displayed no similarity at their insertion sites. This suggests that both the insertion and elimination of active *hobo*s with a length of 1,406 bp most frequently occur without recombinase with the help of transposase. Most likely, old *hobo*s are eliminated mainly due to recombinations. In this process, not only *hobo*s but also the adjacent repeats can recombine. The involvement of recombinases in the destruction of *hobo* elements is confirmed by the data on locus-specific instability of the *yellow* gene from the Uman' population (Zakharenko et al. 2000, 2004). In this study, the authors used Southern blot hybridization to demonstrate that the number and size of the genomic fragments containing chimeric *hobo* elements (*hobo* + a fragment of *yellow* gene regulatory region) decrease with time. This is a very rapid process. The two tens of derivatives of one of the *hobo*-determined y2 strains from the Uman' population obtained during one year contained no strains displaying the identical *yellow* gene structure with *hobo* inserted in the regulatory region. A locus-specific instability in the *yellow* gene in this strain was determined by *hobo* duplications (including chimeric variants), inversions–reinversions of the sequences located between *hobo*s and recombinations between them.

7.3.3 Analysis of the Number of Hypothetical Hobo Insertion Sites Confirms the Suggestion on a Recent Invasion of Hobo in D. melanogaster

The human and mouse TEs that became integral elements of the genomes are present at the same frequencies on both DNA strands; however, the most active TE, *Alu* repeat, is most frequently met in the antisense strand (Sela et al. 2007). The *hobo* sequences in *D. melanogaster* strain *y cn bw sp* are approximately equally distributed in the plus and minus strands: 30 *hobo* sequences in the plus strand (totally, 28,272 nucleotide pairs) and 29 sequences in the minus strand (totally, 28,753 nucleotide pairs). The region of *hobo* insertion is enriched with AAA and TTT sequences. The content of A and T amounts to 61.3 $\pm$ 9.3% of the DNA located at a distance of 20 bp from the insertion site, 60.0 $\pm$ 2.7% of the DNA located at 1 kbp and 57.1 $\pm$ 2.1% of the DNA at 100 kbp. A direct repeat with a length of 8 bp is formed at the *hobo* insertion site (Saville et al. 1999). No consensus was detected at the *hobo* insertion site, except for T at position 2 relative to the insertion site in 19 of the 20 sequences and A at position 7 in 15 of the 20 sequences.

As the genome sequence of strain *y cn bw sp* is almost completely known and as a direct repeat with a characteristic structure remains after *hobo* excision, we attempted to estimate the number of *hobo* insertion sites with the structure nTnnnnAnnTnnnnAn in the studied genome. The overwhelming majority of the sites with the structure in question were in the AT-rich sequences. On average, the analyzed repeat is met in every 8,000 bp. However, none of the real *hobo* insertion sites consists in only A and T nucleotides. The minimal number of G or C per *hobo* insertion site is two (or four per overall direct repeat). If we consider only the sites that contain less than four G or C per overall repeat (25% GC of all the nucleotides of the repeat), their number will be about 10,000 per genome. However, the average ratio of AT to GC in the insertion sites is close to 50%, i.e., the actual number of insertion sites will be significantly lower. For eight G or C nucleotides of 16 with the structure nTnnnnAnnTnnnnAn, this constitutes 424 potential *hobo* insertion sites. The minimal *hobo* transposition frequency in this strain is 10^{-2} per site per genome per generation, i.e., one transposition per two generations occurs in the case of about 50 *hobo* sites in the genome. Assuming that the G and C content in the repeat is 50% or 8 of 16 nucleotides, the maximal number of generations since the moment of *hobo* insertion into the genome of the studied strain is 424/0.5 = 848 generations or 84 years taking into account that 10 *Drosophila* generations pass each year. This estimate matches in the order of magnitude the assumed time of the last *hobo* invasion into the genome of *D. melanogaster* (mid-last century) obtained by analyzing the strains isolated at different times from natural populations (Bonnivard et al. 2000).

Despite that the *hobo* insertion sites are nonrandom, the used analytical algorithms failed to find common characteristics. An original method SITECON, involving detection of conserved DNA properties in samples of aligned sequences, was used to analyze the *hobo* insertion sites (Oshchepkov et al. 2002a).

7.3.4 The New Hobo Sequences Are Evenly Distributed in the Genome, Whereas the Old Hobo Sequences Tend for Pericentromeric Regions

We have earlier analyzed the transposition rate of the *hobo* element in the strain *y cn bw sp* by FISH (Zakharenko et al. 2007). The sensitivity of FISH is insufficient to detect short heterogeneous sequences; therefore, we detected only the long sequences with a high degree of homology to the complete *hobo* sequence used as a probe. According to the FISH data, the number of *hobo* sequences doubled since the time after sequencing the strain *y cn bw sp* (Fig. 7.4). The excisions of *hobo* elements were rarer than their insertions (Fig. 7.4). All the deleted *hobo* sequences (regions 19C, 19E5, 36C2, 36D2, 36E2, 45E1, 95E1, 99B9 and 99D5)

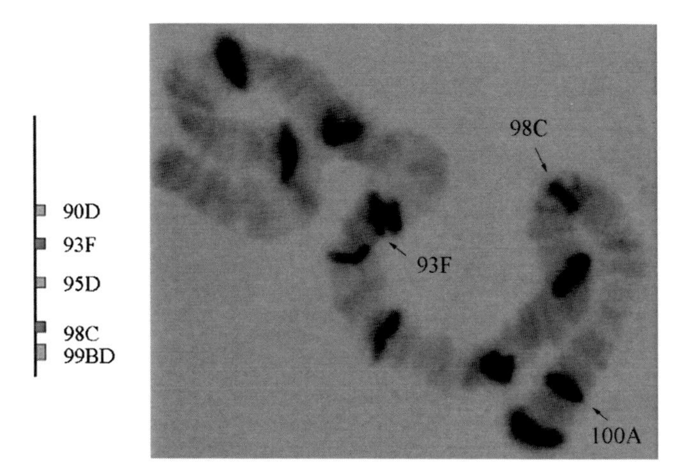

Fig. 7.4 The distribution of stable *mdg1* (93F and 98C) and unstable *hobo* (all other sites) on the larval salivary gland polytene chromosomes of *Drosophila melanogaster* strain *y cn bw sp* according to FISH (*right*) in comparison with *in silico* data (*left*). FISH was performed 15 years after *D. melanogaster* genome sequencing

were 1,406-bp long except for the sequence localized to the region 80D4 (1150 bp). Possibly, this short sequence did not disappear from the genome but only failed to hybridize with the *hobo* element due to its low homology with this element. The sequence with a length of 1,406 bp is the most widespread type of defective *hobo* sequences in the strain in question detectable with FISH and retaining the terminal inverted repeats. Nine of the 20 *hobo* sites with a length of 1,406 bp are localized to genes (in all cases, the insert is in the intron); two, to mobile elements; and the remaining cases, to the sequences with unknown function or intergenic space (FlyBase). This strain has no *hobo* insertions into one another, although *hobo* insertions were detected in other mobile elements (roo and Quasimodo). We found no tandemly arranged *hobo* sequences located immediately near one another. The minimal distance between the neighboring *hobo*s was 15 bp and the maximal, 13,100,981 bp (Fig. 7.5). During 15 years after sequencing, 10 of the 24 *hobo* sites annotated in 2003 were lost; five of them were located in introns.

According to *in silico* data, the *hobo* sequences are nonuniformly distributed over the chromosomes. The distribution density of *hobo* increased with the decrease in the distance to the chromocenter mainly due to short defective *hobo* sequences with a low homology to the complete copy. Two thirds of all *hobo* sequences are localized to the proximal third of the chromosomes adjacent to the chromocenter. The *hobo* sequences are least abundant in the distal third of the chromosomes adjacent to telomeres (Table 7.1) except for the 3R chromosome, where *hobo* elements are distributed relatively uniformly. However, the *hobo* sites that appeared during the time since the genome sequencing were mainly located, according to FISH data, in the central chromosome regions and their distal parts, whereas the disappearance of *hobo* was observed mainly in the chromosome regions adjacent to

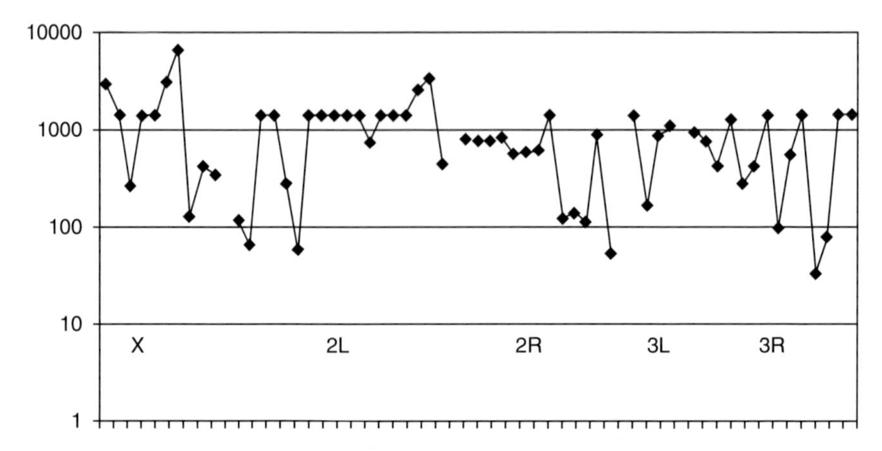

Fig. 7.5 The distance between the neighboring *hobos* according to *in silico* data in the genome *Drosophila melanogaster* strain of *y cn bw sp*

Table 7.1 *hobo* sites number (%) according to *in silico* and FISH data in the different thirds of the chromosomes of *Drosophila melanogaster* strain *y cn bw sp*

	Number of the *hobo* sites (%)		
Type of experiments	Telomere adjacent parts	Central parts	Chromocenter adjacent parts
In silico	10(17)	11 (19)	38 (64)
FISH	12(35)	18(53)	4(12)

the chromocenter (Fig. 7.4). According to the Fisher criterion, the significance of nonuniform *hobo* distribution is 0.001 in the case of *in silico* data and 0.05 in the case of FISH data. Similar results were obtained by Yang et al. (2006) (Yang and Barbash 2008); they demonstrated that the younger sequences of *DINE1* nonautonomous mobile element are more uniformly distributed over the *Drosophila* genome as compared with the older sequences, which tend for the region adjacent to the chromocenter. The most active human TE, *Alu* repeat, is also more uniformly distributed over the genome than the less active TEs (Sela et al. 2007). Presumably, there is a common pattern in the distribution of new and old TEs. Then we can withdraw the contradiction between the data of Rizzon et al. (2002) and the data of Hoogland and Biemont (1996) obtained for *D. melanogaster*. In the first case, the authors analyzed *in silico* mainly the defective TEs, as they are more numerous than the complete copies. In the second case, the data were obtained by *in situ* hybridization, which detected only highly homologous relatively long TE sequences that preserved the similarity to the complete copy.

Thus, the distribution of TEs depends on the situation that has formed in an analyzed strain. If TEs are unable to transpose, they gradually diverge and eliminate, presumably, via recombination. If TEs are able to transpose, their number does not grow infinitely, as the increase in the copy number is accompanied by their excision by both transposase and recombination. The genomes with a large copy number can also be subject to a negative selection. The old defective sequences possibly destroyed due to ectopic recombination are accumulated in the chromosome regions with a decreased recombination frequency, whereas the active copies, moved mainly by transposases, do not follow this rule. Nevertheless, in spite of high *hobo* transposition and recombination frequency there is not one detected *hobo* dependent genetic event in the *D. melanogaster y cn bw sp* genome.

Acknowledgements This work was partially funded by "Biodiversity and Gene Pool Dynamics" grant from Russian Academy of Sciences. 23.29 and grant RFBR #09-04-00213a

References

Bergman CM, Quesneville H, Anxolabehere D, Ashburner M (2006) Recurrent insertion and duplication generate networks of transposable element sequences in the *Drosophila melanogaster* genome. Genome Biol 7:R112; doi:10.1186/gb-2006-7-11-r112

Blackman RK, Grimaila R, Koehler MM, Gelbart WM (1987) Mobilization of *hobo* elements residing within the *decapentaplegic* gene complex: suggestion of a new hybrid dysgenesis system. Cell 49:497–505

Bonnivard E, Bazin C, Denis B, Higuet D (2000) A scenario for the *hobo* transposable element invasion, deduced from the structure of natural populations of *Drosophila melanogaster* using tandem TPE repeats. Genet Res 75:13–23

Duret L, Marais G, Biemont C (2000) Transposons but not retrotransposons are located preferentially in regions of high recombination rate in *Caenorhabditis elegans*. Genetics 156:1661–1669

Eggleston WB, Rim NR, Lim JK (1996) Molecular characterization of *hobo*-mediated inversions in *Drosophila melanogaster*. Genetics 144:647–656

Hoogland C, Biermont C (1996) Chromosomal distribution of transposable elements in *Drosophila melanogaster*: test of the ectopic recombination model for maintenance of insertion site number. Genetics 144:197–204

Fontanillas P, Hartl DL, Reuter M (2007 November) Genome organization and gene expression shape the transposable element distribution in the *Drosophila melanogaster* euchromatin. PLoS Genet. 3(11):e210; doi:10.1371/journal.pgen.0030210

Innan H (2008) Coevolution of duplicated genes. The Sixth International Conference on Bioinformatics of Genome Regulation and Structure. Thesis, p100

Lipatov M, Lenkov K, Petrov DA, Bergman CM (2005) Paucity of chimeric gene-transposable element transcripts in the *Drosophila melanogaster* genome. BMC Biol 3:24; doi:10.1186/1741-7007-3-24

Mariner PD, Walters RD, Espinoza CA, Drullinger LF, Wagner SD, Kugel JF, Goodrich JA (2008) Human *Alu* RNA is a modular transacting repressor of mRNA transcription during heat shock. Mol Cell 29:499–509

Medstrand P, van de Lagemaat LN, Mager DL (2002) Retroelement distributions in the human genome: variations associated with age and proximity to genes. Genome Res 12:1483–1495

Oshchepkov DYu, Furman DP, Katokhin AV, Katokhina LV (2002a) Detection of conservative conformational properties of insertion sites for Drosophila. Proceedings of the third international conference of bioinformatics of genome regulation and structure. Novosibirsk, Russia, Vol. 1. 153–156

Oshchepkov DYu, Turnaev II, Vityaev EE (2002b) SITECON: a method for recognizing transcription factor binding sites basing on analysis of their conservative physicochemical and conformational properties. Proceedings of the third international conference of bioinformatics of genome regulation and structure, vol 1. Novosibirsk, Russia, pp 42–44

Rizzon C, Marais G, Gouy M, Biémont C (2002) Recombination rate and the distribution of transposable elements in the *Drosophila melanogaster* genome. Genome Res 12:400–407

Saville KJ, Warren WD, Atkinson PW, O'Brochta DA (1999) Integration specificity of the *hobo* element of *Drosophila melanogaster* is dependent on sequences flanking the integration site. Genetica 105:133–147

Sela N, Mersch B, Gal-Mark N, Lev-Maor G, Hotz-Wagenblatt A, Ast G (2007) Comparative analysis of transposed element insertion within human and mouse genomes reveals *Alu*'s unique role in shaping the human transcriptome. Genome Biol 8:R127; doi:10.1186/gb-2007-8-6-r127

Thibault ST, Singer MA, Miyazaki WY, Milash B, Dompe NA, Singh CM, Buchholz R, Demsky M, Fawcett R, Francis-Lang HL, Ryner L, Cheung LM, Chong A, Erickson C, Fisher WW, Greer K, Hartouni SR, Howie E, Jakkula L, Joo D, Killpack K, Laufer A, Mazzotta J, Smith RD, Stevens LM, Stuber C, Tan LR, Ventura R, Woo A, Zakrajsek I, Zhao L, Chen F, Swimmer C, Kopczynski C, Duyk G, Winberg ML, Margolis J (2004) A complementary transposon tool kit for *Drosophila melanogaster* using *P* and *piggyBac*. Nat Genet 36:283–287

Whitfield J (2008) Across the curious parallel of language and species evolution. PLoS Biol 6:e186; doi:10.1371/journal.pbio.0060186

Wright SI, Agrawal N, Bureau TE (2003) Effects of recombination rate and gene density on transposable element distributions in *Arabidopsis thaliana. Genome Res* 13:1897–1903

Yang HP, Barbash DA (2008) Abundant and species-specific *DINE-1* transposable elements in 12 *Drosophila* genomes. Genome Biol 9:R39; doi:10.1186/gb-2008-9-2-r39

Yang HP, Hung TL, You TL, Yang TH (2006) Genomewide comparative analysis of the highly abundant transposable element *DINE-1* suggests a recent transpositional burst in *Drosophila yakuba*. Genetics 173(1):189–196

Zakharenko LP, Zakharov IK, Romanova OA, Voloshina MA, Gracheva EM, Kochieva EZ, Golubovskiĭ MD, Georgiev PG (2000) "Mode for mutation" in the natural population of *Drosophila melanogaster* from Uman is caused by distribution of a *hobo*-induced inversion in the regulatory region of the *yellow* gene. Genetika 36:740–748 (Russian)

Zakharenko LP, Zakharov IK, Voloshina MA, Romanova OM, Alekseenko AA, Georgiev PG (2004) Cause for maintaining high instability at the *yellow* gene in *Drosophila melanogaster* lines, isolated during the "mode for mutation" period in Uman populations. Genetika 40:316–321 (Russian)

Zakharenko LP, Kovalenko LV, Mai S (2007) Fluorescence *in situ* hybridization analysis of *hobo, mdg1* and *Dm412* transposable elements reveals genomic instability following the *Drosophila melanogaster* genome sequencing. Heredity 99:525–530

Chapter 8
Long-Term Evolution of Histone Families: Old Notions and New Insights into Their Mechanisms of Diversification Across Eukaryotes

José M. Eirín-López, Rodrigo González-Romero, Deanna Dryhurst, Josefina Méndez, and Juan Ausió

Abstract In eukaryotes and some archaebacteria, DNA is found associated with histones in a nucleoprotein complex called chromatin, which allows for a high extent of compaction of genomic DNA within the limited space of the nucleus. Early studies led to the notion that histones exhibit a conserved structural gene organization and limited protein diversity. However, research has been progressively accumulating to demonstrate that the structure, configuration and copy number of histone genes varies widely across organisms as a result of a long-term evolutionary process that promotes genetic variation. This genetic diversity is mirrored by the structural and functional diversity exhibited by the protein members of the different histone families that is, in most instances, concomitant with the complexity of the organism. The present chapter is aimed at providing a comprehensive review of the most recent information on the origin of eukaryotic histone multigene families. Particular attention is paid to the structural and functional constraints acting on histones and their relevance for the progressive diversification of histone variants during evolution, especially as it pertains to histone gene organization and expression.

R. González-Romero and J. Méndez
Departamento de Biología Celular y Molecular, Universidade da Coruña, E15071, A Coruña, Spain
D. Dryhurst and J.Ausió
Department of Biochemistry and Microbiology, University of Victoria, V8W 3P6 Victoria, Canada
J.M. Eirín-López
Departamento de Biología Celular y Molecular, Facultade de ciencias, Universidade da Coruña, Campus de A Zapateria s/n, E15071 A Coruña, Spain
e-mail: jeirin@udc.es

P. Pontarotti (ed.), *Evolutionary Biology: Concept, Modeling, and Application*,
DOI: 10.1007/978-3-642-00952-5_8,

8.1 Introduction

In eukaryotes and some archaebacteria, DNA is found associated with histones in a nucleoprotein complex called chromatin. Chromatin allows for a high extent of compaction of genomic DNA within the limited space of the nucleus and also provides the support on which most DNA metabolic functions (i.e., replication, transcription and repair) take place. The repetitive subunit of chromatin, the nucleosome, consists of an octamer of core histones (two of each H2A, H2B, H3 and H4) around which two left handed superhelical turns of DNA are wrapped (van Holde 1988). The nucleosome core particles (NCPs) are joined together in the chromatin fiber by short stretches of linker DNA that, interact with linker H1 histones, resulting in an additional folding of the chromatin fiber. Although all the domains in eukaryotic chromatin share a common nucleosomal structure, the dynamic processes responsible for the local heterogeneity observed across the genome are regulated by three principal mechanisms: the replacement of canonical histones with specialized variants that have dedicated functions (Malik and Henikoff 2003), the occurrence of histone posttranslational modifications (Jenuwein and Allis 2001) and the association with remodeling complexes responsible for nucleosome mobilization (Owen-Hughes 2003). The wide range of possible configurations that facilitate different chromatin metabolic processes are the result of the synergistic action of the aforementioned mechanisms.

Early studies on histone genes led to the formulation of several general dogmas about their structure: (a) histone genes were considered to be intronless, encoding highly conserved proteins; (b) they existed in multiple copies closely clustered in the genome and organized into tandemly repeated blocks; (c) except for the case of relatively few replacement histone variants, the mRNAs transcribed from the genes of the canonical histones were not polyadenylated and their expression was largely coupled to S phase of the cell cycle; and (d) the evolution of the histone gene families was thought to be subject to a process of concerted evolution through rapid interlocus recombination or gene conversion (Hentschel and Birnstiel 1981; Kedes 1979; Maxson et al. 1983a; Ohta 1983). However, research has been progressively accumulating to demonstrate that the organization and copy number of eukaryotic histone genes varies widely across organisms. Additional evidence that argues against the assumptions made by these dogmas was found in the case of specialized histone variants. These proteins are encoded by intron-containing genes and are expressed constantly at basal levels throughout the cell cycle (Ausió 2006). These diverse patterns of organization and expression differ even among closely related organisms and most likely reflect the presence of different regulatory mechanisms and a complex mode of gene family evolution (Doenecke et al. 1997; Khochbin 2001; Wells and Kedes 1985). The high degree of structural and functional diversification observed across different histone types provides a compelling argument against the notion of concerted evolution (homogenization) as the major force guiding long-term histone evolution.

The present chapter is aimed at providing a comprehensive review of the most recent information on the origin of eukaryotic histone multigene families and the subsequent mechanisms guiding their long-term evolution. Special attention is paid to the structural and functional constraints acting on histone proteins and their relevance to the progressive diversification of histone variants during evolution, especially as they pertain to histone gene organization and expression. Finally, a model summarizing the process of histone diversification and differentiation from their archael origin to the wide diversity of specialized variants in eukaryotes is presented.

8.2 Histone Genes Display Highly Heterogeneous Organization Patterns Across Eukaryotic Genomes

Histone gene organization has been studied for more than 30 years (Hentschel and Birnstiel 1981; Isenberg 1979; Kedes 1979; Maxson et al. 1983a) and during most of this early period, the concept of a homogeneous gene organization was believed to be the normal arrangement in the genomes of most model organisms studied at the time, especially *Drosophila* and sea urchin (Lifton et al. 1977; Maxson et al. 1983b). However, with the progressive characterization of histone genes in a broader range of organisms, a more complex picture started to emerge showing a dispersal and diversification of the genes that was apparently concomitant with the position of the different organisms on the phylogenetic tree (Doenecke et al. 1997). A great diversity of histone gene organization patterns was observed, however, it was not until histone variants were first identified and their specific functions progressively deciphered that such an extensive heterogeneity was definitively determined to be an intrinsic feature of the different histone families. In sharp contrast to canonical histones, the histone variants have a unique genomic organization (solitary genes), copy number (single-copy), gene structure (presence of introns) and regulation (basal continuous expression throughout the cell cycle) (Ausió 2006). Clearly there is no doubt that the diversification process experienced by histone genes has allowed for the progressive differentiation of histone variants with dedicated functions and this has shaped the complex, efficient and tightly controlled mechanisms of DNA packaging and regulation of chromatin dynamics in the cell nucleus. In order to illustrate the relevance of such diversity, we will next summarize the major modes of histone gene organization across representative groups of eukaryotes.

8.2.1 Prokaryotic Chromatin and the Origin of Histones

Most studies on prokaryotic chromatin have initially focused on the eubacteria *Escherichia coli*, which has a 4.6 Mb circular chromosome organized into supercoiled domains (Worcel and Burgi 1972). The HU proteins are the most abundant proteins in the nucleoid of this organism (Murphy and Zimmerman 1997) and DNA compaction is achieved as a result of the exogenous pressure of macromolecular crowding from the nucleoplasm and from supercoiling introduced by architectural proteins and topoisomerase activity (Sandman et al. 1998). The absence of a nucleosome-based organization of bacterial chromatin contrasts with that observed in archaebateria, which contain histones as well as other chromatin associated proteins (Grayling et al. 1996; Sandman and Reeve 2005). The phylum *Euryarchaea* (Fig. 8.1) is characteristic among *Archaea* in having histone-like proteins that are relatively small in size, do not have N- or C-terminal unstructured tails comprising only a histone-fold domain (Arents and Moudrianakis 1995) and appear to form analogous or perhaps homologous structures to those formed by the eukaryotic $(H3 + H4)_2$ histone tetramers (Pereira et al. 1997). These proteins bind and wrap DNA into nucleosomal-like structures protecting about 60 bp of DNA and can induce DNA supercoiling (Sandman et al. 1998). Histone-encoding genes have also been identified in marine *Crenarchaea*, giving strong support for the argument that histones evolved very early, after the divergence of *Bacteria-Archaea* but before the separation of the archaeal and eukaryotic lineages (Cubonova et al. 2005). The ancestry shared by archaeal and eukaryotic core histones is indeed

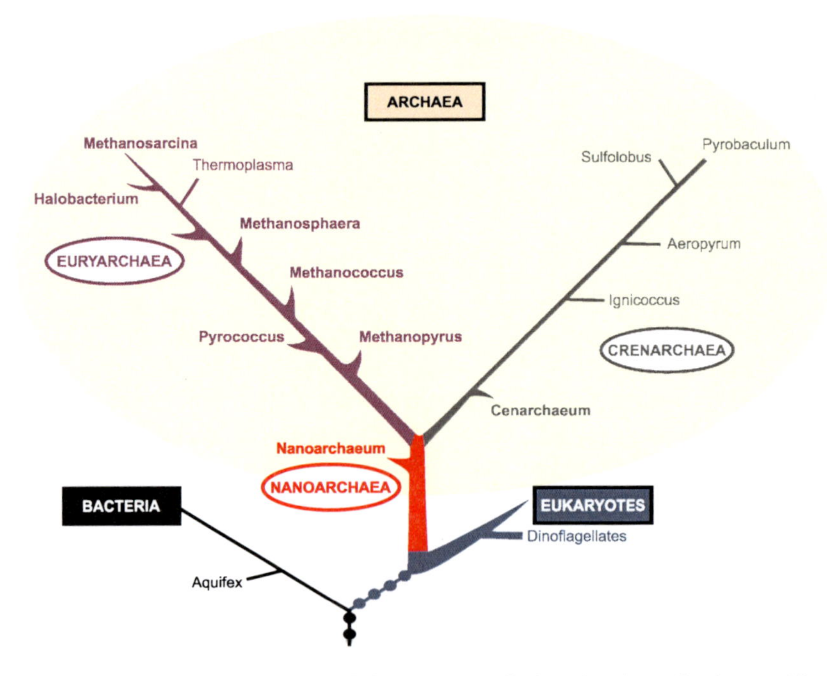

Fig. 8.1 Universal tree of life based on rRNA sequences of selected Archaea (Sandman and Reeve 2006). Different domains as well as archaeal phyla are indicated in the diagram. Thick fancy lines account for those lineages in which histone-fold sequences have been identified while dashed lines indicate uncertainty regarding the time of histone-fold origin

manifested in their conserved histone-fold region as well as in the amino acid residues that are important for histone–DNA and histone dimer–dimer interactions. Furthermore, archaeal histone sequences exhibit a degree of variation that provides clues to the molecular requirements of the evolutionary transition from prokaryotic to eukaryotic chromatin (Sandman and Reeve 2006; Zlatanova 1997).

8.2.2 The Transition Toward the Eukaryotic Cell and the Appearance of Pluricellularity in Light of Histone Diversification

Histones have been extensively characterized in eukaryotic genomes where, with the only exception of some protozoa such as dinoflagellates (Herzog and Soyer 1981), they organize chromatin into a repetitive nucleosome structure. Among lower eukaryotes, yeast is unique in its small number of histone genes, having only two gene copies of each of the four core histones. The H2A and H2B genes are adjacent to one another, they are divergently transcribed and they exist as two genetically unlinked copies encoding two structural variants. This is in contrast to the H3 and H4 genes that encode identical products (van Holde 1988). There is also a candidate yeast linker histone (Hho1p) encoded by a gene that is co-expressed with the core histone genes in S phase (Landsman 1996). This protein has an unusual tertiary structure with two regions (GI and GII) homologous to the single globular domain of the linker histone of higher eukaryotes (Ali and Thomas 2004). The organization of histone genes is also known in other lower eukaryotes such as the ciliates *Stylonychia lemnae*, *Tetrahymena thermophila* and closely related species, where the histone genes are unclustered in the genome and the H1 genes are apparently absent in the micronucleus (Allis et al. 1979; Prescott 1994). Also, single-copy H1 genes independently organized from core histone genes were described in *Volvox* (Lindauer et al. 1993).

Despite the classical notion of histone genes being clustered and tandemly repeated, many arrangements are observed in metazoan genomes as indicated in Fig. 8.2. For instance, in the cnidarian *Acropora formosa* (coral) histone genes consist in clusters of the four core histones (quartets) that are reiterated about 150 times (Miller et al. 1993). A similar organization has also been described in the nematode *Caenorhabditis elegans* that has independently organized single-copy H1 genes (Sanicola et al. 1990). Similarly, the histone genes of annelids are organized into clusters containing H1 and the four core histone genes (quintets) but quartets of core histones that are reiterated about 600–650 times in the genome are also present (del Gaudio et al. 1998; Sellos et al. 1990). However, molluscs represent the paradigm of diversity among metazoans showing up to three different arrangements of histone genes even within a single species. This is the case with the mussel *Mytilus galloprovincialis* (Eirín-López et al. 2004b), which displays clustered quintets linked to 5S rRNA genes, quartets of core histones and independent clusters containing replication independent H1 genes

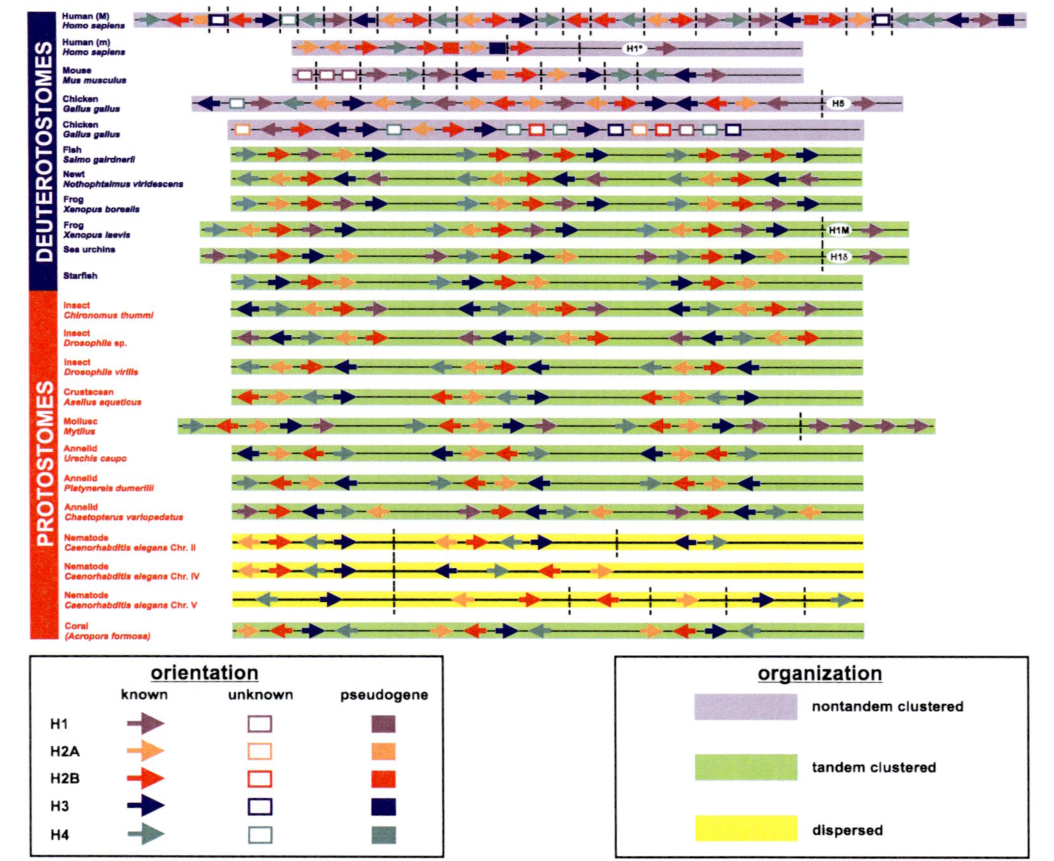

Fig. 8.2 Genomic organization of genes encoding histones across different metazoan lineages highlighting the many possible arrangements in nontandemly clustered, dispersed and tandemly clustered repeats. Directions of transcription are indicated when known, as well as the presence of pseudogenes for the five histone families and some of the replication-independent H1 variants. Both the major (M) and minor (m) human histone clusters are represented, located on chromosomes 6 and 1, respectively. Although two clusters also exist in mice, only a portion of the major cluster is shown. Two histone clusters are also present in the chicken genome but it is not certain whether these are on the same chromosome or not

(Drabent et al. 1999; Eirín-López et al. 2002). In insects, the clustered quintets containing H1 genes are present in *Drosophila* (Kremer and Henning 1990; Lifton et al. 1977; Tsunemoto and Matsuo 2001) and in the midge *Chironomus* (Hankeln and Schmidt 1991). However, clustered quartets of core histones and independent H1 histone genes are present in *Drosophila virilis* (Domier et al. 1986; Schienman et al. 1998) and *Chironomus thummi* (Hankeln and Schmidt 1993).

Echinoderms are probably one of the groups that have been more extensively analyzed among deuterostomes. The sea urchins *Strongylocentrotus purpuratus* and *Psammechinus miliaris* display around 700 repetitions of quintets tandemly arranged and expressed at early embryonic stages (Sures et al. 1978), which coexist in the genome with additional clusters without any regular organization that are expressed later on during development (Maxson et al. 1983b). However, only clustered quartets with variable arrangements have been reported in the starfish species *Pisaster ochraceus*, *P. brevispinus* and *Dermasterias imbricata* (Cool et al. 1988). Although histone genes still remain organized in clusters in many vertebrate genomes, such regular organization appears to be gradually lost during the course of evolution. For instance, amphibians display histone genes organized either in clustered quartets (newts) or quintets (*Xenopus*) with variable copy number (Ruberti et al. 1982; Stephenson et al. 1981; Turner and Woodland 1983; Turner et al. 1983; van Dongen et al. 1981), as do fishes such as *Salmo gairdnerii*, which have quintets repeated about 150 times in their genome (Connor et al. 1984). Although chickens also contain clustered histone quintets repeated about 10 times, they also have solitary and replication independent H1-type genes encoding the specialized H5 histone typical of nucleated erythrocytes (D'Andrea et al. 1985; Scott and Wells 1976). The tandem arrangement of histone genes is definitively lost in mammals. Histones are organized in either a single cluster (located on chromosome 17 in the rat [Walter et al. 1996]) or in two physically independent major and minor clusters on chromosomes 6 and 3 and chromosomes 13 and 3 in humans and mice, respectively (Albig et al. 1997a, b; Wang et al. 1997). In addition, solitary single-copy replication independent H1-type genes are also present in both human and murine genomes, encoding histone $H1^0$ that has dedicated functions in terminally differentiated cellular systems (Doenecke and Alonso 1996).

8.3 Histone Variants Impart Specific Functions to Nucleosomes

8.3.1 Linker Histones

The histone H1 family encompasses one of the largest numbers of isoforms among histones. This diversity is especially evident in mammals where 11 different linker histones have been identified: 7 somatic (H1.1–H1.5, $H1^0$ [Ausió and Abbott 2004; Parseghian and Hamkalo 2001], and H1x [Happel et al. 2005]), 3 sperm-specific (H1t [Seyedin and Kistler 1979], H1t2 [Martianov et al. 2005] and HILS1 [Iguchi et al. 2003; Yan et al. 2003]); and the oocyte-specific variant H1oo (Tanaka et al. 2001).

Other cleavage-specific H1 variants homologous to H1oo have also been described in amphibians (embryonic linker histone H1M or B4 from *Xenopus* [Cho and Wolffe 1994]) and echinoderms (histone CS from sea urchin [Mandl et al. 1997]).

The evolution of H1 has favored the differentiation of other highly specialized isoforms such as histone H5 (Ruiz-Carrillo et al. 1983), a replication-independent H1 variant restricted to the nucleated erythrocytes of birds, which also appears to be present in amphibians (Khochbin 2001) and reptiles. Therefore, H5 is structurally related to $H1^0$, a histone that replaces somatic H1 isoforms in terminally differentiated cells of many vertebrates (Panyim and Chalkey 1969). These variants have been extensively studied and they share characteristic conserved elements in their promoter regions, which are involved in their replication-independent pattern of expression. These include a UCE element (Upstream Conserved Element), an H1 box followed by a G/C rich segment and an H4 box (Eirín-López et al. 2005; Schulze and Schulze 1995). Although both the $H1^0$ and H5 genes encode polyadenylated mRNAs, the H5 transcript contains two additional stem-loop signals in the 3′ UTR region (Doenecke and Alonso 1996). Until recently, the occurrence of replication independent H1 variants had been restricted exclusively to deuterostomes. Although different studies postulated the existence of H1 histone variants in several protostomes (Ausió 1999; Barzotti et al. 2000; del Gaudio et al. 1998; Eirín-López et al. 2002, 2004b; Hankeln and Schmidt 1993) the presence of replication-independent H1 forms has only been studied in detail in molluscs (Eirín-López et al. 2005).

8.3.2 Core Histones

Since histones must be synthesized in stoichiometric amounts for the assembly of chromatin onto newly replicated DNA, transcription of histone genes is tightly regulated during the cell cycle and the bulk of their translation is coordinated with DNA replication during S phase (Marzluff 1992). A unique feature of these replication-dependent histone mRNAs is their lack of polyadenylated tails (replaced by a stem-loop signal) and the regulation of their expression at three different levels including transcriptional, mRNA processing and mRNA stability (Doenecke et al. 1997). Although these regulatory mechanisms are characteristic of most histone genes, there is a small fraction of histones encoded by solitary single-copy genes whose expression prevails in nonproliferating cells, the so-called replacement histones or histone variants (Henikoff and Ahmad 2005; Smith et al. 1984). Histone variant genes exhibit constant basal replication-independent expression throughout the cell cycle, their mRNAs contain long 3′ UTR regions as well as polyadenylated tails that bind the poly(A) binding protein that increases their stability (Marzluff 1992).

Although core histones are far more conserved than H1 histones (Isenberg 1979), this does not preclude the existence of a marked functional differentiation among their variants. This is particularly noticeable in the case of the H2A family

(Ausió and Abbott 2002), which includes the heavily studied variants H2A.X and H2A.Z. These two variants are involved in the maintenance of genome integrity and in the regulation of chromatin dynamics. Histone H2A.X is present in almost all eukaryotes and is particularly expressed in germinal cells where it has functions related to DNA repair and chromosome condensation (Li et al. 2005). This variant is encoded by intron-less genes that are expressed as two different types of transcripts: a short replication-dependent type with a stem-loop signal and a longer replication-independent type that is polyadenylated (Alvelo-Ceron et al. 2000). Histone H2A.Z has been ascribed multiple functions that may differ among species. A growing body of evidence suggests that it participates in regulation of gene expression (Barski et al. 2007; Bruce et al. 2005) but it also plays an important role in the heterochromatin structure of the centromere (Greaves et al. 2007). In vertebrates, H2A.Z exists as a mixture of two protein forms H2A.Z-1 (previously H2A.Z) and H2A.Z-2 (previously H2A.F/Z or H2A.V) that differ by three amino acids (Coon et al. 2005). These two proteins are encoded by separate genes that contain four introns in humans and are expressed through polyadenylated mRNAs (Hatch and Bonner 1988, 1990). The H2A family also includes another highly specialized variant, macroH2A. This variant is characterized by a long non-histone C-terminal tail and has been shown to be involved in female X chromosome inactivation in mammals and birds (Costanzi and Pehrson 1998; Ellegren 2002). In contrast, histone H2A.Bbd is a highly variable quickly evolving mammalian H2A variant (Eirín-López et al. 2008), which is markedly deficient in the inactive X-chromosome and participates in the destabilization of nucleosomes and in the unfolding of the chromatin fiber (González-Romero et al. 2008b).

Other core histone variants include histone H3.3, CENPA and H3.t of the H3 family. In mammals, H3.3 differs from canonical replication-dependent H3.1 by only four amino acids and is enriched in actively transcribed regions of the genome in somatic cells (Mito et al. 2005). Two identical proteins, H3.3A and H3.3B, which are encoded by two different solitary genes, participate in the transition from histones to protamines during spermiogenesis in mammals (Henning 2003; Doenecke et al. 1997; Wells and Kedes 1985). Centromeric protein A (CENPA) is involved in the packaging of chromatin at eukaryotic centromeres (Govin et al. 2005; Palmer et al. 1987) and a testis-specific H3 histone (H3t) has been described in humans (Witt et al. 1996). With regard to the other histone families, the human H2B.1 gene represents the only replication-independent H2B isoform known and encodes transcripts alternatively processed at 3′ UTR regions, yielding replication-dependent and replication-independent mRNAs (Collart et al. 1992). Although no replacement subtypes have been described for the most highly conserved family of histones, the H4 family, an H4 protein with replication-independent properties has been described in *Drosophila* (Akhmanova et al. 1996).

In striking contrast to the animal kingdom, plant histones provide a very different example of histone gene structure and regulation (Kanazin et al. 1996). This is demonstrated by the existence of an intron containing H3 genes in soybean, barley and wheat, which were found to be expressed in different plant organs in a relatively replication-independent fashion. Plant histones do not have stem-loop signals and are transcribed into polyadenylated mRNAs (reviewed by Chaboute et al. [1993]) suggesting that the mechanisms regulating histone expression are very different between the plant and animal kingdoms. The regulation of histone production in plants would essentially occur at the transcriptional level.

8.4 Eukaryotic Histones Arose from Archaeal Histones Following a Recurrent Gene Duplication Process

The evolutionary origin of eukaryotic histones can be traced back to prokaryotes. Histones resembling H2A and H4, as well as the DNA topoisomerase V (a prokaryotic counterpart of eukaryotic topoisomerase I) were found in the hyperthermophilic archaebacteria *Methanopyrus kandleri* (Slerasev et al. 1984) and also archaeal RNA polymerases show common features with eukaryotic RNA polymerases (Reeve et al. 1997). In addition, archaeal histones exhibit some similarities with the central domain of the CBF-A eukaryotic transcription factor subunit (CCAAT-binding factor subunit A), suggesting that the eukaryotic modes of transcription and DNA packaging may have originated before the eukaryotes themselves (Ouzounis and Kyrpides 1996). Consistent with this observation, phylogenetic trees reconstructed from rRNA data place *Archaea* and *Eukarya* on the same branch, indicating the existence of a common ancestor shared by both groups after the divergence of *Archaea* and *Eubacteria* (Sandman and Reeve 1998, 2006).

The presence of histone-like genes in *Euryarchaeota* was explained by Sandman and Reeve (1998) in terms of the 'hydrogen hypothesis for the first eukaryote' (Martin and Müller 1998). Accordingly, the origin of the eukaryotic cell was proposed to have been derived from a symbiotic association between a methanogen archaebacterium and a proteobacterium. Thus, while the eukaryotic nucleus would have been derived from the archaebacterium (including the proteins involved in DNA packaging), most of the cellular metabolism would have been contributed by the proteobacterium. Consequently, the evolution of histones and DNA condensation into nucleosomes apparently occurred in the euryarchaeotal lineage, before the archaeal-eukaryal divergence, facilitating the genome expansion and the development of *Eukarya* (Sandman and Reeve 2005, 2006).

It appears that DNA duplication has been the basis underlying the evolutionary mechanisms driving histone evolution from the early appearance of these proteins in *Archaea* and the evolutionary process shaping this variability can be summarized in three major stages as indicated in Fig. 8.3. Firstly, the evolution of histones in *Archaea* would have been based on their generic role in packaging the genomic DNA. In this context, the first important event of gene duplication would have resulted in at least two histone-like genes in *Archaea* (such as HMfA and HMfB genes identified in *Methanothermus fervidus*, Fig. 8.3a) capable of forming tetrameric complexes to efficiently compact DNA (Sandman et al. 1990; Starich et al. 1996). The next critical duplication event would be represented by the intragenic

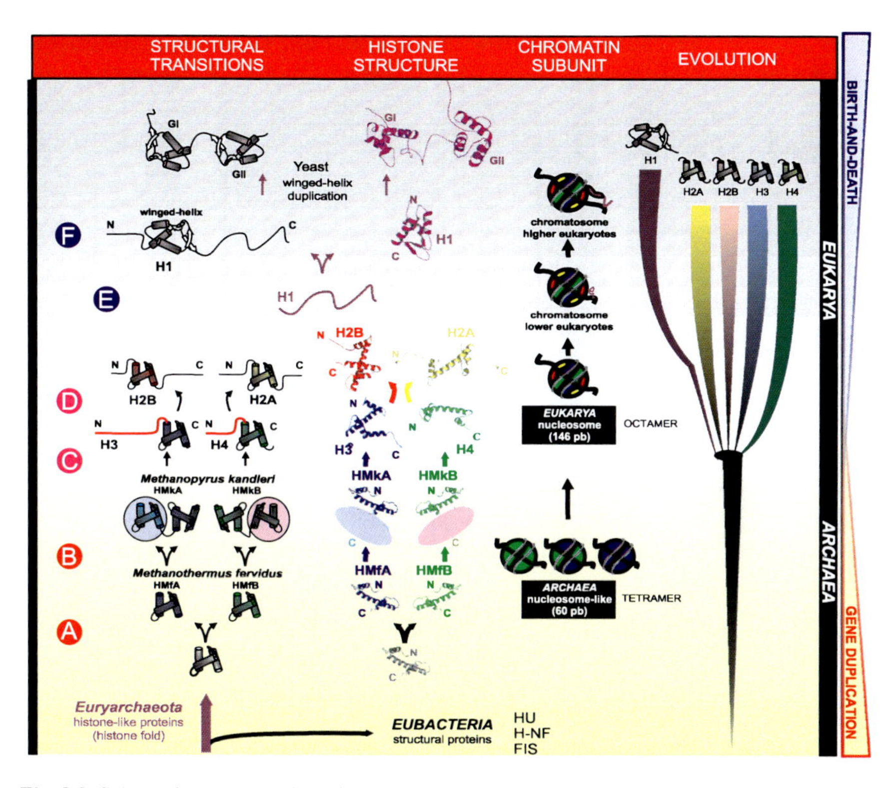

Fig. 8.3 Schematic representation of histone evolution since their origin in archaebacteria through recurrent gene duplication steps, which represent the underlying mechanisms driving histone evolution and the selective mechanisms shaping histone variability. Transitions in histone gene structure were immediately mirrored by the configuration of the nucleosome core particle (NCP) regarding its ability to package DNA. The progressive evolutionary diversification and specialization of core histones is indicated in the right margin, concomitantly with the incorporation of linker histones into the chromatin structure

duplication giving rise to the doublet histones (Fig. 8.3b) such as those identified in *M. kandleri* (Malik and Henikoff 2003). In this scenario it would be plausible that while one of the two histone folds retained its critical role in DNA packaging, the additional one would have been available to provide a substrate for selection experiments, eventually resulting in the further functionalization of the histone N- and C- terminal tails (Fig. 8.3c).

In the second stage, the histone doublet structure resulted in the formation of an asymmetric dimer that would have been the direct predecessor of the cannonical H3-H4 dimers. The evolutionary constraints generated in the transition towards a eukaryotic-specific mitosis would have been strong enough to force the shift towards an octamer structure by incorporating additional H2A–H2B dimers (Fig. 8.3d), a process that may have facilitated the genome expansion and the development of *Eukarya* (Malik and Henikoff 2003). Linker histones were likely the last component to join this structure, providing the maximum level of compaction of DNA. However, it is still not clear how this process took place. The C-terminal tail of eukaryotic H1 histones is able to compact chromatin by itself and this domain constitutes the whole protein in the case of early ancestral eukaryotes such as trypanosomes (Grüter and Betschart 2001). This supports the hypothesis that, in contrast to core histones, ancestral H1 histones were composed of only a C-terminal region (Fig. 8.3e) and the core domain containing the winged-helix motif would have been acquired later in evolution as indicated in Fig. 8.3f (Kasinsky et al. 2001).

In the third stage, the differentiation of the five metazoan histones (H1, H2A, H2B, H3 and H4) would have marked the beginning of the final stage in histone evolution. Although gene duplication has prevailed as the major mechanism in providing the eukaryotic cell with the required amounts and diversity of histones (Malik and Henikoff 2003), it has been shown that concerted evolution does not play a major role in their evolution (Eirín-López et al. 2004a; Eirín-López et al. 2005; Piontkivska et al. 2002; Rooney et al. 2002). Crucial to this stage is the diversification as a result of recurrent gene duplications followed by a strong purifying selection process acting at the protein level. This process is known as birth-and-death evolution (Nei and Hughes 1992; Nei et al. 1997) and represents the major mode of long-term evolution in eukaryote histone families as well as in many other gene families (Nei and Rooney 2006), as discussed in detail in the following section. The evolutionary refinement resulting from such diversification would have been determined by the cellular specialization associated with the appearance of multicellular organisms. Histone variability in these organisms is required in order to accommodate the different packing needs and regulation of gene expression in different cell types and developmental stages.

8.5 The Long-Term Evolution of Histone Genes Is Guided by a Birth-and-Death Process That Promotes Genetic Diversity

In contrast to the notion of divergent evolution, until around 1990 most multigene families were thought to be subject to concerted evolution (Fig. 8.4). Concerted evolution involves a process where a mutation occurring in a repeat spreads all through the gene family members by repeated occurrence of unequal crossover or gene conversion (Arnheim 1983; Smith 1974). The validity of this model, in the case of histones, was further reinforced by the general view that a gene family that produces a large amount of gene products is subject to concerted evolution in order to maintain the homogeneity of the protein product (Baldo et al. 1999; Coen et al. 1982; Kedes 1979; Liao 1999; Matsuo and Yamazaki 1989; Thatcher and Gorovsky 1994). However, as more amino acid and DNA sequence information became available, some serious conceptual concerns arose when trying to apply the

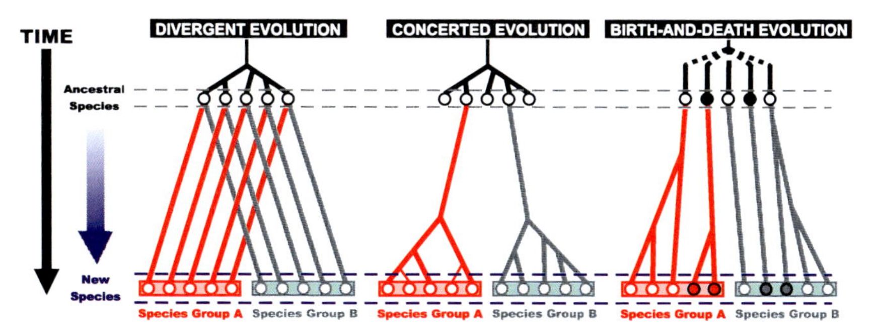

Fig. 8.4 Schematic representation of the three major models of multigene family evolution proposed for histone gene families since their discovery (Nei and Rooney 2006). Divergent evolution was the first mechanism proposed in order to explain the long-term evolution of hemoglobin α, β, γ and δ chains and myoglobin, whose encoding genes have diverged gradually as the duplicate genes acquired new functions. However, the close intraspecific relationship among ribosomal RNA genes was very difficult to reconcile with the aforementioned mechanism, leading to an idea based on a process of unequal crossover or gene conversion responsible for the homogenization of the family members, known as concerted evolution. Later on, the availability of DNA and protein sequences allowed to revisit the mechanisms guiding the evolution of some gene families such as histones, questioning the applicability of concerted evolution and suggesting a model called birth-and-death, based on recurrent gene duplications (open and solid circles indicate active genes and pseudogenes in this model, respectively)

concepts of concerted evolution to some gene families. These include genes involved in immune systems and disease-resistance (Hughes and Nei 1990; Ota and Nei 1994; Zhang et al. 2000), as well as highly conserved gene families such as ubiquitins and histones (Eirín-López et al. 2004a, 2005; González-Romero et al. 2008a; Nei et al. 2000; Piontkivska et al. 2002; Rooney et al. 2002).

As mentioned in the previous section, the differentiation of the five eukaryotic histone families, together with the subsequent specialization of the histone variants, represented an evolutionary breakthrough that allowed for a maximal level of chromatin compaction and structural and functional diversification. The broad gene and protein diversity of histones is seemingly contradictory to what would be expected from a concerted evolution model since such a model would predict close intraspecific relationships among histone genes. Several studies have dealt with this paradigm during recent years by analyzing protein and nucleotide variation levels within the different histone gene families across different groupings of eukaryotes. However, none of these studies found any support for a process of homogenization being involved in histone evolution and there are at least three major lines of evidence that argue against such a process. Firstly, phylogenetic inference of the evolutionary history of histone genes revealed that different histone isoforms cluster by type instead of by species in the phylogenetic trees. Furthermore, the analyses of nucleotide sequences failed to detect any signal of homogenization among histone genes. For instance, comparisons between human and mouse H1 genes showed that paralogous genes are not more closely related than orthologous H1 genes from both species, indicating that the functional differentiation of these genes is most likely due to a process involving selection rather than homogenization (Eirín-López et al. 2004a). A second line of evidence arose from the analyses of the nature of nucleotide variation occurring among histone genes, which indicated that the synonymous nucleotide divergence was always significantly larger than the nonsynonymous variation (Eirín-López et al. 2004a; González-Romero et al. 2008a; Piontkivska et al. 2002; Rooney et al. 2002). This suggests the absence of a concerted evolution process and is again consistent with a selective process acting at the protein level. Finally, pseudogene evolution provides a powerful tool for examining the presence of homogenization across histone repeats, given that they are expected to have a lower level of divergence compared to active genes. In this regard, all the studies focused on the long-term evolution of histone families have detected significant levels of divergence in pseudogenes (Eirín-López et al. 2004a; González-Romero et al. 2008a; Piontkivska

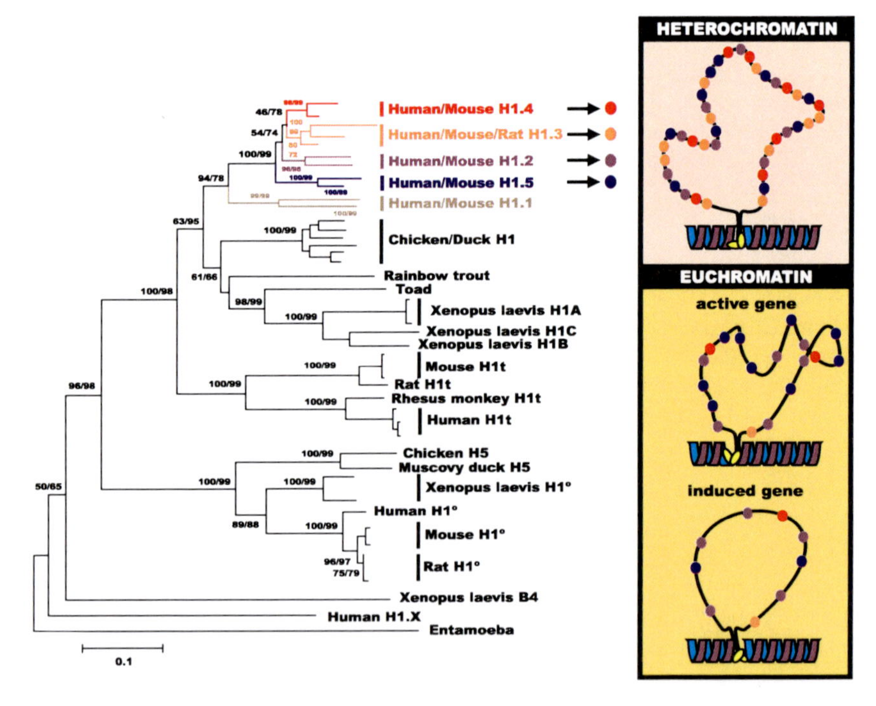

Fig. 8.5 Phylogenetic relationships among protein sequences of members of the H1 family in chordates showing a functional clustering pattern in which orthologs are more closely related among them than to their respective paralogs (Eirín-López et al. 2004a). Although histone H1 somatic types from mammals (H1.1–H1.5) were thought to be functionally redundant, the topology depicts a process of functional differentiation leading to different H1 identities. This notion has been indeed supported by experiments revealing a nonrandom distribution of H1 subtypes in the chromatin fiber, related to different transcriptional states (Parseghian et al. 2000). Numbers in internal nodes indicate confidence intervals calculated as bootstrap and internal brach-test, respectively

et al. 2002; Rooney et al. 2002), discarding any significant effect resulting from concerted evolution.

All the above mentioned studies proposed that the results fit better with the birth-and-death model of evolution (Nei and Hughes 1992) in which the diversification of the family members is the result of a recurrent process of gene duplication (mutation) followed by the inactivaction of some of the newly generated genes (selection, see Fig. 8.4 for details). Homogeneity is maintained by the effect of a strong purifying selection at the protein level and thus, DNA sequences of different gene family members can be very divergent both within and between species (Nei and Hughes 1992; Nei et al. 1997). A good example of this mode of long-term evolution is best illustrated by the histone H1 family, where extensive knockout experiments have demonstrated the existence of a large amount of functional redundancy among somatic replication-dependent H1 subtypes. However, there is now evidence indicating that the four histone H1 subtypes that are present in all mammalian somatic cells are not randomly distributed in chromatin (Fig. 8.5). This provides evidence for different roles for the different somatic H1 subtypes, in agreement with the evolutionary picture of the functional differentiation of H1 histones (Eirín-López et al. 2004a; Parseghian et al. 2000).

The birth-and-death model of evolution, as opposed to the gene homogenization process that would result from concerted evolution, promotes genetic variation. Therefore, it provides quite a reasonable mechanism for explaining the long-term evolution of gene families with high levels of diversity among their members, such as histones. This does by no means imply that gene conversion or unequal crossover does not occur but it strongly suggests that their contribution to the diversification of multigene families is relatively minor.

8.6 Replication-Dependent Histone Variants Are Derived from a Common Replication- Independent Ancestor

Different histone variants are expressed in a tissue- and developmental stage-dependent manner. During the course of evolution, the origin of these variants could have arisen by two different ways: by a gradual process from an ancient differentiation event, or through multiple independent events. An 'orphon' origin was initially proposed to explain the evolutionary origin of replication-independent histone variants followed by a process of concerted evolution. The isolation of these genes from the main histone repeats of replication-dependent variants would account for the divergent, solitary, single-copy nature of their genes (Drabent et al. 1999; Eirín-López et al. 2002, 2004b; Schulze and Schulze 1995). The discovery that long-term histone evolution occurs by a birth-and-death mechanism (Eirín-López et al. 2004a, 2005; González-Romero et al. 2008a; Piontkivska et al. 2002; Rooney et al. 2002) forced a revision of the 'orphon' hypothesis for the origin of the replication-independent histone variants. The process of purifying selection acting at the protein level would have preceded the split between protostomes and

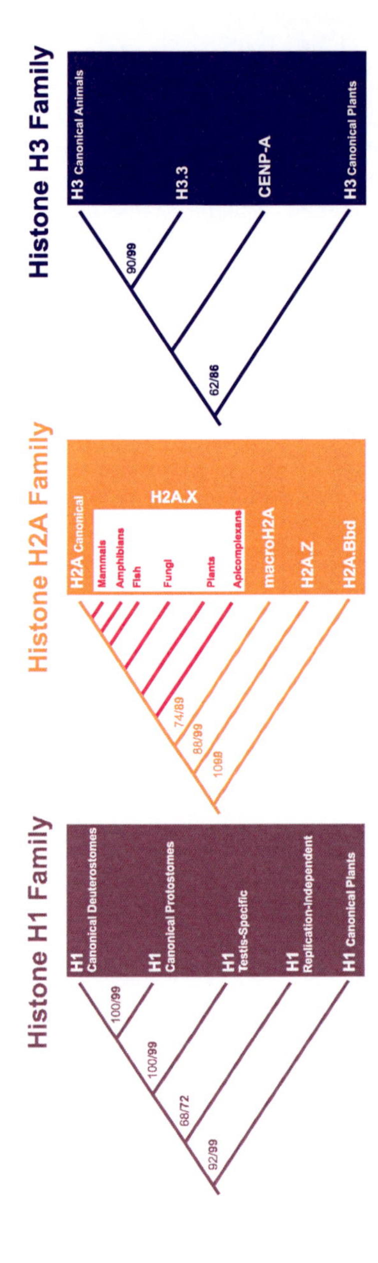

Fig. 8.6 Evolution of the major histone variant lineages within the histone H1, H2A and H3 families depicting an ancient origin except for the case of histone H2A.X. Numbers in internal nodes are indicated as in the case of Fig. 8.5

deuterostomes allowing for the subsequent transposition of replication-independent histone variants to solitary locations in the genome, where they would gradually continue their evolution (Eirín-López et al. 2004a, 2005).

Different histone families seem to be subject to different rates of 'birth-and-death' evolution as indicated by the high levels of diversity exhibited by members of the H1 and H2A families in contrast to the H2B, H3 and H4 families. Although most replication-independent histone variants have an ancient origin, there are notable exceptions to this rule, such as the case of histone H2A.X (Fig. 8.6). It is very likely that H2A.X genes arose separately and recurrently during evolution having totally replaced canonical H2A in organisms such as *Saccharomyces cerevisiae* while being completely absent in *C. elegans* (Thatcher and Gorovsky 1994). Less is known about the H3 genes, where a replication independent gene was initially proposed to be the progenitor of all H3 genes through a single differentiation event that took place early in evolution (Wells et al. 1986). This hypothesis is supported by a study that suggests that a gene similar to that of histone H3 from the protist *Phreatamoeba*, the closest relative to animal and plant H3 genes, may have been the ancestor of the animal, plant and fungal H3 sequences. Nevertheless, the appearance of H3 variants independently in animals, plants and *Tetrahymena* was also taken as evidence for the multiple origin of H3 variants (Thatcher and Gorovsky 1994).

When considering long-term histone evolution it is important to bear in mind the large differences exhibited by the plant and animal kingdoms, which clearly reflect the different evolutionary strategies followed by different organisms despite having all gone through the same histone gene duplication and selection mechanisms. Plant histones show very unique features that clearly differentiate them from the replication-dependent histones from animals but they are very closely related to animal replication-independent histone variants with which they share common traits such as the presence of introns and expression through polyadenylated transcripts (Chaboutè et al. 1993). The most plausible explanation to account for this relies on the fact that plant cells exhibit a much longer cell cycle than animal cells. Therefore, a rapid change in the levels of histone gene expression during S-phase is no longer needed and transcription control has a predominant role over posttranscriptional regulation. All this raises the question of whether ancestral histones were expressed through polyadenylated transcripts. Different authors have suggested that the major plant histone genes evolved from a common polyadenylated ancestor prior to the differentiation between plants and animals (Chaboutè et al. 1993) and that animal histone genes would have acquired specific posttranslational regulatory mechanisms (necessary to ensure the rapid histone biosynthesis in rapidly dividing cells) later on during evolution. This hypothesis is further supported by the polyadenylated nature of different histone transcripts in ancestral eukaryotes preceding the differentiation of the metazoan variants. A good example of this is provided by the histone H1 from trypanosomes (Grüter and Betschart 2001).

The mechanisms of transcriptional regulation of histone genes play a critical role in the specialization of histone variants in different tissues and developmental stages. The bulk of histone mRNA translation is coordinated with DNA replication during S phase of the cell cycle and this process is mediated by the presence of

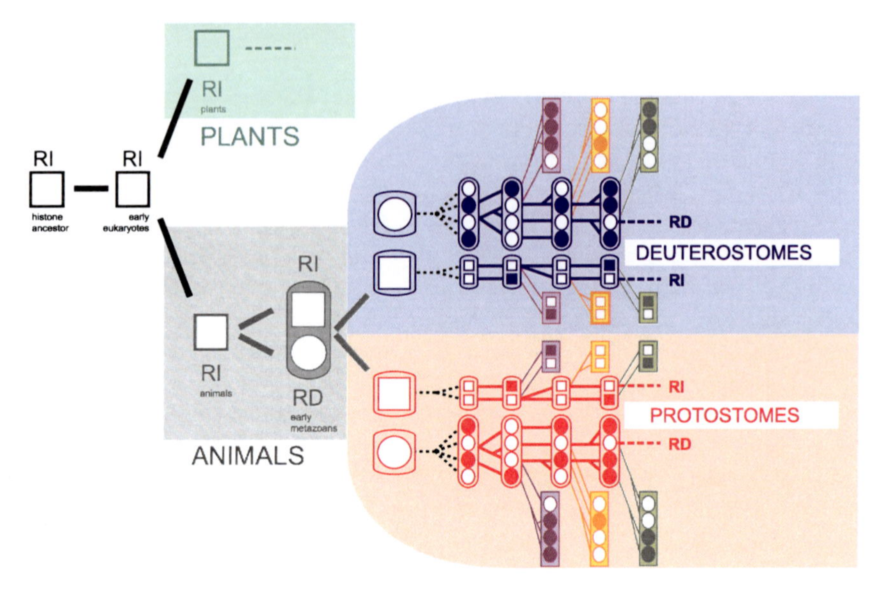

Fig. 8.7 Representation of the evolutionary process followed by an ancestral replication-independent histone leading to an initial differentiation between plant and animal histones (replication independent), which was subsequently followed by the segregation of the two animal histone lineages (replication independent, RI; replication dependent, RD) early in metazoan evolution. Both RI and RD lineages followed a parallel process of birth-and-death evolution across protostomes and deuterostomes leading to the differentiation of the canonical histones (RD) and histone variants (RI) in different taxonomic groups

a stem-loop signal in the 3′ UTR region of the transcript that is unique to replication-dependent histones in animals. In contrast, the expression of all other histone types is mediated by a polyadenylation signal that confers stability to the mRNAs, as is the case for almost any other type of gene in the genome. The fact that the 'canonical' replication-dependent histones in animals represent the only examples of genes lacking polyadenylated transcripts and given the early differentiation exhibited by the replication-independent histone variants, the notion that the primordial genes of eukaryotic histones were also transcribed through polyadenylated transcripts is supported (Fig. 8.7).

8.7 Conclusions

The old idea of histones being small basic proteins that provide a scaffold for DNA in the chromatin fiber and whose genes evolve through concerted evolution is only accurate in describing histones as small and basic proteins. As discussed throughout this work, histones have an evolutionary origin that can be traced back to archaebateria. It involves a progressive diversification and differentiation of the four core histone families through a mechanism of recurrent gene duplication, which eventually facilitated DNA compaction in the transition towards the eukaryotic cell.

The extraordinary structural and functional diversity observed among members of the different histone families is, in most instances, concomitant with the complexity of the organism, owing to the critical role histones play in the evolution of biological systems. Such diversity provides compelling evidence in support of a mechanism directing the long-term evolution of histone families that is geared towards the generation of genetic diversity (birth-and-death), rather than one that induces gene homogenization (concerted evolution). Two major histone lineages must have already been differentiated very early in eukaryotic evolution, one leading to the canonical replication-dependent histones and the other to the replication-independent histone variants. Both lineages appear to share a common replication-independent ancestor containing introns, which most likely was the preferred histone choice in plants. It is important to bear in mind that the loss of introns and the appearance of a replication-dependent expression must have had critical consequences for the evolutionary constraints acting upon the proteins. It probably involved a dramatic temporal switch towards strong positive selection of the replication-dependent histone genes that would result in an oddly similar signature to that arising from purifying selection. The current availability of genome sequences for a broad range of eukaryotic organisms opens a door to further examine the implications of such an intriguing phenomenon, especially as it pertains to the distribution and evolution of introns in early eukaryotes and early metazoans in relation to the selective pressures acting on histones.

Acknowledgments This work was funded in part by the Canadian Institutes of Health Research (CIHR) Grant MOP-57718 to Juan Ausió and by the Marie Curie Outgoing International Fellowship (MOIF-CT-2005–021900) within the 6th Framework Programme (European Union) and by a contract within the Isidro Parga Pondal Program (Xunta de Galicia) to José M. Eirín-López. Rodrigo González-Romero is the recipient of a fellowship from the Diputacion de A Coruña.

References

Akhmanova A, Miedema K, Henning W (1996) Identification and characterization of the Drosophila histone H4 replacement gene. FEBS Lett 388:219–222

Albig W, Meergans T, Doenecke D (1997a) Characterization of the H1.5 genes completes the set of human H1 subtype genes. Gene 184:141–148

Albig W, Kioschis P, Poutska A, Meergans T, Doenecke D (1997b) Human histone gene organization: non-regular arrangement within a large cluster. Genomics 40:314–322

Ali T, Thomas JO (2004) Distinct properties of the two putative “globular domains” of the yeast linker histone, Hho1p J Mol Biol 337:1123–1135

Allis CD, Glover CVC, Gorovsky MA (1979) Micronuclei of Tetrahymena contain two types of histone H3. Proc Natl Acad Sci USA 76:4857–4861

Alvelo-Ceron D, Niu L, Collart DG (2000) Growth regulation of human variant histone genes and acetylation of the encoded proteins. Mol Biol Rep 27:61–71

Arents G, Moudrianakis E (1995) The histone fold: a ubiquitous architectural motif utilized in DNA compaction and protein dimerization. Proc Natl Acad Sci USA 92:11170–11174

Arnheim N (1983) Concerted evolution of multigene families. In: Nei M, Koehn RK (eds) Evolution of Genes and Proteins. Sinauer, Sunderland, MA, pp 38–61

Ausió J (1999) Histone H1 and evolution of sperm nuclear basic proteins. J Biol Chem 274:31115–31118
Ausió J (2006) Histone variants: the structure behind the function. Brief Funct Genomic Proteomic 5:228–243
Ausió J, Abbott DW (2002) The many tales of a tail: carboxyl-terminal tail heterogeneity specializes histona H2A variants for defined chromatin function. Biochemistry 41:5945–5949
Ausió J, Abbott DW (2004) The role of histone variability in chromatin stability and folding. In: Zlatanova J, Leuba SH (eds) Chromatin Structure and Dynamics: State-of-the-Art. Elsevier, New York, pp 241–290
Baldo AM, Les DH, Strausbaugh LD (1999) Potentials and limitations of histone repeat sequences for phylogenetic reconstruction of *Sophophora*. Mol Biol Evol 16:1511–1520
Barski A, Cuddapah S, Cui K, Roth TY, Schones DE, Wang Z, Wei G, Chepelev I, Zhao K (2007) High-resolution profiling of histone methylations in the human genome. Cell 129:823–837
Barzotti R, Pelliccia F, Bucchiarelli E, Rocchini A (2000) Organization, nucleotide sequence, and chromosomal mapping of a tandemly repeated unit containing the four core histone genes and a 5S rRNA gene in an isopod crustacean species. Genome 43:341–345
Bruce K, Myers FA, Mantouvalou E, Lefevre P, Greaves I, Bonifer C, Tremethick DJ, Thorne AW, Crane-Robinson C (2005) The replacement histone H2A.Z in hyperacetylated form is a feature of active genes in chicken. Nucleic Acids Res 33:5633–5639
Chabouté ME, Chaubet N, Gigot C, Phillips G (1993) Histones and histone genes in higher plants: Structure and genomic organization. Biochimie 75:523–531
Cho H, Wolffe AP (1994) Xenopus laevis B4, an intron-containing oocyte-specific linker histone-encoding gene. Gene 143:233–238
Coen E, Strachan T, Dover GA (1982) Dynamics of concerted evolution of ribosomal DNA and histone gene families in the melanogaster species subgroup of Drosophila. J Mol Biol 158:17–35
Collart D, Pockwinse S, Lian JB, Stein JL, Stein GS, Romain PL, Pilapil S, Heubner K, Cannizzaro LA, Croce CM (1992) A human histone H2B.1 Variant gene, located on chromosome 1, utilizes alternative 3 end processing. J Cell Biochem 50:374–385
Connor W, Mezquita J, Winkfein RJ, States JC, Dixon GH (1984) Organization of the histone genes in the rainbow trout (*Salmo gairdnerii*). J Mol Evol 20:272–285
Cool D, Banfield D, Honda BM, Smith MJ (1988) Histone genes in three sea star species: Cluster arrangement, transcriptional polarity and analysis of the flanking regions of H3 and H4 genes. J Mol Evol 27:36–44
Coon JJ, Ueberheide B, Syka JE, Dryhurst DD, Ausió J, Shabanowitz J, Hunt DF (2005) Protein identification using sequential ion/ion reactions and tandem mass spectrometry. Proc Natl Acad Sci USA 102:9463–9468
Costanzi C, Pehrson JR (1998) Histone macroH2A1 is concentrated in the inactive X chromosome of female mammals. Nature 393:599–601
Cubonova L, Sandman K, Hallam SJ, DeLong EF, Reeve JN (2005) Histones in *Crenarchaea*. J Bacteriol 187:5482–5485
D'Andrea R, Coles LS, Lesnikowski C, Tabe L, Wells JRE (1985) Chromosomal organization of chicken histone genes: preferred associations and inverted duplications. Mol Cell Biol 5:3108–3115
del Gaudio N, Potenza N, Stefanoni P, Chiusano ML, Geraci G (1998) Organization and nucleotide sequence of the cluster of five histone genes in the polichaete worm Chaetopterus variopedatus: first record of a H1 histone gene in the phylum annelida. J Mol Evol 46:64–73
Doenecke D, Alonso A (1996) Organization and expression of the developmentally regulated $H1^0$ histone gene in vertebrates. Int J Dev Biol 40:423–431
Doenecke D, Albig W, Bode C, Drabent B, Franke K, Gavenis K, Witt O (1997) Histones: genetic diversity and tissue-specific gene expression. Histochem Cell Biol 107:1–10
Domier LL, Rivard JJ, Sabatini LM, Blumenfeld M (1986) *Drosophila virilis* histone gene clusters lacking H1 coding segments. J Mol Evol 23:149–158

Drabent B, Kim J-S, Albig W, Prats E, Cornudella L, Doenecke D (1999) *Mytilus edulis* histone gene clusters containing only H1 genes. J Mol Evol 49:645–655
Eirín-López JM, González-Tizón AM, Martínez A, Méndez J (2002) Molecular and evolutionary analysis of mussel histone genes (*Mytilus* spp.): possible evidence of an "orphon origin" for H1 histone genes. J Mol Evol 55:272–283
Eirín-López JM, González-Tizón AM, Martínez A, Méndez J (2004a) Birth-and-death evolution with strong purifying selection in the histone H1 multigene family and the origin of *orphon* H1 genes. Mol Biol Evol 21:1992–2003
Eirín-López JM, Ruiz MF, González-Tizón AM, Martínez A, Sánchez L, Méndez J (2004b) Molecular evolutionary characterization of the mussel Mytilus histone multigene family: first record of a tandemly repeated unit of five histone genes containing an H1 subtype with "orphon" features. J Mol Evol 58:131–144
Eirín-López JM, Ruiz MF, González-Tizón AM, Martínez A, Ausió J, Sánchez L, Méndez J (2005) Common evolutionary origin and birth-and-death process in the replication-independent histone H1 isoforms from vertebrate and invertebrate genomes. J Mol Evol 61:398–407
Eirín-López JM, Ishibashi T, Ausió J (2008) H2A.Bbd: a quickly evolving hypervariable mammalian histone that destabilizes nucleosomes in an acetylation-independent way. FASEB J 22:316–326
Ellegren H (2002) Dosage compensation: do birds do it as well? Trends Genet 18:25–28
González-Romero R, Ausió J, Méndez J, Eirín-López JM (2008a) Early evolution of histone genes: prevalence of an 'orphon' H1 lineage in protostomes and birth-and-death process in the H2A family. J Mol Evol 66:505–518
González-Romero R, Méndez J, Ausió J, Eirín-López JM (2008b) Quickly evolving histones, nucleosome stability and chromatin folding: All about histone H2A.Bbd. Gene 413:1–7
Govin J, Caron C, Rousseaux S, Khochbin S (2005) Testis-specific histone H3 expression in somatic cells. Trends Biochem Sci 30:357–359
Grayling RA, Sandman K, Reeve JN (1996) Histones and chromatin structure in hyperthermophilic Archaea. FEMS Microbiol Rev 18:203–213
Greaves IK, Rangasamy D, Ridgway P, Tremethick DJ (2007) H2A.Z contributes to the unique 3D structure of the centromere. Proc Natl Acad Sci USA 104:525–530
Grüter E, Betschart B (2001) Isolation, characterisation and organisation of histone H1 genes in African trypanosomes. Parasitol Res 87:977–984
Hankeln T, Schmidt ER (1991) The organization and nucleotide sequence of the histone genes of the midge *Chironomus thummi*. Chromosoma 105:25–31
Hankeln T, Schmidt ER (1993) Divergent evolution of an "orphon" histone gene cluster in *Chironomus*. J Mol Biol 234:1301–1307
Happel N, Schulze E, Doenecke D (2005) Characterization of human histone H1x. Biol Chem 386:541–551
Hatch CL, Bonner WM (1988) Sequence of cDNAs from mammalian H2A.Z, an evolutionarily diverged but highly conserved basal histone H2A isoprotein species. Nucleic Acids Res 16:1113–1124
Hatch CL, Bonner W (1990) The human histone H2A.Z gene. Sequence and regulation. J Biol Chem 265:15211–15218
Henikoff S, Ahmad K (2005) Assembly of variant histones into chromatin. Annu Rev Cell Dev Biol 21:133–153
Henning W (2003) Chromosomal proteins in the spermatogenesis of Drosophila. Chromosoma 111:489–494
Hentschel CC, Birnstiel ML (1981) The organization and expression of histone gene families. Cell 25:301–313
Herzog M, Soyer MO (1981) Distinctive features of dinoflegellate chromatin. Absence of nucleosomes in a primitive species *Prorocentrum micans*. Eur J Cell Biol 23:295–302
Hughes AL, Nei M (1990) Evolutionary relationships of class II major-histocompatibility complex genes in mammals. Mol Biol Evol 7:491–514

Iguchi N, Tanaka H, Yomogida K, Nishimune Y (2003) Isolation and characterization of a novel cDNA encoding a DNA-binding protein (Hils1) specifically expressed in testicular haploid germ cells. Int J Androl 26:354–365

Isenberg I (1979) Histones. Annu Rev Genet 48:159–191

Jenuwein T, Allis CD (2001) Translating the histone code. Science 293:1074–1080

Kanazin V, Blake T, Shoemaker RC (1996) Organization of the histone H3 genes in soybean, barley, and wheat. Mol Gen Genet 250:137–147

Kasinsky HE, Lewis JD, Dacks JB, Ausió J (2001) Origin of H1 histones. FASEB J 15:34–42

Kedes L (1979) Histone genes and histone messengers. Annu Rev Biochem 225:501–510

Khochbin S (2001) Histone H1 diversity: bridging regulatory signals to linker histone function. Gene 271:1–12

Kremer H, Henning W (1990) Isolation and characterization of a *Drosophila hydei* histone DNA repeat unit. Nucleic Acids Res 18:1573–1580

Landsman D (1996) Histone H1 in *Saccharomyces cerevisiae* – a double mystery solved. Trends Biochem Sci 21:287–288

Li A, Eirín-López JM, Ausió J (2005) H2AX: tailoring histone H2A for chromatin-dependent genomic integrity. Biochem Cell Biol 83:505–515

Liao D (1999) Concerted evolution: molecular mechanism and biological implications. Am J Hum Genet 64:24–30

Lifton RP, Goldberg ML, Karp RW, Hogness DS (1977) The organization of the histone genes in *Drosophila melanogaster*: functions and evolutionary implications. Cold Spring Harbor Symp Quant Biol 42:1047–1051

Lindauer A, Müller K, Schmidt R (1993) Two histone H1-encoding genes of the green alga *Volvox carteri* with features intermediate between plant and animal genes. Gene 239:15–27

Malik HS, Henikoff S (2003) Phylogenomics of the nucleosome. Nat Struct Biol 10:882–891

Mandl B, Brandt WF, Superti-Furga G, Graninger PG, Birnstiel M, Busslinger M (1997) The five cleavage-stage (CS) histones of the sea urchin are encoded by a maternally expressed family of replacement histone genes: functional equivalence of the CS H1 and frog H1M (B4) proteins. Mol Cell Biol 17:1189–1200

Martianov I, Brancorini S, Catena R, Gansmuller A, Kotaja N, Parvinen M, Sassone-Corsi P, Davidson I (2005) Polar nuclear localization of H1T2, a histone H1 variant, required for spermatid elongation and DNA condensation during spermiogenesis. Proc Natl Acad Sci USA 102:2808–2813

Martin W, Müller M (1998) The hydrogen hypothesis for the first eukaryote. Nature 392:37–41

Marzluff WF (1992) Histone 3′ ends: essential and regulatory functions. Gene Express 2:93–97

Matsuo Y, Yamazaki T (1989) Nucleotide variation and divergence in the histone multigene family in *Drosophila melanogaster*. Genetics 122:87–97

Maxson R, Cohn R, Kedes L, Mohun T (1983a) Expression and organization of histone genes. Annu Rev Genet 17:239–277

Maxson R, Mohun T, Gormezano G, Childs G, Kedes L (1983b) Distinct organizations and patterns of expression of early and late histone gene sets in the sea urchin. Nature 301:120–125

Miller DJ, Harrison PL, Mahony TJ, McMillan JP, Miles A, Odorico DM, ten Lohuis MR (1993) Feature of histone gene structure and organization are common to diploblastic and tripoblastic metazoans. J Mol Evol 37:245–253

Mito Y, Henikoff JG, Henikoff S (2005) Genome-scale profiling of histone H3.3 replacement patterns. Nat Genet 37:1090–1097

Murphy LD, Zimmerman SB (1997) Stabilization of compact spermidine nucleoids from *Escherichia coli* under crowded conditions: implications for in vivo nucleoid structure. J Struct Biol 119:336–346

Nei M, Hughes AL (1992) Balanced polymorphism and evolution by the birth-and-death process in the MHC loci. In: Tsuji K, Aizawa M, Sasazuki T (eds) 11th Histocompatibility Workshop and Conference. Oxford University Press, Oxford/England, pp 27–38

Nei M, Rooney AP (2006) Concerted and birth-and-death evolution in multigene families. Annu Rev Genet 39:121–152

Nei M, Gu X, Sitnikova T (1997) Evolution by the birth-and-death process in multigene families of the vertebrate immune system. Proc Natl Acad Sci USA 94:7799–7806

Nei M, Rogozin IB, Piontkivska H (2000) Purifying selection and birth-and-death evolution in the ubiquitin gene family. Proc Natl Acad Sci USA 97:10866–10871

Ohta T (1983) On the evolution of multigene families. Theor Popul Biol 23:216–240

Ota T, Nei M (1994) Divergent evolution and evolution by the birth-and-death process in the immunoglobulin V_H gene family. Mol Biol Evol 11:469–482

Ouzounis CA, Kyrpides NC (1996) Parallel origin of the nucleosome core and eukaryotic transcription from archaea. J Mol Evol 42:234–239

Owen-Hughes T (2003) Colworth memorial lecture. Pathways for remodelling chromatin. Biochem Soc Trans 31:893–905

Palmer DK, O'Day K, Wener MH, Andrews BS, Margolis RL (1987) A 17-kD centromere protein (CENP-A) copurifies with nucleosome core particles and with histones. J Cell Biol 104:805–815

Panyim S, Chalkey R (1969) High resolution acrylamide gel electrophoresis of histones. Arch Biochem Biophys 63:265–297

Parseghian MH, Hamkalo BA (2001) A compendium of the H1 family of somatic subtypes: an elusive cast of characters and their characteristics. Biochem Cell Biol 79:289–304

Parseghian MH, Newcomb RL, Winokur ST, Hamkalo BA (2000) The distribution of somatic H1 subtypes is nonrandom on active vs. inactive chromatin: distribution in human fetal fibroblasts. Chromosome Res 8:405–424

Pereira SL, Grayling RA, Lurz R, Reeve JN (1997) Archaeal nucleosomes. Proc Natl Acad Sci USA 94:12633–12637

Piontkivska H, Rooney AP, Nei M (2002) Purifying selection and birth-and-death evolution in the histone H4 gene family. Mol Biol Evol 19:689–697

Prescott DM (1994) The DNA of ciliated protozoa. Microbiol Rev 58:233–267

Reeve JN, Sandman K, Daniels CJ (1997) Archaeal histones, nucleosomes, and transcription initiation. Cell 89:999–1002

Rooney AP, Piontkivska H, Nei M (2002) Molecular evolution of the nontandemly repeated genes of the histone 3 multigene family. Mol Biol Evol 19:68–75

Ruberti I, Fragapane P, Pierandrei-Arnaldi P, Beccari E, Arnaldi F, Bozzoni I (1982) Characterization of histone genes isolated from Xenopus laevis and Xenopus tropicalis genomic libraries. Nucleic Acids Res 10:1544–1550

Ruiz-Carrillo A, Affolter M, Renaud J (1983) Genomic organization of the genes coding for the main six histones of the chicken: complete sequence of the H5 gene. J Mol Biol 170:843–859

Sandman K, Reeve JN (1998) Origin of the eukaryotic nucleus. Science 281:501–503

Sandman K, Reeve JN (2005) Archaeal chromatin proteins: different structures but common function? Curr Opin Microbiol 8:656–661

Sandman K, Reeve JN (2006) Archaeal histones and the origin of the histone fold. Curr Opin Microbiol 9:520–525

Sandman K, Krzycki JA, Dobrinski B, Lurz R, Reeve JN (1990) HMf, a DNA-binding protein isolated from the hyperthermophilic archaeon *Methanothermus fervidus*, is most closely related to histones. Proc Natl Acad Sci USA 87:5788–5791

Sandman K, Pereira SL, Reeve JN (1998) Diversity of prokaryotic chromosomal proteins and the origin of the nucleosome. Cell Mol Life Sci 54:1350–1364

Sanicola M, Ward S, Childs G, Emmons SW (1990) Identification of a *Caenorhabditis elegans* histone H1 gene family. Characterization of a family member containing an intron and encoding poly(A) + mRNA. J Mol Biol 212:259–268

Schienman JE, Lozovskaya ER, Strausbaugh LD. (1998) *Drosophila virilis* has atypical kinds and arrangements of histone repeats. Chromosoma 107:529–539

Schulze E, Schulze B (1995) The vertebrate linker histones $H1^0$, H5, and H1M are descendants of invertebrate "orphon" histone H1 genes. J Mol Evol 41:833–840

Scott AC, Wells JR (1976) Reiteration frequency of the gene for tissue-specific histone H5 in the chicken genome. Nature 259:635–638

Sellos D, Krawetz SA, Dixon GH (1990) Organization and complete nucleotide sequences of the core-histone-gene cluster of the annelid *Platynereis dumerilii*. Eur J Biochem 190:21–29
Seyedin SM, Kistler WS (1979) H1 histone subfractions of mammalian testes. 1. Organ specificity in the rat. Biochemistry 18:1371–1375
Slerasev AI, Belova GI, Kozyavkin SA, Lake JA (1984) Evidence for an early prokaryotic origin of histones H2A and H4 prior to the emergence of eukaryotes. Nucleic Acids Res 26:427–430
Smith BJ, Harris MR, Sigournay CM, Mayes ELV, Bustin M (1984) A survey of $H1^0$- and H5-like protein structure and distribution in lower and higher eukaryotes. Eur J Biochem 138:309–317
Smith GP (1974) Unequal crossover and the evolution of multigene families. Cold Spring Harbor Symp Quant Biol 38:507–513
Starich MR, Sandman K, Reeve JN, Summers MF (1996) NMR Structure of the HMfB from the hyperthermophile, *Methanothermus fervidus*, confirms that this archaeal protein is a histone. J Mol Biol 255:187–203
Stephenson E, Erba H, Gall J (1981) Characterization of a cloned histone gene cluster of the newt *Notophtalmus viridescens*. Nucleic Acids Res 9:2281–2295
Sures I, Lowry J, Kedes L (1978) The DNA sequence of the sea urchin (*S. purpuratus*) H2A, H2B and H3 histone coding and spacer regions. Cell 15:1033–1044
Tanaka M, Hennebold JD, MacFarlane J, Adashi EY (2001) A mammalian oocyte-specific linker histone gene H1oo: homology with the genes for oocyte-specific cleavage stage histones (cs-H1) of sea urchin and the B4/H1M histone of the frog. Development 128:655–664
Thatcher TH, Gorovsky MA (1994) Phylogenetic analysis of the core histones H2A, H2B, H3, and H4. Nucleic Acids Res 22:174–179
Tsunemoto K, Matsuo Y (2001) Molecular evolutionary analysis of a histone gene repeating unit from *Drosophila simulans*. Genes Genet Syst 76:355–361
Turner PC, Woodland HR (1983) Histone gene number and organisation in *Xenopus*: *Xenopus borealis* has a homogeneous major cluster. Nucleic Acids Res 11:971–986
Turner PC, Aldridge TC, Woodland HR, Old RW (1983) Nucleotide sequences of H1 histone genes from *Xenopus laevis*. A recently diverged pair of H1 genes and an unusual H1 pseudogene. Nucleic Acids Res 11:4093–4106
van Dongen W, de Laaf L, Zaal R, Moorman AF, Destree O (1981) The organization of the histone genes in the genome of *Xenopus laevis*. Nucleic Acids Res 25:2297–2311
van Holde KE (1988) Chromatin. Springer-Verlag, New York
Walter L, Klinga-Levan K, Helou K, Albig W, Drabent B, Doenecke D, Günther E, Levan G (1996) Chromosomal mapping of rat histone genes H1fv ($H1^0$), H1d, H1t, Th2a and Th2b. Cytogenet Cell Genet 75:136–139
Wang ZF, Sirotkin AM, Buchold GM, Skoultchi AI, Marzluff WF (1997) The mouse histone H1 genes: gene organization and differential regulation. J Mol Biol 271:124–138
Wells JRE, Kedes L (1985) Structure of a human histone cDNA: evidence that basally expressed histone genes have intervening sequences and encode polyadenylated mRNAs. Proc Natl Acad Sci USA 82:2834–2838
Wells JRE, Bains W, Kedes L (1986) Codon usage of histone gene families in higher eukaryotes reflects functional rather than phylogenetic relationships. J Mol Evol 23:224–241
Witt O, Albig W, Doenecke D (1996) Testis-specific expression of a novel human H3 histone gene. Exp Cell Res 229:301–306
Worcel A, Burgi E (1972) On the structure of the folded chromosome of *Escherichia coli*. J Mol Biol 71:127–147
Yan W, Lang M, Burns KH, Matzuk MM (2003) HILS1 is a spermatid-specific linker histone H1-like protein implicated in chromatin remodeling during mammalian spermiogenesis. Proc Natl Acad Sci USA 100:10546–10551
Zhang J, Dyer KD, Rosenberg HF (2000) Evolution of the rodent eosinophil-associated RNase gene family by rapid gene sorting and positive selection. Proc Natl Acad Sci USA 97:4701–4706
Zlatanova J (1997) Archaeal chromatin: virtual or real? Proc Natl Acad Sci USA 94:12251–12254

Chapter 9
Masculinization Events and Doubly Uniparental Inheritance of Mitochondrial DNA: A Model for Understanding the Evolutionary Dynamics of Gender-Associated mtDNA in Mussels

Donald T. Stewart, Sophie Breton, Pierre U. Blier, and Walter R. Hoeh

Abstract Bivalved mollusks have an unusual system of mitochondrial DNA transmission referred to as "Doubly Uniparental Inheritance" or DUI. Species with DUI are characterized by the presence of two gender-associated mitochondrial DNA lineages that are inherited through males (male-transmitted or M types) or females (female-transmitted or F types), respectively. In marine mussels (Mytiloida), an M type and an F type occasionally recombine. Complete mtDNA genome sequencing studies have demonstrated that new, recombinant M types are occasionally produced that are primarily composed of an F type protein coding genes, rRNA genes, tRNA genes, F type control region and, significantly, an M type control region. We have studied various properties of one of these novel recombinant M types (i.e., a "recently masculinized male-transmitted genome or RM type") and compared its performance with an evolutionary older "standard" M type (i.e., an SM type) and found evidence for faster swimming speed and more efficient mitochondrially-encoded enzyme activity of the RM type. These studies lead us to a model to explain the evolutionary dynamics and phylogenetic relationships of the various male and female-transmitted mtDNA genomes found in marine mussels (Mytiloida).

D.T. Stewart
Department of Biology, Acadia University, Wolfville, Nova Scotia, B4P 2R6, Canada
e-mail: don.stewart@acadiau.ca
S. Breton and W.R. Hoeh
Department of Biological Sciences, Kent State University, Kent, Ohio 44242, USA
P.U. Blier
Département de Biologie, Université du Québec à Rimouski, 300 Allée des Ursulines, Rimouski, Québec, G5L 3A1, Canada

P. Pontarotti (ed.), *Evolutionary Biology: Concept, Modeling, and Application*,
DOI: 10.1007/978-3-642-00952-5_9,

9.1 Doubly Uniparental Inheritance of Mitochondrial DNA in Bivalves – An Overview

Marine mussels (Mytiloida), marine clams (Veneroida) and freshwater mussels (Unionoida) have a unique mode of mitochondrial DNA transmission called "Doubly Uniparental Inheritance" or DUI. This system is fundamentally different from strict maternal inheritance that is found in other animals. Unique aspects of DUI include: (1) the presence of distinct gender-associated mtDNA lineages that are transmitted through sperm or eggs (i.e., males or females, respectively), (2) periodic "role reversals" of female-transmitted mtDNA genomes such that they are subsequently transmitted through sperm, (3) recombination of male (M) and female (F) transmitted mtDNA genomes, (4) an accelerated rate of molecular evolution of both the M and F genomes but particularly M genomes and (5) the presence of a unique extension of the cytochrome *c* oxidase subunit II gene in the M (but not the F) genome in freshwater mussels (see Breton et al. 2007 and references therein). Because of its novelty, bivalve species possessing DUI constitute an excellent model system for studying basic aspects of mitochondrial DNA inheritance and the evolution of mtDNA genomes in general. Because DUI is the exception to the rule, understanding why bivalves evolved distinct male and female mtDNA lineages can provide us with insights into the evolutionary forces that maintain strictly maternal inheritance in most animals.

9.2 Details of DUI and Variations on the Basic Model

Species exhibiting DUI, such as the blue mussel *Mytilus edulis*, are characterized by the presence of distinct gender-associated mitochondrial DNA genomes that are inherited either through eggs (female-transmitted or F types) or through sperm (male-transmitted or M types). Females tend to be homoplasmic for the F type, although traces of M type have been reported in female somatic tissues and even female gonad tissue and eggs (e.g., Obata et al. 2006; Sano et al. 2007). In males, the F type is predominant in somatic tissues but the M type is predominant in gonad tissue samples (Garrido-Ramos et al. 1998) and appears to be the exclusive type in sperm (Venetis et al. 2006). The M type is actively transcribed in both gonadal and somatic tissues of males (Dalziel and Stewart 2002). A striking difference between strictly maternal inheritance and DUI is that in theory, DUI allows selection to act directly on the M type; unlike most animals, male mussels do not represent an evolutionary dead-end for mtDNA. Consequently, microevolutionary forces such as drift and selection can act directly on M type polymorphisms.

The evolutionary dynamics of M and F types differ markedly between freshwater mussels and marine bivalves. The M and F mtDNA lineages have been distinct and generally stable for over 200 MY in freshwater mussels (Hoeh et al. 2002; but see Walker et al. 2006). In contrast, marine mussels (family Mytilidae) and marine clams (order Veneroida) have periodically experienced "role-reversal events" in which an F type becomes "masculinized" and invades the male route of transmission (Hoeh et al. 1997; Passamonti and Scali 2001; Everett et al. 2004; Knock et al. 2005). A fascinating hypothesis proposed recently by Theologodis et al. (2007) is that recombination of an F type and a so-called "Standard Male" or SM type may be the first step in the masculinization process. Several sequencing studies have demonstrated that all "recently masculinized" or RM types examined to date are recombinants (e.g., Quesada et al. 2003; Burzynski et al. 2003, 2006; Cao et al. 2004; Mizi et al. 2005; Rawson 2005; Breton et al. 2006). The RM genomes sequenced to date are primarily composed of an F genome's protein coding genes, rRNA genes, tRNA gene, control region and, significantly, an additional M type genome's control region.

9.3 Phylogenetic Patterns

DUI is found exclusively within the bivalve mollusks (Breton et al. 2007). Phylogenetic relationships among the M and F types within and between species have been examined in several studies. Hoeh et al. (1996, 1997) examined partial sequences from the mitochondrial 16S rRNA and cytochrome c oxidase subunit I (*CoxI*) genes to reconstruct evolutionary relationships between the M and F types that had been detected in gonad tissues of male and female freshwater mussels (family Unionidae) and marine mussels (family Mytilidae). They initially concluded that DUI has arisen independently at least three times in bivalves; once in the common ancestor to the three species of freshwater mussels they examined (*Pyganodon grandis, P. fragilis* and *Fusconaia flava*), once in the common ancestor of *M. trossulus* and *M. edulis* and once in *Geukensia demissa* (see Fig. 1, Hoeh et al. 1997). However, subsequent sequencing and phylogenetic analysis of a few relatively rare M types in *Mytilus* demonstrated that some of these male transmitted genomes segregating in male *M. edulis* and *M. trossulus* mussels in Nova Scotia, Canada, were more closely related to F types in their respective species (at least for the genes sequenced) than they were to other M types segregating in those populations. This led Hoeh et al. (1997) to reject the hypothesis of multiple origins of DUI in these marine and freshwater mussels and to accept instead the hypothesis of periodic "role reversal" events, which was first proposed by Hoeh et al. 1996. Specifically, the latter hypothesis suggested that occasionally an F type switches from the female route of inheritance into the male route of inheritance. This hypothesis could explain the observed polymorphism among M types within the species *M. edulis* and *M. trossulus*. Furthermore, this hypothesis could be extended to explain why the M and F types of *G. demissa* are more closely related to one another than to the M or F types, respectively, of other marine mussels. Hoeh et al. (1997) also extended the hypothesis of role reversals to all bivalve species with distinct M and F transmitted mitochondrial DNA lineages (i.e., those species with DUI) and suggested that the phenomenon of DUI evolved

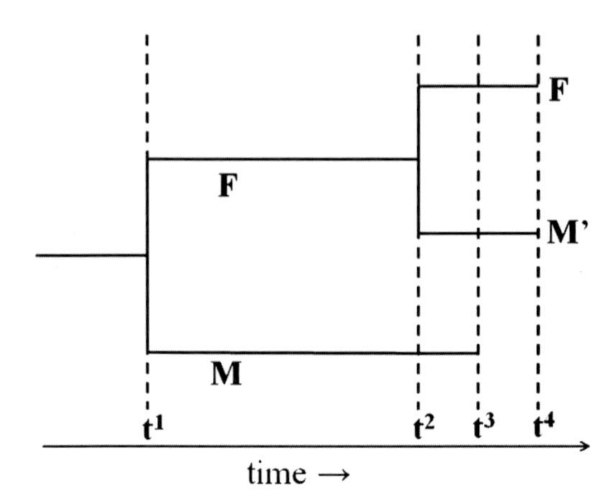

Fig. 9.1 Simplified version of a role reversal or masculinization event by which a genome that consists primarily of F type genes (plus a recombined M control region) invades the male route of inheritance. (i) At some point in time (t^1), DUI originated in some ancestral bivalve species resulting in distinct male transmitted (M) and female transmitted (F) types. (ii) At some subsequent point in time, (t^2), an F type switches into the male route of transmission resulting in two male types segregating within the population; the original M type and the newly masculinized M type (M'). (iii) At a third point in time, (t^3), the original M type goes to extinction (either as a result of stochastic or deterministic factors; see discussion herein). (iv) It is important to note that if this species is assayed at t^4, based on the amount of sequence divergence between F and M', one might infer that DUI had originated at time t^2, rather than at t^1

once early on in the history of bivalves prior to the divergence of the freshwater mussels and the marine bivalves (i.e., several hundred million years ago) but role reversal events periodically reset the amount of divergence between the M and F mtDNA genomes within an organismic lineage back to zero that destroyed the record of the previous (i.e., evolutionary older) M lineage (see Fig. 9.1).

DUI has subsequently been documented, based on the presence of distinct M and F types in male and female gonad tissues, respectively, in several additional bivalve taxa. For example, Hoeh et al. (2002) analyzed M and F types in six additional freshwater species from the superfamily Unionoidea and one species from the order Trigonioida. To date, all M and F unionoidean sequences examined phylogenetically are, unlike the situation in mytilids, reciprocally monophyletic (i.e., the M types form one monophyletic group and the F types form another monophyletic group). These data indicate that role reversal events either do not occur in freshwater mussels, or at the very least, that any newly masculinized M types do not persist in these taxa (Hoeh et al. 2002). A unique coding feature of the cytochrome c oxidase subunit II gene in the unionoidean M types (but not the F types) is an extension of up to several hundred additional nucleotides that are transcribed and translated, which suggests that the M genome has become so specialized in these bivalves that recombination, leading to role reversals, is no longer possible (Curole and Kocher 2002; Chakrabarti et al. 2007). Passamonti and Scali (2001) and Passamonti (2007) have demonstrated distinct M and F types in the venerid clam *Tapes phillipinarum* and another species of Mytilid, *Musculista senhousia*. The M and F types within each of these species are distinct from one another but are also each other's closest relatives at least based on the species of mussels that have been analyzed to date. These phylogenetic patterns have been interpreted as evidence of role reversal events in these taxa as well (Passamonti and Scali 2001; Passamonti 2007).

The outstanding question pertaining to DUI is, of course, why did it evolve in the first place; i.e., why were distinct M and F types "tolerated" in some ancestral bivalve species when selection for strict maternal inheritance appears to be the norm in animals? As mentioned above, DUI should allow selection on the M mitochondrial genome to optimize male associated functions such as sperm motility where selective pressures on, for example, sperm mitochondria, might differ from selective pressures on egg mitochondria and female somatic tissues (Zeh and Zeh 2005; Burt and Trivers 2006). While the reciprocal monophyly of M and F types in freshwater mussels is consistent with this hypothesis, the frequency of role reversal events in marine mussels suggests that a simple model of DUI persistence based on selection for favorable mitochondrial mutations associated with male associated functions is too simplistic. Data from recent experiments conducted on *M. edulis* sperm may help us construct a model to explain the evolutionary dynamics of M and F types in marine mussels and to suggest a research program that will help evaluate this model.

9.4 Functional Studies of M Type Polymorphisms in *Mytilus*

In *Mytilus* mussels, protein coding genes and RNA genes of recently masculinized and standard M types exhibit up to 20% divergence in nucleotide sequence (Breton et al. 2006). These differences include non-synonymous substitutions resulting in an amino acid sequence divergence of 8–9% between these polymorphic M types (e.g., Stewart et al. 1996). Presumably this level of amino acid substitution could affect enzyme efficiency of the ATP production machinery of the sperm mitochondria that could, in turn, affect sperm motility parameters (Everett et al. 2004).

The first attempt to assess functional differences in sperm motility between sperm containing the so-called standard male (SM) or recently masculinized (RM) male types in *M. edulis* was conducted by Everett et al. (2004). For this study, *Mytilus* mussels were obtained from Lunenburg Bay, Nova Scotia, which is known to contain both *M. edulis* and *M. trossulus*. Mussels were induced to spawn by injecting them with 0.5 ml of 0.5 M KCL and by thermal shock (i.e., placing them in seawater warmed to 20°C) and their sperm motility was videotaped and analyzed by manually tracking sperm movement every frame (1/30th second in length) through three full oscillations. For each spawned male, 20 individual sperm were tracked. The mussels were sacrificed post-spawning and genotyped to confirm species (i.e., *M. edulis* or *M. trossulus*) and mitotype (i.e., SM or RM). In total, 14 SM males and 4 RM males were analyzed. In this initial study, no significant differences in sperm motility parameters such as curvilinear velocity or average path velocity were observed. However, because (a) this initial study was limited to a few individuals ($n = 18$), (b) the number of RM males was particularly low ($n = 4$), (c) the mussels examined were spawned and analyzed over several months and (d) a relatively unsophisticated sperm analysis system was used, the question was revisited using larger numbers of samples spawned over a 3-day period and analyzed using a much more sophisticated computer assisted sperm analysis (i.e., CASA) system. Specifically, for the second study (see Jha et al. 2008 for details), a CEROS 12.1 sperm analyser with CEROS Animal Motility Software was used. Mussels were again collected from the southern coast of Nova Scotia, Canada, from an area where both *M. edulis* and *M. trossulus* mussels have been documented. Over 250 individuals of both sexes and both species were induced to spawn by a combination of physical agitation and thermal shock. Sperm samples were loaded into chambered slides with a 20 μm depth and a total of ten fields of view were analyzed per slide. The CEROS system tracks sperm movement every 1/60th of a second so 30 frames of sperm motility data were collected in 0.5 s. Whereas Everett et al. (2004) were limited to 20 individual spermatozoa per individual that also swam in a more or less progressive linear direction, the CEROS software enabled a much more accurate examination of many more sperm per individual (frequently hundreds per individual). Because we were able to analyze considerably more individuals and to analyze individual sperm motility parameters based on far more sperm per individual, we are confident that the second set of results is in fact more objective and more accurate than the first set.

To summarize the results of the second study, the sperm of RM males was found to exhibit faster average curvilinear velocity ($n = 8$; mean $VCL_{RM} = 127.27 \pm 2.15$ μm/s) than the SM males ($n = 60$; mean $VCL_{SM} = 117.77 \pm 1.43$ μm/s). Similarly, the RM males also exhibited faster average path velocities (mean $VAP_{RM} = 114.21 \pm 2.15$ μm/s) than the SM males (mean $VAP_{SM} = 103.62 \pm 1.76$ μm/s) (Jha et al. 2008).

In a set of parallel experiments, we also explored the molecular basis of the observed difference in sperm motility by testing for differences in mitochondrial respiratory chain enzyme activities between RM and SM mtDNA bearing-sperm. To this end, we measured the activity of the major mitochondrial respiratory chain complexes (complexes II, I + III, and IV) as well as the activity of citrate synthase in gonad samples of the same individuals of *M. edulis* as were analyzed in Jha et al. (2008). Our study found that the mitochondrial subunits encoded by the RM types were associated with higher enzymatic activities than the gene products of the SM type genome.

To briefly summarize the methods, samples of gonad tissues were standardized across individuals and homogenized in a buffer consisting in 50 mM imidazole, 2 mM $MgCl_2$ and 5 mM EDTA, pH 7.5. Homogenates were centrifuged at 500 *g* for 5 min at 4°C and the supernatant was used to analyze the following enzyme activity levels. Citrate synthase (CS), succinate dehydrogenase (complex II or SDH), electron transport system (ETS), which comprises NADH dehydrogenase and cytochrome *c* reductase (complexes I and III) and cytochrome *c* oxidase (complex

IV or COX) activities were measured using a UV/VIS spectrophotometer. For details of the assay conditions, see Breton et al. (2009).

The results of these enzyme activity assays were revealing. Mean activity for *COX* (complex IV) was significantly greater in RM gonad samples, whereas there was no significant difference between SM and RM male mussels in mean activity of other complexes or citrate synthase (CS). The ratio of *COX* activity to CS was also significantly greater in RM gonad samples, which controls for electron transport chain activity as a function of the corresponding mitochondrial mass (Dalziel et al. 2005). Although higher ratios were also observed for RM gonad samples than SM samples for other comparisons of CS-corrected activities, the differences were not statistically significant.

9.5 A (Primarily) Deterministic Model for Periodic Replacement of SM Types by RM Types

The results of these two studies suggest a deterministic model for the periodic replacement of the evolutionarily older SM types by relatively younger, recombinant RM types in the marine mussel *Mytilus*. We hypothesize that faster swimming speed, which may be a direct consequence of more efficient *COX* activity levels, may confer a fitness advantage to the RM types over the SM types under certain ecological parameters. This result appears to run counter to the hypothesis that DUI may be adaptive because it allows selection to act directly on the M genome and optimize it for male associated functions such as sperm motility. If this were the only microevolutionary force acting on the M genomes, then one would expect the evolutionarily older SM types to outcompete a nascent, recently masculinized type that is primarily an F genome (with a small piece of recombined SM genome including the control region), which should be primarily adapted to functioning in female gonad tissues and somatic tissues. A complete model may be, not surprisingly, more complex than this.

Several factors involving selection at various levels likely play a role in determining the relative fitnesses of SM and RM types. First, because of the apparent dynamic nature of DUI in the marine mussels, the descriptor "standard male" or SM genome may be something of a misnomer. The SM genomes in all marine mussels examined to date are themselves likely the product of role reversal events at some earlier point in time in the evolutionary history of the species. Therefore, when we compare the functional properties of RM to SM genomes, we are really comparing *two* recently masculinized types; the RM types are simply more recently masculinized than the SM types but both are the product of the same process. Perhaps *previously masculinized* or PM type may be more appropriate than the SM type but we will continue to use the acronym SM herein.

Second, comparative studies of the rate of molecular evolution of the M types in bivalves generally indicate that they have a more rapid rate of evolution than the

F types (e.g., Stewart et al. 1996 but see Passamonti 2007 for a possible exception to this general rule). It has been suggested that because the M types are predominant in fewer cellular "arenas" than the F types (i.e., in fewer tissue types), that selection is relatively relaxed on the M types (Stewart et al. 1996). This relaxation of selection may be one additional factor that contributes to the relative fitness of SM vs. RM types over time. If selection is relatively relaxed, then M types will gradually accumulate more slightly deleterious mutations than F types. As an M type becomes evolutionarily older, these slightly deleterious mutations will begin to accumulate a significant genetic load. We propose that as an M type becomes more and more divergent, the probability of a nascent RM type being able to function more efficiently than the older SM type will gradually increase. In other words, while a brand new RM type may not be more fit than a slightly evolutionarily older SM type, that same RM type may be more fit than a more divergent SM type with a greater mutational load.

Third, layered on top of these factors are considerations based on the complexities of calculating fitness factors for alternative mitochondrial polymorphisms in "real life" settings. While our results from the sperm motility experiments and the sperm enzyme activity assays suggest that the RM types (a) swim faster than the SM types and (b) exhibit more efficient *COX* activity than the SM types, this does not necessarily lead to the conclusion that the RM types are more fit than the SM types. As Levitan (2000) has pointed out with respect to alternative mitochondrial DNA polymorphisms in sea urchins, the relative fitness of a particular mitotype will depend on the environment. In a high density population of broadcast spawners, faster sperm may have an advantage over slower sperm simply because they can reach their targets (i.e., eggs) more quickly. However, in a lower density population, sperm that can swim at a somewhat slower speed but do so over a longer distance may be at an advantage. This may be a factor in explaining the current maintenance of the RM and the SM polymorphisms in *M. edulis* in Nova Scotia, Canada. Given that the RM type is approximately 1.5–2% divergent from the F type for *CoxI* for example, it has clearly been present and segregating in the population for a significant period of time. Despite its faster swimming speed, it has not replaced the SM type in this population and is, in fact, still the minority mitotype at ~10–15% frequency along the southern coast of Nova Scotia. Swimming speed is clearly only one of several factors that combine to produce the fitness of a particular mitochondrial genome.

A fourth consideration that must be taken into account when constructing a model for the periodic replacement of one M type by another is that stochastic factors (i.e., drift) could play a role in fluctuations and fixation of allele frequencies just as they do for any molecular polymorphism. If the RM and SM types are similar enough in their selection coefficients that they essentially represent neutral polymorphisms, then drift may result in a mitotype going to fixation. The consequence of drift on the evolutionary dynamics of RM and SM types is, however, asymmetrical. While new RM types may be produced regularly by recombination of an F type and an M type in the population, SM types cannot be reproduced by this process. Once an SM type is lost either by stochastic or deterministic forces, its relatively divergent coding sequences become extinct and cannot be recreated.

9.6 Future Research Opportunities

We are still far from fully understanding the adaptive significance of doubly uniparental inheritance of mitochondrial DNA in bivalve mollusks. Similarly, there is still much work to be done to understand the roles of recombination, selection and drift on the evolutionary dynamics of M genomes in the marine mussels. We have initiated a variety of studies on mussel sperm to better understand the functional properties of alternative M types (i.e., RM vs. SM polymorphisms) in *M. edulis*. Similar baseline studies of sperm motility and enzyme activities levels in closely related species such as *M. edulis*, *G. demissa*, and other species in the family Mytilidae will be very informative. However, more realistic studies of the relative fitness of these mitochondrial DNA polymorphisms will involve analyzing sperm motility, longevity and fertilization success of alternative polymorphisms under a variety of conditions (e.g., time post-spawning, various densities of eggs, various temperature and salinity regimes, etc.) and investigating the coevolution interactions of mitochondrial and nuclear genes in both RM and SM males.

Acknowledgments This work was funded by Discovery grants from the Natural Sciences and Engineering Research Council of Canada (NSERC) to D. Stewart and P. Blier and by a National Science Foundation grant to W.R. Hoeh. S. Breton was funded by an NSERC postgraduate scholarship and is currently funded by an NSERC postdoctoral fellowship. We acknowledge members of the Blier and Stewart labs for assistance with various aspects of this study.

References

Breton S, Burger G, Stewart DT, Blier PU (2006) Comparative analysis of gender-associated complete mitochondrial genomes in marine mussels (*Mytilus* spp.). Genetics 172:1107–1119

Breton S, Doucet Beaupré H, Stewart DT, Hoeh WR, Blier PU (2007) The unusual system of doubly uniparental inheritance of mtDNA: Isn't one enough? TIG 23:464–475

Breton S, Stewart DT, Blier PU (2009) Role-reversal of gender-associated mitochondrial DNA affects mitochondrial function in *Mytilus edulis* (Bivalvia: Mytilidae). J Exp Zool B (Mol Dev Evol) 312B:108–117

Burt A, Trivers RL (2006) Genes in conflict. Belknap Press of Harvard University Press, Boston, MA

Burzyński A, Zbawicka M, Skibinski DOF, Wenne R (2003) Evidence for recombination of mtDNA in the marine mussel *Mytilus trossulus* from the Baltic. Mol Biol Evol 20:388–392

Burzyński A, Zbawicka M, Skibinski DOF, Wenne R (2006) Doubly uniparental inheritance is associated with high polymorphism for rearranged and recombinant control region haplotypes in Baltic *Mytilus trossulus*. Genetics 174:1081–1094

Cao LQ, Kenchington E, Zouros E, Rodakis GC (2004) Evidence that the large noncoding sequence is the main control region of maternally and paternally transmitted mitochondrial genomes of the marine mussel (*Mytilus* spp.). Genetics 167:835–850

Chakrabarti R, Walker JM, Chapman EG, Shepardson SP, Trdan RJ, Curole JP, Watters GT, Stewart DT, Vijayaraghavan S, Hoeh WR (2007) Reproductive function for a C-terminus extended, male-transmitted cytochrome c oxidase subunit II protein expressed in both spermatozoa and eggs. FEBS Lett 581:5213–5219

Curole JP, Kocher TD (2002) Ancient sex-specific extension of the cytochrome c oxidase II gene in bivalves and the fidelity of doubly uniparental inheritance. Mol Biol Evol 19:1323–1328

Dalziel AC, Stewart DT (2002) Tissue-specific expression of male-transmitted mitochondrial DNA and its implications for rates of molecular evolution in *Mytilus* mussels (Bivalvia: Mytilidae). Genome 45:348–355

Dalziel AC, Moore SE, Moyes CD (2005) Mitochondrial enzyme content in the muscles of high performance fish: Evolution and variation among fiber types. Am J Physiol Regul Integr Comp Physiol 288:R163–R172

Everett EM, Williams P, Gibson G, Stewart DT (2004) Mitochondrial DNA polymorphisms and sperm motility in *Mytilus edulis* (Bivalvia: Mytilidae). J Exp A (Comp Exp Biol) 301A:906–910

Garrido-Ramos MA, Stewart DT, Sutherland BW, Zouros E (1998) The distribution of male-transmitted and female-transmitted mitochondrial DNA types in somatic tissues of blue mussels: Implications for the operation of doubly uniparental inheritance of mitochondrial DNA. Genome 41:818–824

Hoeh WR, Stewart DT, Sutherland BW, Zouros E (1996) Multiple origins of gender-associated mitochondrial DNA lineages in bivalves (Mollusca: Bivalvia). Evolution 50:2276–2286

Hoeh WR, Stewart DT, Saavedra C, Sutherland BW, Zouros E (1997) Phylogenetic evidence for role-reversals of gender-associated mitochondrial DNA in *Mytilus* (Bivalvia: Mytilidae). Mol Biol Evol 14:959–967

Hoeh WR, Stewart DT, Guttman SI (2002) High fidelity of mitochondrial genome transmission under the doubly uniparental mode of inheritance in freshwater mussels (Bivalvia: Unionoidea). Evolution 56:2252–2261

Jha M, Côté J, Hoeh WR, Blier PU, Stewart DT (2008) Sperm motility in *Mytilus edulis* in relation to mitochondrial DNA polymorphisms: implications for the evolution of doubly uniparental inheritance in bivalves. Evolution 62:99–106

Knock E, Petersen SD, Stewart DT (2005) Differential display reverse transcription PCR applied to male *Mytilus edulis* mussels with two distinct mitochondrial DNA types. Biochem Syst Ecol 33:715–724

Levitan DR. (2000) Sperm velocity and longevity trade off each other and influence fertilization in the sea urchin *Lytechinus variegatus*. Proc R Soc Lond B Biol Sci 267:531–534.

Mizi A, Zouros E, Moschonas N, Rodakis GC (2005) The complete maternal and paternal mitochondrial genomes of the Mediterranean mussel *Mytilus galloprovincialis*: Implications for the doubly uniparental inheritance mode of mtDNA. Mol Biol Evol 22:952–967

Obata M, Kamiya K, Kawamura K, Komaru A (2006) Sperm mitochondrial DNA transmission to both male and female offspring in the blue mussel *Mytilus galloprovincialis*. Dev Growth Differ 48:253–261

Passamonti M, Scali V (2001) Gender-associated mitochondrial DNA heteroplasmy in the venerid clam *Tapes philippinarum* (Mollusca Bivalvia). Curr Genetics 39:117–124

Passamonti M (2007) An unusual case of gender-associated mitochondrial DNA heteroplasmy: the mytilid *Musculista senhousia* (Mollusca Bivalvia). BMC Evol Biol 7(Suppl 2):S7 doi:10.1186/1471-2148-7-S2-S7

Quesada H, Stuckas H, Skibinski DOF (2003) Heteroplasmy suggests paternal co-transmission of multiple genomes and pervasive reversion of maternally into paternally transmitted genomes of mussel (*Mytilus*) mitochondrial DNA. J Mol Evol 57:S138–S147

Rawson PD (2005) Nonhomologous recombination between the large unassigned region of the male and female mitochondrial genomes in the mussel, *Mytilus trossulus*. J Mol Evol 61:717–732

Sano N, Obata M, Komaru A (2007) Quantitation of the male and female types of mitochondrial DNA in a blue mussel, *Mytilus galloprovincialis*, using real-time polymerase chain reaction assay. Dev Growth Diff 49:67–72

Stewart DT, Kenchington E, Singh R, Zouros E (1996) Degree of selective constraint as an explanation of the different rates of evolution of gender-specific mitochondrial DNA lineages in the mussel *Mytilus*. Genetics 143:1349–1357

Theologidis I, Saavedra C, Zouros E (2007) No evidence for absence of paternal mtDNA in male progeny from pair matings of the mussel *Mytilus galloprovincialis*. Genetics 176:1367–1369

Venetis C, Theologidis I, Zouros E, Rodakis GC (2006) No evidence for presence of maternal mitochondrial DNA in the sperm of *Mytilus galloprovincialis* males. Proc R Soc Lond B Biol Sci 273:2483–2489

Walker JM, Bogan AE, Garo K, Soliman GN, Hoeh WR (2006) Hermaphroditism in the Iridinidae (Bivalvia: Etherioidea). J Molluscan Stud 72: 216–217

Zeh JA, Zeh DW (2005) Maternal inheritance, sexual conflict and the maladapted male. TIG 21:281–286

Chapter 10
Missing the Subcellular Target: A Mechanism of Eukaryotic Gene Evolution

S.A. Byun McKay, R. Geeta, R. Duggan, B. Carroll, and S.J. McKay

Abstract Gene duplication is a critical force in genome evolution. By serving as the raw material for new genetic functions, duplications are believed to lead to the evolution of new genes and genetic innovations. The increasing availability of genomic data reveals that gene duplication through polyploidization or other mechanisms is prevalent in genomes from across all three domains of life. Elucidating the mechanisms by which duplicated genes acquire novel functions is a key question in genome evolution. We propose that one such mechanism is protein subcellular relocalization (PSR), where small changes to N-terminal signal peptides lead to differential targeting of otherwise identical duplicated proteins. We argue that changing the location of duplicate proteins within the eukaryotic cell can lead to novel functions. If these new functions are advantageous, these duplicates would then be retained as novel genes. We suggest that PSR is an important evolutionary mechanism and that it has played an influential role in the origin of new eukaryotic genes.

10.1 Introduction

For nearly 40 years, gene duplication has been considered to be one of the most central evolutionary processes involved in the generation of new genes. With the

R. Duggan and B. Carroll
Fairfield University, Dept. of Biology, 1073 North Benson Road, Fairfield CT, 06824, USA

S.A. Byun McKay
Fairfield University, Dept. of Biology, 1073 North Benson Road, Fairfield CT, 06824, USA
e-mail: sbyun@mail.fairfield.edu

R. Geeta
Stony Brook University, Dept. of Ecology and Evolution, 650 Life Sciences Building, Stony Brook NY 11794, USA

S.J. McKay
Cold Spring Harbor Laboratory, 1 Bungtown Rd, Cold Spring Harbor, NY, 11724, USA

P. Pontarotti (ed.), *Evolutionary Biology: Concept, Modeling, and Application*,
DOI: 10.1007/978-3-642-00952-5_10,

increased availability of genomic data over the last 15 years, greater effort has been given towards understanding the various possible mechanisms by which duplicate genes can evolve novel functions (Hughes 1994; Long et al. 2003; Zhang 2003; Taylor and Raes 2004). The classic idea of how duplication results in novel genes, coined neofunctionalization, was first proposed by Ohno (1970). Neofunctionalization is based primarily on the premise that, initially, duplicate genes are functionally redundant. This redundancy, which results in a relaxation of selection, allows duplicate genes the freedom to accumulate mutations without compromising the original gene function. Although most mutations are deleterious, rare beneficial mutations can impart a new or modified function to a duplicated protein. Purifying selection could then launch the paralogue that encodes the neofunctionalized protein onto its own evolutionary trajectory, ultimately to become a new gene (Ohno 1970). However, a major problem with neofunctionalization is the rarity of beneficial mutations. As such, it is not certain how common neofunctionalization is and exactly how influential it has been in producing new genes following duplication.

A widely discussed and well-substantiated mechanism associated with duplicate genes is subfunctionalization (Force et al. 1999, 2005). A key component of subfunctionalization is its reliance on deleterious mutations rather than on beneficial ones. Through the accumulation of deleterious mutations, which occur in both duplicates, the original gene function becomes partitioned such that each duplicate retains a subset of that function. Because both copies are now required to carry out the original function, both duplicates are retained in the genome and do not deteriorate into pseudogenes. Because the mechanism of subfunctionalization is based on accumulation of much more common deleterious, rather than advantageous mutations, it would be a more frequently occurring and arguably a more plausible evolutionary mechanism for retention of duplicated genes. However, though subfunctionalization can explain how a duplicate gene can be retained within the genome through purifying selection, it does not necessarily account for how a duplicate gene can ultimately evolve novel functions without beneficial mutations, which would enhance or modify function. Other mechanisms such as adaptive conflict (Des Mariais and Rauscher 2008), epigenetic complementation (Rodin et al. 2005) and dosage compensation (Papp et al. 2003) have also been discussed. All of these mechanisms have likely played roles in the evolution of new genes to some extent.

In this chapter, we discuss another mechanism that may have had an influential role in the evolution of eukaryotic genes. We propose that protein subcellular relocalization (PSR) (Byun et al. 2007) may be a significant mechanism that could lead to the evolution of new gene functions following gene duplication. Through changes in the N-terminal peptide of duplicate proteins by deletions, insertions and/or point mutations, the duplicate can be redirected to a new location within the eukaryotic cell. As a consequence of its new biochemical surroundings, the duplicate protein can potentially take on new or modified roles and if advantageous, will be retained as a new gene. We will summarize the major concepts of PSR, discuss some of the evidence for this mechanism and supplement the discussion with some preliminary data from our current work. We argue that the evidence collectively suggests that PSR has played a significant role in the origin of eukaryotic genes.

10.2 Protein Subcellular Relocations (PSR): A Mechanism for the Evolution of New Genes and Gene Functions

10.2.1 Protein Subcellular Localization through the N-Terminal Peptide

The central dogma of molecular biology is that genes typically produce proteins. These proteins, which are translated in the cytosol, either remain there or are subsequently redirected to various locations within the cell. Subcellular localization is determined through a variety of signals such as C-terminal motifs and nuclear localization signals. One of the best understood mechanisms of protein subcellular localization is the N-terminal peptide. The N-terminal peptide is a sequence of about 13–30 amino acids located at the N-terminus of a protein. Depending on the sequence, the protein attached to the N-terminal peptide will either remain in the cytosol or be directed to a membrane bound organelle such as the endoplasmic reticulum, the mitochondria, or the chloroplast. Once the protein reaches its final destination, the N-terminal peptide is typically cleaved off and degraded. As such, the N-terminal peptide does not usually participate directly in protein function (reviewed in Lodish et al. 2000; Bannai et al. 2002 and references therein).

A key feature of N-terminal peptides is their high degeneracy. For example, Kaiser et al. (1987) demonstrated that about 20% of random human genomic sequences could all direct an invertase protein to the endoplasmic reticulum in *Saccharomyces cervisiae*. The degeneracy of the N-terminal peptide sequence results in a relatively faster rate of evolution than that of the adjacent peptide. However, that does not necessarily mean that the N-terminal peptide region is not subjected to purifying selection. In an examination of 76 mouse-rat orthologues, the N-terminal peptide sequence, although evolving at a rate of about two to five times faster than the rest of the protein, was still subject to purifying selection (Williams et al. 2000). Our analysis of *Arabidopsis thaliana*, *Trypanosoma cruzi* and *Homo sapiens* paralogues suggests that the N-terminal peptide also evolves more quickly relative to the rest of the proteins (e.g., Fig. 10.1) (Byun McKay et al. 2009, unpublished data).

Because many random alternatives can direct proteins to the same subcellular location, N-terminal peptides are not characterized by a specific sequence of amino acids. Rather the specific target of an N-terminal peptide seems to be determined by the presence of certain features such as hydrophobic cores, α -helices and the relative abundance of acidic amino acids. Consequently, changes in N-terminal peptide targeting do not necessarily require a large number of amino acid substitutions. In fact, a single amino acid substitution is potentially sufficient to

```
tc1     MSYTHIIMEQTVNTVKPFPTFFCSKTTEVTPPQNAAGGSVAGGNAESGEEASAQSIAGLG
tc2     MMTTYRLLCALLVLALCCCSSVCVTATGEMPPQNAAGVSVTGGNAESGEEASAQSIAGLG
        *  *: ::   :  .    : .* .:*   ******* **:*******************

tc1     GEGSGEPSDDGSAGRLAHGSEGSGSGRSKEDEGKAEELNQQGGAPPSPLPLPPAPTGPIN
tc2     GEGSGEPSDDCSAGRLAHGSGGSGSGRSKEDEGKAEELNQQGGAPPSPLPIPPAPTGPIN
        ********** ********* *****************************:*********

tc1     LPRSAEPGEPATSGVTTEPQSAQQLQSQLPAAGNQGVTGDGSPEPGTAAASGATRSTTAG
tc2     LPRSAEPGEPATSGVTTEPQSAQQQQSQLPAAGNQGVTGDGSPEPGTAAASGATRSTPAW
        ************************ ********************************.*

tc1     GDAEPTSPSPGGQAAASGPGENSAAEGTPNGTPPSAAAVTHDDNTNTTDTATSERNSTAP
tc2     GDAEPTSPSPGGQAAASGPGENSAAEGTPNGTPPSAAAVTHDDNTNTTDTATSERNSTAA
        ***********************************************************.

tc1     GMPAPLSSAPRTKALESGVGNDACFHDTRLRAPLLLALAALAYAALG
tc2     GMLAPLSSAPRTKALESGVGNDACFHDTRLHAPLLLALAALAYAALG
        ** ****************************:***************
```

Fig. 10.1 Alignment of two paralogues from *Trypanosoma cruzi*. Paralogues tc1 and tc2 have an amino acid identity of 87.1% overall and are predicted to localize to two different locations within the cell. Note that the majority of amino acid substitutions (28/36) occur in the first 30 residues of the N-terminal region (tc1 – XM 804024.1 tc2 – XM 799964.1)

Table 10.1 NTP sequences and their putative subcellular targets

NTP sequence	Putative subcellular target
MSSSTLLHLLELSSLLFSCLPNAKPQQAED	cytosol
MSSSTLLHLLLLSSLLFSCLPNAKPQQAED	endoplasmic reticulum
MSSSTLLHLLRLSSLLFSCLPNAKPQQAED	mitochondria
MSSSTLLHLLHLSSLLFSCLPNAKPQQAED	chloroplast

The NTP sequences above are identical except for the single amino acid substitutions outlined by the box. Based on iPSORT predictions, the single amino acid substitution is sufficient for PSR. Modified from Byun McKay and Geeta, 2007

redirect the protein to a new subcellular location, illustrating how quickly PSR could act (Table 10.1).

Clearly, point mutations are not the only change in the N-terminal peptide that could ultimately result in PSR. Another type of change is an incomplete duplication that results in an insertion or deletion in the N terminus of a duplicate gene. For example, acetyl-CoA carboxylase (ACCase) in *Brassica napus* is grouped into two classes, class I and Class II. Class II is represented by a single gene localized to the cytosol while Class II consists of two genes both of which are characterized by an additional exon at the 5' end. This addition alters the N terminus of the resulting protein, which apparently retargets the protein from the cytosol to the chloroplast (Schlute et al. 1997). Based on a broad survey of *A. thaliana*, *H. sapiens*, *Chlamydomonas reinhardtii* and *Tetrahymena thermophila* genomes, deletions/insertions in the N-terminal peptide region appeared to be a common occurrence (e.g., Fig. 10.2) (Byun McKay et al. 2009, unpublished data). Such changes are particularly interesting because these deletions/insertions have the potential to instantaneously alter subcellular localization and as such, provide a novel context for new or altered functions to emerge.

```
cr1    MAPRIRAAMGLALLVALAATVCDARRELSMMKVMPRAVYTQTNDPAGNQIVAIMFNASSG
cr2    ---------------------------------------------------MFDASSG
                                                           **:****

cr1    MLVGNMTTRTPTGGMGAAGLGANGQPVTADGLFSQGAVAVSGSWLLAVNGGSNTLSLFRI
cr2    MLVGNMTTRTPTGGMGAAGLGANGQPVTADGLFSQGAVAISGSWLFAVNGGSNTLSLFRI
       ***************************************:*****:**************

cr1    SDSDPTSLMLVGSPVDTLGDFPVSVTYSAKCKTACVLNGGKRDGVSCFAVSATGLKPLDA
cr2    SDSDPTTLMLVGTPVDTLGDFPVSVTYSAKCKTACVLNGGKRDGVSCFAVSATGLKPLDA
       ******:*****:***********************************************

cr1    SPRPVGLGQSANPPTGPPSTASMIAFNPKGDILAVTIKGNLNDNRLGYIGLYRVSKTGMV
cr2    SPRPVGLGQSANPPTGPPSTASMIAFNPKGDILAVTIKGNLNDNRLGYIGLYRVSKTGMV
       ************************************************************

cr1    SRTQTRAYSIFTSQPVGAGASMIGFLPFGFAWVDDMTLSISDPSVGAVTLCISSRDPMLA
cr2     SRTQSRAYSIFTSQPVGAGASMIGFLPFGFAWVDDMTLSISDPSVGAVTLCISSRDPMLA
       ****:*******************************************************

cr1    SIPVYKTDMAGGYAPAVFTLPEAAPCWAAYSDASGCAYIADAATGTLVEIMPKSGGKLRS
cr2    SIPVYKTDMAGGYAPAVFTLPEAAPCWAAYSDASGCAYIADAATGTLVEIMPKSGGKLRS
       ************************************************************

cr1    VFNVTSAFGITLGSTPANTYGLLDTAAAGDLLFSVSPRSGSVVVVDVSGGSMAPRQLLAA
cr2    VFNVTSAFGITLGSTPANTYGLLDTAAAGDLLFSVSPRSGSVVVVDVSGGSMAPRQLLAA
       ************************************************************

cr1    GLPVSSMGLAVAC
cr2    GLPVSSMGLAVAC
       *************
```

Fig. 10.2 Alignment of two paralogues from *Chlamydomonas reinhardtii*. Paralogues, cr1 and cr2, have an amino acid identity of 98.4% and are predicted to localize to the ER and chloroplast respectively. The change in subcellular location is due to the large deletion/insertion located in the NTP region. (1 – XM 001696505.1 and 2 – XM 001696498.1)

Given the variety of ways in which subcellular targeting can be altered through the N-terminal peptide and the potential rapidity of such changes, we would expect protein subcellular relocalization (PSR) to be a common feature in diverse gene families.

10.2.2 PSR is Widespread in Gene Families

Recent large-scale genomic analyses have suggested that subcellular relocalization is a common feature amongst paralogues. For example, an analysis of *A. thaliana* revealed that at least 239 gene families have at least two members localized to different subcellular locations (Heilmann et al. 2004). Similarly, a recent examination of duplicate gene pairs created by a whole genome duplication in *S. cervisae* suggested that 24–37% of the pairs were localized to different subcellular compartments (Marques et al. 2008).

There are abundant data on subcellular targeting within specific gene families as well. For example, two gene families FK506 binding proteins and cyclophilin within the green algae *Chlamydomonas* have been empirically determined to have members localized to a variety of subcellular locations including the nucleus, cytosol, chloroplast, mitochondria and ER. α-carbonic anhydrase is another example of a gene family that is apparently localized to various subcellular locations. Members of this gene family function primarily in the reversible hydration of carbon dioxide, a process essential for sequestering CO_2 and facilitating its transport into the cell. Although much of what we know about this gene family is derived from studies in animal models, α-carbonic anhydrase is also found in plants, where its function is generally unknown. In *Dioscorea* (yams) certain storage proteins were determined to be α-carbonic anhydrases (Conlan et al. 1995; Hewett-Emmett and Tashian 1996). Predictions of subcellular localization within this gene family in a range of different plants such as *Arabidopsis*, rice and tobacco suggest multiple subcellular targets, consistent with the predications of PSR (McKay and Geeta 2009, unpublished data).

10.2.3 Changes in Subcellular Location Can Alter Protein Function

The subcellular location of a protein plays a critical role in its function. Each compartment of the cell has a unique metabolic microenvironment, which can alter or modify a protein's activity. As discussed earlier, α-carbonic anhydrase within the genus *Dioscorea* is believed to have additional functions, such as storage. It has also been suggested that within *Dioscorea*, α-carbonic anhydrase also has monodehydroascobate reductase activity, trypsin inhibition, chitinase activity and possibly lectin/mannose binding ability. Interestingly, some members of the α-carbonic anhydrase family in this genus has been localized to the vacuoles rather than its typical subcellular location, the cytosol (Conlan et al. 1995; Hou et al. 1999; Gaidamashvil et al. 2004).

A change in the function of a protein as a consequence of its subcellular location is well documented in the literature. For example, phospholipase D (PLD) is a gene family known to localize to multiple subcellular compartments such as the nucleus, the golgi apparatus, mitochondria and cytoskeleton. In each of these locations, PLD activities differ as a result of being stimulated by different signaling molecules (Liscovitch et al. 1999). The change in subcellular location can also result in dramatically different enzymatic activities, as is the case for insulin degrading enzyme (IDE). IDE is a widely expressed zinc metallopeptidase, typically localized to the cytosol. In the cytosol, its primary function is to regulate levels of cerebral amyloid β-peptide and plasma insulin. However, when IDE was experimentally redirected to the mitochondria, it degraded additional, non-insulin peptides, an activity unknown to the cytosolic form (Lessring et al. 2004).

Changes in subcellular location can also modify an existing function. The *Arabidopsis* desaturase gene family consists in members that catalyze the desaturation of fatty acids by inserting covalent double bonds in specific locations along the fatty acid chain. When desaturase proteins were experimentally relocalized to the cytoplasm and chloroplast, a change in the location of double bond insertion within the fatty acid chain was observed. Altering the location of covalent double bond insertion usually requires at least two to six key point mutations in the mature desaturase protein. However, similar changes in enzymatic function were evidently achieved through subcellular relocalization (Heilmann et al. 2004). While most changes in function have been demonstrated within gene families that encode proteins involved in metabolic processes, changes in function due to different subcellular targeting has also been documented in a family of transcription factors known as LEAFY. Although these proteins are typically localized to the nucleus, in some cases, LEAFY is localized to the cytoplasm where it was observed to undergo intercellular movement and possibly participate in cell to cell signaling (Sessions et al. 2000; Wu et al. 2003). The above examples clearly illustrate how changes in a protein's subcellular location can have a dramatic influence on its function.

10.3 Conclusion

Based on cumulative evidence that (a) subcellular relocalization is a common feature within gene families (b) that a change in subcellular location can alter the function of a duplicate protein (c) that changes in subcellular targeting is potentially rapid and relatively easy to achieve, we suggest that protein subcellular relocalization is an important mechanism of duplicate gene diversification and that this mechanism has played a key role in the evolution of new eukaryotic genes (Byun Mckay and Geeta 2007).

The hypothesis of PSR as a mechanism of new gene evolution is based on changes that occur in subcellular targeting. We focused specifically on the N-terminal peptide because most information about subcellular targeting is based on this region. However, we consider any change in the subcellular location through the N-terminal peptide, C-terminal motifs or nuclear localization signals as falling under the mechanism of PSR.

One of the advantages of PSR is the relative ease by which proteins can be relocalized within the cell. Such relocalization can occur through the relatively rapid accumulation of mutations in the N-terminal peptide region compared to the rest of the protein or potentially as a consequence of an imperfect duplication of the 5' end of the gene. Because the N-terminal peptide is removed when the protein reaches its destination, any change in protein function as a consequence of relocalization is achieved without a mutation occurring in the mature protein sequence. Thus PSR differs from neofunctionalization in that the affected part of the protein is removed and does not directly impact the protein's function. Because PSR does not necessarily involve degenerative mutations and partitioning of genetic functions, it is also distinct from subfunctionalization.

Changing the subcellular target of a duplicated protein can happen relatively quickly. Deletions and insertions that can occur through an incomplete duplication can instantaneously result in relocalization. Point mutations, which tend to accumulate more rapidly in the N-terminal peptide can also result in PSR. As little as a single amino acid substitution within the N-terminal peptide is apparently sufficient to alter the subcellular location of a protein. Such changes in subcellular targeting of a duplicate protein can result in a change or modification of function. Once the duplicate protein reaches its new subcellular location, it may be non-functional, have unaltered function, or may take on a modified or even entirely new function. If such functional changes are advantageous, these duplicate genes will be subjected to purifying selection and ultimately evolve into a new gene. PSR represents one of the few mechanisms that outline how duplicated genes are not only retained within a genome but how it can actually evolve a novel function. We suggest that PSR is an important mechanism through which genes functionally diversified following duplication and that this model is worth pursuing as a fundamental mechanism of eukaryotic genome evolution.

References

Bannai H, Tamada Y, Maruyama O, Nakai K, Miyano S (2002) Extensive feature detection of N-teminal protein sorting signals. Bioinformatics 18:298–305

Byun McKay SA, Geeta R (2007) Protein subcellular relocalization: a new perspective on the origin of novel genes. TREE 22(7):338–344

Conlan RS, Griffiths IA, Napier JA, Shewry PR, Mantell S, Ainsworth C (1995) Isolation and characterization of cDNA clones representing the genes encoding the major tuber storage protein (dioscorin) of yam (*Dioscorea cayenensis* Lam). Plant Mol. Biol. 28:369–380

Des Marais DL, Rausher MD (2008) Escape from adaptive conflict after duplication in an anthocyanin pathway gene. Nature 454:762–765

Force A, Lynch M, Bryan Pickett F, Amores A, Yan Y, Postlewait J (1999) Preservation of duplicate genes by complementary, degenerative mutations. Genetics 15:1531–1545

Force A, Cresko WA, Bryan Pickett F, Proulx SR, Amemiya C, Lynch M (2005) The origin of subfunctions and modular gene regulation. Genetics 170:443–446

Gaidamashvil M, Ohizumi Y, Iijima S, Takayama T, Ogawa T, Muramoto, K (2004) Characterization of the yam tuber storage proteins from *Dioscorea batatas* exhibiting unique lectin activities. J Biol Chem 279:26028–26035

Heilmann I, Pidkowich MS, Girke T, Shanklin J (2004) Switching desaturase enzyme specificity by alternate subcellular targeting. Proc Natl Acad Sci USA 101:10266–10271

Hewett-Emmett D, Tashian, RE (1996) Functional diversity, conservation, and convergence in the evolution of the alpha-, beta-, and gamma-carbonic anhydrase gene families. Mol Phylogenet Evol 5:50–77

Hou W, Liu JS, Chen HJ, Chen TE, Chang CF, Lin YH (1999) Dioscorin, the major tuber storage protein of yam (*Dioscorea batatas* Decne) with carbonic anhydrase and trypsin inhibitor activities. J Agric Food Chem 47:2168–2172

Hughes A (1994) Gene duplication and the origin of novel proteins. Proc Natl Acad Sci USA 102:8791–8792
Kaiser CA, Preuss D, Grisafi P, Botstein D (1987) Many random sequences functionally replace the secretion signal sequence of yeast invertase. Science 235:312–317
Lessring MA, Farris W, Wu Xining, Christodoulou DC, Haigis MC, Guarente L, Selkoe DJ (2004) Alternative translation initiation generates a novel isoform of insulin-degrading enzyme targeted to mitochondria. Biochem J 383:439–446
Liscovitch M, Czarny M, Fiucci G, Lavie Y, Tang X (1999) Localization and possible functions of phospholipase D isozymes. Biochim Biophys Acta 1439:245–263
Lodish H, Berk A, Zipursky SL, Matsudaira P, Baltimore D, Darnell JE (2000) Molecular cell biology, 4th edn. W.H. Freeman, New York
Long M, Betrán E, Thornton K, Wang W (2003) The origin of new genes: glimpses from the young and old. Nat. Rev. Genet. 4:865–875
Marques AC, Vickenbosch N, Brawand D, Kaessmann H (2008) Functional diversification of duplicate genes through subcellular adaptation of encoded proteins. Genome Biol 9(3):R54
Ohno S (1970) Evolution by gene duplication. Springer-Verlag, New York
Papp B, Pál C, Hurst LD (2003) Dosage sensitivity and the evolution of gene families in yeast. Nature 424:194–197
Rodin SN, Parkhomchuk DV, Rodin AS, Holmquist GP, Riggs AD (2005) Repositioning dependent fate of duplicate genes. DNA Cell Biol 24:529–542
Schulte W, Töpfer R, Stracke R, Schell J, Norbert M (1997) Multi-functional acetyl-CoA carboxylase from Brassica napus is encoded by a multi-gene family:indication for plastidic localization of at least one isoform. Proc Natl Acad Sci USA 94:3465–3470
Sessions A, Yanofsky MF, Weigel D (2000) Cell-cell signaling and movement by the floral transcription factors LEAFY and APETALA1. Science 289:779–781
Taylor JS, Raes J (2004) Duplication and divergence: The evolution of new genes and old ideas. Ann Rev Genet 28:615–643
Williams EJ, Pal C, Hurst LD (2000) The molecular evolution of signal peptides. Gene 253:313–322
Wu X, Dinneny JR, Crawford KM, Rhee Y, Citovsky V, Zambryski PC, Weigel D (2003) Modes of intercellular transcription factor movement in the Arabidopsis apex. Development 130:3735–3745
Zhang J (2003) Evolution by gene duplication: An update. Trends Ecol Evol 18:292–298

Chapter 11
The Evolution of Functional Gene Clusters in Eukaryote Genomes

Takashi Makino and Aoife McLysaght

Abstract It is increasingly clear that eukaryotic gene order is nonrandom, being constrained in some instances by expression patterns, expression levels, protein interactions or epistatic effects. The relationship between gene order and function is patchy (not every gene is in a functional cluster) and the factors that influence it are numerous and interconnected. Not surprisingly, then, our knowledge of genome structure is still growing. Here we review reported gene clusters in eukaryote genomes and obstacles such as leaky expression and tandem duplication effects for identification of functional gene clusters. In particular, we show interacting gene clusters, which are identified by protein–protein interactions (PPIs), are robust against the problems. Furthermore, we emphasize that evolutionary analyses of functional gene clusters are very important to assess their biological meaning.

11.1 Physically and Functionally Linked Gene Clusters

11.1.1 Operons: Typical Gene Clusters in Prokaryote Genomes

We define a gene cluster as two or more genes having both functional links (interaction of the gene products) and physical links (close location of the genes) in an organism. Prokaryotes possess characteristic gene clusters known as operons, where several genes that function together in the same biochemical pathway are located close together in the genome and their expression is coordinated. There are more than 600 operons in *Escherichia coli* (Salgado et al. 2000). For example, there are five genes in the tryptophan operon in *E. coli* that are polycistronically transcribed in a single mRNA. The five kinds of proteins translated from the mRNA

T. Makino
Smurfit Institute of Genetics, University of Dublin, Trinity College, Dublin 2, Ireland
e-mail: makinot@tcd.ie

P. Pontarotti (ed.), *Evolutionary Biology: Concept, Modeling, and Application*,
DOI: 10.1007/978-3-642-00952-5_11, © Springer-Verlag Berlin Heidelberg 2009

work together to synthesize tryptophan from chorismic acid. Operon-like structures have been observed in a few eukaryotes: 15% of genes in *Caenorhabditis elegans* (Blumenthal et al. 2002), 20% of genes in *Ciona intestinalis* (Satou et al. 2008) and several genes in *Saccharomyces cerevisiae* (David et al. 2006; He et al. 2003) are in operon-like structures. However these are rare exceptions and operons are not a general feature of eukaryotic genomes (Table 11.1).

11.1.2 Gene Clusters in Eukaryote Genomes

Although almost all known eukaryote genomes do not have operon structures, there are many examples of nonrandom gene order (Hurst et al. 2004; Sproul et al. 2005; Table 11.1). Housekeeping or highly expressed genes are clustered together in the same genomic regions in eukaryotes (Craig and Bickmore 1994; Lercher et al. 2002; Singer et al. 2005; Williams and Hurst 2002). Approximately 80% of imprinted genes are clustered on the genome (Reik and Walter 2001). Hox, β-globin, epidermal differentiation complex (EDC) and major histocompatibility complex (MHC) gene clusters, which are mainly derived from tandem gene duplication events, are found in mammalian genomes (Sproul et al. 2005). In addition, closely linked bidirectional gene pairs have been identified in yeast (Kruglyak and Tang 2000) and mammals (Adachi and Lieber 2002; Franck et al. 2008; Koyanagi et al. 2005; Li et al. 2006; Takai and Jones 2004; Trinklein et al. 2004). The gene pairs possibly share promoter regions and therefore their expression patterns are often correlated (Li et al. 2006; Trinklein et al. 2004). Genes in the same pathway are often clustered in eukaryote genomes (Lee and Sonnhammer 2003). Interacting gene clusters have also been found in yeast (Poyatos and Hurst 2006; Teichmann and Veitia 2004) and human (Makino and McLysaght 2008). Using protein-protein interactions for identifying gene clusters can minimize detection of artificial gene clusters as discussed in detail below.

Table 11.1 Functional gene clusters in eukaryote genomes

Functional gene cluster	Data source	Example
Operon-like structure	Structure of transcripts	Operon-like structures in yeast, worm, and ascidian
Co-expressed gene cluster	Expression patterns	Housekeeping gene clusters
Imprinted gene cluster	Epigenetic feature	The Beckwith-Wiedemann syndrome cluster
Tandemly duplicated gene cluster	Genomic structure	Hox gene clusters
Bidirectional gene cluster	Genomic structure	Bidirectional gene clusters in mammals
Gene cluster in the same pathway	Metabolic pathways	Glycolysis pathway
Interacting gene cluster	Protein–protein interactions	Immune-related gene clusters

11.2 Leaky Gene Expression

Many reported gene clusters in eukaryotes were identified from gene expression data (Hurst et al. 2004), however most reports did not examine or discuss the possibility of leaky expression caused by expression of a neighboring gene (Spellman and Rubin 2002; Fig. 11.1a). Liao and Zhang showed that co-expression between neighboring genes is negatively correlated with the physical distance between them (Liao and Zhang 2008). Furthermore, the correlation between transcription and translation is weak in yeast (Ghaemmaghami et al. 2003). The coincidental transcripts are likely to be suppressed. Thus it is not clear that there is a biological or functional significance to the co-expression of closely located genes.

Ebisuya and colleagues recently reported a "ripple effect" of transcription (Ebisuya et al. 2008). In particular, the authors observed that transcription of immediate-early genes (IEGs) up-regulated by addition of fibroblast growth factor (FGF) activated transcription of their neighbors. The shorter a distance between IEG and its neighbor, the more quickly transcription of the neighbor is activated (Ebisuya et al. 2008). Therefore, we cannot easily distinguish functional

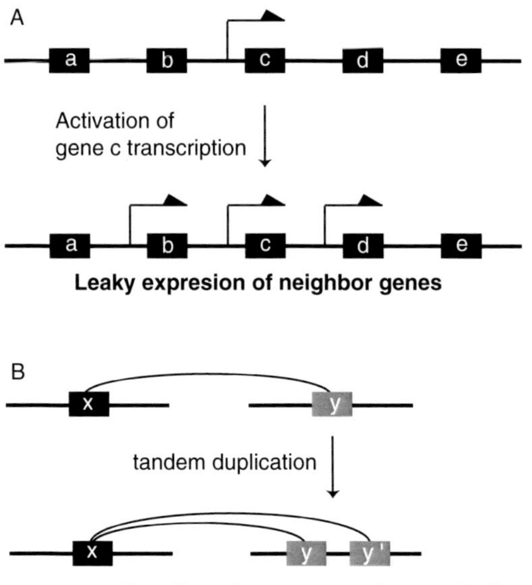

Fig. 11.1 Leaky gene expression and tandemly duplicated genes. Rectangles, horizontal lines, horizontal arrows and curved lines represent genes, chromosomes, gene expression and protein–protein interactions, respectively. A character on the rectangle indicates a gene name. (**a**) Leaky gene expression. When transcription of a gene "c" is activated, that of its neighbors may also be activated without biological relevance. (**b**) Duplication of a gene "y" interacting with another gene "x" on a different chromosome. There is not an interacting gene cluster after tandem gene duplication. On the other hand, a co-expressed gene cluster y–y' may be created by the gene duplication, when regulatory regions of the gene are also duplicated tandemly with the gene itself

co-expression from leaky expression using only gene expression data (Fig. 11.1a). We emphasize the need for other biological information, such as functional relationships, pathway information or a physical interaction between gene products, to identify functional gene clusters.

11.3 Tandemly Duplicated Genes

One third of all duplicated genes have been derived from tandem gene duplication in human, mouse and rat genomes (Shoja and Zhang 2006). If regulatory regions of a gene are also duplicated tandemly with the gene itself, the duplicated copies behave like a co-expressed gene cluster though the clustering is a historical artifact of the mechanism of duplication rather than a biologically significant arrangement. We know that there are functional tandemly-duplicated gene clusters such as *Hox* gene clusters (Sproul et al. 2005) but the most physical proximity between tandemly duplicated genes is not biologically relevant. Many early reports about co-expressed gene clusters did not control for the tandem duplicate effect (Hurst et al. 2004; Fig. 11.1b). One proposed operational definition of duplicated genes is hits from an all-against-all BLASTP similarity search with an E-value <0.2, because 90% of gene pairs classified in different gene families had E-value ≥ 0.2 (Lercher et al. 2002). Many subsequent studies have followed this definition to minimize the tandem duplication effect (Fukuoka et al. 2004; Lercher et al. 2003; Liao and Zhang 2008; Makino and McLysaght 2008; Semon and Duret 2006). However, a recent study reported co-expressed gene clusters in plants with a very conservative definition of tandemly duplicated genes (Mentzen and Wurtele 2008). The authors only included highly similar sequences as duplicated genes (all-against-all BLASTP search, E-value $<10^{-20}$), which only detects approximately 44% of the duplicated genes under the more conventional threshold (E-value <0.2; data not shown). We would like to stress that the strict elimination of tandemly-duplicated genes is important for the identification of functional gene clusters.

11.4 Interacting Gene Clusters

Identification of co-expressed gene clusters must exclude leaky expression and tandem gene duplication effects. On the other hand, the identification of gene clusters using protein–proteins interaction (PPIs) minimizes tandem duplication effects and is not affected by leaky expression (Makino and McLysaght 2008; Fig. 11.1b). Teichmann and Veitia have reported interacting gene clusters among subunits of a protein complex in yeast (Teichmann and Veitia 2004). The authors observed co-localization of genes encoding subunits of a protein complex within

10–30 kb in the yeast genome. In addition, neighboring genes on the yeast genome are likely to be neighbors in the yeast protein interaction network (Poyatos and Hurst 2006).

It had been unclear whether there are interacting gene clusters on a human genome, which is more complex and larger than the yeast genome (Hurst and Lercher 2005; Hurst et al. 2004; Poyatos and Hurst 2006). We showed the presence of interacting gene clusters within 1 Mbp on the human genome, even after the strict removal of tandem duplicated genes (Makino and McLysaght 2008). The tandem duplicated genes were excluded not only by blast but also by sensitive homology search PSI-blast. In addition, we also removed gene pairs sharing protein domains in InterPro (http://www.ebi.ac.uk/interpro/) or SCOP (http://scop.mrc-lmb.cam.ac.uk/scop/) as tandem duplicated genes (Makino and McLysaght 2008). The identified human interacting gene clusters do not belong to previously known functional gene clusters such as housekeeping genes or pairs in the same protein complexes but they are significantly enriched for response to stimulus related genes, particularly immune system related genes.

We give several examples of immune related interacting gene clusters shown in Makino and McLysaght (2008). Two interacting pairs are ATP-binding cassette transporter 1 (TAP1)—TAP binding protein (TAPBP) and ATP-binding cassette transporter 2 (TAP2)—TAPBP that mediate interactions between newly assembled MHC class I molecules and the transporter associated with antigen processing (Sadasivan et al. 1996). We found genes related to not only MHC class I but also MHC class II in the interacting gene clusters. While MHC class I molecules are present on almost all nucleated cells, MHC class II molecules are restricted to antigen-presenting cells such as macrophages, dendritic cells, B-cells and T-cells. MHC class II regulatory factor RFX1 binds to the X-boxes of MHC class II genes and is essential for their expression (Reith et al. 1990). An interacting pair, the high affinity immunoglobulin epsilon receptor (FCER1G)—the low affinity immunoglobulin gamma receptor (FCGR3A), is present on the surface of natural killer cells and macrophages to recognize immunoglobulin (Huizinga et al. 1990; Le Coniat et al. 1990). Recombination activating gene 1 (RAG1) and recombination activating gene 2 (RAG2) act together to activate immunoglobulin V-D-J recombination (Oettinger et al. 1990). An interacting pair macrophage stimulating protein 1 (MST1)—MST1 receptor (MST1R) is related to macrophage stimulation (Sakamoto et al. 1997). Two interacting pairs, complement component 4B (C4B)—complement component 2 (C2) and C4B—complement factor B (CFB) are related to innate immunity and have function in antigen–antibody complex and other complement components (Janeway et al. 2004).

Interestingly, these immune-related interacting gene clusters are located not only in previously reported large gene clusters such as the MHC on human chromosome 6 but also in other local genomic regions (Makino and McLysaght 2008). These results indicate that the organization of co-localized interacting genes is a biologically significant feature for many components of the immune system.

11.5 Evolution of Functional Gene Clusters

The co-localization of interacting genes is likely to be important for co-regulation. As mentioned above, there are bidirectional promoters in yeast (Kruglyak and Tang 2000) and mammals (Adachi and Lieber 2002; Franck et al. 2008; Koyanagi et al. 2005; Li et al. 2006; Takai and Jones 2004; Trinklein et al. 2004). Closely linked genes often share enhancers (West et al. 2002) and are under chromatin-mediated regulation (Finnegan et al. 2004; Hurst et al. 2004; Robyr et al. 2002). Furthermore, epistatic interactions between products of genes at linked loci will be preserved the closer the loci are located to each other (Nei 1967). Co-regulation and epistatic effects both predict functional constraints and thus evolutionary conservation of the interacting clusters. In addition, we may estimate the timing of gene cluster creation through comparative genomics studies.

11.5.1 Evolution of Bidirectional Promoters in Eukaryote Genomes

In yeast, a single regulatory system may control a gene pair (Kruglyak and Tang 2000), however there is no evidence for functional constraint on transcriptional orientation for co-expressed gene pairs (Huynen et al. 2001; Seoighe et al. 2000). On the other hand, closely linked head-to-head (h2h; divergent orientation) gene pairs are enriched in mammalian genomes (Adachi and Lieber 2002; Franck et al. 2008; Koyanagi et al. 2005; Li et al. 2006; Takai and Jones 2004; Trinklein et al. 2004), are often co-expressed in human and their h2h orientations have been conserved in the mouse genome (Trinklein et al. 2004). Koyanagi and her colleagues have also shown that the orientation of the h2h gene pairs has arisen recently in the mammalian lineage (Koyanagi et al. 2005). In further analyses, Li and his colleagues showed that expression patterns between the h2h gene pairs are correlated and the gene pairs are functionally correlated (Li et al. 2006). Furthermore, h2h gene pairs have been conserved during vertebrate evolution (Li et al. 2006) and are older—many originating in the tetrapod ancestor—than other gene pair orientations (Franck et al. 2008). These observations strongly suggest that bidirectional promoters exist in at least mammalian genomes.

11.5.2 Evolution of Co-expressed Gene Clusters

S. cerevisiae gene adjacencies are frequently conserved in *Candida albicans* and these conserved gene pairs are significantly co-expressed in yeast (Hurst et al. 2002; Huynen et al. 2001). The same trend was observed in vertebrate genomes (Semon and Duret 2006; Singer et al. 2005). Co-expressed and housekeeping gene clusters are not likely to contain chromosomal breakpoints between human and mouse genomes (Singer et al. 2005). In addition, co-expressed gene clusters have often been conserved between human and chicken genomes during vertebrate evolution, although there was not a large difference in the fraction of co-expressed gene clusters between observation and simulation and some co-expressed gene clusters probably arose under neutral evolution (Semon and Duret 2006). Interestingly, Liao and Zhang found that linkage of human co-expressed genes is less conserved among mammalian genomes than that of non-co-expressed genes and suggested that co-localization of co-expressed genes is disadvantageous (Liao and Zhang 2008).

The relationship of co-localization and co-expression is complex. Rapid evolutionary rates of gene expression itself must be one of the factors that makes a clear pattern difficult to discern (Khaitovich et al. 2004; Makova and Li 2003; Yanai et al. 2004).

11.5.3 Evolution of Interacting Gene Clusters

It has also been shown that interacting gene clusters have been conserved during vertebrate evolution (Makino and McLysaght 2008). In particular, the difference between the observed level of conservation of the interacting gene pair and expectation in mammals was the greatest among examined evolutionary categories (Makino and McLysaght 2008). On the other hand, the interacting gene clusters were rarely conserved in the ascidian genome, although about 40% of orthologous genes in the gene clusters were identified (Makino and McLysaght 2008). We also found several examples of the construction of interacting gene clusters from non-clustered genes within the mammalian lineage. These results indicate that substantial interacting gene clusters have been created in the vertebrate lineage, particularly the mammalian lineage.

Evolution of interacting gene clusters has not been studied well, however protein–protein interactions (PPIs) are likely to be one of the many factors affecting the evolution of genome organization.

11.6 Concluding Remarks

Physical proximity of genes facilitates co-regulation and co-inheritance of allelic combinations. This is beneficial for an organism when those genes are also functionally related. In complex organisms with many tissues and developmental stages, functional relationships can be defined in many diverse and overlapping ways. Perhaps for this reason the biological significance of the organization of eukaryotic genomes took some time to be discovered. Confounding factors also arise from the nature of the mechanisms of gene duplication that commonly results in duplicate, thus functionally related, genes arranged in an array in one genomic locality. However, it is now clear that functional gene clusters exist in eukaryotic genomes. There is not one single character that typifies these clusters but they are governed by a compendium of regulatory, epistatic and interaction effects. Comparative genomics, which exploits the natural experimentation of evolution over long periods of time, has been instrumental in uncovering the functional constraints on the organization of eukaryotic genomes, which has led directly to a greater understanding of genomic structure.

Acknowledgment This work is supported by the Science Foundation Ireland.

References

Adachi N, Lieber MR (2002) Bidirectional gene organization: a common architectural feature of the human genome. Cell 109:807–809

Blumenthal T, Evans D, Link CD, Guffanti A, Lawson D, Thierry-Mieg J, Thierry-Mieg D, Chiu WL, Duke K, Kiraly M, Kim SK (2002) A global analysis of *Caenorhabditis elegans* operons. Nature 417:851–854

Craig JM, Bickmore WA (1994) The distribution of CpG islands in mammalian chromosomes. Nat Genet 7:376–382

David L, Huber W, Granovskaia M, Toedling J, Palm CJ, Bofkin L, Jones T, Davis RW, Steinmetz LM (2006) A high-resolution map of transcription in the yeast genome. Proc Natl Acad Sci USA 103:5320–5325

Ebisuya M, Yamamoto T, Nakajima M, Nishida E (2008) Ripples from neighbouring transcription. Nat Cell Biol 10:1106–1113

Finnegan EJ, Sheldon CC, Jardinaud F, Peacock WJ, Dennis ES (2004) A cluster of Arabidopsis genes with a coordinate response to an environmental stimulus. Curr Biol 14:911–916

Franck E, Hulsen T, Huynen MA, de Jong WW, Lubsen NH, Madsen O (2008) Evolution of closely linked gene pairs in vertebrate genomes. Mol Biol Evol 25:1909–1921

Fukuoka Y, Inaoka H, Kohane IS (2004) Inter-species differences of co-expression of neighboring genes in eukaryotic genomes. BMC Genomics 5:4

Ghaemmaghami S, Huh WK, Bower K, Howson RW, Belle A, Dephoure N, O'Shea EK, Weissman JS (2003) Global analysis of protein expression in yeast. Nature 425:737–741

He F, Li X, Spatrick P, Casillo R, Dong S, Jacobson A (2003) Genome-wide analysis of mRNAs regulated by the nonsense-mediated and 5′ to 3′ mRNA decay pathways in yeast. Mol Cell 12:1439–1452

Huizinga TW, Kuijpers RW, Kleijer M, Schulpen TW, Cuypers HT, Roos D, von dem Borne AE (1990) Maternal genomic neutrophil FcRIII deficiency leading to neonatal isoimmune neutropenia. Blood 76:1927–1932

Hurst LD, Lercher MJ (2005) Unusual linkage patterns of ligands and their cognate receptors indicate a novel reason for non-random gene order in the human genome. BMC Evol Biol 5:62

Hurst LD, Pal C, Lercher MJ (2004) The evolutionary dynamics of eukaryotic gene order. Nat Rev Genet 5:299–310

Hurst LD, Williams EJ, Pal C (2002) Natural selection promotes the conservation of linkage of co-expressed genes. Trends Genet 18:604–606

Huynen MA, Snel B, Bork P (2001) Inversions and the dynamics of eukaryotic gene order. Trends Genet 17:304–306

Janeway C, Travers P, Walport M, Sholomchik M (2004) Immunobiology, 6th edn. Garland Science, New York/London

Khaitovich P, Muetzel B, She X, Lachmann M, Hellmann I, Dietzsch J, Steigele S, Do HH, Weiss G, Enard W, Heissig F, Arendt T, Nieselt-Struwe K, Eichler EE, Paabo S (2004) Regional patterns of gene expression in human and chimpanzee brains. Genome Res 14:1462–1473

Koyanagi KO, Hagiwara M, Itoh T, Gojobori T, Imanishi T (2005) Comparative genomics of bidirectional gene pairs and its implications for the evolution of a transcriptional regulation system. Gene 353:169–176

Kruglyak S, Tang H (2000) Regulation of adjacent yeast genes. Trends Genet 16:109–111

Le Coniat M, Kinet JP, Berger R (1990) The human genes for the alpha and gamma subunits of the mast cell receptor for immunoglobulin E are located on human chromosome band 1q23. Immunogenetics 32:183–186

Lee JM, Sonnhammer EL (2003) Genomic gene clustering analysis of pathways in eukaryotes. Genome Res 13:875–882

Lercher MJ, Blumenthal T, Hurst LD (2003) Coexpression of neighboring genes in Caenorhabditis elegans is mostly due to operons and duplicate genes. Genome Res 13:238–243

Lercher MJ, Urrutia AO, Hurst LD (2002) Clustering of housekeeping genes provides a unified model of gene order in the human genome. Nat Genet 31:180–183

Li YY, Yu H, Guo ZM, Guo TQ, Tu K, Li YX (2006) Systematic analysis of head-to-head gene organization: evolutionary conservation and potential biological relevance. PLoS Comput Biol 2:e74

Liao BY, Zhang J (2008) Coexpression of linked genes in mammalian genomes is generally disadvantageous. Mol Biol Evol 25:1555–1565

Makino T, McLysaght A (2008) Interacting gene clusters and the evolution of the vertebrate immune system. Mol Biol Evol 25:1855–1862

Makova KD, Li WH (2003) Divergence in the spatial pattern of gene expression between human duplicate genes. Genome Res 13:1638–1645

Mentzen WI, Wurtele ES (2008) Regulon organization of Arabidopsis. BMC Plant Biol 8:99

Nei M (1967) Modification of linkage intensity by natural selection. Genetics 57:625–641

Oettinger MA, Schatz DG, Gorka C, Baltimore D (1990) RAG-1 and RAG-2, adjacent genes that synergistically activate V(D)J recombination. Science 248:1517–1523

Poyatos JF, Hurst LD (2006) Is optimal gene order impossible? Trends Genet 22:420–423

Reik W, Walter J (2001) Genomic imprinting: parental influence on the genome. Nat Rev Genet 2:21–32

Reith W, Herrero-Sanchez C, Kobr M, Silacci P, Berte C, Barras E, Fey S, Mach B (1990) MHC class II regulatory factor RFX has a novel DNA-binding domain and a functionally independent dimerization domain. Genes Dev 4:1528–1540

Robyr D, Suka Y, Xenarios I, Kurdistani SK, Wang A, Suka N, Grunstein M (2002) Microarray deacetylation maps determine genome-wide functions for yeast histone deacetylases. Cell 109:437–446

Sadasivan B, Lehner PJ, Ortmann B, Spies T, Cresswell P (1996) Roles for calreticulin and a novel glycoprotein, tapasin, in the interaction of MHC class I molecules with TAP. Immunity 5:103–114

Sakamoto O, Iwama A, Amitani R, Takehara T, Yamaguchi N, Yamamoto T, Masuyama K, Yamanaka T, Ando M, Suda T (1997) Role of macrophage-stimulating protein and its receptor, RON tyrosine kinase, in ciliary motility. J Clin Invest 99:701–709

Salgado H, Moreno-Hagelsieb G, Smith TF, Collado-Vides J (2000) Operons in *Escherichia coli*: genomic analyses and predictions. Proc Natl Acad Sci USA 97:6652–6657

Satou Y, Mineta K, Ogasawara M, Sasakura Y, Shoguchi E, Ueno K, Yamada L, Matsumoto J, Wasserscheid J, Dewar K, Wiley GB, Macmil SL, Roe BA, Zeller RW, Hastings KE, Lemaire P, Lindquist E, Endo T, Hotta K, Inaba K (2008) Improved genome assembly and evidence-based global gene model set for the chordate *Ciona intestinalis*: new insight into intron and operon populations. Genome Biol 9:R152

Semon M, Duret L (2006) Evolutionary origin and maintenance of coexpressed gene clusters in mammals. Mol Biol Evol 23:1715–1723

Seoighe C, Federspiel N, Jones T, Hansen N, Bivolarovic V, Surzycki R, Tamse R, Komp C, Huizar L, Davis RW, Scherer S, Tait E, Shaw DJ, Harris D, Murphy L, Oliver K, Taylor K, Rajandream MA, Barrell BG, Wolfe KH (2000) Prevalence of small inversions in yeast gene order evolution. Proc Natl Acad Sci USA 97:14433–14437

Shoja V, Zhang L (2006) A roadmap of tandemly arrayed genes in the genomes of human, mouse, and rat. Mol Biol Evol 23:2134–2141

Singer GA, Lloyd AT, Huminiecki LB, Wolfe KH (2005) Clusters of co-expressed genes in mammalian genomes are conserved by natural selection. Mol Biol Evol 22:767–775

Spellman PT, Rubin GM (2002) Evidence for large domains of similarly expressed genes in the Drosophila genome. J Biol 1:5

Sproul D, Gilbert N, Bickmore WA (2005) The role of chromatin structure in regulating the expression of clustered genes. Nat Rev Genet 6:775–781

Takai D, Jones PA (2004) Origins of bidirectional promoters: computational analyses of intergenic distance in the human genome. Mol Biol Evol 21:463–467

Teichmann SA, Veitia RA (2004) Genes encoding subunits of stable complexes are clustered on the yeast chromosomes: an interpretation from a dosage balance perspective. Genetics 167:2121–2125

Trinklein ND, Aldred SF, Hartman SJ, Schroeder DI, Otillar RP, Myers RM (2004) An abundance of bidirectional promoters in the human genome. Genome Res 14:62–66

West AG, Gaszner M, Felsenfeld G (2002) Insulators: many functions, many mechanisms. Genes Dev 16:271–288

Williams EJ, Hurst LD (2002) Clustering of tissue-specific genes underlies much of the similarity in rates of protein evolution of linked genes. J Mol Evol 54:511–518

Yanai I, Graur D, Ophir R (2004) Incongruent expression profiles between human and mouse orthologous genes suggest widespread neutral evolution of transcription control. Omics 8:15–24

Chapter 12
Knowledge Standardization in Evolutionary Biology: The Comparative Data Analysis Ontology

Francisco Prosdocimi, Brandon Chisham, Enrico Pontelli, Arlin Stoltzfus, and Julie D. Thompson

Abstract In this chapter we describe the development of a new biomedical ontology in the context of the modern knowledge representation research field. We also present the modeled concepts and their relevance in the light of the history of evolutionary biology. CDAO stands for "Comparative Data Analysis Ontology" and allows the representation of data produced in evolutionary biology studies in the form of a set of well-defined concepts and the relationships among them. CDAO is not intended to be a glossary or a simple taxonomy of evolution-related terminology. Since evolutionary theory provides a broad framework for almost all fields of biology, the concepts in CDAO reflect a rich history of controversies stressed by academics in philosophical analyses of the whole field of biology. The concept of an evolutionary tree to represent relationships between organisms is credited to Darwin. However, the nature of species and the operational taxonomic units (OTU) used in evolutionary analysis are still a matter of controversy among scholars. The same can be said for a number of other concepts modeled in CDAO. For instance, the choice of a methodological basis for evolutionary analysis is still a matter of debate: should researchers use simple "non-theory" based comparative approaches to analyze their data? Should they assume parsimony to adapt phylogeny towards a popperian concept of science? Should they consider likelihood or Bayesian methods as more appropriate to their endeavor? In the first part of this chapter we try to understand the role of a knowledge representation task in the context of modern

F. Prosdocimi and J.D. Thompson
Institut de Génétique et de Biologie Moléculaire et Cellulaire (IGBMC), Department of Structural Biology and Genomics, Strasbourg, France. 1 rue Laurent Fries/BP 10142/67404 Illkirch Cedex, France
email: fpros@igbmc.fr
B. Chisham and E. Pontelli
Department of Computer Science, New Mexico State University, P.O. Box 30001, MSC CS Las Cruces, NM 88003, USA
A. Stoltzfus
Center for Advanced Research in Biotechnology, University of Maryland Biotechnology Institute, 9600 Gudelsky Drive, Rockville, MD 20850, USA

P. Pontarotti (ed.), *Evolutionary Biology: Concept, Modeling, and Application*,
DOI: 10.1007/978-3-642-00952-5_12,

research in biology. The second part is devoted to the presentation of the concepts modeled in CDAO and their specification using standard ontology descriptors. Finally, the third part of the chapter deals with historical discussions in evolutionary biology that influenced the genesis and development of CDAO's formalized concepts. This approach will be extended to show how evolutionary data can be represented in ontologies in order to cope with the multiplicity of approaches and philosophical backgrounds used in this endeavor. CDAO will prove to be a very pluralistic ontology, allowing the representation of evolutionary data in a number of different theoretical backgrounds normally assumed by evolutionary biologists. Although further discussions and new software will be needed, we believe that CDAO will become the standard way for describing evolutionary biology concepts in the near future. This statement is based on the fact that (1) CDAO is highly theoretical and describes the most relevant evolutionary biology concepts; and (2) it is flexible enough to be able to represent, inter-relate and allow further reasoning over the flood of data from the modern petabyte era of biological research.

12.1 Introduction

12.1.1 Knowledge Representation as a Positive Heuristic in Biomedicine

At the intersection of information, computation and biological sciences, there exists an interdisciplinary research field that has been growing steadily over the last decade. This field is related to new techniques of ontology production for biomedical science. The term ontology is inherited from its original application in philosophy and is frequently said to be related *to the study of the nature of being, existence and reality*. From the point of view of information science (or knowledge representation), this new task can be understood as the formalization of the structure of knowledge in some area of research. This formalization is frequently performed through the definition of fixed words linked to rigorous concepts that researchers consider to be relevant in some specific area (Gruber 1995). Furthermore, by providing a common representation of interesting research topics, ontologies permit data and knowledge to be integrated, reused and shared easily by both researchers and computers (Harris 2008). If everybody uses the same terms and relations to describe their data, they can understand each other better, exchange results, integrate them in a common perspective and write algorithms to analyze the organized information in a large-scale fashion.

One of the advantages of these ontologies inspired by information science, lies in the fact that they allow the construction of structures similar to phrases by which concepts link to each other using relations. The semantics appear and defined entities are related to each other using explicit verbs or verbal-derived forms, such as *is_a*, *has*, *contains*. The creation of an ontology may be compared to the description of a new language – in the sense that symbols (nouns or sets-of-nouns) are defined to describe events and entities in the real world and other symbols (actions or verbs) are created to relate these entities based on some theoretical background. The relationships between the terms or symbols are generally built using some kind of pre-defined formal rules. In a classical paper called *Empiricism, semantics and ontology*, the German philosopher Rudolf Carnap considered mathematics as a sort of ontology and suggests that mathematicians speak about "symbols and formulas manipulated according to given formal rules" (Carnap 1950). From a modern knowledge representation point of view, the symbols described by Carnap can be understood as the ontology concepts and the formal rules are represented by the relations between the terms and some restrictions to these relations. An example from molecular biology is the Gene Ontology in which the relationships allow the hierarchical description of gene functions, such as: "glycolysis" (GO:0006096) is part_of "hexose metabolic process" (GO:0019320) (Ashburner et al. 2000). This concept-and-relations schema allows ontologies to formalize and create meaning through such semantic sentences.

Although long-standing classical studies in epistemology and philosophy of science have clearly demonstrated that this task of knowledge conceptualization can never be consensual among all researchers in some area, well-designed formal representation of natural, real-world entities can advance human knowledge and generate new research fields. The Hungarian philosopher of science Imre Lakatos in his classical book *Methodology of scientific research programs* argues that a research (scientific) program is growing when it generates what he calls *positive heuristics*, i.e., when it generates even more research and becomes more studied and specialized over time (Lakatos 1978). Figure 12.1 demonstrates the exponential growth in the study of biomedical ontologies, represented by the number of papers with the word *ontology* in their title or abstract.

The data suggest that ontologies are deemed useful by the scientific community and that the consideration of conceptualization issues in biological sciences brings positive heuristics to this research field. In other words, the production of biomedical ontologies is a fertile research program that should promote the development of the field.

12.1.2 Ontologies in the Petabyte Era of Biological Research

When initially describing some conceptual system, all the terms and relations must be explicitly described as clearly as possible in order to solicit criticisms from the community. These criticisms should generate discussions about the formalized concepts in order to make them as clear, broad and consensual as possible. This agreement among researchers is of critical importance to the general acceptance of the ontology and thus, most scientific ontologies in use today are the result of

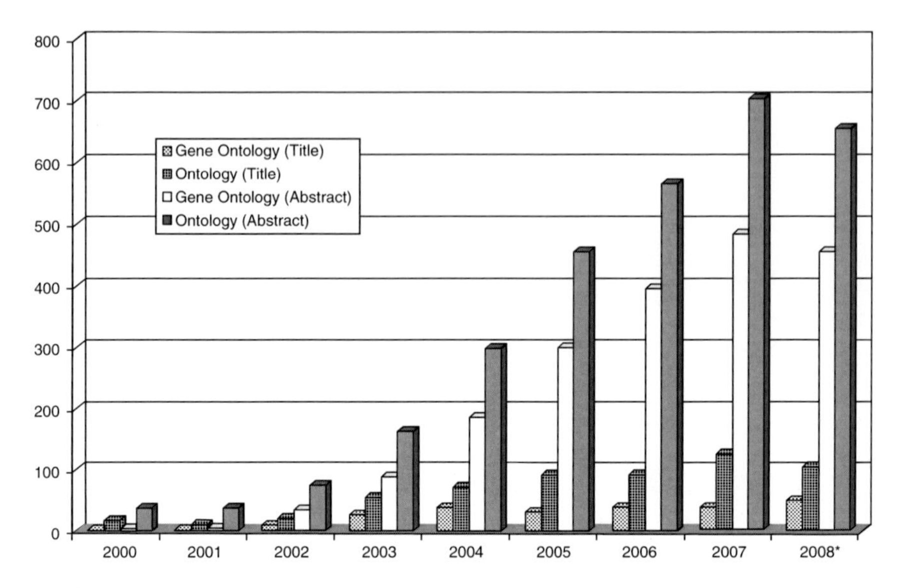

Fig. 12.1 Growth in ontology usage in biomedical science research measured by Pubmed searches for terms "ontology" and "gene ontology" in abstracts and titles of papers since 2000

collaborative scientific research efforts (Rodrigues et al. 2006; Leontis et al. 2006; Gene Ontology Consortium 2001).

As a consequence, once a system of concepts is artificially created to represent some area of human knowledge, empirical data should be described using this prototype ontology. This in turn, allows the use of computational techniques, such as the vast array of expert algorithms (Sirin et al. 2007; Tsarkov and Horrocks 2006) developed to read large amounts of empirical data and to inter-relate and interpret them under some rational perspective. These algorithms, which extract knowledge from ontology-based resources, are frequently called *reasoners*. Computer reasoning allows the production of higher-level information from large volumes of raw data – for example, derived from hard-science molecular experiments – and facilitates subsequent analyses and interpretation by humans. The extraction of knowledge from high-throughput data clearly cannot be performed by a single individual, or even by a trained group of experts. In order to develop modern science, new computer-readable standards are urgently needed in a number of research areas, including evolutionary biology.

In a recent set of reports published in Nature (Issue 7209 dated 4 September 2008) about "big data," the editors argue that biology will arrive soon in what they have called the "Petabyte era" (the petabyte is a value corresponding to 1,024 terabytes that, in turn, represents about 1,000,000 megabytes or 10^{15} bits of information). An entirely new generation of super-fast DNA sequencers will soon be up and running all over the world (Rothberg and Leamon 2008; Shendure and Ji 2008) and modern DNA sequencers such as Illumina's Solexa, the Applied Biosystems' Sequencing by Oligonucleotide Ligation and Detection (SOLiD) technology and the GS FLX instruments from Roche/454 Life Sciences will soon revolutionize the study of life sciences (Marguerat et al. 2008). Some of these sequencers are capable of producing the information necessary to build an entire human genome (3Gb) in about a week (Dohm et al. 2008) but we still lack the informational and computational infrastructure to actually analyze all these sequences, in order to understand the biological and evolutionary perspectives. Bioinformatics resources have begun to respond to the challenges posed and are now addressing the issues of storage, availability, reasoning and interpretation of this explosion of information (Cochrane et al. 2009). The next decade will certainly see the rise of new ways of working in the so-called "big science" field and it is probably not too risky to suppose that the graph shown in Fig. 12.1 represents only the start of a more accentuated growth in the usage of ontologies and ontology-based studies over the next decade.

12.1.3 The Central Role of Evolutionary Biology

The modern version of Darwinian evolutionary theory could be said to be the most unifying conceptual system in the whole field of biology. "*Nothing in biology makes sense except in the light of evolution*," says one of most notorious biologists from the 20th century and one of the architects of *synthetic theory of evolution*, or the modern synthesis, the Russian geneticist Theodosius Dobzhansky (Dobzhansky 1973). If we assume that evolution is the explicative point of convergence of the biological sciences, it follows that researchers in this crucial field must adapt their methodology to today's petabyte era. However, the long tradition of evolutionary studies has highlighted a number of conceptual and philosophical discussions on its main problems, concepts and data analysis paradigms. Recurrent academic discussions in the evolutionary field concern, among others, (1) the nature of species, (2) the strength of natural selection and the relevance of random processes in evolution (selectionism versus neutralism), (3) the origin of life, (4) the rate of genomic and anatomic modification over time (gradualism, punctuated equilibrium and saltationism) and (5) the different empirical approaches to study and comprehend evolutionary data (phenetics, cladistics, parsimony, likelihood, distance metrics, etc.). Thus, in order to formalize the basic knowledge in the evolutionary biology research field, most of these ideas, discussions and paradigms must be taken into account. As we will see, most of these discussions are particularly relevant when choosing and naming fundamental concepts in the evolutionary biology research field and defining their scope and the relations between them. The terminology must be exact and precise and must allow the multiplicity of approaches to be represented, so that every evolutionary biologist will be able to describe his data using his preferred methodology. The goal of comparative data analysis ontology (CDAO) is exactly this: to model a limited number of important concepts in the field in such a way that researchers will feel free to describe their data from their own perspective.

12.1.4 Current Biomedical Ontologies

Many ontologies have been developed recently in the biomedical field to represent specific aspects of the real world, such as anatomical parts (Gaudet et al. 2008; Maglia et al. 2007; Larson et al. 2007; Trelease 2006; Bard 2005; Lee and Sternberg 2003), development, disease, or DNA sequences (Segerdell et al. 2008; Schulz et al. 2007, 2008; Chabalier et al. 2007; Bard et al. 2005; Jaiswal et al. 2005). A significant number of these ontologies are housed on the Open Biomedical Ontologies (OBO) web site, created by Ashburner and Lewis in 2001 as an umbrella body for the developers of life science ontologies (Smith et al. 2007). Probably, the most well-known and widely used ontology in the biomedical field is the Gene Ontology (GO), which consists in three separate parts, describing different cellular structures, molecular functions and biological processes of genes. GO has been very successful as the first set of concepts to be actually applied in a large number of genome projects, e.g., FlyBase (*Drosophila*), the *Saccharomyces* Genome Database (SGD), the Mouse Genome Database (MGD) and modern scientific studies all over the world.

Although GO allows a clear and concise representation of the human knowledge associated with gene functions, it contains very simple semantic relationships among the concepts described therein. The concepts are structured in a hierarchy ranging from the most general to the most specific (Ashburner et al. 2000) and specific sub-concepts, such as "DNA recombination" (GO:0006310) or "DNA repair" (GO:0006281), are related to their parents, such as "DNA metabolism" (GO:0006259), using the *is_a* or *part_of* semantic relationships (http://wiki.geneontology.org/index.php/Relation_composition). Most of the other accepted ontologies in the biomedical field are also conservative in the use of new relations among entities, illustrated by the fact that the Relation Ontology (Smith et al. 2005) contained only about a dozen relations in its latest release. However, it is clear that some fields of knowledge cannot be represented by such a narrow set of relations and a dedicated discussion list on the subject has recently been set up (http://www.obofoundry.org/ro/) with proposals for a number of new relations from scientific research organizations, such as the UCDHSC (University of Colorado at Denver and Health Sciences Center) and the OBI (Open Biomedical Investigations).

As we argued before, modern high-throughput methods will require that data is described in terms of standard vocabularies and today's substantial efforts to produce consistent ontologies will be valuable for twenty-first-century biology. Nevertheless, it is clear that ontology concepts that are clearly correlated to the physical world (such as anatomical parts or even DNA-based descriptors) are less interpretative in fashion than more general or abstract concepts, such as the OTU or character in evolutionary theory. This makes the production of broad theory ontologies a different task, requiring distinct formalisms, structures and relations.

12.2 The Comparative Data Analysis Ontology

12.2.1 History of CDAO Development

Taking into account the necessity to analyze modern high-throughput biology data under an evolutionary perspective, the production of an ontology was envisioned by the members of the Evolutionary Informatics group at NESCent (http://evoinfo.nescent.org). The Evolutionary Informatics group includes many of the most prominent world-leaders in software development for evolutionary analysis. The members of the group have been meeting at Durham, NC biannually since early 2007 to discuss future infrastructure requirements in this research area. The idea of developing an ontology to formalize the knowledge in the field and allow automatic reasoning came in the early days. Although everyone had agreed that concepts in evolutionary biology could be interpreted differently depending on the various perspectives – as we shall discuss – a number of these experts decided to launch this ambitious project of creating a broad-range ontology to describe this central notion in biology: the evolutionary theory. The comparative data analysis ontology (CDAO) represents the first step in this process.

CDAO was developed using the OWL language using the Stanford University ontology editor Protégé. OWL (Web-Ontology Language) was chosen as the most modern, flexible and robust language in the field, being the standard recommended by the World Wide Web consortium (W3C) for ontology implementation. Although recent and still under development, the ontology editor Protégé has been shown to be a user-friendly, robust and powerful tool to build OWL ontologies. With more than 100,000 registered users, this software is supported by numerous academic, government and corporate users (http://protege.stanford.edu/).

The overall structure and the main concepts described in CDAO are shown in Fig. 12.2.

12.2.2 CDAO: Some Evaluation Considerations

The main purpose of the first prototype release of CDAO is to describe the most basic structure of knowledge underlying the key entities in evolutionary analysis. Thus, it does not provide an extensive terminological coverage (e.g., covering the complete set of concepts identified in the controlled vocabulary at https://www.nescent.org/wg_evoinfo/ConceptGlossary). This distinguishes CDAO from ontologies like the Gene Ontology – where there is a more limited emphasis on structuring knowledge (e.g., by using only a very basic set of relations between terms) but an extensive informational coverage, mostly expressed in terms of a deep taxonomy.

Several schemes have been proposed to provide a classification of ontologies. According to such schemes, CDAO can be classified as follows:

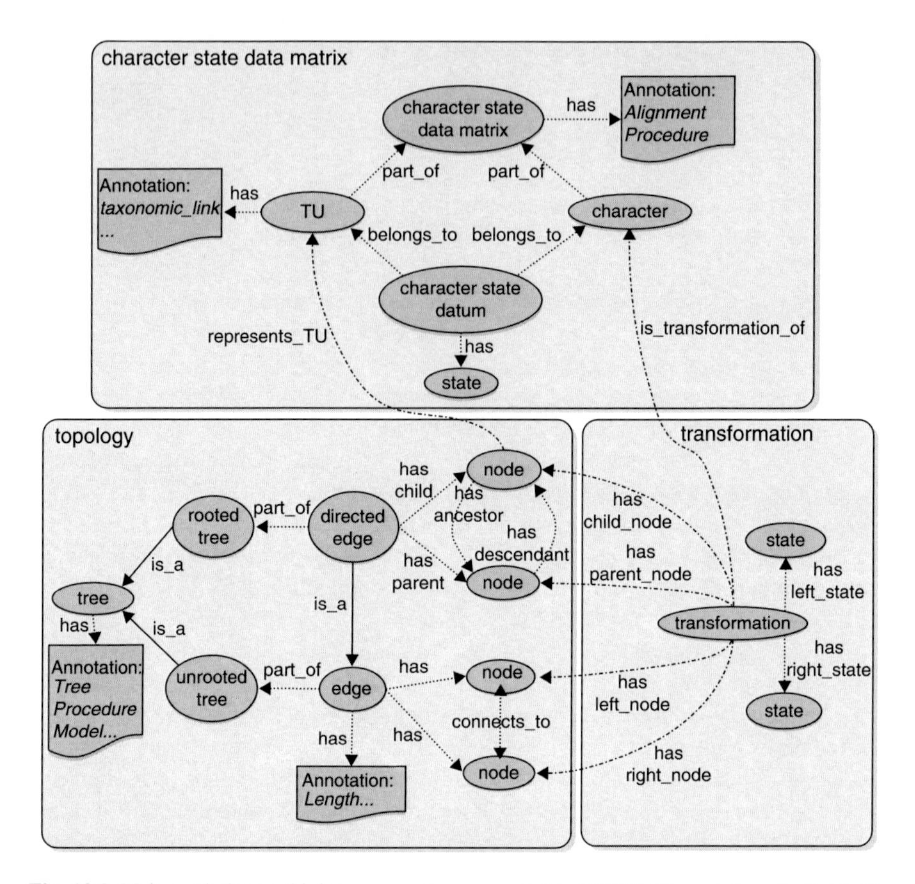

Fig. 12.2 Main evolutionary biology concepts represented in CDAO. The ontology is divided in three inter-related parts: (1) the character-state data matrix, representing data conceptualization and description; (2) the tree topology, representing the ancestral relationships among the groups analyzed; and (3) the transformation, representing the step-wise modification in characters along evolutionary time. *2008 data collected on November 18, 2008

- Following the richness classification (Lassila and McGuiness 2001), CDAO is at one of the highest levels of complexity, as it includes general logical constraints and disjoint classes.
- According to the subject-based classification (Gomes-Perez et al. 2004), CDAO can be viewed as a domain-task ontology, as it describes concepts and relations valid across one specific (although broad) domain: evolutionary analysis, with a focus on specific tasks within the domain (e.g., phylogenetic inference and description).
- Van Heijst et al. (1997) provides two dimensions for the classification of ontologies. The first dimension measures the amount of structure present in the ontology; in this regard, CDAO can be viewed as a knowledge modeling ontology, having less of a focus on extensive terminological coverage and

deep hierarchical classification and a greater emphasis instead on the basic conceptual structure of knowledge in the domain. The second dimension classifies ontologies based on the subject of the conceptualization; in this regard, CDAO is a domain ontology (like the majority of ontologies in the OBO foundry).

- Mizoguchi and Vanwelkenhuysen (1995) classify ontologies based on their use. The four top levels of the classification distinguish between Content, Communication, Indexing and Meta ontologies. CDAO is clearly a Content ontology; its aim is to enable the reuse of knowledge related to evolutionary analysis across agents and applications (e.g., stages of an analysis pipeline). It is important to observe that CDAO is not a Communication ontology, as the purpose is to emphasize the *structure of knowledge*. Content ontologies are further classified as Workplace, Task, Domain and General ontologies. CDAO is a Domain ontology, which is Task-independent (not being tied to one specific task), Activity-related (since CDAO emphasizes, at this time, the representation of knowledge related to phylogenetic analysis) and it represents an Object ontology (since it describes structure, behavior and function of entities).

12.2.3 CDAO Version 2.0

Ontology development is a continuous, iterative process and work on version 2 of CDAO has already begun. This version aims to provide a more complete structure or classification of the concepts defined in CDAO 1.0, in order to facilitate more complex querying of CDAO data sets. As an example, we have added a richer set of classes to describe the topology of trees. Using this richer annotation set, higher-order concepts such as "Fully Resolved Tree" or "Polytomy" can be easily deduced, either by a researcher or by an automatic reasoning software system. Previously, these concepts could be inferred from the topology of a tree, if the reasoning system had information concerning the definition of a polytomy. Our goals in including these definitions in CDAO are (1) to standardize the definition of these concepts and (2) to allow the use of general reasoners that have no specific evolutionary knowledge. The terminological/informational component of CDAO is also being expanded to provide a more extensive coverage of evolutionary concepts. For instance, CDAO 2.0 includes imports for MyGrid (http://www.mygrid.org.uk/) terms, in order to facilitate its use within web services. The additional terms specify a large number of phylogenetic file formats, repositories and web services in common use. This change will hopefully contribute to early efforts to develop a MIAPA (*Minimal Information About a Phylogenetic Analysis*) standard (Leebens-Mack et al. 2006).

All the documents published and other relevant information can be found on our web site (http://www.evolutionaryontology.org). Please refer to these documents when searching for the development of CDAO-based applications.

12.3 Conceptual Revolutions in Evolutionary Biology and Historical Analysis of CDAO Concepts

In this section our aim is to make a very brief recapitulation of the history of evolutionary biology regarding some conceptual modifications in this discipline since the work of Charles Darwin. We will discuss how these concepts have evolved and how they are currently represented in CDAO. We hope to provide a theoretical analysis of our conceptualization work, illustrating the theoretical plasticity of our ontology and how it facilitates more practical tasks related to data annotation and representation of trees and character matrices, for example.

The concepts formalized in CDAO represent a vast and rich history of controversies. The theory of evolution, as the broadest conceptual model in biology, has been the target of a number of theoretical revisions mainly during the twentieth century, resulting in today's rich version that encompasses and links together many diverse research fields, including biochemistry, genetics, genomics, ecology, zoology, botany, microbiology, development, physiology and medicine. Every single data collection made in biology can be viewed from an evolutionary perspective and our goal is to represent them in a modern and scalable way.

12.3.1 Conceptual Reformulations in Evolutionary Biology Since Darwin

The years that followed Darwin's publication *The Origin of Species* (1859) revealed a number of misunderstandings about the meaning of his global theory. Although evolution, as opposed to creationism and common ancestry were soon accepted by most evolutionists in the nineteenth century, the other parts of the theory were still seen skeptically. The first point of theoretical disagreement was resolved in the nineteenth century by the German biologist August Weissmann in his book *On Heredity*, finally bringing to an end the Lamarckian idea of the inheritance of acquired characteristics and the use and disuse of characters (Lamarck 1809). Weissmann advocated the "germ plasm" theory, in which inheritance could only take place by modifications in germ cells, since the other cells of the body do not influence future generations and therefore could not act as agents of heredity (Weissmann 1889). The term **Neodarwinism** is frequently used to refer to Weissmann's view of evolutionary theory.

During the first decade of the twentieth century, the works of Mendel were rediscovered, although the interpretation of his works, made mainly by the Dutch botanist Hugo De Vries, turned this theory into a mutation or saltationistic view of evolution as opposed to Darwinian gradualism (Jacob 1970). Further development in this field in Britain, made mainly by William Bateson, led to the birth of genetics as a scientific discipline and the proliferation of saltationism. Nevertheless, it was only during the 1930s and 1940s that a new conceptual revolution in evolutionary biology would take place, based on the work of Thomas Morgan in New York. His studies on heredity and evolution were mainly developed in fruit-flies of the genus *Drosophila*, making this insect one of the most studied model organisms in the whole field of biology. According to Mayr (1991), the studies of Morgan and his students indicated that small mutations could allow a gradual modification in populations; the sudden leaps predicted by saltation theory were no longer needed to explain evolution. The subsequent revolution, integrating gradualism, Mendelism and the recent population genetics from a Darwinistic perspective was named the **Fisherian synthesis** (Mayr 2004). The next conceptual integration brought a common perspective to population genetics, biodiversity, spatial-temporal patterns and evolutionary theory. While population geneticists were interested in explaining the evolution and allelic changes inside a population group (anagenesis), naturalists were interested in how a population could differentiate into two groups, leading to a new species (cladogenesis). The publication of *Genetics and the Origin of Species* by Dobzhansky in 1937 opened the doors for the integration of these two previously inconsistent points-of-view into a single evolutionary framework. Julian Huxley named this the **synthetic theory of evolution,** or new synthesis and it was better understood and elaborated after the publication of books by Mayr, Simpson, Huxley and Stebbins.

Finally, the last great conceptual revolution in evolutionary biology resulted from the development of molecular biology. Crick's central dogma, predicting that the flux of genetic information was unidirectional, finally explained the molecular basis for why the information from the environment could not directly influence DNA, corroborating Weissmann's studies on heredity (Mayr 1991). The overall similarity of the genetic code among all life forms could be seen as a final proof corroborating all evolutionary theory. Moreover, the astonishing similarities amongst the sequences of genes and proteins involved in the most basic metabolic pathways between bacteria, archaeabacteria, protists, fungi, plants and animals may be seen as the ultimate attestation that evolution occurs. The advent of molecular biology, together with previous theoretical synthesis, has led to our modern evolutionary theory, that Mayr suggests should be simply called **Darwinism**.

CDAO can only be envisioned today thanks to all of these theorists, who have provided us with a solid theoretical basis. In the following sections, we will discuss the history of some specific concepts formalized in CDAO.

12.3.2 History of CDAO Concepts: The Taxonomic Unit

A widely used term in the evolutionary analysis field is "OTU," which stands for "*Operational Taxonomic Unit*." OTU generally refers to the organisms or group of organisms the researcher is working on when he begins his studies. The problem of the OTU concept is clearly related to the so-called "*species problem*." For a long time, species were considered to be biological organisms sharing some arbitrarily-defined morphological characteristics. However, with the development of biology in the nineteenth and twentieth centuries, a number of obstacles were identified, such as (1) the description of a number of cryptic species, where organisms very similar in form have completely different evolutionary origins; and (2) the enormous amount of variation observed in organisms from the same species, linked to sex, age, or particular environments. In the early twentieth century, the concept of *reproductive isolation* was proposed and species came to be considered as closed genetic groups changing gene alleles inside their reproductive group. But this concept has rapidly presented problems too. How should this be applied to asexual species? And how does this explain the formation of hybrids? No general consensus about the species problem has been achieved yet and pragmatically speaking, taxonomists still use morphological and physiological characters to describe new species.

Recently, *molecules* have been widely used as units for comparison. Although molecules can be representative of species or organism groups, researchers must take into account the fact that the evolution of a gene or protein may be biased by a number of factors. The evolutionary history of a gene family can be compared to the evolutionary history of the organisms harboring these genes but the evaluation of a number of different genes would be necessary to actually infer a phylogeny of species based on molecular data.

12.3.2.1 TU@CDAO

In the context of an ontology representation of evolutionary data, all these philosophical problems concerning the definition of biological groups have been, of course, inherited. It is clear that the definition of any entity, or "taxonomic unit," to be compared is entirely under the responsibility of the researcher and CDAO should allow the representation of entities at the level of molecules, gene/protein families, organisms, populations, species, or any other higher taxonomic level.

One particular problem that we encountered in the formalization of the taxonomic unit concept in CDAO is that the term "OTU" is often used in phylogenetic studies to indicate present-day entities (species, populations, individuals, molecules, etc.) under study, while a different term "HTU," or hypothetical taxonomic unit, is used to refer to the ancestral entities. In some specialized phylogenetic studies, for example in viral studies, the distinction between present-day entities and ancestors is less clear, since it is possible that the ancestors of these rapidly evolving organisms still exist. As a consequence, we decided to define a single concept for both present-day and ancestral entities: the TU, or taxonomic unit.

12.3.3 History of CDAO Concepts: The Character and the Character-State Data Matrix

Once the taxonomic units to be studied have been clearly defined, the next step in an evolutionary analysis is the choice of the specific traits or characters to be used to classify these TUs. Until the theory of descent was actually accepted in biology and used as a criterion for classification, morphological traits were employed as characters to distinguish species by naturalists. David Hull credits to Hennig the first serious attempt to claim that systematic classifications should be made using evolutionary homologies (Hull 1988) and this has become one of the most relevant concepts in evolution: the concept of *character homology*. Two characters are said to be homologous if they descend from the same ancestral character present in an ancestral organism. This argument was so strong that it was subsequently incorporated by most evolutionary biologists and it could be said that systematics today cannot be considered as separate from phylogenetics.

Maureen Kearney however points out that the identification and definition of characters in species have led to the formation of two different schools in evolutionary analysis and taxonomy (Kearney 2007). The first school was formed by the *pheneticists* who claimed that both evolutionary analysis and systematics should be done in a "theory-free" context. They claimed that a homology assumption would create some sort of circularity in the methodology of evolution: if evolutionists want to use phylogenetic trees to test hypotheses about evolution they should not use evolutionary homology assumptions to build the trees. The second school of *cladists* argues that both the evolutionary analysis of organism groups and the biological nomenclature of them should be based on Darwinism, i.e., that characters should be analyzed under an evolutionary context *if and only if* they can be assumed to be homologous. In response to the tautology accusation made by pheneticists, cladists advocate the philosophical principle of parsimony and the general scientific principle known as the Occam's razor normally understood as "*Pluralitas non est ponenda sine necessitate*" ("plurality should not be posited without necessity"). The cladists claim that their theory is closer to the popperian philosophy of science in the sense that the most parsimonious tree is the representation of the evolutionary history that could not be falsified by the available data in the light of modern evolutionary theory.

12.3.3.1 Character@CDAO

In the face of this plurality, CDAO allows the user to define their characters independently of any supposed homology among them. Although it has now become almost a consensus that classification should be done using homologous characters, whole-genome and multiple alignment comparisons are capable of producing characters that seem more phenetic than cladistic. Once the user has defined the characters, they can be annotated as being homologous or not.

Depending on the particular entities (TUs) under study, widely different characters may be used to classify them, including morphological traits, molecular characters, such as nucleotide residues or amino acids, gene functions, cellular localizations or expression levels. The diverse types of data are taken into account in CDAO by defining sub-classes of the more general character concept, namely discrete_character, continuous_character, categorical_character and compound_character. The compound_character provides a mechanism for combining individual characters into a single character, for example, to allow the definition of different parts of a molecule as special characters in an evolutionary study.

For a given set of TUs, the state of each of the selected characters is entered into a matrix, known as a "character-state data matrix," where the rows of the matrix correspond to TUs and the columns represent the characters. The "character-state" data model is generic and can be applied to most data types, e.g., in a protein sequence alignment, the TU represents a protein, the character represents a column in the alignment and the character state is either a residue or a gap. The states of high-level biological characters (morphology, development, anatomy, behavior) are also typically encoded as discrete states. Many of these character states are defined in existing bio-medical ontologies, often in OWL format and can be used in conjunction with CDAO to annotate specific characters and character states.

Missing data and absent features, common in biological data, may be treated as an extra state. Thus, a feature that is found only in some TUs, such as an intron, can be represented by a binary character with "presence" and "absence" states. This concept will become increasingly important with the more widespread use of automatic phylogenetic inference approaches, as defined by (Eisen 1998).

12.3.4 History of CDAO Concepts: The Tree

The fact that biological organisms relate to each other following a tree-like schema was originally described by Darwin in *The Origin of Species* (1859). One of the biggest pieces of scientific work in all the history of science, this paradigm-shift book – in the words of the philosopher of science Thomas Kuhn (Kuhn 1962) – was published exactly 150 years ago and contained only a single illustration. This picture is the ancestor of all modern studies in evolution and opened the way for a field mainly developed during the 1950s by Willi Hennig: cladistics (or phylogenetics). Figure 12.3 shows the original picture in Darwin's publication and a modern phylogenetic tree of life.

According to Mayr (2004), none of the Darwinian sub-theories was accepted so enthusiastically as the so-called theory of *common ancestry*. This common ancestry insight gave birth to the idea of building a tree to relate modern organisms back to ancestors in the past. In his book *What makes biology unique?*, Mayr states: well-known "*similarities, such as the cord in tunicates and the brachial arcs in fishes and terrestrial vertebrates were completely disconcerting until they were interpreted as vestiges of a common past.*"

Since then, the common ancestry idea has been referred to as the *theory of descent* (Hennig 1950; Weissmann 1889) and formed the basis for the initial formalization of methods in evolutionary biology. The methods for the description of an evolutionary tree evolved relatively slowly during the twentieth century, until recently when it converged on the graph theory originally developed in the computer sciences. The origin of graph theory is frequently attributed to a paper of the

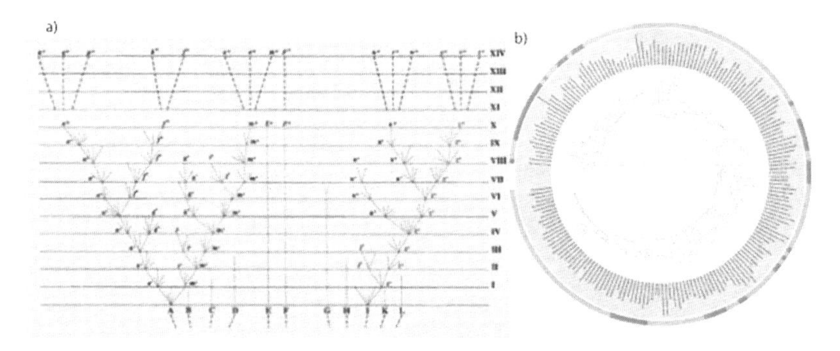

Fig. 12.3 The original and modern stages in phylogenetic tree building. (**a**) The very first phylogenetic tree of organisms contained in the *Origin of Species* (Darwin 1859); (**b**) a modern phylogenetic tree of life

Swiss mathematician Leonhard Euler, more than one century before Darwin. With the multidisciplinary research fields growing at the end of the twentieth century, information theorists started to work with evolutionary biology data and standard methods to treat and represent tree-structured data in computer and information sciences were introduced into evolutionary biology. The most widely used format existing today to represent an evolutionary tree is the so-called Newick standard – which makes use of the correspondence between nodes in a tree and nested parentheses. This format was adopted in evolutionary biology in 1986 during an informal meeting of the Society for the Study of Evolution in Durham, New Hampshire.

12.3.4.1 Tree@CDAO

The traditional binary tree, where extant species evolved from a common ancestor, is now being replaced by a more general representation, a phylogenetic network, for example, in the case where horizontal gene transfer events produce complex trees with criss-crossing branches. CDAO uses a semantic-based structure to represent both networks and trees, with nodes and branches defined as separate concepts. The trees can be either rooted or unrooted and branches in a rooted tree can be assigned a "direction," indicating the parent-child relationships between the associated nodes.

The structure defined in CDAO allows researchers to annotate both branches and nodes with specific information and in this way, character states can be associated with specific tree lineages, present-day or ancestral TUs. Similarly, genetic events (duplication, lateral transfer, inversion, transposition, deletion, insertion, etc.), leading to modifications or transformations in the state of a character, can be localized at specific branches of the tree. The existence of dedicated concepts associated with ancestral TUs in an evolutionary tree allows the representation of the most likely status of characters in putatively ancestral organisms and can help researchers to formulate better hypotheses about both living and non-living organisms.

12.4 Conclusions

Most of the discussions in evolutionary biology made during the twentieth century have produced what we now call modern Darwinism or, simply, Darwinism. Many ancient controversies have been solved and the ones that have not been solved do not seem to pose any practical problem to the modern evolutionary biologist. Darwinism has been completely integrated theoretically with other broad fields in biology, such as genetics and molecular biology. If an ontology can now be envisaged to represent and annotate data in evolutionary biology, it is certainly based on the work of numerous biologists and philosophers towards developing unifying concepts in evolution. Even inside Darwinism, the controversies are now well-understood and evolutionary biologists can work together, even though they base their premises in different traditional schools. Modern evolutionists can switch between theories, such as cladistics or phenetics, choosing the most pertinent framework, without being accused of philosophical betray. In the twenty-first century era of integrative biology, researchers have understood that no method is superior in all cases and there will always be a number of variables or conditions that will make one or another methodology more suitable for a particular case. And although a number of controversies still exist, such as the nature of species (or TUs), the best method to analyze data or the fact that characters should or should not be homologies, evolutionary biologists can understand the limitations of their methods and concepts in order to produce high-quality scientific understanding.

As a consequence, any modern method that aspires to unite evolutionary biologists must be able to cope with this diversity of definitions. CDAO was developed with these discussions in mind and it should permit the representation of data in any theoretical background currently available. Although general enough to allow the representation of different paradigmatic views in evolution, CDAO represents a well-defined structure of the knowledge in evolutionary biology that will facilitate automatic reasoning by algorithms.

In conclusion, CDAO is the result of initial efforts made by members of the NESCent Evolutionary Informatics consortium to conceptualize and define the most pertinent concepts used in a comparative data analysis that focuses on the evolution of biological organisms. Therefore CDAO allows a formal representation of evolutionary data, where annotations of any sort can be linked to the TUs, characters, character-state data matrix and the branches of a tree. Thanks to the formalisms defined in CDAO, most of this information can be subsequently parsed and analyzed by automatic reasoning algorithms, with the goal of producing new and high-level information. CDAO is intended to provide the basis for a standardized file format for evolutionary data, promoting data reuse and data interoperability. More importantly, it is hoped that it will eventually offer a complete framework for both computer and human expert analyses.

12.5 Future Developments

CDAO is an ongoing project and we plan to add more specific concepts as new versions of the ontology are built in the near future. Next steps on CDAO development will include the addition of still missing standard evolutionary concepts, such as (1) the ones used in systematics; (2) sister group relations; (3) standard homology relationships, including molecular-based ones; and (4) homoplasy and evolutionary convergence annotation. Additionally, since organism can be better understood in ecological contexts, we also intend to associate CDAO with some environmental ontology, such as the one provided at http://www.environmentontology.org/ has been envisioned. Initially developed behavioral ontologies (Midford 2004) shall also be used in association with CDAO instances to allow better description of evolutionary features of complex organisms.

User-friendly algorithms to transform standard format of character-state data matrices in CDAO formatted instances will soon be produced to facilitate the representation of evolutionary data in the CDAO format. The production of these algorithms for data representation is strikingly relevant to approximate the evolutionary biologist in his lab to this new technology of data representation.

CDAO also aims to be integrated into a workflow-pipeline of evolutionary software on which the users will be able to input their data, click in some buttons to activate the methods they want to use and then run these methods on a grid machine. The results shall be stored in the tree-structure part of CDAO and they will be able to be analyzed automatically by a reasoner, returning the analyzed data to the user for interpretation. Moreover, once CDAO stores the raw data, there is the possibility of associating a number of different tree-topologies (and the methods associated to them) for the same data set. Moreover, the statistical methods of re-sampling will also be able to be annotated and values of confidence will be associated on tree-branches. These developments however will depend on the development of MIAPA and further submission of projects aiming to provide the informatics structure for workflow development.

Furthermore, new digital representation of data matrices on evolutionary biology has been produced and Ramírez et al. (2007) describes a data base of digital images and character data matrices information that could be automatically transformed into CDAO instances. The usage of CDAO as the standard file format for storing, annotating and sharing evolutionary information may ease communication in the near future, although standard algorithms and converting tools must be made available as soon as possible.

At last, considering that we envision the possibility of getting together a number of CDAO-annotated data sets produced by different research teams all over the world, we strongly encourage evolutionary biology researchers to provide a complete and accurate description of their represented data. The possibility of joining a number of evolutionary data produced everywhere in the globe opens a wonderful possibility of producing complete and powerful evolutionary information using thousands of diverse characters, from the molecular to the morphological, behavioral and ecological. To the best of our knowledge and belief that sort of global, heterogeneous, joined data set information could never be achieved without making use of a standard well-defined knowledge representation vocabulary, as the one described on CDAO.

Acknowledgments We thank NESCent for financing our group meetings and the members of the Evolutionary Informatics team for their stimulus and support. The French ANR (EvolHHuPro: BLAN07-1_198915) are gratefully acknowledged for financial support. Francisco Prosdocimi and Julie D. Thompson are supported by institute funds from the Institut National de la Santé et de la Recherche Médicale, the Centre National de la Recherche Scientifique and the Université Louis Pasteur de Strasbourg. Enrico Pontelli has been supported by National Science Foundation grants HRD-0420407 and CNS-0220590. Brandon Chisham is supported by an IGERT fellowship from NSF grant DGE-0504304. The identification of specific commercial software products in this paper is for the purpose of specifying a protocol and does not imply a recommendation or endorsement by the National Institute of Standards and Technology.

References

Ashburner M, Ball CA, Blake JA, Botstein D, Butler H, Cherry JM, Davis AP, Dolinski K, Dwight SS, Eppig JT, Harris MA, Hill DP, Issel-Tarver L, Kasarskis A, Lewis S, Matese JC, Richardson JE, Ringwald M, Rubin GM, Sherlock G (2000) Gene ontology: tool for the unification of biology. The Gene Ontology Consortium. Nat Genet 25(1):25–29

Bard J, Rhee SY, Ashburner M (2005) An ontology for cell types. Genome Biol 6(2):R21

Bard JB (2005) Anatomics: the intersection of anatomy and bioinformatics. J Anat 206(1):1–16

Carnap R (1950) Empiricism, Semantics, and Ontology. Revue Internationale de Philosophie 4:20–40. On the web: http://www.ditext.com/carnap/carnap.html

Chabalier J, Mosser J, Burgun A (2007) Integrating biological pathways in disease ontologies. Stud Health Technol Inform 129(Pt 1):791–795

Cochrane G, Akhtar R, Bonfield J, Bower L, Demiralp F, Faruque N, Gibson R, Hoad G, Hubbard T, Hunter C, Jang M, Juhos S, Leinonen R, Leonard S, Lin Q, Lopez R, Lorenc D, McWilliam H, Mukherjee G, Plaister S, Radhakrishnan R, Robinson S, Sobhany S, Hoopen PT, Vaughan R, Zalunin V, Birney E (2009) Petabyte-scale innovations at the European Nucleotide Archive. Nucleic Acids Res 37(Database issue):D19–25

Darwin CR (1859) On the origin of species by means of natural selection, or the preservation of favoured races in the struggle for life, 1st edn. John Murray, London. On the web: http://darwin-online.org.uk/content/frameset?itemID = F373& viewtype = text&pageseq = 1

Dobzhansky T (1973) Nothing in biology makes sense except in the light of evolution. Am Biol Teacher 35:125–129

Dohm JC, Lottaz C, Borodina T, Himmelbauer H (2008) Substantial biases in ultra-short read data sets from high-throughput DNA sequencing. Nucleic Acids Res 36(16):e105

Eisen JA (1998) Phylogenomics: improving functional predictions for uncharacterized genes by evolutionary analysis. Genome Res 8:163–167

Gaudet P, Williams JG, Fey P, Chisholm RL (2008) An anatomy ontology to represent biological knowledge in Dictyostelium discoideum. BMC Genom 18(9):130

Gene Ontology Consortium (2001) Creating the gene ontology resource: design and implementation. Genome Res 11(8):1425–1433

Gomes-Perez A, Fernando-Lopez M, Corcho O (2004) Ontological engineering: theoretical foundations of ontologies. Springer Verlag, New York

Gruber TR (1995) Toward principles for the design of ontologies used for knowledge sharing. International Journal of Human-Computer Studies 43(4–5):907–928

Harris MA (2008) Chapter 5: developing an ontology. In: Jonathan M, Keith (ed) Bioinformatics data, sequence analysis and evolution. Humana Press, New York, pp 111–124

Hennig W (1950) Phylogenetic systematics. (trans. D. Davis and R. Zangerl). University of Illinois Press, Urbana 1966, reprinted 1979

Hull D (1988) Science as Process. University of Chicago Press, Chicago, IL

Jacob F (1970) La logique du vivant: une histoire de l'heredité. Gallimard, Paris

Jaiswal P, Avraham S, Ilic K, Kellogg EA, McCouch S, Pujar A, Reiser L, Rhee SY, Sachs MM, Schaeffer M, Stein L, Stevens P, Vincent L, Ware D, Zapata F (2005) Plant ontology (PO): a controlled vocabulary of plant structures and growth stages. Comp Funct Genom 6(7–8): 388–397

Kearney M (2007) Phylosophy and phylogenetics: historical and current connections. In: David L Hull, Michael Ruse (eds) The Cambridge companion to the philosophy of biology. Cambridge University Press, Cambridge, pp 211–232

Kuhn TS (1962) The structure of scientific revolutions, 1st edn. University of Chicago Press, Chicago, IL, p 168

Lakatos I (1978) The methodology of scientific research programmes: philosophical papers, vol 1. Cambridge University Press, Cambridge

Lamarck chevalier de (Jean-Baptiste Pierre Antoine de Monet) (1809) Philosophie Zoologique. On the web: http://fr.wikisource.org/wiki/Philosophie_zoologique

Larson SD, Fong LL, Gupta A, Condit C, Bug WJ, Martone ME (2007) A formal ontology of subcellular neuroanatomy. Front Neuroinform 1:3

Lassila O, McGuinness DL (2001) The role of frame-based representation on the semantic web. Linköping Electron Art Computer Inform Sci 6:5

Leebens-Mack J, Vision T, Brenner E, Bowers JE, Cannon S, Clement MJ, Cunningham CW, dePamphilis C, deSalle R, Doyle JJ, Eisen JA, Gu X, Harshman J, Jansen RK, Kellogg EA, Koonin EV, Mishler BD, Philippe H, Pires JC, Qiu YL, Rhee SY, Sjölander K, Soltis DE, Soltis PS, Stevenson DW, Wall K, Warnow T, Zmasek C (2006) Taking the first steps towards a standard for reporting on phylogenies: minimum information about a phylogenetic analysis (MIAPA). OMICS 10(2):231–237

Lee RY, Sternberg PW (2003) Building a cell and anatomy ontology of caenorhabditis elegans. Comp Funct Genom 4(1):121–126

Leontis NB, Altman RB, Berman HM, Brenner SE, Brown JW, Engelke DR, Harvey SC, Holbrook SR, Jossinet F, Lewis SE, Major F, Mathews DH, Richardson JS, Williamson JR, Westhof E (2006) The RNA Ontology Consortium: an open invitation to the RNA community. RNA 12(4):533–541

Maglia AM, Leopold JL, Pugener LA, Gauch S (2007) An anatomical ontology for amphibians. Pac Symp Biocomput 367–378

Marguerat S, Wilhelm BT, Bähler J (2008) Next-generation sequencing: applications beyond genomes. Biochem Soc Trans 36(Pt 5):1091–1096

Mayr E (1991) One long argument: Charles Darwin and the genesis of modern evolutionary thought (questions of science). Harvard University Press, Cambridge

Mayr E (2004) What makes biology unique?: considerations on the autonomy of a scientific discipline. Cambridge University Press, Cambridge

Midford PE (2004) Ontologies for behavior. Bioinformatics 20:3700–3701

Mizoguchi R, Vanwelkenhuysen J (1995) Task ontology for reuse of problem solving knowledge. Proceedings of KB & KS, pp 46–59

Ramírez MJ, Coddington JA, Maddison WP, Midford PE, Prendini L, Miller J, Griswold CE, Hormiga G, Sierwald P, Scharff N, Benjamin SP, Wheeler WC (2007) Linking of digital images to phylogenetic data matrices using a morphological ontology. Syst Biol 56:283–294

Rodrigues JM, Rector A, Zanstra P, Baud R, Innes K, Rogers J, Rassinoux AM, Schulz S, Trombert Paviot B, ten Napel H, Clavel L, van der Haring E, Mateus C (2006) An ontology driven collaborative development for biomedical terminologies: from the French CCAM to the Australian ICHI coding system. Stud Health Technol Inform 124:863–868

Rothberg JM, Leamon JH (2008) The development and impact of 454 sequencing. Nat Biotechnol 26(10):1117–1124

Schulz S, Markó K, Hahn U (2007) Spatial location and its relevance for terminological inferences in bio-ontologies. BMC Bioinform 8:134

Schulz S, Stenzhorn H, Boeker M (2008) The ontology of biological taxa. Bioinformatics 24(13): i313–321

Segerdell E, Bowes JB, Pollet N, Vize PD (2008) An ontology for Xenopus anatomy and development. BMC Dev Biol 8:92

Shendure J, Ji H (2008) Next-generation DNA sequencing. Nat Biotechnol 26(10):1135–1145

Sirin S, Parsia B, Grau BC, Kalyanpur A, Katz Y (2007) Pellet: a practical OWL-DL reasoner. Web Semantics 5:51–53

Smith B, Ceusters W, Klagges B, Köhler J, Kumar A, Lomax J, Mungall C, Neuhaus F, Rector AL, Rosse C (2005) Relations in biomedical ontologies. Genome Biol 6(5):R46

Smith B, Ashburner M, Rosse C, Bard J, Bug W, Ceusters W, Goldberg LJ, Eilbeck K, Ireland A, Mungall CJ; OBI Consortium, Leontis N, Rocca-Serra P, Ruttenberg A, Sansone SA, Scheuermann RH, Shah N, Whetzel PL, Lewis S (2007) The OBO Foundry: coordinated evolution of ontologies to support biomedical data integration. Nat Biotechnol 25(11):1251–1255

Trelease RB (2006) Anatomical reasoning in the informatics age: Principles, ontologies, and agendas. Anat Rec B New Anat 289(2):72–84

Tsarkov D, Horrocks I (2006) FaCT + + description logic reasoner: system description. In: Furbach U, Shankar N (eds) IJCAR 2006, LNAI 4130, pp 292–297

van Heijst G, Schreiber AT, Wielinga BJ (1997) Roles are not classes: a reply to Nicola Guarino. Int J Hum-Comput Stud 46(2):311–318

Weissmann A (1889) Essays upon heredity, vols 1 and 2. Clarendon, Oxford. On the web: http://www.esp.org/books/weismann/essays/facsimile/

Part II
Modeling

Chapter 13
Large-Scale Analyses of Positive Selection Using Codon Models

Romain A. Studer and Marc Robinson-Rechavi

Abstract Positive selection is the mechanism of adaptation to the environment, as well as the main source of novelty in evolution and thus it is of great interest to find its trace in genomes. During the last decade, different evolutionary models have been developed to detect positive selection at the gene level, based on divergence between species. Most recently, these models have been applied to large-scale comparisons of genomes. We present in this chapter some strengths and limitations of such genomic scans for positive selection and discuss the main recent large-scale studies, as well as relevant databases. We particularly discuss our recent results concerning the impact of genome duplication in vertebrate evolution and our related database Selectome.

13.1 Introduction

13.1.1 Positive Selection as a Mechanism of Adaptation

A fundamental concept in evolution is selective pressure from the environment. Mutations in genomes occur at random. These mutations could have an impact in the phenotype by modifying, e.g., biochemical function, or the expression of genes affected. Most mutations have a negative effect on the fitness (they are deleterious) and thus almost all genes are under purifying selection to preserve their function.

R.A. Studer
Department of Ecology and Evolution, Biophore, Lausanne University, CH-1015, Lausanne, Switzerland; Swiss Institute of Bioinformatics, CH-1015 Lausanne, Switzerland

M. Robinson-Rechavi
Department of Ecology and Evolution, Biophore, Lausanne University, CH-1015, Lausanne, Switzerland; Swiss Institute of Bioinformatics, CH-1015 Lausanne, Switzerland
email: marc.robinson-rechavi@unil.ch

P. Pontarotti (ed.), *Evolutionary Biology: Concept, Modeling, and Application*,
DOI: 10.1007/978-3-642-00952-5_13,

Some mutations have no effect on fitness and may be fixed under the neutral process of drift. Others will have a beneficial effect and they will be kept in the genome by the process of positive selection (also called Darwinian, adaptive, or directional selection). This positive selection is the mechanism of adaptation to the environment and thus it is of great interest to find its trace in genomes. During the last decade, many molecular analyses have found different categories of genes preferentially affected by adaptive selection (Yang 2006).

13.1.2 Functional Categories of Genes

The most representative categories of genes under positive selection are involved in arm-race adaptation. Virus and bacteria are rapidly mutating due to the absence of control in replication and to large population size, respectively. These mutations could affect various genes like HIV proteins (env [Nielsen and Yang 1998], gag, pol) or the wsp protein in the outer membrane in Wolbachia bacteria (Jiggins, Hurst et al. 2002). These genes are on perpetual adaptation against drugs and immune systems. Similarly, other organisms have to continually optimize their defenses in order to counter these attacks. The Major Histocompatibility Complex (MHC) classes I and II are evolving to recognize any kind of external peptides and are subject to positive selection in both classes (Hughes and Nei 1988, 1989; Hughes et al. 1994). Many other immunity genes are also under positive selection, like Glycophorin A (Baum et al. 2002), CD4 glycoprotein (Zhang et al. 2008) or TRIMα (Sawyer and Wu 2005).

Others main categories are genes involved in sexual reproduction, like the sperm lysin in abalone (Lee et al. 1995) or the protamine P1 in primates (Rooney and Zhang 1999). Genes of perception are also found to be under positive selection (i.e., Olfactory Receptor OR5I1 [Moreno-Estrada et al. 2008]). Cases of genes involved in digestion have also been reported, notably the Lysosyme in primates (Messier and Stewart 1997). This could be explained by a change in diet.

13.1.3 The Case of Duplicated Genes

Another group of genes, which are frequently reported to evolve under positive selection, are not related to a particular functional category: duplicated genes. After a duplication event, a gene will be present twice in the genome, implying a cost of redundancy, which may be balanced by a selective gain in function. Different theoretical models have been elaborated to predict the fate of these duplicated genes (see review [Zhang 2003]). The main fate is simply loss of one copy, by **nonfunctionalization**. Assuming that duplicates are deleterious (in terms of stoichiometry or cost of expression), most additional copies will be rapidly erased from the genome and become pseudogenes. It has been estimated that around 60–80% of copies are lost after a whole genome duplication (Brunet et al. 2006; Semon and

Wolfe 2007). But what is the fate of genes that stay in two copies in the genome? The first predicted fate, of special interest for studies of positive selection, is **neofunctionalization (NF)** (Ohno 1970; Force et al. 1999). In its simplest version, one copy will keep the ancestral function and the other will rapidly acquire mutations during a period a relaxation of selection. These mutations at strategic positions will promote a new function. Function is a sometimes ambiguous term and can be defined notably as the biochemical function of the protein, its interaction partners, or the spatiotemporal expression of the gene. This model invokes positive selection to fix advantageous mutations (Kondrashov and Kondrashov 2006; Shiu et al. 2006). An alternative to neofunctionalization is **subfunctionalization (SF)**. It assumes that the ancestral gene has several functions. After the duplication event, each copy will lose one or more functional parts and these genes will be complementary. For example, the developmental gene *Xhox3* of the Xenopus frog is expressed in the tail, the analia–genitalia and the nervous system. In the zebrafish, there are two genes homologous to *Xhox3:* the gene *evx1*, which is expressed in the analia–genitalia and the nervous system and its duplicate *eve1,* which is expressed in the tail and during gastrulation (Avaron et al. 2003). The best known version of this model is the Duplication–Degeneration–Complementation (DDC) model (Force et al. 1999). Importantly, it does not involve positive selection, *in contrario* to the neofunctionalization model. The **sub-neofunctionalization (SNF)** model (He and Zhang 2005) is a mix between the two others models, in which duplicated genes may have a rapid subfunctionalization step followed by a longer neofunctionalization step. Other models have also been proposed but we will not describe them here (Conant and Wolfe 2008). We just would like to mention that it is usually difficult to attribute the evolutionary pattern of a pair of genes exactly to one model.

We can now use whole genome sequences to test hypotheses related to the occurrence of positive selection, such as what makes the differences between chimpanzees and us, or the level of implication of positive selection in the preservation of duplicate genes. We will present in this chapter different analyses of positive selection across evolution of genomes. The emphasis will be on problems of scaling codon based models; excellent discussion of methods for single gene studies and/or polymorphism based methods, can be found elsewhere (Eyre-Walker 2006; Anisimova and Liberles 2007).

13.2 Which Codon Model for Which Problem?

The selective pressure, which occurred during the divergence between two homologous genes, can be measured at the nucleotide level by computing the d*N*/d*S* ratio (ω). d*N* is defined as the number of non-synonymous mutations per non-synonymous site and indicates in first approximation the substitutions that are generated by mutation and fixed by a combination of drift and selection (on the function of the protein). d*S* is the number of synonymous mutations per synonymous site and indicates in first approximation the substitutions that are generated by mutation and fixed by drift alone. If a gene is under purifying selection for the function of the encoded protein, this ω ratio is expected to be lower than 1, since most amino acid changes will be rejected; this is the case for most genes. If a gene is evolving without selective constraints (neutral evolution), the ω ratio is expected to be equal to 1, with no impact of either synonymous or amino acid changes on fitness. Finally, if there is positive selection to change the structure or function of the encoded protein, it is possible that ω will be higher than 1: amino acid changes are selected for their new role in the protein and thus kept in the genome more frequently than expected under mutation + drift. These expectations mean that ω can be used to estimate the direction and intensity of selection on protein coding genes.

The most popular package for such an estimation of evolutionary pressure is PAML (Phylogenetic Analysis by Maximum Likelihood) (Yang 1997, 2007). For an overview of the other methods, we refer to Anisimova and Liberles (2007). In the following, we present the main codon models of the PAML package.

13.2.1 Pairwise Estimate of *d*N/*d*S

The pairwise measure consists in simply estimating the d*N* and d*S* values between two genes. This method is interesting for genes that are closely related, when no additional information is available. This is typically the case when comparing orthologs from two closely related genomes. But it can be problematic when measuring more divergent genes. Notably this is a risk of saturation in the estimate of d*S*, when a synonymous site has multiple substitutions. We discuss this in Section 13.3.3

13.2.2 Branch Models

The "branch models" (Yang 1998) were the first codon models to be implemented in PAML. They take advantage of a multiple sequence alignment to estimate d*N* and d*S* on specific branches using a mixture of Markov process and maximum likelihood inference. Depending on the model used, it can assign different d*N*/d*S* values to all branches, or to a few categories of branches. These models are useful to detect genes that have undergone strong positive selection, which has modified deeply and rapidly the amino acid sequences. The basic model is model 0, or "one-ratio" model. It assumes only one d*N*/d*S* for all branches of the tree but the d*N* and d*S* could vary for each branch. At the opposite, the "free-ratio" model estimates one d*N*/d*S* for each branch. It is in principle useful when we have no a-priori hypothesis but the result should be taken with caution, due to a wide number of free parameters. Between these two extremities, the "two-or-more ratios" models can be used to test an a-priori hypothesis, by specifying the branch or branches that are thought to be under positive selection. We will obtain one d*N*/d*S* for the branch(es) of interest and another for all other branches. We can then construct a likelihood ratio test (LRT) by contrasting this model against a simpler model, such as the one-ratio model. This test is more precise than the free ratio model. But one potential disadvantage in all the "branch models" is that dN/dS values are averaged over all positions in the alignment.

13.2.3 Site Models

It is obvious that not all sites in a protein are equivalent and it may not be a reasonable assumption to expect positive selection to act on all the protein sequence. Thus the second type of codon models are the "site models" (Yang et al. 2000). These models are quite useful to detect specific amino acids that are continuously under positive selection in a gene family. This is expected to be the case in arm-races such as experience by HIV proteins (Yang et al. 2000) or the MHC (Yang and Swanson 2002), or in sexual conflict, as identified in the mollusk abalone sperm lysin (Yang et al. 2000). The classic usage of these models is to perform an LRT contrasting the positive selection model M2a, which assigns sites into three different dN/dS classes ($\omega_0 < 1$, $\omega_1 = 1$ and $\omega_2 \geq 1$), to a nearly-neutral model M1a that assigns sites into only two classes ($\omega_0 < 1$, $\omega_1 = 1$). Other LRTs can be constructed using model M3 (3 classes of sites under no constraints [ω_0, ω_1, ω_2]) vs. M0 (1 class of site with ω_0), or model M8 (10 classes + 1 class $\omega \geq 1$) vs. model M7 (10 classes, no positive selection allowed). These models need at least six sequences to be reasonably powerful (see Section 13.3). Moreover, they will only detect positive selection that acts over long periods of time (relative to the sequence sampling). In practice, this makes the corresponding tests very conservative and adapted only to detecting the most extreme examples of positive selection, such as HIV proteins.

13.2.4 Branch-Site Models

The most recent class of models are "branch-site" models (Yang and Nielsen 2002; Zhang et al. 2005). They present an interesting mix of the two previous models. These models allow an estimation of the proportion of sites under positive selection (if any) during a specific evolutionary time (determined by a specific phylogenetic branch). This model is intuitively appealing, as positive selection could be expected to affect only some sites, during a limited time of functional change (Fig. 13.1). It is also consistent with theoretical expectations from realistic models of molecular evolution, such as the model of episodic selection (Gillespie 1991). If only a few substitutions occurred during the change of function, a simple branch model will fail to detect them, because the dN/dS per branch is averaged over all positions and most of these remain under purifying selection. Thus branch-site models are best

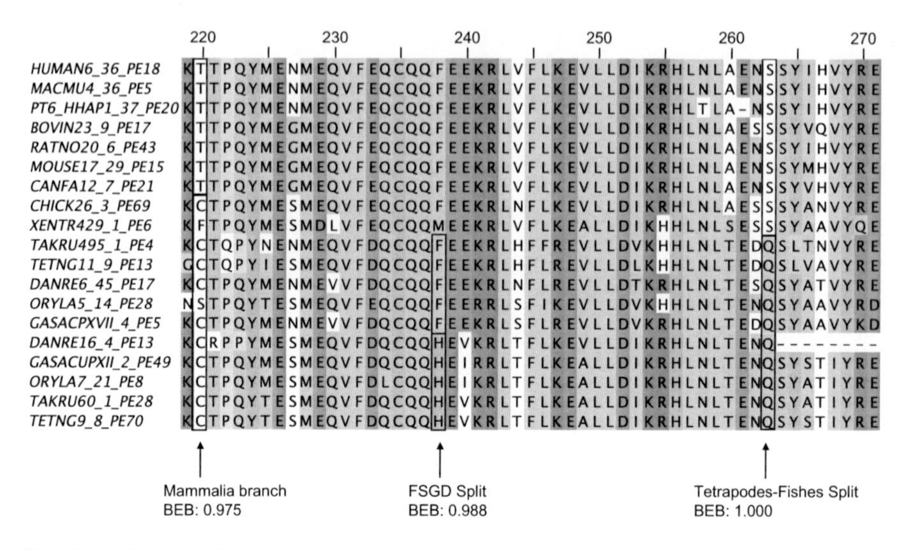

Fig. 13.1 The gene family "Protein kinase C and casein kinase substrate in neurons protein (SwisProt:PACN1_HUMAN)" (code HBG059468 in Homolens release 3) presents positive selection at three different evolutionary times, as revealed by the branch-site model of PAML (Yang and Nielsen 2002; Zhang et al. 2005). We highlighted three different sites: (**a**) site 220 was selected for a threonine in the Mammalia branch, (**b**) site 238 for a histidine in the longest branch in fishes and (**c**) in site 263 for a serine in vertebrates against a glutamine in fishes. These sites are above a cut-off of 95% in the Bayes Empirical Bayes analysis from CodeML (Yang et al. 2005)

suited to identifying mutations that fine tune proteins, as in the case of adaptation of plant photosynthesis with the optimization of the RubisCO enzyme (Christin et al. 2008).

In branch-site models, a branch of interest must be defined as the foreground, while all the other branches are defined as background. Positive selection is excluded on the background branches, while it may be allowed on the foreground. The branch-site model A will estimate three different d*N*/d*S* ratios (ω_0, ω_1, and ω_2) and assign sites into four different classes: class K0 sites are under purifying selection ($0 \leq \omega_0 \leq 1$) on all branches (foreground and background); class K1 sites are under neutral evolution ($\omega_1 = 1$) on all branches; class K2a sites may be under positive selection ($\omega_2 \geq 1$) on the foreground branch but under purifying selection ($0 \leq \omega_0 \leq 1$) on background branches; and finally class K2b sites may be under positive selection ($\omega_2 \geq 1$) on the foreground branch but under neutral evolution ($\omega_1 = 1$) on background branches.

There are two different likelihood ratio-tests (LRT) to infer the significance of this model. The original version compared the positive selection model A against the nearly-neutral site model M1a (Yang and Nielsen 2002). In this case, relaxation of purifying selection on the foreground branch could be wrongly interpreted as significant positive selection (Zhang 2004). The improved version (Zhang et al. 2005) contrasts the branch-site model A with positive selection ($\omega_2 \geq 1$) against a constrained branch-site model A where sites can be only under purifying or neutral evolution (ω_2 fixed to 1). Thus the LRT is significant only if positive selection on the foreground branch is a better explanation of the data than a possible relaxation of purifying selection on the foreground branch. When the test is significant, a Bayes Empirical Bayes (BEB) prediction (Yang et al. 2005) can identify sites under positive selection, according to their posterior probability (PP) (Fig. 13.1). Of note, positive selection may be supported by the LRT even in cases where the BEB has insufficient power to predict specific sites.

13.3 Issues in Deep and Large-Scale Analysis

13.3.1 Sampling

Different problems could appear when analyzing large data sets of genes using codon models but the main one is probably sequence sampling. Simulations suggest a minimum number of six sequences in the alignment to have enough power and accuracy (Anisimova et al. 2001, 2002; Anisimova and Yang 2007). This is in itself a major issue for genomic studies involving too few species (e.g., two or three primates). Moreover, sequences that are too close will not contain enough information for reliable estimation of d*N* and d*S*, while sequences that are too divergent may be difficult to align (Section 13.3.2) or have issues of saturation of d*S* (Section 13.3.3). Good sampling can help resolve the latter problem (Section 13.3.3).

13.3.2 Alignment Quality

The quality of the multiple sequences alignment is critical, as in many comparative analyses (e.g., phylogeny, molecular modeling). If the alignment is of poor quality, the final result will also be of poor quality ("Garbage In, Garbage Out") (Landan and Graur 2007; Wong, Suchard et al. 2008). In most phylogenetic software, any column with at least a gap will be removed from the alignment, mainly because interpretation of gaps in evolutionary context is still poorly understood. But the residues immediately surrounding gaps are often difficult to align. The GBLOCK method (Castresana 2000) has been developed to extract the best parts of a multiple sequence alignment, based on gap patterns and can be used automatically in large-scale analyses.

13.3.3 Saturation of d*S*

When sequence divergence increases, so does the probability that each synonymous site has undergone multiple substitutions. This leads to the classical issue in molecular evolution of saturation, whence it can becomes difficult or impossible to estimate the number of substitutions. Pairwise analysis of sequence is very sensitive to this issue. But, like in phylogeny, an appropriate sampling scheme can "break" long branches and improve the estimation of d*S* (and d*N*) values. For example in a Putative RNA methylase family, the pairwise method of PAML will estimate a d*S* of 0.43 between the human gene ENSG00000066651 and its mouse ortholog ENSMUSG00000019792, which is below saturation. But with the longer divergence between this human gene and its zebrafish ortholog ENSDARG00000040033, the pairwise d*S* is estimated at 13, clearly saturated. Using the corresponding Ensembl gene tree adds 13 orthologs to these two genes. Then the one-ratio branch model estimates a maximum d*S* value of 1.5 for any one branch and a total of 3.4 if we sum the d*S* values of all branches separating the two original genes (Fig. 13.2). While distance based estimates of d*S* may saturate

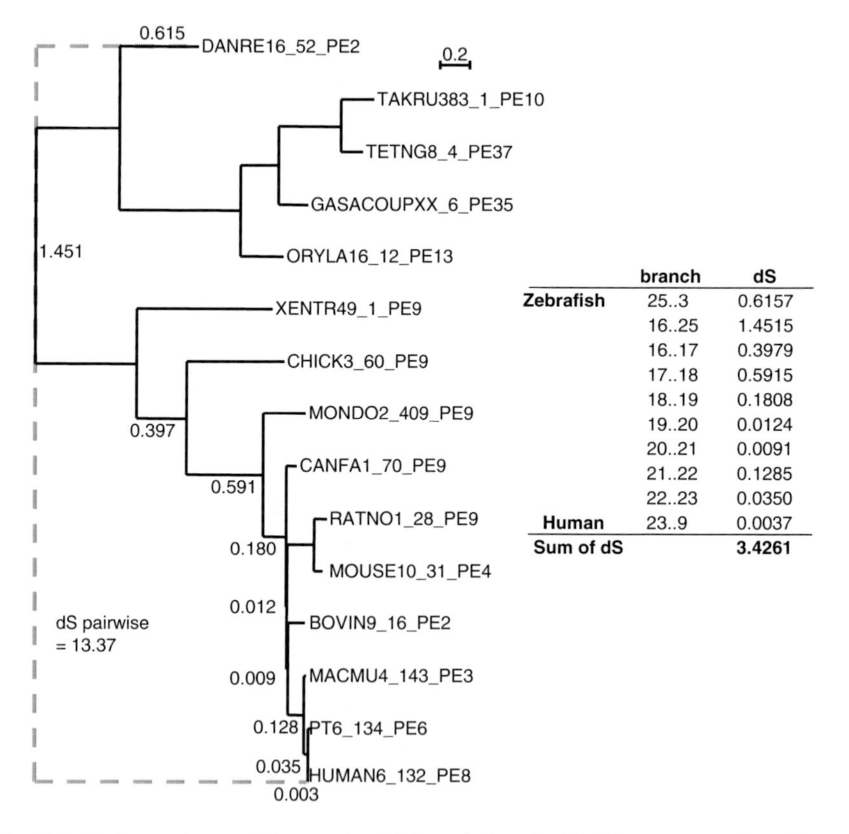

Fig. 13.2 Phylogenetic tree of the putative RNA methylase family. The tree comes from Homolens release 3 (code: HBG000007). The pairwise analysis of human and zebrafish genes results in a d*S* of 13.37, which is saturated. Adding all others species and computing under the "one-ratio" model of PAML, the codon method takes advantage of breaking the branches and results in more precise d*S* estimation, with a maximum of 1.45 and a sum of 3.43 for all branches separated in zebrafish and human branches

relatively early ($dS \geq 1$), simulations indicate that maximum likelihood methods are more robust against saturation problems (Anisimova et al. 2001).

13.3.4 False Discovery Rate

A large-scale scan for positive selection will encounter the same problem as any other large scan, namely test repetition. Both testing different branches for one gene tree and testing many different genes, may lead to false positive results. A simulation study suggested that the q-value (Storey and Tibshirani 2003) provides a good compromise between power and specificity in the case of multiple testing in one gene tree (Anisimova and Yang 2007). This method evaluates the proportion of false positives according to the global distribution of all p-values (Storey and Tibshirani 2003). Since the q-value is also a method of choice for genomic scans (indeed was developed for genomic data), it should also be appropriate for the second issue of testing many genes. Indeed, our simulations found that q-value correction over p-values from multiple branches and multiple genes was both powerful and specific (Studer et al. 2008).

13.4 Large-Scale Studies

We present here some selected whole genome scans for positive selection. More examples are reviewed elsewhere (Biswas and Akey 2006; Eyre-Walker 2006).

13.4.1 General Scans for Positive Selection

One of the first large-scale scans for positive selection was performed by Endo et al. (1996). They scanned the DDBJ/EMBL/Genbank database and classified more than 24,000 sequences into 3,595 groups of homologous genes. They used pairwise computations of dN/dS to estimate the level of positive selection and they found 17 (0.48%) groups susceptible to be under positive selection, of which 9 are proteins from parasites or viruses. They also used a window analysis to find regions under selection in each gene. This study was made before the rise of a more sophisticated codon model. It should be noted that problems of false positives have recently been reported for the sliding window approach (Schmid and Yang 2008).

A more recent study searched for genes evolving under positive selection in *Escherichia coli* (Petersen et al. 2007). The authors used site models (Nielsen and Yang 1998) to scan 3,757 genes from strain K12 with at least two other orthologs in other strains. They found positive selection is present in eight gene categories, based on the EcoCyc Database, especially in genes encoding cell surface proteins. High incidence of positive selection in *E. coli* is consistent with expectations from large population size.

13.4.2 From Human—Chimpanzee Comparisons to a Study of Vertebrates

In recent years, several scans of positive selection have been performed with a special focus on the human and other primate lineages. Most recently, these approaches have been generalized to mammalian and vertebrate gene trees.

The simplest approach is to perform a pairwise d*N*/d*S* analysis between all orthologs of two genomes. Thus Nielsen et al. (2005) scanned more than 13,000 genes between human and chimpanzee. The aim was to identify genes that experimented positive selection in either or both lineages; 733 genes were reported but only 35 have a *p*-value under 5%. They found over-representation of immune-defense-related genes and sensory perception, as well as genes on the X chromosome; and under-representation of genes expressed in the brain.

Tests for positive selection can be made directional by including a third genome but usually at the cost of analyzing fewer genes. Clark et al. (2003) searched for positive selection between human and chimpanzee, using the mouse as an outgroup. They analyzed 7,645 ortholog groups of human–chimpanzee–mouse. They used two models: the branch model (Yang 1998) and the original version of the branch-site model (Yang and Nielsen 2002). They checked for a number of potential biases, such as GC content, repeat density, local recombination and segmental duplication. The scan identified positive selection on the human branch for 1,547 genes and on the chimpanzee branch for 1,534 genes. They found an overrepresentation of genes present in OMIM (Hamosh et al. 2005), as well as an overrepresentation of olfactory genes in human. Interestingly, some genes involved in speech were found under positive selection in human. More surprisingly, they found genes involved in the metabolism of amino acids. This has been interpreted as linked to a change in diet, since humans but not chimpanzees eat meat. Finally, Clark et al. (2003) propose that, as suggested by King and Wilson (1975), most differences between chimpanzees and humans may be due to regulatory changes. Jorgensen et al. (2005) used the three different types of codon model (branch, site, and branch-site (original version)) to infer positive selection in 1,120 trios of human–mouse–pig genes. They found only one gene with the branch model but 3.0% of genes in the human lineage show positive selection with the branch-site model, relative to 2.0% for pig and 2.2% for mouse. However, Jorgensen et al. (2005) are cautious due to the small taxon sampling in their study and suggest using it only as a first step for further analyses. The most recent gene trio study scanned 13,888 genes using the macaque as an outgroup to identify branches leading to human or chimpanzee lineages (Bakewell et al. 2007). A notable advance over previous studies is the use of the improved branch-site model (Zhang et al. 2005). They found more genes with positive selection in chimpanzee evolution than in human evolution and indeed after correcting for multiple testing identified only two genes in the human lineage (59 in the chimpanzee lineage). Even so, it has been suggested that the small sequence sampling used induced false-positives (Suzuki 2008).

The recent availability of multiple genome sequences has allowed for studies that combine a large number of genes with a phylogenetic framework. Arbiza et al. (2006) thus investigated more than 13,000 genes with orthologs in human, chimpanzee, mouse, rat and dog, to estimate positive selection, relaxation, or evolutionary rate acceleration in the human lineage. They used relative-rate tests (Robinson-Rechavi and Huchon 2000) and branch-site models (original and improved versions) to estimate divergence between human and chimpanzee genes. They found positive selection associated with Gene Ontology terms such as sensory perception, GPCR signaling pathway, immune response, DNA/RNA metabolism and transcription. An investigation in the mammalian lineage used the power of six different species to analyze approximately 16,500 human genes, with at least two orthologs in another mammalian (Kosiol et al. 2008). All duplicated genes were excluded. All the six species are represented in 42% of gene families. They used the site model to look over the entire tree and the branch-site model (modification of the improved version) to infer lineage specific positive selection. Again, they found over-representation in processes of immunity/defense and sensory perception. They also found several pathways containing large numbers of genes under positive selection, such as the complement immunity system and the FAS/p53 apoptotic pathway. Some differences were found between primates and rodents: perception is overrepresented in primates and immunity in rodents. The analysis of microarray expression data indicated that genes under positive selection are less expressed and more tissue-specific than other genes. A more focused study investigated the molecular cause of species differences in disease (Vamathevan et al. 2008). In this study, they analyzed 3,079 orthologs genes of the same five mammalian genomes as Arbiza et al. (2006), with a special focus on human and primate diseases. They first looked for positive selection using the free-ratio model in order to have an overview of the evolutionary rate. They found an ω between 0.14 to 0.20, depending on the branch tested. Using the branch-site model, they found 511 genes under positive selection, with an excess in the chimpanzee branch (162 in chimp. vs 52 in human), as seen in Bakewell et al. (2007). They confirmed overrepresentation in nucleic acid metabolism, neuronal activities and immunity/ defense. Of special interest, genes under positive selection are more likely to be present in OMIM. Finally genes under positive selection appear to interact more often with other positively selected genes than expected by chance. Finally, we have investigated positive selection using the improved branch-site test in 884 gene trees, including at least four mammals, chicken, Xenopus and five fishes (Studer et al. 2008). We found evidence for positive selection, after correcting for multiple testing, in a surprising 77% of gene trees. This appears to be due to an increased power of the test with more species and greater divergence between sequences. We discuss this study in more detail in the next section.

13.4.3 What is the Effect of Genome Duplication on the Incidence of Positive Selection?

As discussed in Section 13.1, positive selection has been suggested to be important in the evolution of duplicated genes. To test this, we studied positive selection in vertebrate gene trees, which were constrained to include only specific duplication patterns. During vertebrate evolution, three different rounds of whole genome duplication occurred: two rounds before the split tetrapodes-teleost fishes (known as "2R" [Putnam et al. 2008]) and a third before the diversification of teleost fishes ("3R" or "FSGD" [Jaillon, Aury et al. 2004]). The availability of many vertebrate sequenced genomes, especially the five teleost fishes provides us with enough power to test the hypothesis of positive selection as a retention mode of duplicated genes. We found that positive selection is pervasive along the vertebrate tree but is surprisingly independent of the evolutionary event, speciation, or duplication (Studer et al. 2008).

To perform a large-scale analysis with constraints on gene phylogeny, we used the database HomolEns version 3 (http://pbil.univ-lyon1.fr/databases/homolens.html), based on Ensembl release 41 (Oct 2006). HomolEns is built on the same model as Hovergen (Duret et al. 1994) or Hobacgen (Perriere et al. 2000). All predicted peptides are compared to each other with BLASTP2 (BLOSSUM62 as substitution matrix and e-value cut-off = 10^{-4}), followed by a transitive clustering of genes (cluster size varies from 1 to more than 1,000). For each family, a multiple sequences alignment is built with MUSCLE (Edgar 2004) and a maximum-likelihood phylogenetic tree is estimated, on conserved blocks of the alignments selected with GBLOCKS, with PhyML (substitution model = JTT, estimated proportion of invariable sites, four categories, estimated gamma, initial tree with BIONJ) (Guindon and Gascuel 2003). For each family, tree reconciliation is performed using a taxonomic reference tree (Dufayard et al. 2005; Kuzniar et al. 2008). Each node is a defined either as a speciation event (separating orthologs) or a duplication event (separating paralogs). Importantly, the query tool FamFetch includes the TreePattern editor (Dufayard et al. 2005), which allows searching for families according to specific tree topologies. In our study, we used three different topologies (Fig. 13.3), which all share strong constraints on species sampling and phylogenetic relations among vertebrates: (a) gene trees where no retention of duplicates is allowed, after the fish specific genome duplication or otherwise ("singleton") and (b) and (c) gene trees where duplicated genes after the FSGD are retained in all fishes, forbidding or not other duplications in the tree. The genes with retention after FSGD provide the data set to test the impact of duplication and the singleton genes provide a control in the absence of duplication. In addition, we performed a search for genes retained in duplicate after 2R, to control for the eventual impact of this event on evolution of the vertebrate genes. It should be noted that our stringent selection procedure enriched our data set in basic cell processes, which are usually underrepresented in reports of positive selection.

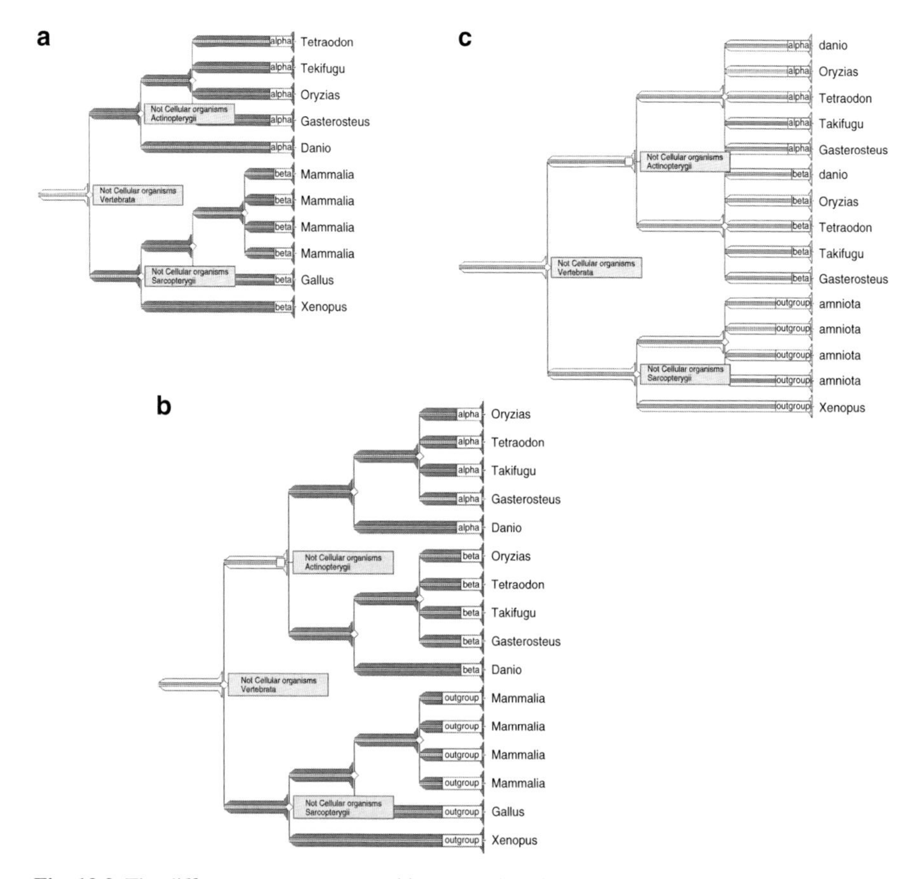

Fig. 13.3 The different tree patterns used in our study. Diamonds represent speciation nodes and boxes represent duplication nodes. Clear branches have no constraint whereas duplication is forbidden on dark gray branches. (**a**) Singleton topology: no duplication allowed, the gene tree must follow the expected species tree. (**b**) Conservation of paralogs from the fish specific duplication enforced in all five fishes; no other duplication allowed and the gene tree must follow the expected species tree. (**c**) Same as B but other duplications are allowed and there might be slight differences between the gene tree and the species tree

For all our analyses, we used CodeML 3.15 from the PAML package (Yang 1997). We first tried the **"branch model"** to estimate d*N*/d*S* among branches. As we said before, this model is interesting to detect branches under strong positive selection but suffers from two major problems in regards of our data set: the number of parameters to estimate (one d*N*/d*S* per branch) and the long evolutionary time. Taken together, these problems generated important convergence problems. We next tried the **"site model"** to estimate the selective pressure acting on specific sites. However, we found no significant results. Although this model can be useful on very fast evolving genes such as MHC or HIV proteins, it fails to detect weaker or transient positive selection on genes, which mostly evolve under strong purifying selection. Thus neither of these models were used in the final analysis. Finally, we implemented the improved **"branch-site"** model (Zhang et al. 2005) in a bioinformatic pipeline for large gene trees. Of note, the original branch-site model was used in one of the rare studies of ancient positive selection, on 2R duplication in the Troponin C gene family (Bielawski and Yang 2003). We corrected for test repetition using the q-value method (Storey and Tibshirani 2003), with $q = 10\%$. A specificity of results from the improved branch-site test is the bimodal distribution of p-values, due to many cases in which the best fit of the alternative model is identical to the null model and thus $p = 1$. Because of this, we use the *bootstrap* option to evaluate π_0 in the R package QVALUE.

The first striking result is that more than one third of FSGD duplicated genes (36%) have experienced positive selection shortly after duplication. Although this seems a high figure, it should be noted that (i) positive selection only concerns on average 3.0% of sites and (ii) we lack an expectation of the amount of positive selection on such genes with this method and sampling. To solve the second point, we focused on other vertebrate branches. Surprisingly, in the singleton data set, we found larger proportions of genes having experienced positive selection (Fig. 13.4). As reported above, we found that on our total data set (884 families), 77% of gene families have at least one branch with significant results. It must be noted again that most positive selection concerns a few sites, depending on the branch tested: only between 0.9% to 4.7% of sites are affected by positive selection on average. It is probable that this low percentage, over one branch, would be not detected by a simpler evolutionary model.

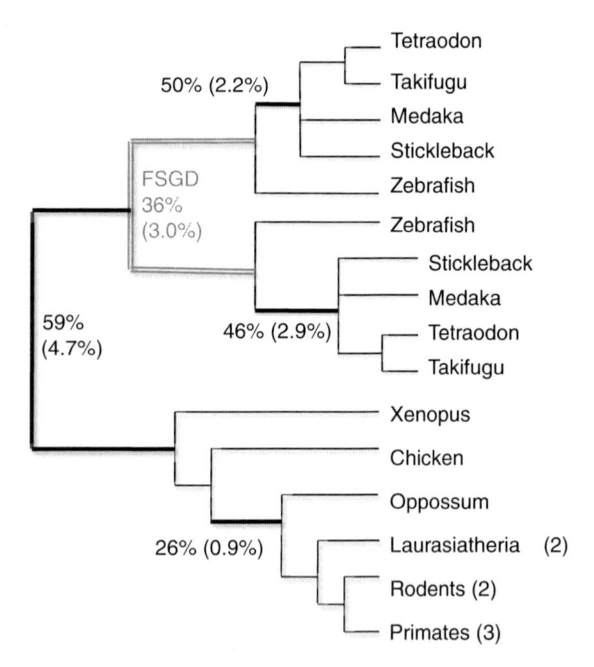

Fig. 13.4 The taxonomic tree of species used in our analysis of positive selection (Studer et al. 2008). The numbers in brackets after lineage names are the number of species in taxa represented by more than one species. The doubled branches are the split between duplicates from the fish specific genome duplication. The speciation branches analyzed are in bold. The values on branches represent the proportion of cases where positive selection has been detected on a branch, with the mean proportion of sites involved in brackets

We checked further the influence of whole genome duplication. Using Wilcoxon tests on different model parameters, no significant differences in the incidence of positive selection are found comparing either fish duplicate branches against others in the same gene tree, FSGD topology genes against singleton genes, or families with vertebrate 2R detected against no 2R detected.

As our data set is based on specific patterns, the multiple alignments are generally good, due to the tree constraints we imposed. To remove any doubt, we also performed additional analyses with GBLOCK and with MAFFT (Katoh et al. 2005). In all cases, the results were consistent, with small differences between alignment methods.

As said previously, a common criticism in analyses involving estimation d*N* and d*S* is the possibility of d*S* saturation. To ensure this is not the case in our study, we performed simulations using the same global parameters as the real data for each gene: number of sequences, sequence length, tree topology, branch lengths, d*N*/d*S* ration ω, transition/transversion ratio κ and codon usage. This procedure guaranties that the simulated data set has the same distribution of parameters as the real data set, including potential confounding factors such as codon usage bias or long branches. These results and those of Anisimova et al. (2002; Anisimova and Yang 2007) showed that the maximum likelihood estimate is robust to d*S* saturation, even for large divergence as shown by our simulations with doubled branch lengths. This is probably due to the use of more sequences, which "break" the long branches of the gene tree.

Finally, the conclusion of our study is threefold: (i) positive selection affects diverse phylogenetic branches and diverse gene categories during vertebrate evolution; (ii) it concerns only a small proportion of sites (1–5%); and (iii) whole genome duplication had no detectable incidence on the prevalence of this positive selection.

13.5 Selectome, A Database of Branch-Site Positive Selection

Most of these different methods of positive selection are time-consuming and can be difficult to use, thus it is of interest to have precomputed results available in a database. One of the first such databases is TAED (The Adaptive Evolution Database), which contains gene sequences, multiple alignment and trees from "higher plants" and chordates (Roth et al. 2005). For each branch of the tree, a d*N*/d*S* estimation is performed in a pairwise manner, using ancestral sequences reconstruction. Another database is the HumanPAML browser, which presents results of positive selection specific to human genes (Nickel et al. 2008). The computation of d*N*/d*S* is performed in $\sim$14,000 human genes with mammalian orthologs. Several models from PAML are used, including branch models, site models and branch-site models (original and improved). When a branch has to be specified for the model, results are computed only for the human branch. In our study of vertebrate genomes (Studer et al. 2008), we developed a fully automatic procedure to detect positive selection in any branch of phylogenetic trees, using the branch-site test. The main core consists in formatting the multiple alignment sequences of nucleotides, preparing the tree file by specifying automatically each branch of interest (coded with a #1), launching CodeML and finally retrieving the data and statistically analyzing them. In that study, we focused only on some specific patterns of evolution. But it should be of interest to have an exhaustive catalog of branches under positive selection, using the same methodology. Thus, we then developed Selectome (Proux et al. 2008). We used an improved version of the previous bioinformatic pipeline to scan all vertebrate gene families in the TreeFam database (Ruan et al. 2008). Release 1 focused only on the manually curated part (TreeFam-A). The database, built in MySQL, can be explored through a web interface at http://bioinfo.unil.ch/selectome. The user can perform searches using different criteria, such as gene name, gene ID, keywords, or by selecting specific branches of the phylogeny, with or without duplication. The result is a display of relevant sub-families, with annotated phylogenetic trees. Duplicated nodes are colored in red and branches under positive selection are in green boxes, with the associated p-value indicated. The user can also view the protein multiple sequences alignment with the Jalview applet (Clamp et al. 2004). For each significant branch, bars identify sites detected by BEB methods (Yang et al. 2005), color coded according to level of posterior value. We think this could be very useful not only in evolutionary biology but also in molecular biology studies to identify, e.g., candidates for site-directed mutagenesis.

13.6 Conclusion

In the studies presented here, different models have been used to infer evolutionary rate. An important point is that these models do not always detect the same pattern of positive selection. Pairwise estimation or branch models are useful to detect positive selection between two closely related genes, or in a specific lineage in a well-sampled family. Site models are more specific to detect amino acids that make rapid adaptation to external factors. And the branch-site models can be used to detect ancient or specific adaptive mutational events. Another point to note in conclusion is that there is of necessity a trade-off between the number of genes analyzed and phylogenetic sampling. Thus pairwise comparisons of closely related genomes will always include more genes than studies spanning many species. Obviously, the information from both types of studies is useful and complete each other.

We have seen through different analyses that various classes of genes could undergo positive selection. The genes for arm-race, for sexuality, for sensorial perception are the most dominant but positive selection can also be detected in housekeeping genes. Duplicated genes somewhat surprisingly do not appear to experience more positive selection than other genes. It will be of special interest in the future to be able to combine increasingly complex and realistic models of codon evolution (e.g., Mayrose et al. 2007; Yang and Nielsen 2008) with large amounts of data and good species sampling.

References

Anisimova M, Bielawski JP, et al (2001) Accuracy and power of the likelihood ratio test in detecting adaptive molecular evolution. Mol Biol Evol 18(8):1585–1592

Anisimova M, Bielawski JP, et al (2002) Accuracy and power of bayes prediction of amino acid sites under positive selection. Mol Biol Evol 19(6):950–958

Anisimova M, Liberles DA (2007) The quest for natural selection in the age of comparative genomics. Heredity 99(6):567–579

Anisimova M, Yang Z (2007) Multiple hypothesis testing to detect lineages under positive selection that affects only a few sites. Mol Biol Evol 24(5):1219–1228

Arbiza L, Dopazo J, et al (2006) Positive selection, relaxation, and acceleration in the evolution of the human and chimp genome. PLoS Comput Biol 2(4):e38

Avaron F, Thaeron-Antono C, et al (2003) Comparison of even-skipped related gene expression pattern in vertebrates shows an association between expression domain loss and modification of selective constraints on sequences. Evol Dev 5(2):145–156

Bakewell MA, Shi P, et al (2007) More genes underwent positive selection in chimpanzee evolution than in human evolution. Proc Natl Acad Sci USA 104(18):7489–7494

Baum J, Ward RH, et al (2002) Natural selection on the erythrocyte surface. Mol Biol Evol 19 (3):223–229

Bielawski JP, Yang Z (2003) Maximum likelihood methods for detecting adaptive evolution after gene duplication. J Struct Funct Genomics 3(1–4):201–212

Biswas S, Akey JM (2006) Genomic insights into positive selection. Trends Genet 22(8):437–446

Brunet FG, Crollius HR, et al (2006) Gene loss and evolutionary rates following whole-genome duplication in teleost fishes. Mol Biol Evol 23(9):1808–1816

Castresana J (2000) Selection of conserved blocks from multiple alignments for their use in phylogenetic analysis. Mol Biol Evol 17(4):540–552

Christin PA, Salamin N, et al (2008) Evolutionary switch and genetic convergence on rbcL following the evolution of C4 photosynthesis. Mol Biol Evol 25(11):2361–2368

Clamp M, Cuff J, et al (2004) The Jalview Java alignment editor. Bioinformatics 20(3):426–427

Clark AG, Glanowski S, et al (2003) Inferring nonneutral evolution from human-chimp-mouse orthologous gene trios. Science 302(5652):1960–1963

Conant GC, Wolfe KH (2008) Turning a hobby into a job: How duplicated genes find new functions. Nat Rev Genet 9(12):938–950

Dufayard JF, Duret L, et al (2005) Tree pattern matching in phylogenetic trees: automatic search for orthologs or paralogs in homologous gene sequence databases. Bioinformatics 21 (11):2596–2603

Duret L, Mouchiroud D, et al (1994) HOVERGEN: a database of homologous vertebrate genes. Nucleic Acids Res 22(12):2360–2365

Edgar RC (2004) MUSCLE: multiple sequence alignment with high accuracy and high throughput. Nucleic Acids Res 32(5):1792–1797

Endo T, Ikeo K, et al (1996) Large-scale search for genes on which positive selection may operate. Mol Biol Evol 13(5):685–690

Eyre-Walker A (2006) The genomic rate of adaptive evolution. Trends Ecol Evol 21(10):569–575

Force A, Lynch M, et al (1999) Preservation of duplicate genes by complementary, degenerative mutations. Genetics 151(4):1531–1545

Gillespie JH (1991) The causes of molecular evolution. Oxford University Press, New York

Guindon S, Gascuel O (2003) A simple, fast, and accurate algorithm to estimate large phylogenies by maximum likelihood. Syst Biol 52(5):696–704

Hamosh A, Scott AF, et al (2005) Online Mendelian Inheritance in Man (OMIM), a knowledgebase of human genes and genetic disorders. Nucl Acids Res 33(Suppl_1):D514–517

He X, Zhang J (2005) Rapid subfunctionalization accompanied by prolonged and substantial neofunctionalization in duplicate gene evolution. Genetics 169(2):1157–1164

Hughes AL, Hughes MK, et al (1994) Natural selection at the class II major histocompatibility complex loci of mammals. Philos Trans R Soc Lond B Biol Sci 346(1317):359–366; discussion 366–367

Hughes AL, Nei M (1988) Pattern of nucleotide substitution at major histocompatibility complex class I loci reveals overdominant selection. Nature 335(6186):167–170

Hughes AL, Nei M (1989) Nucleotide substitution at major histocompatibility complex class II loci: evidence for overdominant selection. Proc Natl Acad Sci USA 86(3):958–962

Jaillon O, Aury JM, et al (2004) Genome duplication in the teleost fish Tetraodon nigroviridis reveals the early vertebrate proto-karyotype. Nature 431(7011):946–957

Jiggins FM, Hurst GD, et al (2002) Host-symbiont conflicts: positive selection on an outer membrane protein of parasitic but not mutualistic Rickettsiaceae. Mol Biol Evol 19(8):1341–1349

Jorgensen FG, Hobolth A, et al (2005) Comparative analysis of protein coding sequences from human, mouse and the domesticated pig. BMC Biol 3:2

Katoh K, Kuma K, et al (2005) MAFFT version 5: improvement in accuracy of multiple sequence alignment. Nucleic Acids Res 33(2):511–518

King MC, Wilson AC (1975) Evolution at two levels in humans and chimpanzees. Science 188 (4184):107–116

Kondrashov FA, Kondrashov AS (2006) Role of selection in fixation of gene duplications. J Theor Biol 239(2):141–151

Kosiol C, Vinar T, et al (2008) Patterns of positive selection in six Mammalian genomes. PLoS Genet 4(8):e1000144

Kuzniar A, van Ham RC, et al (2008) The quest for orthologs: finding the corresponding gene across genomes. Trends Genet 24(11):539–551

Landan G, Graur D (2007) Heads or tails: a simple reliability check for multiple sequence alignments. Mol Biol Evol 24(6):1380–1383

Lee YH, Ota T, et al (1995) Positive selection is a general phenomenon in the evolution of abalone sperm lysin. Mol Biol Evol 12(2):231–238

Mayrose I, Doron-Faigenboim A, et al (2007) Towards realistic codon models: among site variability and dependency of synonymous and non-synonymous rates. Bioinformatics 23 (13):i319–327

Messier W, Stewart CB (1997) Episodic adaptive evolution of primate lysozymes. Nature 385 (6612):151–154

Moreno-Estrada A, Casals F, et al (2008) Signatures of selection in the human olfactory receptor OR5I1 gene. Mol Biol Evol 25(1):144–154

Nickel GC, Tefft D, et al (2008) Human PAML browser: a database of positive selection on human genes using phylogenetic methods. Nucleic Acids Res 36(Database issue):D800–D808

Nielsen R, Bustamante C, et al (2005) A scan for positively selected genes in the genomes of humans and chimpanzees. PLoS Biol 3(6):e170

Nielsen R, Yang Z (1998) Likelihood models for detecting positively selected amino acid sites and applications to the HIV-1 envelope gene. Genetics 148(3):929–936

Ohno S (1970) Evolution by gene duplication. Springer, New York

Perriere G, Duret L, et al (2000) HOBACGEN: database system for comparative genomics in bacteria. Genome Res 10(3):379–385

Petersen L, Bollback JP, et al (2007) Genes under positive selection in *Escherichia coli*. Genome Res 17(9):1336–1343

Proux E, Studer RA, et al (2008) Selectome: a database of positive selection. Nucleic Acids Res 37:D404–D407

Putnam NH, Butts T, et al (2008) The amphioxus genome and the evolution of the chordate karyotype. Nature 453(7198):1064–1071

Robinson-Rechavi M, Huchon D (2000) RRTree: Relative-Rate Tests between groups of sequences on a phylogenetic tree. Bioinformatics 16(3):296–297
Rooney AP, Zhang J (1999) Rapid evolution of a primate sperm protein: relaxation of functional constraint or positive Darwinian selection? Mol Biol Evol 16(5):706–710
Roth C, Betts MJ, et al (2005) The Adaptive Evolution Database (TAED): a phylogeny based tool for comparative genomics. Nucleic Acids Res 33(Database issue):D495–D497
Ruan J, Li H, et al (2008) TreeFam: 2008 Update. Nucleic Acids Res 36(Database issue):D735–D740
Sawyer SL, Wu LI, et al (2005) Positive selection of primate TRIM5alpha identifies a critical species-specific retroviral restriction domain. Proc Natl Acad Sci USA 102(8):2832–2837
Schmid K, Yang Z (2008) The trouble with sliding windows and the selective pressure in BRCA1. PLoS ONE 3(11):e3746
Semon M, Wolfe KH (2007) Consequences of genome duplication. Curr Opin Genet Dev 17 (6):505–512
Shiu SH, Byrnes JK, et al (2006) Role of positive selection in the retention of duplicate genes in mammalian genomes. Proc Natl Acad Sci USA 103(7):2232–2236
Storey JD, Tibshirani R (2003) Statistical significance for genomewide studies. Proc Natl Acad Sci USA 100(16):9440–9445
Studer RA, Penel S, et al (2008) Pervasive positive selection on duplicated and nonduplicated vertebrate protein coding genes. Genome Res 18(9):1393–1402
Suzuki Y (2008) False-positive results obtained from the branch-site test of positive selection. Genes Genet Syst 83(4):331–338
Vamathevan JJ, Hasan S, et al (2008) The role of positive selection in determining the molecular cause of species differences in disease. BMC Evol Biol 8:273
Wong KM, Suchard MA, et al (2008) Alignment uncertainty and genomic analysis. Science 319 (5862):473–476
Yang Z (1997) PAML: a program package for phylogenetic analysis by maximum likelihood. Comput Appl Biosci 13(5):555–556
Yang Z (1998) Likelihood ratio tests for detecting positive selection and application to primate lysozyme evolution. Mol Biol Evol 15(5):568–573
Yang Z (2006) Computational molecular evolution. Oxford University Press, New York
Yang Z (2007) PAML 4: phylogenetic analysis by maximum likelihood. Mol Biol Evol 24 (8):1586–1591
Yang Z, Nielsen R (2002) Codon-substitution models for detecting molecular adaptation at individual sites along specific lineages. Mol Biol Evol 19(6):908–917
Yang Z, Nielsen R (2008) Mutation-selection models of codon substitution and their use to estimate selective strengths on codon usage. Mol Biol Evol 25(3):568–579
Yang Z, Nielsen R, et al (2000) Codon-substitution models for heterogeneous selection pressure at amino acid sites. Genetics 155(1):431–449
Yang Z, Swanson WJ (2002) Codon-substitution models to detect adaptive evolution that account for heterogeneous selective pressures among site classes. Mol Biol Evol 19(1):49–57
Yang Z, Swanson WJ, et al (2000) Maximum-likelihood analysis of molecular adaptation in abalone sperm lysin reveals variable selective pressures among lineages and sites. Mol Biol Evol 17(10):1446–1455
Yang Z, Wong WS, et al (2005) Bayes empirical bayes inference of amino acid sites under positive selection. Mol Biol Evol 22(4):1107–1118
Zhang J (2004) Frequent false detection of positive selection by the likelihood method with branch-site models. Mol Biol Evol 21(7):1332–1339
Zhang J, Nielsen R, et al (2005) Evaluation of an improved branch-site likelihood method for detecting positive selection at the molecular level. Mol Biol Evol 22(12):2472–2479
Zhang JZ (2003) Evolution by gene duplication: an update. Trends Ecol Evol 18(6):292–298
Zhang ZD, Weinstock G, et al (2008) Rapid evolution by positive Darwinian selection in T-cell antigen CD4 in primates. J Mol Evol 66(5):446–456

Chapter 14
Molecular Coevolution and the Three-Dimensionality of Natural Selection

Mario A. Fares and Christina Toft

Abstract Natural selection is the force driving evolution and therefore much effort has been invested in the deciphering and understanding of the main mechanisms underlying such a force. Because of the main implications of identifying selective processes in proteins and the perspectives for defining functional/structural amino acid sites in protein structures, many models have been devised in order to search for shifts in the selective constraints throughout evolution. However, most of these models suffer from simplistic assumptions that deem results inconclusive or ambiguous. In this chapter we discuss the importance of natural selection in the emergence and generation of evolutionary novelties and the many different approaches built to identify the forces shaping the evolution of proteins. We also highlight the fact that, despite the plethora of new mathematical/statistical methods to identify selection at the molecular level, much remains to be done to build more realistic models. Molecular coevolution is among the most promising approaches to tackle the simplistic assumptions of linearity of protein sequence evolution but the field in this respect remains in its infancy. Here we discuss in depth the marriage between the linear sequence analysis of selection and the three-dimensionality of proteins through coevolutionary analyses and urge researchers to account for amino acid dependencies when looking for Darwin thoughts swamped in a sea of neutral evolution and pre-existing finite mutational and fitness landscape.

M.A. Fares and C. Toft
Department of Genetics, University of Dublin, Trinity College, Lincoln Place, Dublin 2, Dublin, Ireland
e-mail: faresm@tcd.ie

P. Pontarotti (ed.), *Evolutionary Biology: Concept, Modeling, and Application*,
DOI: 10.1007/978-3-642-00952-5_14,

14.1 Natural Selection and the Neutral Theory of Molecular Evolution

DNA mutation is an inherent phenomenon inextricably linked to replication. Mutations occur stochastically at the DNA level and their fixation at the protein level is responsible for the morphological, behavioral and physiological variation within populations. Ultimately, such variation leads to functional innovation and species divergence. The mechanisms whereby such mutations are fixed in genes and genomes have remained the subject of intense debate ever since the formulation of the neutral theory of molecular evolution by Kimura and later by King and Jukes in the late 1960s (Kimura 1968; King and Jukes 1969). Despite the apparent challenge of this theory to one of the main pillars of neo-Darwinism, this does not attempt to undermine the role of natural selection in the shaping of morphological variation. Rather it claims that fixation of selectively advantageous variants occurs at a very low frequency at the molecular level, making it irrelevant on a large-scale. Under the neutral theory of molecular evolution the fitness landscape associated to the mutational background is very smooth, so most of the mutations fixed have little to no effects on fitness. According to this theory, evolution of genes occurs due to the fixation of neutrally selected mutations by genetic drift (Kimura 1983). Because proteins are expected to have undergone optimization throughout evolution, most of the newly arising mutations are likely to be deleterious. These mutations will be selected against and removed by selection (purifying selection) (Li 1997). In exceedingly rare cases, new mutations may increase the fitness of its carriers in which case such mutations will be advantageous and subjected to positive selection. Implicit in the neutral theory of evolution is the assumption of homogeneity of substitution rates along time (Ohta and Kimura 1971; Bastolla et al. 2003).

The arrival of new molecular data intensified dramatically the debate about the relative importance of genetic drift and selection in driving evolution (Kimura and Ohta 1974). This data challenged in different ways the overwhelming acceptance of the neutral theory as the sole explanation for evolution. First, assuming that synonymous substitutions (those causing no amino acid replacements) that were fixed neutrally could be used as a benchmark to determine whether fixation of amino acid replacing mutations were favored or hindered by selection (Miyata and Yasunaga 1980; Li et al. 1985; Nei and Gojobori 1986). This assumption permitted the identification of hitherto increasing list of genes under positive selection. Second, theoretical studies have shown that the variances of substitution rates deviate from their expected Poisson distribution (Langly and Fitch 1974; Gillespie 1989). These deviations were also observed at the protein structural level, with proteins in different structural locations showing different rates of evolution (Bastolla et al. 2000; Bastolla et al. 2001). Finally, large-scale analyses of single genes and protein families demonstrated the correlation between the succession of events of adaptive evolution and the relative contribution of the protein to the biological fitness of the individuals (Endo et al. 1996; Roth et al. 2005). Despite these evidences fueling the neutralist-selectionist debate, only a few cases of adaptive evolution have been documented (Yang and Bielawski 2000). The main reason for such poor yield on the detection of proteins having undergone adaptive evolution is the fact that positive selection acts over few amino acid sites and in an episodic fashion being followed by strong purifying selection to preserve the new advantageous state (Gillespie 1991; Sharp 1997). Consequently, the signal for adaptive evolution becomes diluted with that of selection purifying most of the amino acid substitutions along evolution.

Because of the direct link between biological fitness of an organism mediated by protein's performance and adaptive evolution, identifying such events has become one of the main aims of evolutionary biologists and biomedical researchers aimed at detecting important amino acid sites in proteins of pharmacological relevance. In fact, the importance of adaptive evolution to predict the functional relevance of genes has been shown in several studies related to the immune response (Hughes and Nei 1988; Fares et al. 2002; Yang and Nielsen 2002; Yang and Swanson 2002) and in reproductive genes (Swanson et al. 2001a, b, 2003; Tsaur et al. 2001; Torgerson et al. 2002; Civetta 2003). Adaptive evolution has also been detected in genes linked to breast cancer in the human lineage (Fleming et al. 2003) and in the protein G6PD that confers resistance to malaria (Tishkoff et al. 2001). Because of this, a plethora of probabilistic and non-parametric methods has been developed aimed at detecting adaptive evolution at both the population and species level. The increasing sensitivity of such methods to identify adaptive evolution at single amino acid sites and in a nearly episodic fashion has heralded new avenues in basic and biomedical research. We are only beginning to scratch the surface of the mechanism whereby selection acts and more remains to be done as to understand the basic mutational dynamics underlying protein's evolution.

14.2 Measuring the Intensity of Selection

Protein-coding genes are the portions of DNA that have attracted most attention from evolutionary and molecular biologists owing to their functional and medical importance. From this perspective, proteins contribute in relative quantities to the biological fitness of the individual carrying them. Some proteins may hence be absolutely essential for the organism to work, others may be less essential under normal environmental conditions. Protein's function depends primarily on its three-dimensional structure, with only native structures being functional while unfolded ones being usually inactive. The folding of proteins is a complex problem yet to be resolved but the contribution of the amino acid composition of proteins to their appropriate folding and three-dimensional configuration is widely accepted. In dealing with protein-coding nucleotide sequences it is important to discriminate between substitutions that do not change the amino acid composition of the protein (called synonymous substitutions) and those that do change the amino acids (also called non-synonymous replacements). Because synonymous substitutions do not contribute to the evolution of the amino acid protein composition, they are assumed to accumulate neutrally and to be, on average, proportional to the time since the divergence of the species being compared (Miyata and Yasunaga 1980).

Even though this assumption may hold for most of the protein-coding genome of organisms, synonymous substitutions may well present evidence of being under selective constraints in some other organisms, such as RNA viruses whose secondary structure stability plays an essential role in the virus fitness (Chamary et al. 2006; Parmley et al. 2006; Tully and Fares 2009). Other elegant computational studies in HIV-1 viruses also quantified the constraints operating at synonymous sites (Mayrose et al. 2007). Codon bias is also an important factor, disturbing the equality of synonymous nucleotide substitution rates among genes. Conversely, non-synonymous substitutions change the amino acid composition of proteins and are consequently subjected to the evolutionary filtering depending on the selective value of such mutations and the population dynamics of the species. Non-synonymous substitutions thus change from gene to gene due to purifying selection, the extent of which also varies among genes (Kimura 1983).

Several case studies have demonstrated that with some genes non-synonymous substitutions occur at a higher rate than synonymous substitutions, this being an example of the action of positive Darwinian selection (e.g., Hughes and Nei 1988). Taking into account these parameters, many authors considered the non-synonymous-to-synonymous rates ratio (the standard notation of which is $\omega = d_N/d_S$, with d_N and d_S representing the number of replacements per non-synonymous sites and synonymous sites, respectively) to be a good indicator of the intensity of selection acting on a gene (Sharp 1997; Akashi 1999a, b; Crandall et al. 1999; Yang et al. 2000). There are several methods to estimate d_N and d_S that can be classified into three groups: (1) evolutionary pathway methods, such as the model of Nei and Gojobori (Nei and Gojobori 1986); (2) Kimura's 2-paramater model, such as the methods of Li, Wu and Luo (Li et al. 1985), the method of Pamilo and Bianchi (Pamilo and Bianchi 1993), method of Comeron (Comeron 1995) and the method of Ina (Ina 1995); and (3) maximum-likelihood methods with codon substitution models, such as the model of Goldman and Yang (Goldman and Yang 1994) and the different increasingly complex models developed thereafter (Nielsen 1998; Yang et al. 2000). Methods aimed at detecting selective constraints either at the level of populations or species are however subject to limitations mainly due to the heterogeneous distribution of selective constraints among genes throughout the evolutionary time and among protein regions. Other methods have been recently developed based on the 2-parameter Kimura's models that account for selective constraints operating at single amino acid sites (Suzuki and Gojobori 1999; Suzuki 2004a), at linear sequence window regions (Fares et al. 2002) and at three-dimensional structural window regions (Suzuki 2004b; Berglund et al. 2005).

14.2.1 Heterogeneous Selective Constraints Throughout Time and Sequence Space

The different contribution of genes to the biological fitness of individuals within populations shapes the degree of heterogeneity in the distribution of selective constraints among them. More interesting is the fact that within proteins different amino acid sites are subjected to different selective constraints for various reasons, including structural and functional constraints. Amino acid sites located at highly dense molecules (e.g., the core of the protein) will evolve slower than those located at structurally relaxed protein regions. This means that the greater the importance of the protein's function for the biological performance of the organism the stronger are the selective constraints against amino acid mutations and hence purifying selection acts efficiently at these proteins.

When the rate of evolution differs among amino acid sites, the distribution of these rates generally follows a gamma distribution (Uzzell and Corbin 1971; Gu et al. 1995; Yang 1996). If there is no rate shift throughout the evolutionary history of a gene then this distribution is called a homogenous gamma distribution. The shape of such a distribution is generally represented by the shape parameter (α), with small values ($\alpha < 2$) indicating a strongly skewed distribution (high variance in the substitution rates among sites), while large values ($\alpha > 20$) suggest almost Poisson distribution of substitution rates among amino acid sites. Selection shifts on specific amino acid sites throughout the evolutionary time of a gene deem the homogeneous gamma distribution model inappropriate to test for selection and other models accounting for selection shifts along time may have to be considered (Gu 1999; Gu 2001; Wang and Gu 2001; Fares and Travers 2006; Ruano-Rubio and Fares 2007). In these methods the violation of two of the assumptions of the homogeneous gamma model are accounted for and parameterized, including the variation of evolutionary rates along time and the covariation between amino acid sites. Regarding variation of substitution rates among sites under a gamma distribution, several empirical models of amino acid substitutions such as JTT can be combined with the gamma distribution, leading to models such as JTT + Γ (Yang 1994). However, based on the assumption that models of amino acid transitions may vary among sites, different authors developed site-specific transition models and models allowing few classes of sites to evolve under different Markov-chain models (Thorne et al. 1996; Goldman et al. 1998).

Genes that are under the continuous pressure of the environment to change and diversify, such as in the case of genes that are involved in the host–pathogen antagonism, may present diversifying adaptive evolution during most of their evolutionary time. In these genes, application of site-specific or branch-specific models to identify adaptive Darwinian selection is reasonable since such models average ω along the tree or the sequence space, respectively. Generally however, genes only undergo adaptive evolution punctually, by such events being mostly followed by bursts of purified selective events that swamp and dilute episodic positive selection phenomena signals. In these cases, models combining a variation of selective constraints among lineages and amino acid sites may be more sensitive to identify punctual selective events. Several models have been developed for such purposes, including maximum and Bayesian based approaches (Yang and Nielsen 2002; Zhang 2004; Zhang et al. 2005). Other models were based on more simple assumptions and were developed to identify selective constraints at single gene regions and branches based on the assumption that adjacent sites in the sequence or the structure may correlate in their selective constraints (Fares et al. 2002; Suzuki 2004b; Berglund et al. 2005). Despite such improvements, these methods suffer from the unaccounted correlated variation among amino acid sites, something sharply governed by the functional and structural parameters of proteins.

14.3 Structural Constraints and Molecular Coevolution

Not all the amino acid sites are born equal but rather some are devoted to form catalytic pockets, others are building the core of the proteins, few are generally involved in the establishment of protein–protein or subunit–subunit interfaces and others however have no apparent structural or functional role. This plethora of possible roles and structural categories an amino acid belongs to pinpoints the cause of the high variance in the evolutionary rates among them. For instance, analysis of 5,000 protein structures downloaded from databases such as NCBI supports this claim. Amino acids close in the three-dimensional structure present high correlation in their molecular weights (Fig. 14.1a) as evidenced by the lower difference in these weight differences at lower atomic distances in the structures (Fig. 14.1a). Moreover, when we build multiple sequence alignments for each one of these protein structures and measure the average Poisson-corrected amino acid distances at each amino acid site we see that such evolutionary distances decrease with increasing atomic densities (here atomic densities were measured as the number of amino acids surrounding a particular amino acid in the three-dimensional protein structure). The evolutionary rate is therefore negatively correlated with the number of atomic interactions a particular amino acid site establishes with the structural neighborhood (Fig. 14.1b). The more complex the network of such interactions the greater the selection coefficient may be against amino acid mutations because small mutations may lead to magnified destabilizing structural events. Grouping amino acid sites for selective constraints analyses is hence instrumental in our understanding of the processes of molecular evolution and the discovery of the main physico-chemical components shaping the evolution of molecules.

Analyses of correlated changes among sites were first attempted by ad hoc methods that based the grouping of such amino acids on non-statistical criteria (Hughes and Nei 1988; Clark and Kao 1991). These models have put forward the conclusion that indeed amino acid sites do not evolve independently, albeit several theoretical problems remained unresolved. For example, methods that use linear windowing to identify selective constraints ignore linearly distant amino acids that are three-dimensionally proximal in the structure. Conversely, methods that take into account three-dimensional proximity of amino acid sites do not account for correlated variation between distantly located amino acid sites (Fares and Travers 2006). To monitor correlated variation among amino acid sites, several methods

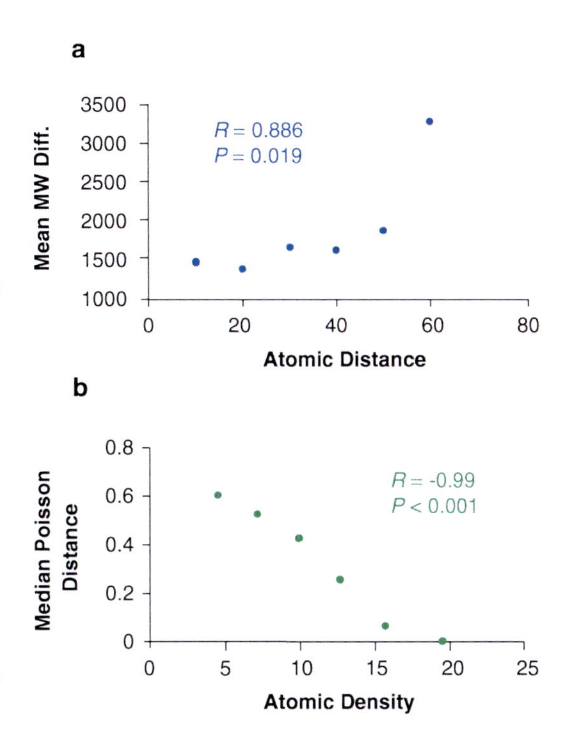

Fig. 14.1 Dependency between physico-chemical features of amino acids and their evolutionary rates. The three-dimensional distance between amino acid sites in the crystal structure (atomic distance) correlates positively with the difference in the molecular weight of the amino acids considered (**a**). The evolutionary rate of amino acids measured using the Poisson-corrected amino acid distances correlates negatively with the atomic interactions that amino acid establishes with the surrounding structural atomic environment (**b**). In this figure, the x-axis represents the different categories (ranging between 3 and 18) of atomic densities of amino acids measured as the average number of amino acids surrounding a particular residue in the crystal protein structure. The analyses presented in this figure are based on the study of 5,000 non-redundant protein structures downloaded from the National Centre for Biotechnology Information (NCBI) in the form of PDB files

have been developed and later improved for such purpose. The underlying complexity of co-mutating processes at the molecular level has compromised the sensitivity of such methods to identify real coevolutionary processes. This difficulty lies in the fact that coevolution between two amino acid sites can be decomposed into different parameters that describe the functional, structural, physical, phylogenetic and stochastic inter-link between the evolutionary processes of these sites (Fig. 14.2) (Codoner and Fares 2008). Different parametric and non-parametric methods have been developed to remove the phylogenetic and stochastic components from the coevolutionary estimates, although disentangling this new filtered coevolution into the components due to structural, functional, or physical covariation has proved difficult and remains yet to be achieved.

14.3.1 Methods to Measure Correlated Variation in Proteins

Many of the methods that have been developed during the last years attempted to resolve many of the problems that could not be resolved using standard selective

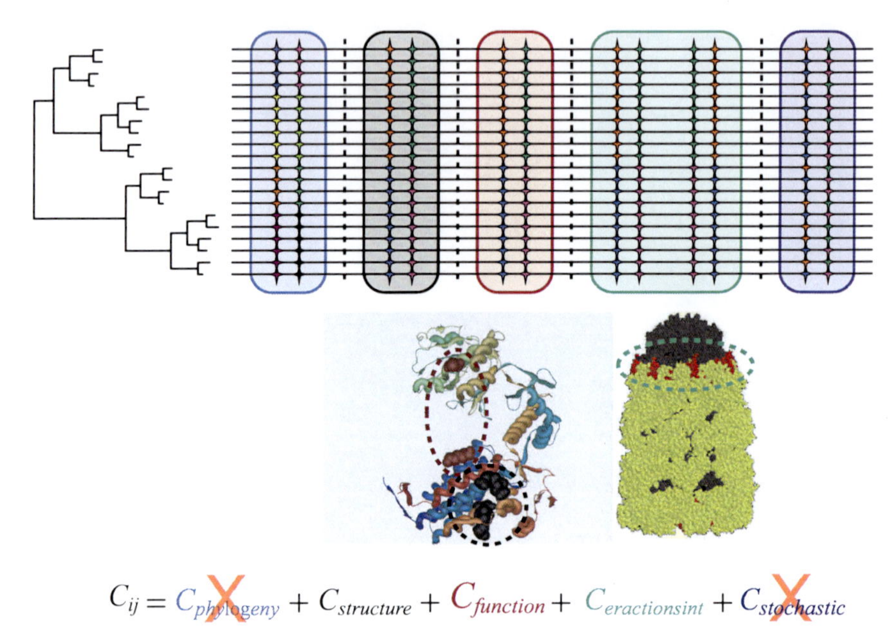

Fig. 14.2 Evolutionary and functional causes of variation dependencies among amino acid sites. Sequences representing different species or individuals within populations are usually historically (phylogenetically) interrelated. Generally, amino acid sites within multiple sequence alignments (symbolized by a group of lines in the figure) may show correlated patterns of variation along evolution (symbolized by the different color codes in the scheme). These patterns may be caused by either structurally constrained dependencies among sites in the protein structure, functional constraints, related interacting functions between amino acid sites or alternatively by historical or stochastic evolutionary processes. Several methods have been developed aimed at removing the stochastic and phylogenetic components

constraints methods. Indeed, models and methods aimed at identifying the selective forces shaping the evolution of proteins generally assume independence among amino acid sites. Such methods however fail to account for the interaction of the mutation effects among amino acid sites. Coevolution entails far more than the mere interaction between two amino acid sites that passively acquiesce to the compensation of one another but it has dramatic consequences in the stability of protein folding. Covariation among amino acid sites can fill the gulf between our knowledge of mutational dynamics and our awareness of protein structure and function and may have pragmatic implication for structure prediction and drug design (Fares 2006).

Coevolution is based on the idea of covariation first proposed by Fitch and Markowitz (Fitch and Markowitz 1970). Following their rationale, throughout the evolution of a gene, regions that were first invariable may become variable and vice versa. As mutations are fixed elsewhere in the sequence, selection constraints on invariable regions may change (Fitch 1971). Later, several authors have used this concept to explain covariation between morphological characters or nucleotide sequences (Schoniger and von Haeseler 1994; Rzhetsky 1995). In general, the scientific community is aware of the gap between the conclusions drawn from the treatment of amino acid sites as independent entities through mutagenesis and those extracted from functional data and coevolution can in principle ameliorate this problem (Gobel et al. 1994; Pazos et al. 1997). As a result many parametric and non-parametric methods have been devised to identify molecular coevolution at the amino acid level. Most of such methods assume that, at the amino acid level, residues under functional or structural constraints exhibit correlated changes (Suel et al. 2003; Gloor et al. 2005; Fares and Travers 2006; Travers and Fares 2007). Besides from the functional coevolution among amino acid sites, dependency of functional domains within proteins may forge strong compensatory co-dependency among three-dimensionally surrounding sites (Taylor and Hatrick 1994; Atwell et al. 1997; Chelvanayagam et al. 1997; Pazos et al. 1997; Martin et al. 2005; Codoner et al. 2006; Kim et al. 2006).

In general terms, methods to identify patterns of molecular amino acid sites covariation can be divided into parametric and non-parametric. Non-parametric methods are based on the information theory. These methods rely on the amount of variability at the amino acid sites and they use the Shanon's entropy as an appropriate approximation to the measurement of variability. Following these methods, detection of coevolutionary amino acid sites is possible by the probability of the different amino acid symbols to occur and co-occur in a particular amino acid site. This likelihood is defined using the Shanon's entropy for the site of interest i (H_i) and measuring its correlation with that of site j (H_j), using the mutual information content MI. This information is calculated as:

$$MI = H_i + H_j - H_{(i,j)}$$

$$H_i = -\sum_{x_i=A,S,L\ldots} P(x_i)\log[P(x_i)]$$

$$H_{(i,j)} = -\sum_{x_i,x_j} P(x_i,x_j)\log[P(x_i,x_j)]$$

Non-parametric methods to detect molecular coevolution are generally subjected to important problems of high false discovery rates. Recent studies however have introduced important corrections that have increased the split between the signals provided by noise and that provided by functional evolutionary dependencies among amino acid sites.

Unlike non-parametric methods, parametric approaches to detect coevolution have not received much attention because of their reliance on *a priori* assumptions. Most of these assumptions contributed substantially to increasing the inaccuracies of the predictions made prompted by our limited knowledge of the relative weight of the different components of coevolutionary parameters. Despite this, much has been achieved by the development of progressively accurate models based on maximum-likelihood (Pollock et al. 1999; Choi et al. 2005), Bayesian probabilities

(Dimmic et al. 2005), sequence divergence (Fares and 2006), continuous-time Markov processes (Yeang et al. 2007) and phylogenetic approaches (Goh et al. 2000; Pazos et al. 2005). Despite the fruitful results gained from such methods, much remains yet to be done to develop more accurate and decomposing methods to unearth the mechanistic reasons for inter-amino acid sites dependencies.

14.3.2 Molecular Adaptive Coevolution and Epistasis

Implicit in the concept of coevolution is the idea that a mutation at each amino acid site of the coevolving pair has a slightly negative effect *per se* over the performance of the protein. The co-occurrence of the two mutations however produces an antagonistic epistasis that improves the biological function of the protein. Examples of such cases have been reported before. For example, in a recent article authors examined the covariation of amino acid sites in functional domains of the heat-shock protein Hsp90 in a wide range of eukaryotic organisms (Fares and Travers 2006). Their results put forward the conclusion that the different functional domains of the Hsp90 communicate through a tightly concerted evolution of their corresponding amino acid sites. This coevolution has also been observed to yield insights into the communication between physically interacting proteins and to even identify residues involved in protein-protein interactions (Fig. 14.3; Travers and Fares 2007). The probability for the fixation of slightly deleterious mutations however depends upon the population-genetics parameters of organisms. For example, the greater the effective population size the lower is the probability for a slightly deleterious mutation to become fixed because the selection coefficient is directly proportional to the population size. Slightly deleterious mutations can only drift to fixation in qualitatively small populations such as eukaryotes as opposed to prokaryotic and viral populations. Endosymbiotic bacteria that live housed in specialized host insect cells also present small effective population sizes as a result of the strong bottlenecks that their effective population sizes are subjected to during the host inter-generational bacterial transmission. This population dynamics allows the accumulation of slightly deleterious mutations that may lead to functionally advantageous mutations through the fixation of three-dimensionally proximal compensatory mutations (Toft 2008). The adaptive value of such compensatory events has also been highlighted in a recent study showing that such events are inextricably linked to the generation of viral diversity and new epidemiological strategies (Tully and Fares 2009).

In conclusion, epistatic and pleiotropic effects of mutations at inter-dependent sites may generate an astronomical diversity in living forms irrespective of the small fraction of possible genotypes that "descended with modification from a common ancestor." Not only this combination may lead to such diversity under the environmental conditions organisms perpetuate through but also most of the potential evolvability of organisms is coded within such hidden genotypes whose phenotypic manifestations will be triggered by the environmental eventualities.

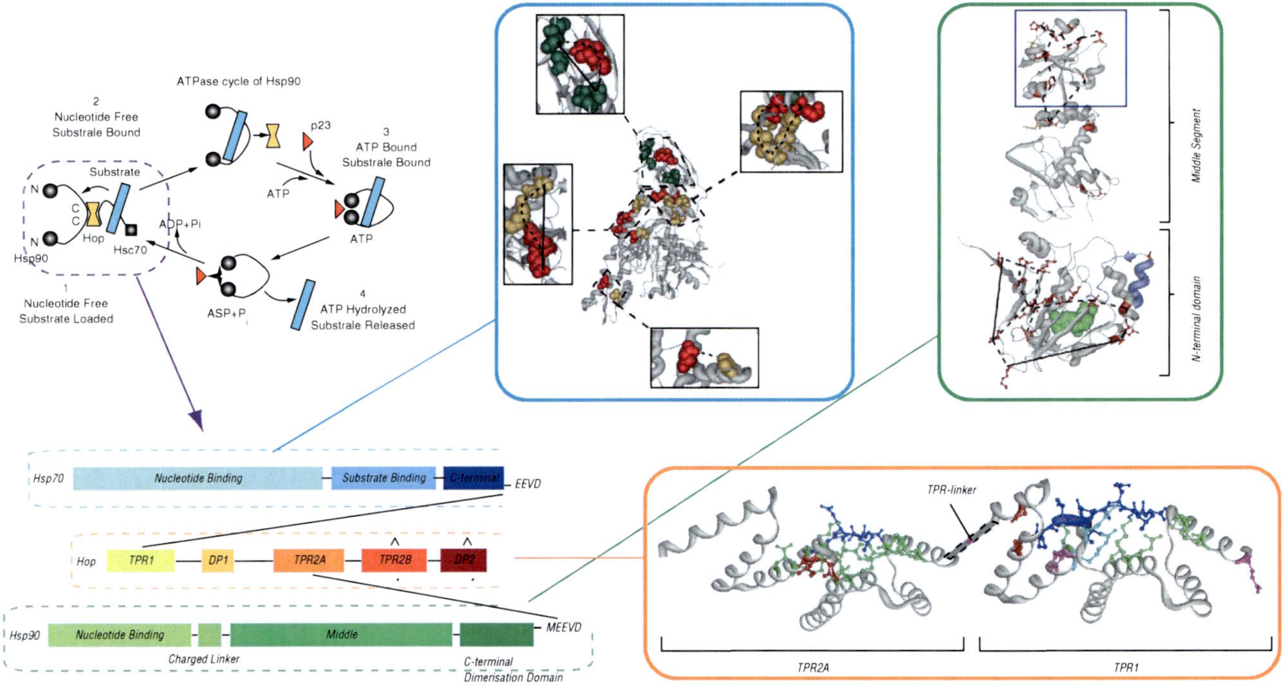

Fig. 14.3 Using coevolutionary approaches to identify residues involved in the interaction surfaces of proteins. In this figure we represent the case of the interaction between the heat-shock proteins Hsp90 and Hsp70 and the heat-organizing protein Hop. Proteins recently synthesized interact with these proteins to acquire their final productive folding in an ATP-dependent manner (top left side of the figure). Hsp70 interacts first with the protein to partially fold it and passes this partially folded protein into Hsp90 through a bridge established by Hop between both these proteins. The interaction between Hsp70 or Hsp90 and Hop has been shown previously to occur between the carboxi-terminal region of these proteins and several domains of Hop (highlighted in solid lines in the bottom left cartoon). Several other domains have been hypothesized to participate in such interactions (symbolized by hat and asterisk symbols). Coevolutionary analyses conducted in Travers and Fares (2007) identifies sites either functionally proposed or hypothesized to be part of such protein-protein interactions (sites shown in the corresponding three-dimensional structures in the right side of the figure)

References

Akashi H (1999a) Inferring the fitness effects of DNA mutations from polymorphism and divergence data: statistical power to detect directional selection under stationarity and free recombination. Genetics 151:221–238

Akashi H (1999b) Within- and between-species DNA sequence variation and the "footprint" of natural selection. Gene 238:39–51

Atwell SM, Ultsch AM, De Vos, Wells JA (1997) Structural plasticity in a remodeled protein-protein interface. Science 278:1125–1128

Bastolla U, Vendruscolo M, Roman HE (2000) Structurally constrained protein evolution: results from a lattice simulation. Eur Phys J B 15:13

Bastolla U, Farwer J, Knapp EW, Vendruscolo M (2001) How to guarantee optimal stability for most representative structures in the Protein Data Bank. Proteins 44:79–96

Bastolla U, Porto M, Roman HE, Vendruscolo M (2003) Statistical properties of neutral evolution. J Mol Evol 57(Suppl 1):S103–119

Berglund AC, Wallner B, Elofsson A, Liberles DA (2005) Tertiary windowing to detect positive diversifying selection. J Mol Evol 60:499–504

Chamary JV, Parmley JL, Hurst LD (2006) Hearing silence: non-neutral evolution at synonymous sites in mammals. Nat Rev Genet 7:98–108

Chelvanayagam G, Eggenschwiler A, Knecht L, Gonnet GH, Benner SA (1997) An analysis of simultaneous variation in protein structures. Protein Eng 10:307–316

Choi SS, Li W, Lahn BT (2005) Robust signals of coevolution of interacting residues in mammalian proteomes identified by phylogeny-aided structural analysis. Nat Genet 37:1367–1371

Civetta A (2003) Positive selection within sperm-egg adhesion domains of fertilin: an ADAM gene with a potential role in fertilization. Mol Biol Evol 20:21–29

Clark AG, Kao TH (1991) Excess nonsynonymous substitution of shared polymorphic sites among self-incompatibility alleles of Solanaceae. Proc Natl Acad Sci USA 88:9823–9827

Codoner FM, Fares MA (2008) Why should we care about molecular coevolution? Evol Bioinform Online 4:9

Codoner FM, Fares MA, Elena SF (2006) Adaptive covariation between the coat and movement proteins of prunus necrotic ringspot virus. J Virol 80:5833–5840

Comeron JM (1995) A method for estimating the numbers of synonymous and nonsynonymous substitutions per site. J Mol Evol 41:1152–1159

Crandall KA, Vasco DA, Posada D, Imamichi H (1999) Advances in understanding the evolution of HIV. AIDS 13(Suppl A):S39–47

Dimmic MW, Hubisz MJ, Bustamante CD, Nielsen R (2005) Detecting coevolving amino acid sites using Bayesian mutational mapping. Bioinformatics 21(Suppl 1):i126–135

Endo T, Ikeo K, Gojobori T (1996) Large-scale search for genes on which positive selection may operate. Mol Biol Evol 13:685–690

Fares MA (2006) Computational and Statistical methods to explore the various dimensions of protein evolution. Curr Bioinform 1:207–217

Fares MA, Travers SA (2006) A novel method for detecting intramolecular coevolution: adding a further dimension to selective constraints analyses. Genetics 173:9–23

Fares MA, Elena SF, Ortiz J, Moya A, Barrio E (2002) A sliding window-based method to detect selective constraints in protein-coding genes and its application to RNA viruses. J Mol Evol 55:509–521

Fitch WM (1971) Rate of change of concomitantly variable codons. J Mol Evol 1:84–96

Fitch WM, Markowitz E (1970) An improved method for determining codon variability in a gene and its application to the rate of fixation of mutations in evolution. Biochem Genet 4:579–593

Fleming MA, Potter JD, Ramirez CJ, Ostrander GK, Ostrander EA (2003) Understanding missense mutations in the BRCA1 gene: an evolutionary approach. Proc Natl Acad Sci USA 100:1151–1156

Gillespie JH (1989) Lineage effects and the index of dispersion of molecular evolution. Mol Biol Evol 6:636–647
Gillespie JH (1991) The cause of molecular evolution. Oxford University Press, Oxford
Gloor GB, Martin LC, Wahl LM, Dunn SD (2005) Mutual information in protein multiple sequence alignments reveals two classes of coevolving positions. Biochemistry 44:7156–7165
Gobel U, Sander C, Schneider R, Valencia A (1994) Correlated mutations and residue contacts in proteins. Proteins 18:309–317
Goh CS, Bogan AA, Joachimiak M, Walther D, Cohen FE (2000) Co-evolution of proteins with their interaction partners. J Mol Biol 299:283–293
Goldman N, Yang Z (1994) A codon-based model of nucleotide substitution for protein-coding DNA sequences. Mol Biol Evol 11:725–736
Goldman N, Thorne JL, Jones DT (1998) Assessing the impact of secondary structure and solvent accessibility on protein evolution. Genetics 149:445–458
Gu, X (1999) Statistical methods for testing functional divergence after gene duplication. Mol Biol Evol 16:1664–1674
Gu, X (2001) Maximum-likelihood approach for gene family evolution under functional divergence. Mol Biol Evol 18:453–464
Gu X, Fu YX, Li WH (1995) Maximum likelihood estimation of the heterogeneity of substitution rate among nucleotide sites. Mol Biol Evol 12:546–557
Hughes AL, Nei M (1988) Pattern of nucleotide substitution at major histocompatibility complex class I loci reveals overdominant selection. Nature 335:167–170
Ina, Y. (1995) New methods for estimating the numbers of synonymous and nonsynonymous substitutions. J Mol Evol 40:190–226
Kim Y, Koyuturk M, Topkara U, Grama A, Subramaniam S (2006) Inferring functional information from domain co-evolution. Bioinformatics 22:40–49
Kimura, M (1968) Evolutionary rate at the molecular level. Nature 217:624–626
Kimura M (1983) The neutral theory of molecular evolution. Cambridge University Press, London
Kimura M, Ohta T (1974) On some principles governing molecular evolution. Proc Natl Acad Sci USA 71:5
King JL, Jukes TH (1969) Non-Darwinian evolution. Science 164:788–798
Langly CH, Fitch WM (1974) An estimation of the constancy of the rate of molecular evolution. J Mol Evol 3:17
Li WH (1997) Molecular evolution. Sinauer, Sunderland, MA
Li WH, Wu CI, Luo CC (1985) A new method for estimating synonymous and nonsynonymous rates of nucleotide substitution considering the relative likelihood of nucleotide and codon changes. Mol Biol Evol 2:150–174
Martin LC, Gloor GB, Dunn SD, Wahl LM (2005) Using information theory to search for co-evolving residues in proteins. Bioinformatics 21:4116–4124
Mayrose I, Doron-Faigenboim A, Bacharach E, Pupko T (2007) Towards realistic codon models: among site variability and dependency of synonymous and non-synonymous rates. Bioinformatics 23:i319–327
Miyata T, Yasunaga T (1980) Molecular evolution of mRNA: a method for estimating evolutionary rates of synonymous and amino acid substitutions from homologous nucleotide sequences and its application. J Mol Evol 16:23–36
Nei M, Gojobori T (1986) Simple methods for estimating the numbers of synonymous and nonsynonymous nucleotide substitutions. Mol Biol Evol 3:418–426
Nielsen, R (1998) Maximum likelihood estimation of population divergence times and population phylogenies under the infinite sites model. Theor Popul Biol 53:143–151
Ohta T, Kimura M (1971) On the constancy of the evolutionary rate of cistrons. J Mol Evol 1:8
Pamilo P, Bianchi NO (1993) Evolution of the Zfx and Zfy genes: rates and interdependence between the genes. Mol Biol Evol 10:271–281
Parmley JL, Chamary JV, Hurst LD (2006) Evidence for purifying selection against synonymous mutations in mammalian exonic splicing enhancers. Mol Biol Evol 23:301–309

Pazos F, Helmer-Citterich M, Ausiello G, Valencia A (1997) Correlated mutations contain information about protein-protein interaction. J Mol Biol 271:511–523

Pazos F, Ranea JA, Juan D, Sternberg MJ (2005) Assessing protein co-evolution in the context of the tree of life assists in the prediction of the interactome. J Mol Biol 352:1002–1015

Pollock DD, Taylor WR, Goldman N (1999) Coevolving protein residues: maximum likelihood identification and relationship to structure. J Mol Biol 287:187–198

Roth C, Betts MJ, Steffansson P, Saelensminde G, Liberles DA (2005) The Adaptive Evolution Database (TAED): a phylogeny based tool for comparative genomics. Nucleic Acids Res 33: D495–497

Ruano-Rubio V, Fares MA (2007) Artifactual phylogenies caused by correlated distribution of substitution rates among sites and lineages: the good, the bad, and the ugly. Syst Biol 56:68–82

Rzhetsky A (1995) Estimating substitution rates in ribosomal RNA genes. Genetics 141:771–783

Schoniger M, von Haeseler A (1994) A stochastic model for the evolution of autocorrelated DNA sequences. Mol Phylogenet Evol 3:240–247

Sharp PM (1997) In search of molecular Darwinism. Nature 385:111–112

Suel GM, Lockless SW, Wall MA, Ranganathan R (2003) Evolutionarily conserved networks of residues mediate allosteric communication in proteins. Nat Struct Biol 10:59–69

Suzuki Y (2004a) New methods for detecting positive selection at single amino acid sites. J Mol Evol 59:11–19

Suzuki Y (2004b) Three-dimensional window analysis for detecting positive selection at structural regions of proteins. Mol Biol Evol 21:2352–2359

Suzuki Y, Gojobori T (1999) A method for detecting positive selection at single amino acid sites. Mol Biol Evol 16:1315–1328

Swanson WJ, Aquadro CF, Vacquier VD (2001) Polymorphism in abalone fertilization proteins is consistent with the neutral evolution of the *egg*'s receptor for lysin (VERL) and positive Darwinian selection of sperm lysin. Mol Biol Evol 18:376–383

Swanson WJ, Yang Z, Wolfner MF, Aquadro CF (2001) Positive Darwinian selection drives the evolution of several female reproductive proteins in mammals. Proc Natl Acad Sci USA 98:2509–2514

Swanson WJ, Nielsen R, Yang Q (2003) Pervasive adaptive evolution in mammalian fertilization proteins. Mol Biol Evol 20:18–20

Taylor WR, Hatrick K (1994) Compensating changes in protein multiple sequence alignments. Protein Eng 7:341–348

Thorne JL, Goldman N, Jones DT (1996) Combining protein evolution and secondary structure. Mol Biol Evol 13:666–673

Tishkoff SA, Varkonyi R, Cahinhinan N, Abbes S, Argyropoulos G, Destro-Bisol G, Drousiotou A, Dangerfield B, Lefranc G, Loiselet J, Piro A, Stoneking M, Tagarelli A, Tagarelli G, Touma EH, Williams SM, Clark AG (2001) Haplotype diversity and linkage disequilibrium at human G6PD: recent origin of alleles that confer malarial resistance. Science 293:455–462

Toft C (2008) Complex evolutionary dynamics in simple genomes: the paradoxical survival of intra-cellular symbiotic bacteria. Ph.D. thesis, University of Dublin, TCD

Torgerson DG, Kulathinal RJ, Singh RS (2002) Mammalian sperm proteins are rapidly evolving: evidence of positive selection in functionally diverse genes. Mol Biol Evol 19:1973–1980

Travers SA, Fares MA (2007) Functional coevolutionary networks of the Hsp70-Hop-Hsp90 system revealed through computational analyses. Mol Biol Evol 24:1032–1044

Tsaur SC, Ting CT, Wu CI (2001) Sex in Drosophila mauritiana: a very high level of amino acid polymorphism in a male reproductive protein gene, Acp26Aa. Mol Biol Evol 18:22–26

Tully DC, Fares MA (2009) Shifts in the selection-drift balance drives the evolution and epidemiology of foot-and-mouth disease virus. J Virol Jan 83(2):781–790. Epub 2008 Nov 12

Uzzell T, Corbin KW (1971) Fitting discrete probability distributions to evolutionary events. Science 172:1089–1096

Wang Y, Gu X (2001) Functional divergence in the caspase gene family and altered functional constraints: statistical analysis and prediction. Genetics 158:1311–1320

Yang Z (1994) Estimation of evolutionary distances between protein sequences. Yi Chuan Xue Bao 21:193–200
Yang Z (1996) Statistical properties of a DNA sample under the finite-sites model. Genetics 144:1941–1950
Yang Z, Bielawski JP (2000) Statistical methods for detecting molecular adaptation. Trends Ecol Evol 15:496–503
Yang Z, Nielsen R (2002) Codon-substitution models for detecting molecular adaptation at individual sites along specific lineages. Mol Biol Evol 19:908–917
Yang Z, Swanson WJ (2002) Codon-substitution models to detect adaptive evolution that account for heterogeneous selective pressures among site classes. Mol Biol Evol 19:49–57
Yang Z, Nielsen R, Goldman N, Pedersen AM (2000) Codon-substitution models for heterogeneous selection pressure at amino acid sites. Genetics 155:431–449
Yeang CH, Darot JF, Noller HF, Haussler D (2007) Detecting the coevolution of biosequences—an example of RNA interaction prediction. Mol Biol Evol 24:2119–2131
Zhang J (2004) Frequent false detection of positive selection by the likelihood method with branch-site models. Mol Biol Evol 21:1332–1339
Zhang J, Nielsen R, Yang Z (2005) Evaluation of an improved branch-site likelihood method for detecting positive selection at the molecular level. Mol Biol Evol 22:2472–2479

Chapter 15
The Evolutionary Constraints in Mutational Replacements

Branko Borštnik, Borut Oblak, and Danilo Pumpernik

Abstract The point mutations in the form of single nucleotide replacements occurring during the human–chimpanzee divergence process are analyzed in order to obtain an insight into the nature of the constraints that are imposed on various levels of cellular machinery upon the mutational events. The replacement patterns were examined separately for CpG islands (CGI), coding DNA regions, repetitive and nonrepetitive regions and for average genomic regions. The replacement counts were processed by the algebra based upon the rate equations. The starting point was the replacement count matrix. On this basis the replacement probability matrices were calculated. A model replacement probability matrix with three free parameters was constructed and the optimization procedure was used to determine the parameters. The replacements within the CGI and non-CGI regions were reproduced to the extent where the correlation coefficients attained values close to $r = 0.9$. For the exonic sequences the relative matrix of mutation rejections was calculated and it was found that the differences in the amino acid polarities and their opposites—the hydrophobicities—correlate in the greatest extent with the mutation rejection probabilities.

15.1 Introduction

Single nucleotide replacements, insertions and deletions are the basic building blocks of molecular evolution and its generating mechanisms are not yet really understood. The question whether the concept of neutral evolution is a solid ground on which the Darwinian principle of "survival of the fittest" can be superimposed is not a trivial one. There is no doubt that the stochasticity of the mutational events is

B. Borštnik, Borut Oblak, and Danilo Pumpernik
National Institute of Chemistry, Hajdrihova 19, SI-1000, Ljubljana, Slovenia
e-mail: branko@hp10.ki.si; borut.oblak@cmm.ki.si; dani.pumpernik@ki.si

P. Pontarotti (ed.), *Evolutionary Biology: Concept, Modeling, and Application*,
DOI: 10.1007/978-3-642-00952-5_15,

a predominant characteristic and that it facilitates the sampling of the sequence space. The exact logic that is governing the acceptance and fixation of the submitted changes is still beyond our comprehension.

The availability of several genomes of related organisms enables us to obtain an insight into the single nucleotide mutational replacements. When two sequences are aligned the directionality of the replacements can not be unveiled. In order to identify the nucleotide at the ancestral site one needs to align more than two closely related nucleotide sequences. In the primate region one can choose human, chimpanzee and macaque genomes. According to our experience the determination of the directionality of the mutations is a rather elusive goal and in our following working hypothesis the evolution is running close to a steady state regime in which the mutational replacements do not alter the compositional characteristics to a significant extent. We realized that the human–chimpanzee comparison is much more acceptable than the human–macaque and chimpanzee–macaque comparisons because 5–7 million years (My) that separate human and chimpanzee is a short enough period in which the number of superimposed mutational events is low enough and the detected picture is clear. The two other comparisons unveil the changes that were accumulated for more than 20 My. The corresponding replacement count matrices exhibit the time convoluted form so that the elementary events can not be easily recovered.

The most numerous forms of point mutations are single nucleotide replacements. The alignments of pairs of primate genomes (Lander et al. 2001; Mikkelsen et al. 2005; Gibbs et al. 2007) reveal several millions of single nucleotide replacements. Also the changes that are deposited in the single nucleotide polymorphism (SNP) databases (Jiang and Zhao 2006a, b; Zhao and Boerwinkle 2002; Zhao and Zhang 2006) represent a rich source of data. The analyses of the available data on the level of single nucleotide replacements reveal the 4×4 nucleotide replacement matrix with 12 off-diagonal elements. The (A/T) and (C/G) symmetry caused by strand complementarity makes the A<=>C replacement equivalent to T<=>G and A<=>G replacement to T<=>C. The replacements can be classified into transitions (A<=>G and C<=>T) that are three to four times more frequent than transversions (A<=>C, T<=>G, C<=>G, and A<=>T). Transversions are of two types—they may change the strength of the hydrogen bonding between the two strands (A<=>C and T<=>G (=W(eak)<=>S(trong)) replacement) or not (C<=>G (= S<=>S) and A<=>T (= W<=>W) replacement).

One should be aware that in spite of the fact that we are dealing with single nucleotide replacements it would be oversimplified if one would try to present the results in the form of a single 4×4 nucleotide replacement matrix, although this is frequently done (Karro et al. 2008). It is necessary to take into account the context dependence of the mutational events (Siepel and Haussler 2004), which means that one should cluster the single nucleotide replacements of various types into subclasses according to the sequence neighborhood. On a large-scale we shall distinguish CpG (or CG) islands (CGIs), protein coding regions, repetitive and nonrepetitive regions and, of course, the average genomic regions. Each single nucleotide mutational replacement is going to be processed in a dinucleotide

(Michel 2007) or trinucleotide context. This means that each replacement is stored while incrementing the dinucleotide and trinucleotide replacement A matrices.

In the protein coding regions (Yampolsky et al. 2005) the trinucleotide codons are the basic building blocks and the condition that protein function may not be impaired represents the main constraint on mutations. In exonic regions the strand complementarity symmetry in nucleotide replacement frequencies is not observed but we shall show that even in this case the global DNA constraints are playing the role and that the preservation of the protein function can be explained to a certain extent by the preservation of amino acid properties.

We shall analyze the mutational replacement data that are generated in the form of dinucleotide and trinucleotide replacement count matrices A, which bear the information about the details of the mutational processes. We construct a simple mutational replacement model. To evaluate the goodness of the model we compare the results of the model with the results obtained by analyzing the natural sequences. Quantification of the differences between the model and the observed results is performed by the correlation coefficient r.

We shall disregard the insertion/deletion component of mutational processes in spite of the fact that the repeat amplification mutational mechanism (Borštnik and Pumpernik 2002, 2005) that modifies short tandem repeats is rather powerful and contributes approximately one half of the DNA modification in the human genome.

It is our aim to reveal the global constraints acting on the genomic sequences. The constraints are mainly imposed by the cellular machinery responsible for the DNA replication, DNA packaging into the nucleosomes and resistance to degrading enzymes. In protein coding regions the mutational constraints are defined by the protein function, by the patterns imposed by the mRNA secondary structure (Shabalina et al. 2006) and by the tRNA availability. The question that we pose is the following: How far is it possible to parametrize the mutational replacement processes in average genomic sequences, in CpG islands and in protein coding regions?

15.2 Methods

Human (hg18)—chimpanzee (panTro2) alignments were taken from http://hgdownload.cse.ucsc.edu. The procedure led us to 2,900 Mbp of alignments that were analyzed and the results were stored in the symmetric dinucleotide and trinucleotide replacement matrices A, separately for nonrepetitive and repeat masked parts of the sequences. The distinction was also made between the CGIs and the remaining parts of the sequences. The CGIs were determined by the CpGIF algorithm (Ye et al. 2008). The aligned sites, where both organisms have identical nucleotides, contribute to the diagonal A_{ii} values and the sites, where one of the sequences was hit by mutation, contribute to the off-diagonal A_{ij} elements. The sites containing the gaps or non-ACGT symbols were ignored.

The A matrix was constructed also for exons. The alignments were taken from the same source as genomic alignments. Locations of exons were inferred from the annotations of the NCBI RefSeq database Human Genome Build 36-1 at ftp://ftp.ncbi.nlm.nih.gov/genomes/H_sapiens where also information about the order and the orientation of the contigs along the chromosome assembly is provided. The number of exons, successfully mapped on the alignment, was 31418.

Several types of A matrices were generated for coding regions: the codon–codon replacement matrices and dinucleotide replacement matrices. The latter were evaluated separately for each pair of positions within the codons 1,2; 2,3 and 3,1.

The A matrices serve as the primary extract of data containing the information about the mutational processes. Let $N = \Sigma_i \Sigma_j A_{ij}$ denote the total number of sequential units (dinucleotides, trinucleotides, or codons). The probability of the occurrence of sequential units equates to

$$f_j = \sum_i A_{ij}/N \tag{15.1}$$

Further, one can define the elements of the mutation probability matrices

$$W_{ij} = A_{ij}/(f_j N) \tag{15.2}$$

In the case that the evolutionary processes are running in a steady state regime the composition of genetic sequences in terms of dinucleotides and trinucleotides are conserved. This conjecture can be written in the following form:

$$[W]f = f \tag{15.3}$$

Equation (15.3) can be exploited when constructing the model of the mutational process. To construct the model means to define a scheme on how the point mutations are running—in other words to define the W matrix. After one constructs the mutation probability matrix elements, equation (15.3) can be treated as a system of linear equations for unknown values of the components of the composition vector. Since (15.3) has an apparent form of the eigenvalue equation, the determinant should be equal to zero for the nontrivial solution, which means that the rank of the matrix is $n - 1$ and the equations are not independent. One can thus remove an arbitrary line from (15.3) and replace it by the equation representing the normalization condition of the dinucleotide population $\Sigma_i f_i = 1$. After solving the system of equations and obtaining the population vector one can calculate the model A matrix and subsequently one can compare the model values with the natural values. The model can be optimized by choosing some criterion to quantify the goodness of the model such as the evaluation of the correlation coefficients. The optimization procedure was carried through by brute force parameter variation procedure.

The model of the dinucleotide mutation probability matrix was constructed with a set of two parameters and a choice of a list of dinucleotide replacements that run at a higher rate with a factor that represents the third variable parameter. The first parameter refers to the transition/transversion ratio. The second parameter refers to the enhanced probability of CpG dinucleotide mutability.

15.3 Results

15.3.1 Dinucleotide Replacements

Our basic results obtained by analyzing the human–chimpanzee alignments are dinucleotide replacement counts A_{ij}. We generated the A matrices for various classes of DNA sequences such as CGI and non-CGI sequences, the repeat sequences and the matrices for exons, separately for the three dinucleotide positions relative to the trinucleotide reading frame, separately for CGI and non-CGI parts of the sequences.

The basic reference matrix is the one corresponding to genomic non-CGI sequences that comprise roughly 97% of all the sequences—roughly 2.9 Gbp. The matrix clearly shows the following characteristic features: (i) The A_{ij} elements fall into three magnitude classes with their maximal values being eight-digit numbers along the diagonal, five-digit numbers if the ith and jth dinucleotide differ at one place only and three-digit numbers if the two dinucleotides differ at both sites. This is consistent with the fact that the human–chimpanzee divergence is roughly 1% and that the probability of a two-nucleotide replacement in dinucleotides is proportional to the square of the probability for one-nucleotide replacement. (ii) The one-nucleotide replacements in the form of transitions are three to four times more frequent than transversions. (iii) The CpG dinucleotides, which are roughly ten times less frequent than other dinucleotides, exhibit the replacement counts that are comparable to other dinucleotides, which means that the probability for their involvement in the mutational process exceeds the mutation probability of non-CpG dinucleotides nearly tenfold.

When the mutational processes that take place outside the CpG islands are subjected to the parameter optimization procedure, the A_{ij} and f_i values are reproduced as depicted in Fig. 15.1 where the horizontal coordinates refer to the natural A_{ij} and f_i values and the vertical coordinates to the values reproduced by the model. Goodness of the model can be measured by the closeness of the points to a straight line—which can be quantified by the value of the correlation coefficient. The distribution of the points in Fig. 15.1 results in r_f = 0.938 and r_A = 0.976. The following parameters emerged when trying to reproduce the f_i and A_{ij} elements: the transition over transversion ratio emerged with a value 3.45; CpG decay enhancement factor with the value 9; the enhancement factor of the dinucleotide decays listed in Table 15.1 emerged with a value of 1.8.

The main difference in the parameterization of mutational processes within the CpG islands is due to the increased CpG dinucleotide content and reduced CpG mutability. In addition to the CpG dinucleotide also the other three SS (S = C or G)

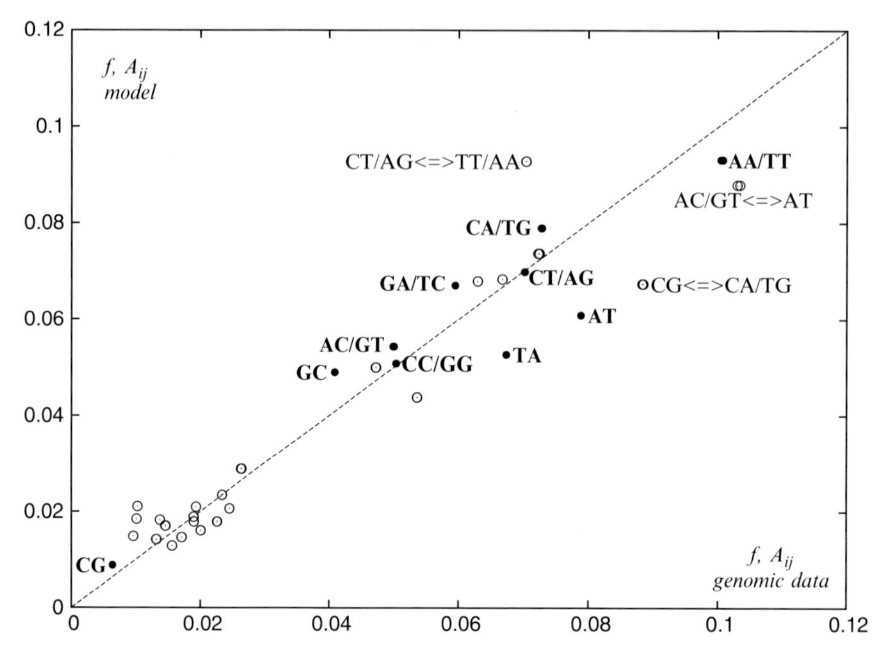

Fig. 15.1 Our model values versus natural values for the probability of occurrence of dinucleotides (*f*, bullets) and replacement count matrix elements (A_{ij}, circles, arbitrary units) for the non-CGI (non-CpG islands) sequences. The numbers along the two axes correspond to the *f* values and the dinucleotides are identified by one-letter symbols. The strand complementarity symmetry causes that the locations of the bullets corresponding to the complementary pair (such as CA/TG) coincide. Also, the A_{ij}values exhibiting the largest discrepancy between the model and natural values are labeled. The replacements where both nucleotides are replaced are rare. They are located very close to the origin of the coordinate system and are not displayed. The group of 16 pairs of transversions is located at the lower left part of the figure. The remaining circles correspond to the eight pairs of transitions

Table 15.1 The list of the enhanced dinucleotide decays outside the CGI. The upper row represents the target dinucleotide and the lower row the mutational product

CC	*GG*	*AC*	*GT*	*GC*	*AT*	*TA*	*AG*	*CA*	*CT*	*TG*
CA,CG, CT	AG,CG,TG	AA,AG,AT	AT,CT,TT	CC,GG	AC,GT	CA,TG	AA	AA	TT	TT

dinucleotides (CC, GG and GC) are more abundant in the average genomic sequences, while the WW (W = A or T) dinucleotides (that is TA, AA, TT and AT) are less abundant. This can be seen in Fig. 15.2 where the CGI dinucleotide abundance is plotted along the vertical axis and the abundance of the remaining part of the genome is plotted along the horizontal axis. The above mentioned SS and WW dinucleotides are located along the off-diagonal lines—the CG, CC, GG and GC dinucleotides at one side and the TA, AA, TT and AT nucleotides on the other

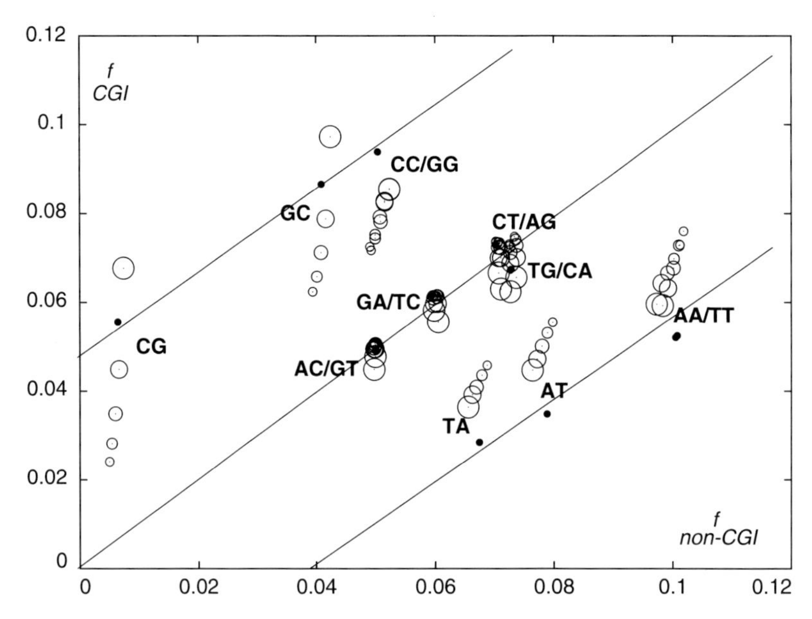

Fig. 15.2 The comparison between the dinucleotide compositions of the CpG islands (CGI) (vertical axis) and the non-CGI sequences (horizontal axis). The bullets represent the results obtained by CpGIF algorithm and the circles represent the result of our own procedure that was run with five sets of parameters. The sizes of the circles are proportional to the CpG density threshold imposed in the process of CGI determination

side of the diagonal. The SW and WS dinucleotides are positioned along the diagonal—they are equally abundant within the CGIs as outside the CGIs. Further, as expected, the CG dinucleotide exhibits very low content outside the CGIs and close to the average content within the CGIs where the TA dinucleotide accepts the status of rareness. In order to see the influence of the parameterization on the determination of the CGIs we have performed the search for CGIs with various stringency criteria for minimal CG dinucleotide density within the CGIs. In Fig. 15.2 it is shown that the stringency criteria do not influence the positioning of mixed dinucleotides with mixed composition, while the pure S-type or W-type dinucleotide composition is drifting away from the diagonal as stringency is increased.

In Fig. 15.3 the elements of the A matrix are plotted in a similar way as the components of the composition vector are plotted in Fig. 15.2. Only 48 elements corresponding to the dinucleotide replacements between the pairs of dinucleotides are plotted where only one nucleotide is replaced. Transversions are located at the lower left corner and are not labeled. Transitions that are 16 in number are clustered in pairs as a consequence of a strand symmetry. The most drastic difference between the CGIs and the remaining part of the genome is expected to take place in the replacements in which CG dinucleotide is involved. Outside the CGIs the CG

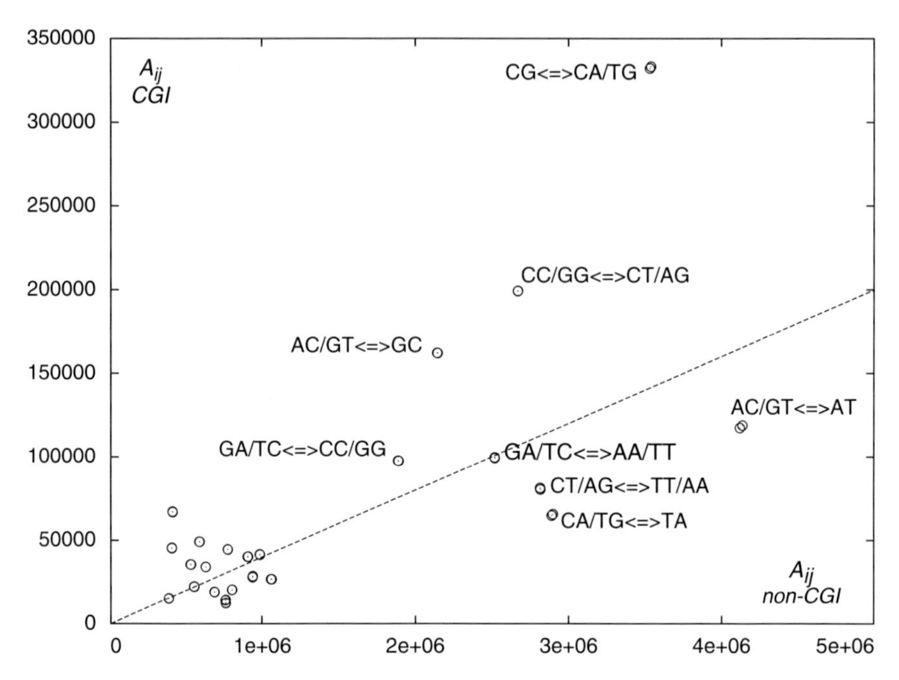

Fig. 15.3 The comparison between the replacement count matrix elements A_{ij} of CpG islands (CGI) (vertical axis) and the non-CGI sequences (horizontal axis) referring to the dinucleotide replacements where only one nucleotide is replaced. The symbols close to the origin correspond to transversions (the exact description of the replacement is not shown), the remaining symbols correspond to transitions

dinucleotide is nearly ten times less frequent and at the same time also ten times more mutable than other dinucleotides, which has for the consequence the fact that the CA/TG<=>CG replacement counts are positioned close to the center of the distribution of A_{ij} values along the horizontal axis. Within the CGIs the CG dinucleotides exhibit close to the average share among all the other dinucleotides and since the CGs are not excessively mutable one would expect that also the A_{ij} element counting the CA/TG<=>CG replacements would lay somewhere in the middle of the distribution along the vertical axis of Fig. 15.3. This is, however, not the case. The CA/TG<=>CG replacements are the most frequent within the CGIs. We think that this is a consequence of the fact that whatever algorithm is used for the CGI retrieval, some methylated—and thus hypermutable CG dinucleotides remain. The residual hypermutable CGs in the CGIs interfere also with the modeling procedure as seen in Fig. 15.4, where one can see that the CA/TG<=>CG replacement count exhibits the strongest discrepancy between the model and natural values. The parameters that optimize the model for CGI sequences are as follows. The transition/ transversion ratio (= 3.45) results in the same value as for the non-CGI sequences. The T<= C transition within the CG dinucleotides is

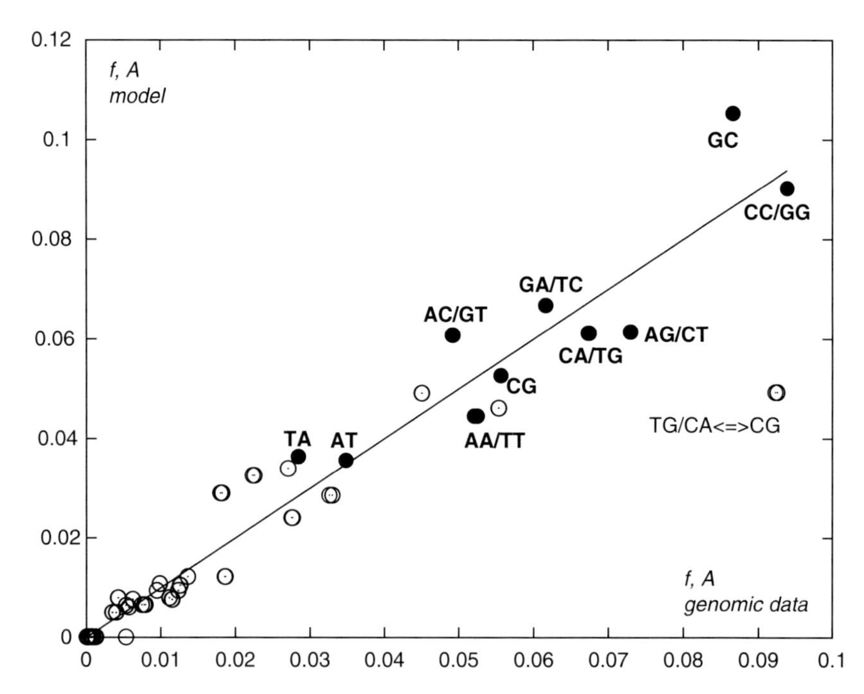

Fig. 15.4 Our model values versus natural values for the probability of the occurrence of the dinucleotides (f, bullets) and replacement count matrix elements (A_{ij}, circles) for the CGI sequences. See also the caption of Fig. 15.1

Table 15.2 The list of the enhanced dinucleotide decays within the CGI. The upper row represents the target dinucleotide and the lower row the mutational product

AT	*TA*	*AC*	*GT*	*CA*	*TG*	*AG*	*CT*	*AA*	*TT*
AC, GT	CA, TG	GC	GC	CG	CG	GG	CC	AG	CT

modeled to occur with double probability in comparison with all other transitions. The enhanced dinucleotide decays listed in Table 15.2 have an enhancement factor equal to 2. The resulting correlation coefficients of f_i and A_{ij} values (see Fig. 15.4) are 0.89 and 0.93, respectively.

The CGIs have interesting properties from the point of view of dinucleotide composition. If one plots the components of the dinucleotide composition vector as a function of the dinucleotide base pairs stacking free energy (Frappat and Sciarrino 2005), one obtains the dependence that is just opposite as in the case when the graph is plotted for the non-CGI sequences, as seen in Fig. 15.5. In spite of the fact that one would expect that the issue of functionality is the main determinant of coding patterns, it turns out that also the energetics of the DNA sequence play a role. The results can be interpreted if considering that the CGIs are generally associated with promoters (Saxonov et al. 2006; Ponger et al. 2001). One could speculate that a

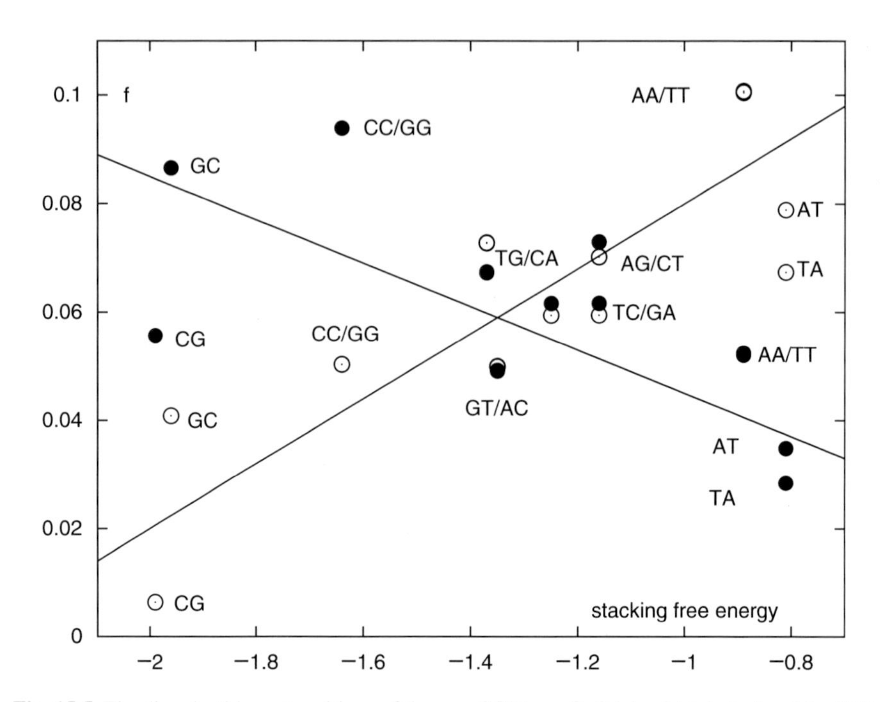

Fig. 15.5 The dinucleotide compositions of the non-CGI (non-CpG islands) (circles) and the CGI (bullets) sequences as a function of the free energy of stacking (in kcal/mol). The two lines are drawn in order to emphasize the fact that within the CGIs the dinucleotides population is a decreasing function of the stacking free energy while in the rest of the genome the situation is just the opposite

strong binding (that is with three hydrogen bonds) between S-type base pairings and more negative stacking free energy makes CGIs suitable for a more important role in promoter regions. On the other hand it is also known that the promoter region also contains the TATA box that can be easily unwound and its constituent dinucleotide is known for its paucity (De Amicis and Marchetti 2000).

15.3.2 Trinucleotide Replacements

Trinucleotide A(64×64) matrix possesses 4,096 matrix elements. Since we work within the approximation of a symmetric A matrix only $64 \times 63/2 = 2{,}016$ elements need to be considered. By taking into account only the replacements where two trinucleotides differ only at one place, the number of relevant matrix elements reduces to 288. We have shown above to what extent the neighboring site is determining the replacement events in dinucleotides. The next question that needs to be answered is whether is it necessary to go further and to construct a new model that will deal with the influence of two neighboring nucleotides on the single nucleotide replacement. We tested the situation by expressing the trinucleotide replacement probability matrix elements in terms of dinucleotide matrix elements. To express the probability that a trinucleotide $\alpha\beta\gamma$ is replaced by a trinucleotide *abc* where only in one of the pairs of the corresponding nucleotides (a,α), (b,β), and (c,γ) the nucleotides are different, one should distinguish two possibilities: In the case that the replacement takes place at the first or the third position of the trinucleotide one can simply use the corresponding dinucleotide replacement probability matrix elements $W_{abc,\alpha\beta\gamma} = W_{ab,\alpha b}$ and $W_{abc,\alpha\beta\gamma} = W_{bc,b\gamma}$, respectively. This means that only the first neighbor was taken into account. In the case that the difference is in the middle position the situation is less straightforward and one should evaluate the importance of the left and right neighbor and weigh the two contributions accordingly. The most efficient approximation is to take into account only the neighbor that exhibits stronger influence on the nucleotide that is being replaced. The matrix element $W_{abc,\alpha\beta\gamma}$ is taken to be equal to $W_{ab,\alpha\beta}$ or $W_{bc,\beta c}$, the one that is farther away from the average value of the dinucleotide transition probabilities.

After the trinucleotide replacement probability matrix is constructed in the above described manner the system of equations for the trinucleotide frequencies was solved and they were compared with the corresponding data from the triple alignments. The following two correlation coefficients were obtained: $r_A = 0.97$ and $r_f = 0.95$. One can conclude that the correlation coefficients are high enough and that only the first neighbor of the replacement site is important in determining the replacement probability.

15.3.3 Codon–Codon Replacements

There are 288 ($= 64 \times 9/2$, since each of the three nucleotides in the triplet can mutate to any of other three nucleotides) single nucleotide replacements between 64 genomic trinucleotides. When counting the single nucleotide replacements between 61 codons coding for 20 amino acids, one should consider that 25 replacements represent so-called nonsense mutations—the replacement of a codon coding an amino acid with *stop* codon. Two replacements represent the interchange between the *stop* codons and what remains are 67 synonymous codon replacements and 196 amino acid replacement mutations. The synonymous codon replacements are the most frequent mutations in the protein coding regions. If all the synonymous codons of a certain amino acid would be equivalent and uniformly used, the synonymous mutations in exons would run with the same pace as the corresponding nucleotide replacements in noncoding regions. The synonymous changes do not correlate perfectly with the corresponding genomic replacements. The corresponding correlation coefficient is $r_A = 0.881$. The situation is to some extent similar with non-synonymous single nucleotide replacements where the correlation is still weaker $r_A = 0.83$. The population of 61 trinucleotide codons correlates with the population of genomic trinucleotides with the correlation coefficient $r_f = 0.45$.

The crucial point we would like to address is the differences between the acceptance rate of the amino acid replacements in protein coding regions of genes and the acceptance rate of the average genomic nonrepetitive sequences. A straightforward proposition would be that the differences between the amino acid properties are the dominant determinants of the amino acid replacements. This proposition can be tested by comparing the amino acid replacement suppression factor versus the difference of the amino acid properties. The amino acid replacement suppression factor was defined in such a way that it was zero if the amino acid replacement probability exhibits the same mutation probability as the genomic groups of trinucleotide, which are equal to the codons coding for the particular amino acid. A strong suppression of the probability of the replacement of one amino acid with another one results in a high positive value of the amino acid replacement suppression factor:

$$s_{ij} = 1 - A_{ij}^{\text{exon}} / A_{ij}^{\text{genome}}. \tag{R1}$$

In the literature one can find an extensive list of amino acid properties (Kawashima et al. 1999). The file with several hundred sets of amino acid values was scanned systematically in order to find the set of properties that demonstrates the strongest correlation between the amino acid properties and the amino acid replacement suppression factor. We found that this is the case with the amino acid polarity and hydrophobicity. However, the correlation between the amino acid replacement suppression factor and the differences of the amino acid polarity values is rather moderate—the correlation coefficient is 0.35.

Another possibility to compare the mutational dynamics of exons with the rest of the genome is to evaluate the similarity indices between the eight mutational replacement matrices that we calculated for the CGI and non-CGI exonic regions and the non-exonic regions. The results are presented in Table 15.3. The highest similarity can be found outside CGI. The 23 and 31 positions within exons is very close to the non-CGI genomic A matrix. The parts of the exons that are located within the CGIs correlate with the genomic CGI A matrix rather poorly—even

Table 15.3 The similarity (quantified by means of the correlation coefficient) between the A matrices corresponding to the eight DNA sequential categories. The 12, 23 and 31 indices refer to the first two and the last two places in codons. The 31 index refers to the third place in the codon and the first place of the next codon

	CGI genome	non-CGI genome
CGI exon 12	0.538	0.164
CGI exon 23	0.373	0.048
CGI exon 31	0.435	0.083
non-CGI exon 12	0.695	0.752
non-CGI exon 23	0.658	0.840
non-CGI exon 31	0.667	0.831
CGI genome	1.000	0.698
non-CGI genome	0.698	1.000

non-CGI exons exhibit stronger resemblance to the CGI genomic A matrix than the CGI exonic A matrices and genomic CGI A matrix.

15.4 Discussion

We have shown that the analyzes of point mutations enable us to obtain an insight into the nature of the mutational processes. The results show that stochasticity is the essential driving force in evolution and that the concept of neutrality deserves recognition. Within a strict Darwinian concept it would be hard to expect that the mutational patterns of the sequential environments as diverse as the protein coding regions and noncoding regions could have a common denominator in the form of replacement matrices that exhibit the correlation coefficient as high as 0.7. This means that there exist global genomic constraints that control the mutational replacements. We tried to obtain an insight into the nature of the mutational patterns of various DNA sequential categories.

Comparing the mutational patterns in CGI and non-CGI environments the main difference is due to CpG mutability (Duret and Galtier 2000; Fryxell and Zuckerkandl 2000; Michel 2007; Ollila et al. 1996; Saxonov et al. 2006). Elaborating on a naíve picture of the CpG islands one could envisage that the CGIs would exhibit a uniform dinucleotide composition, since the main cause of the population nonuniformity—the hypermutability of the CpGs due to the cytosine methylation—is not present in the CGIs. However, such a reasoning is oversimplified because it is estimated (Fryxell and Moon 2005) that hypomethylation of the CGI reduces their mutation rates approximately twofold—thus the hypermutability of the CpGs still persists. Also, our CGI model (see Section 15.3.1) unravels the CpG mutability being for a factor of two above unity, which can be compared with the factor equal to 9 for non-CGI sequences. There are strong indications that the CGIs are involved in the control of the gene expression (Ponger et al. 2001) but it turned out that this is true only for less than three quarters of promoters (Saxonov et al. 2006). Our preliminary results show that the transcription factor binding sites dinucleotide replacement frequency matrix is very close to the corresponding CGI matrix. However, the working definitions of what constitutes a CGI may vary and depends upon the thresholds. It is illustrated in Fig. 15.2 that the threshold variation does not change the qualitative picture as far as the dinucleotide composition of CGI and non-CGI regions is concerned. We can see that the three classes of dinucleotides SS, SW+WS and WW lay along the three parallel straight lines without respect to which CpG threshold densities were used in the calculation. The next step towards the understanding of the mutational patterns and compositional properties of CGI and non-CGI sequences in terms of dinucleotides is to consolidate the information provided by Figs. 15.1 to 15.4 and Tables 15.1 and 15.2. The scattering of the points off the diagonal line in Figs. 15.1 and 15.4 is a measure for the adequateness of our models and the question we can pose is the following: Can one expect a better coverage of the mutational patterns by the model? We modeled non-CGI regions of human—chimpanzee genomic alignment separately for repetitive and nonrepetitive sequences (the UCSC repeat masking result of human genome was used as a source of repeat/no-repeat distinction) and we found that the original model that ignores the repetitiveness fits equally well to both, the repetitive and nonrepetitive part of the genome. This means that the mutational patterns belonging to specific sequence categories such as exons or transcription factor binding sites covering a minor part of genome do not represent a statistically significant contribution to the non-CGI mutational patterns.

Also the result depicted in Fig. 15.5 is relevant to the question whether it is meaningful to construct independent models for CGI and non-CGI parts of the genomic sequences. The fact that the dinucleotide composition of the two sequence categories exhibits meaningful stacking free energy dependence means that the two sequential categories are real entities. We can thus conclude that the partition of genomic sequences into CGI and non-CGI categories is the most meaningful choice when trying to resolve the one-nucleotide replacement pattern of mutational process.

Acknowledgments This work was financed by the Slovenian Research Agency.

References

Arndt PF, Petrov DA, Hwa T (2003) Distinct changes of genomic biases in nucleotide substitution at the time of mammalian radiation. Mol Biol Evol 20:1887–1896

Borštnik B, Pumpernik D (2002) Tandem repeats in protein coding regions of primate genes. Genome Res 12:909–915

Borštnik B, Pumpernik D (2005) Evidence on DNA slippage step-length distribution. Phys Rev E—Stat, Nonlinear, Soft Matter Phys 71:031913/1–031913/7

De Amicic F, Marchetti S (2000) Intercodon dinucleotides affect codon choice in plant genes. Nucleic Acids Res 28:3339–3345

Duret L, Galtier N (2000) The covariation between TpA deficiency, CpG deficiency, and G+C content of human isochores is due to a mathematical artifact. Mol Biol Evol 17:1620–1625

Frappat L, Sciarrino A (2005) Sum rules for free energy and frequency distribution of DNA. dinucleotides. Physica A 351:448–460

Fryxell KJ, Moon W-J (2005) CpG mutation rates in the human genome are highly dependent on local GC content. Mol Biol Evol 22:650–658

Fryxell KJ, Zuckerkandl E (2000) Cytosine deamination plays a primary role in the evolution of mammalian isochores. Mol Biol Evol 17:1371–1383

Gibbs RA, Rogers J, et al (2007) Evolutionary and biomedical insights from the rhesus macaque genome. Science 316:222–234

Jiang C, Zhao Z (2006a) Directionality of point mutation and 5-methylcytosine deamination rates in the chimpanzee genome. BMC Genomics 7:316 doi:10.1186/1471–2164–7–316

Jiang C, Zhao Z (2006b) Mutational spectrum in the recent human genome inferred by single nucleotide polymorphisms. Genomics 88:527–534

Karro JE, Pfeifer M, Hardison RC, Kollmann M, von Gruenberg HH (2008) Exponential decay of GC content detected by strand-symmetric substitution rates influences the evolution of isochore structure. Mol Biol Evol 25:362–374

Kawashima S, Ogata H, Kanehisa M (1999) AAindex: Amino acid index database. Nucleic Acids Res 27:368–369

Lander ES, Linton LM, et al (2001) Initial sequencing and analysis of the human genome. Nature 409:860–921
Michel CJ (2007) Evolution probabilities and phylogenetic distance of dinucleotides. J Theor Biol 249:271–277
Mikkelsen TS, Hillier LW, et al (2005) Initial sequence of the chimpanzee genome and comparison with the human genome. Nature 437:69–87
Ollila J, Lappalainen I, et al (1996) Sequence specificity in CpG mutation hotspots. FEBS Lett 396:119–122
Ponger L, Duret L, Mouchiroud D (2001) Determinants of CpG islands: expression in early embryo and isochore structure. Genome Res, 11:1854–1860
Saxonov S, Berg P and Brutlag DL (2006) A genome-wide analysis of CpG dinucleotides in the human genome distinguishes two distinct classes of promoters. Proc Natl Acad Sci USA 103:1412–1417
Siepel A, Haussler D (2004) Phylogenetic Estimation of Context-Dependent Substitution Rates by Maximum Likelihood. Mol Biol Evol 21:468–488
Shabalina SA, Ogurtsov AY, Spiridonov NA (2006) A periodic pattern of mRNA secondary structure created by the genetic code. Nucleic Acids Res 34:2428–2437
Yampolsky LY, Kondrashov FA, Kondrashov AS (2005) Distribution of the strength of selection against amino acid replacements in human proteins. Human Mol Genetics 14:3191–3201
Ye S, Asaithambi A, Liu Y (2008) CpGIF: an algorithm for the identification of CpG islands. Bioinformation 2:335–338
Zhao Z, Boerwinkle E (2002) Neighboring-nucleotide effects on single nucleotide polymorphisms: A study of 2.6 million polymorphisms across the human genome. Genome Res 12:1679–1686
Zhao Z, Zhang F (2006) Sequence context analysis of 8.2 million single nucleotide polymorphisms in the human genome. Gene 366:316–324

Chapter 16
Why Phylogenetic Trees are Often Quite Robust Against Lateral Transfers

Marc Thuillard

Abstract The circular order of a tree is the order at which the leaves are encountered in a clockwise scanning of a tree. The circular order of a tree is quite robust against lateral transfers. We show that if lateral transfers are only between consecutive nodes, the tree reconstructed with the Neighbor-Joining algorithm furnishes a perfect order of the nodes. The order of the node corresponds to one of the possible orders of the tree prior to lateral transfer. This result permits to understand why phylogenies obtained from molecular data often furnish reasonable trees despite lateral transfers. Using the mathematical framework introduced in the first part of this chapter, new methods to localize lateral transfers are presented. These methods use minimum contradiction matrices to identify lateral transfers. Several examples on real data show the potential of minimum contradiction matrices in phylogenetic studies.

16.1 Introduction

With the advent of molecular biology, phylogenetic trees can be obtained by comparing the distance between genomes at the sequence level. The distance between two taxa can be estimated by aligning proteins or ADN sequences and looking at differences between the sequences with an appropriate model of evolution. Already before the completion of the human genome sequencing in 2001, an article (Doolittle 2000) had summarized the increasing doubts in the scientific community about the possibility of drawing a tree of life based on genome evolution. These doubts arose after the discovery of the importance of lateral transfer in genes evolution. Does the transfer of genetic material destroy the

M. Thuillard
La Colline, Creuze 9, CH-2072 St-Blaise, Switzerland
e-mail: Thuillweb@hotmail.com

P. Pontarotti (ed.), *Evolutionary Biology: Concept, Modeling, and Application*,
DOI: 10.1007/978-3-642-00952-5_16,

original phylogenetic signal? Does lateral transfer lead to trees that cannot be trusted to reconstruct the evolution of species? One is today in a somewhat paradoxical situation. On the one hand, almost nobody disputes the existence and importance of lateral transfer in evolution. On the other hand, there are numerous studies (Fukami-Kobayashi et al. 2007; Dutilh et al. 2007; Kunin et al. 2005b) that show a good level of consistency between phylogenetic trees obtained with whole genomes and accepted phylogenies obtained with different methods when such phylogenies do exist. The first possible answer to this apparent paradox is that the phylogenetic signal is preserved on average and that lateral transfer can be regarded as some overlapping noise. In this chapter we present a different answer, namely that the order of the taxa on a tree is a robust phylogenetic signal. In order to show this, we discuss the effect of lateral transfers on a tree reconstruction algorithm. Provided lateral transfer is only between consecutive leaves the reconstructed tree retains the neighboring relationships between taxa. Obviously there are some limits to the robustness of a tree against lateral transfers. We describe them and present a new approach that has been developed recently to represent phylogenetic information in such cases. This approach based on minimum contradiction matrices has been presented in two publications and the reader is referred to them for technical details on the mathematical aspects (Thuillard 2007, 2008).

16.2 Effect of Lateral Transfer on the Order of the Branches in a Tree

A valued X-tree T is a graph with X as its set of leaves and a unique path between any two distinct vertices x and y, with internal vertices of at most degree 3. A circular order on an X-tree corresponds to an indexing of the n leaves according to a circular (clockwise or anti-clockwise) scanning of the leaves in T (Makarenkov and Leclerc 1997). Figure 16.1 shows a tree and an indexing of the taxa that corresponds to a circular order. For taxa indexed according to a circular order the distance matrix $Y_{i,j}^n$, with $Y_{i,j}^n$ the distance between a reference node n and the path i–j, fulfils the so-called Kalmanson inequalities: $Y_{i,j}^n \geq Y_{i,k}^n, Y_{k,j}^n \geq Y_{k,i}^n (i \leq j \leq k)$. The distance matrix $Y_{i,j}^n$ has the property that the distance diminishes away from the diagonal (Kalmanson 1975). This property is visualized in Fig. 16.1. If the values of the distance matrix are represented by different levels of gray, the level of gray is shading away from the diagonal. This property of the matrix characterizes a Kalmanson matrix and an order satisfying all Kalmanson inequalities is called a perfect order.

A lateral transfer is a process in which the genome of an organism incorporates genetic material from a distantly related organism. The word « lateral » is used to distinguish this process from the vertical evolution in which genetic material is transferred to the offspring by its ancestors. A simple model can be used to understand the effect of lateral transfer. Let us define an α-lateral transfer from

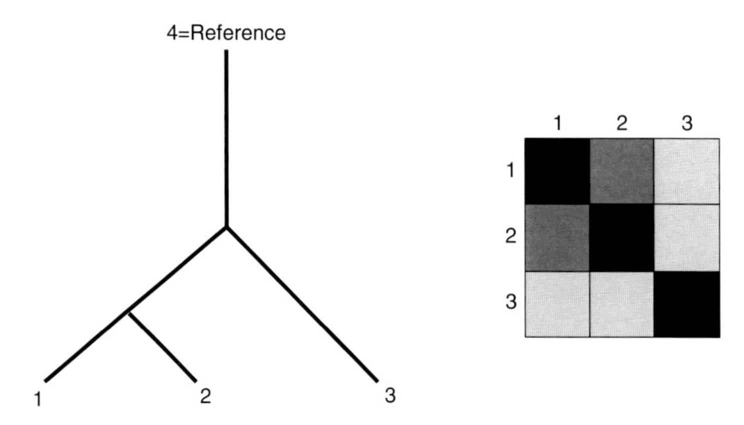

Fig. 16.1 The distance $Y_{i,j}^{\mathrm{Ref}}$ between a reference taxa and a path *i-j* on an X-tree fulfills Kalmanson inequalities. If the values of the distance matrix $Y_{i,j}^{\mathrm{Ref}}$ are coded in a gray scale, the level of gray decreases as one moves away from the diagonal

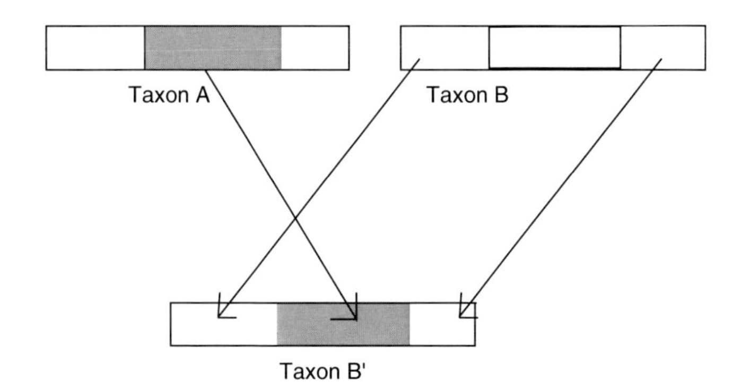

Fig. 16.2 An α-lateral transfer from taxon A to taxon B corresponds to the replacement of some proportion α of taxon B by a segment of A

A to B as the transfer of a proportion α of the genetic material of taxon A to taxon B, as sketched in Fig. 16.2. An α-lateral transfer corresponds to a transformation of the distance matrix. For segments of equal lengths and in the framework of the Jukes-Cantor model the expected value of the distance (Thuillard 2007) between two leaves $\hat{d}_{B,C}$ is given by

$$\hat{d}_{B,C} = \alpha \cdot d_{A,C} + (1 - \alpha) \cdot d_{B,C} \tag{16.1a}$$

$$\hat{d}_{A,B} = (1 - \alpha) \cdot d_{A,B} \tag{16.1b}$$

with $d_{A,B}$ the distance between the sequences A and B in the X-tree (i.e., without lateral transfer).

Despite the fact that a lateral transfer suppresses the X-tree structure, some of the original tree topology can still be retrieved. If a lateral transfer takes place between two consecutive vertices, then the original order is preserved and the inequalities defining perfect order are still fulfilled by the transformed distance matrix. Consider a lateral transfer from one leaf A to another leaf B of a valued X-tree. For any X-tree, there is at least a circular order for which two arbitrary leaves are consecutive. Without restriction, let us choose A and B to be consecutive in a circular order. After lateral transfer, the expected distances are related to the distances on the tree without lateral transfer by the equation: $\hat{Y}^n_{B,C} = \alpha \cdot Y^n_{A,C} + (1-\alpha) \cdot Y^n_{B,C}$ and $\hat{Y}^n_{A,B} = \alpha \cdot Y^n_{A,A} + (1-\alpha) \cdot Y^n_{A,B}$. As $\hat{Y}_{B,C}$ and $\hat{Y}_{A,B}$ are weighted averages on two consecutive sequences, the order is preserved and the distance matrix still satisfies the inequalities characterizing perfect order.

One understands from the previous discussion that an X-tree can accommodate a large number of lateral transfers. A tree has many degrees of freedom that can be used to make any two nodes consecutive. Lateral transfers between consecutive taxa do not destroy the structure of the distance matrix at least in a probabilistic framework. As long as lateral transfers are between consecutive nodes (or consecutive ancestor nodes), perfect order is preserved. The expected values of the distance matrix $Y^n_{i,j}$ decrease away from the diagonal and the expected values of the distance matrix are Kalmanson. In Fig. 16.3 and following, a lateral transfer is represented by an arrow. Provided arrows are only between adjacent nodes, then perfect order is preserved.

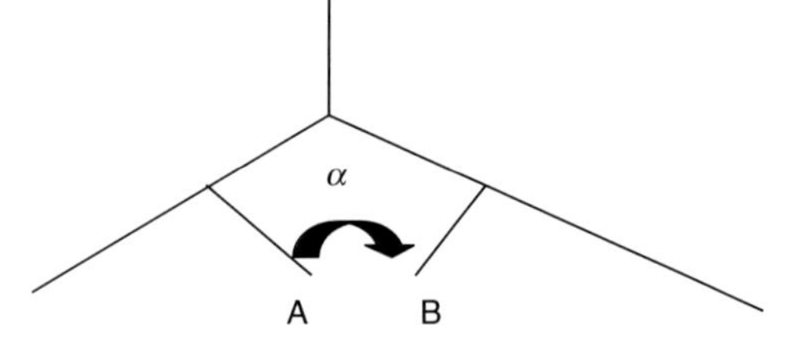

Fig. 16.3 A tree has many degrees of freedom that can be used to make any two nodes consecutive. If a lateral transfer takes place between leaves that are consecutive then the expected distance matrix can be perfectly ordered

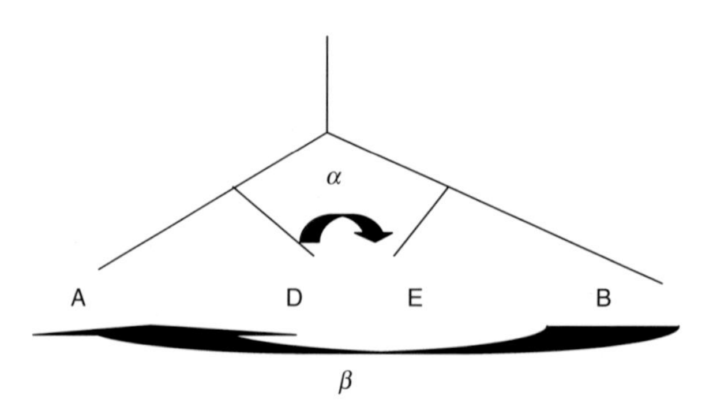

Fig. 16.4 Perfect order cannot be guaranteed if there is no order for which taxa pairs involved in lateral transfer are consecutive

The robustness of the order against lateral transfer has also some limits. Figure 16.4 shows an example of lateral transfers that do not preserve order. This case is characterized by the fact that it is impossible to find an order of the sequences for which all sequence pairs involved in lateral transfers are consecutive leaves.

16.3 Effect of Lateral Transfer on the Reconstructed Tree with the Neighbor-Joining Algorithm (NJ)

The most popular method to reconstruct an evolution tree from distances between sequences is the Neighbor-Joining algorithm (NJ) first developed 30 years ago by Saitou and Nei (1987). The NJ method has become theoretically better understood in very recent years. One can now explain why the method works practically so well and why it is, in several instances, optimal (Gascuel and Steel 2006; Mihaescu et al. 2006).

The NJ algorithm contains two main steps that are repeated iteratively. For an X-tree, the NJ algorithm discovers, at each iteration step, two leaves that form a cherry in the tree. Once the two leaves are discovered, they are replaced by a new leaf. More precisely, in the first step, the two vertices (*i,j*) maximizing $r_i + r_j - (N-2) \cdot d(i,j)$ $(r_i = \sum_{k=1..N} d_{i,k})$ are joined. In the second step the two leaves (*i,j*) are replaced by a new leaf *i-j*. The distance to another leaf k is obtained by averaging the distance between leaves i,j and the leaf *k*: $d_{i-j,k} = 1/2 \cdot (d_{i,k} + d_{j,k})$. The Neighbor-Joining algorithm can be used to obtain an initial "best order." The best order is then refined with more complex algorithms.

The NJ-based ordering algorithm has two main steps.

1. The two taxa *a* and *b* to be joined are determined by the NJ algorithm.
2. Given the leaves $a_1, a_2, ..., a_i$ $(resp.\ b_1, b_2, ..., b_j)$ having the vertex a (resp. b) as first ancestor. The best order of the leaves is chosen so as to minimize the contradiction among four possibilities: $(ab, \bar{a}b, a\bar{b}, \bar{a}\bar{b}$ with ab the order $a_1, a_2, ..., a_i, b_1, b_2, ..., b_j$ and $\bar{a}$ the inversed order $a_i, a_{i-1}, ..., a_1$. The contradiction is summed up over all other nodes taken as reference.

Proposition: Given a perfectly ordered X-tree and a number of α-lateral transfers between consecutive nodes defined by Eq. 16.1, then in the limit of infinitely long sequences, the output of the NJ-based ordering algorithm corresponds to a perfect order satisfying all Kalmanson inequalities.

Proof: The proof uses two known results, namely that (i) The quantity $S_{i,j} = r_i + r_j - (N-2) \cdot d_{i,j}$ with $r_i = \sum_{j=1,...,N} d_{i,j}$ is maximized for two consecutive nodes (Proof in Levy and Pachter 2009). (ii) The NJ algorithm preserves perfect order. When two nodes *i* and *i* + 1 are joined, the distance from the new node is computed by averaging the distance over the two taxa: $d(new)_{i,j} = 1/2 \cdot (d_{i,j} + d_{i+1,j})$. Averaging between neighbors preserves the Kalmanson inequalities. It follows that perfect order is preserved by the NJ algorithm. In order to complete the demonstration it remains to show that the ordering procedure

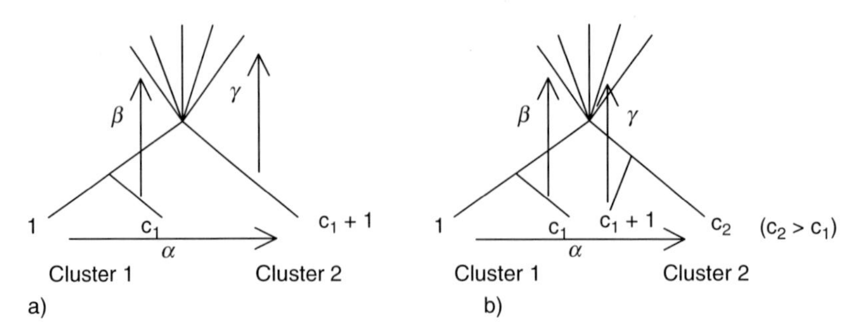

Fig. 16.5 The two configurations (a) and (b) are the two situations that must be detected by the algorithm as they correspond to the wrong order. The two clusters C1 and C2 are the two clusters that are joined by the NJ algorithm and the branches at the top correspond to the remaining nodes that do not belong to clusters C1 and C2

furnishes a perfect order. At each step the NJ joins two nodes C1 and C2 that possibly result from the fusion of a cluster of nodes. At least two of the four possibilities tested by the algorithm correspond to a partial order of the taxa that fulfils all Kalmanson inequalities. If the two nodes form a cherry then any order preserves perfect order. It remains to show that all situations, leading to two nonconsecutive taxa being related by a lateral transfer, correspond to a non-zero contradiction value in step 2 of the ordering algorithm for at least one possible reference node.

Figure 16.5 shows the two cases that must be detected by the algorithm as they correspond to the wrong order. These cases result into non-zero values of the contradiction. Without limiting the generality, let us assume that $Y^n_{1,c1+1} = Y^n_{c1,c1+1}$ as in Fig. 16.5a. With lateral transfers, one has $\hat{Y}^n_{1,c1+1} \geq \hat{Y}^n_{c1,c1+1}$; the equality holding for $\alpha = \beta = 0$. The sense of the arrows in the transfer does not change the above inequalities. It follows that a contradiction is found if $\beta > 0$ or $\alpha > 0$. The same inequalities hold if $\gamma > 0$ and a contradiction is also found when $\beta > 0$ or $\alpha > 0$. Repeating the procedure for the taxa 1, c1+1 and c2 in Fig. 16.5b one obtains similarly that if $\beta > 0$ or $\alpha > 0$ or $\gamma > 0$ then at least a contradiction is registered.

16.4 Representing Phylogenetic Information in Case of Lateral Transfers

In real applications, the distance matrix $Y^n_{i,j}$ often only partially fulfills the inequalities corresponding to a perfect order. The contradiction on the order of the taxa can be defined as

$$C = \sum_{\substack{k>j\geq i \\ i,j,k\neq n}} (\max((Y^n_{i,k} - Y^n_{i,j}), 0))^2 + \sum_{\substack{k\geq j>i \\ i,j,k\neq n}} (\max((Y^n_{i,k} - Y^n_{j,k}), 0))^2. \quad (16.2)$$

The best order of a distance matrix is, per definition, the order minimizing the contradiction. The ordered matrix $Y_{i,j}^n$ corresponding to the best order is defined as the minimum contradiction matrix for the reference taxon n.

For a perfectly ordered X-tree, the contradiction C is zero. A tree with a low contradiction value C is a tree that can be trusted, while a high contradiction value C is the indication of a distance matrix deviating significantly from an X-tree. Depending on the level of (local) contradiction, different representations of the phylogenetic information may be advisable. Kalmanson inequalities permit to relate X-trees, split networks and minimum contradiction matrices.

If the Kalmanson inequalities are fulfilled to a good precision, one may consider either a tree representation or a split network obtained for instance with NJ or NeighborNet (Bryant and Moulton 2004). When large deviations to a perfect order are registered, then a minimum contradiction matrix may be a good alternative representation of the phylogenetic information. Large deviations correspond to taxa whose description with a tree or a split network may be quite problematic. In order to understand the previous statements, one has to recall the relationship between the four-point condition and Kalmanson's inequalities.

A necessary and satisfactory condition for the existence of a unique tree is that the dissimilarity matrix d satisfies the so-called four-point condition. For any four elements in an X-tree the four-point condition requires that

$$d(x_i, x_j) + d(x_{k,} x_n) \leq \max(d(x_i, x_k) + d(x_j, x_n), d(x_i, x_n) + d(x_j, x_k)) \tag{16.3}$$

If a distance matrix d fulfills the four-point inequalities, then the distance matrix can be exactly represented by an X-tree (Bunemann 1971). The Neighbor-Joining algorithm can be used to recover the tree.

The Kalmanson inequalities are equivalent to the following inequalities

$$\begin{aligned} & d(x_j, x_n) + d(x_i, x_k) \geq d(x_k, x_n) + d(x_i, x_j) \\ \text{and} \quad & d(x_j, x_n) + d(x_k, x_i) \geq d(x_i, x_n) + d(x_k, x_j)(i \leq j \leq k < n). \end{aligned} \tag{16.4}$$

These inequalities have a form similar to the four-point condition (2). By comparing Eqs. (16.3) and (16.4), one understands that Kalmanson inequalities are a subset of the four-point conditions limited to ordered i,j,k,n and not to all i,j,k,n as in the four-point condition.

Bandelt and Dress (1992) have shown that if a distance matrix d fulfills Kalmanson inequalities, then the distance matrix can be exactly represented by a split network (Fig. 16.6). The order is well recovered by the NJ algorithm provided lateral transfer is between consecutive nodes. Kalmanson inequalities are at the center of a number of important results relating convexity, phylogenetic trees and networks and the traveling salesman problem (Deineko et al. 1995; Christopher et al. 1996; Dress and Huson 2004). These results show that a perfect order corresponds to a solution of the traveling salesman problem (TSP). Let us recall that the traveling salesman problem is a fundamental problem in computer

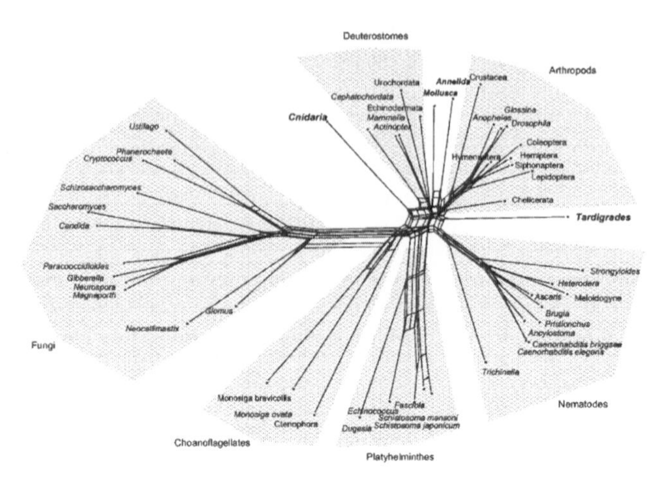

Fig. 16.6 Example of a split network, a generalization of phylogenetic trees that can accommodate some incompatibilities between data such as those caused by lateral transfers between consecutive nodes. (Reprinted with permission from Huson and Bryant 2006)

Table 16.1 Phylogenetic Representations Question: are the indents in the text of the table voluntary?

Relations between taxa	Description		Result with NJ algorithm
Four-point condition		Tree	Correct tree
Kalmanson = four-point condition for consecutive taxa		Split network	See Chap. 3
Contradictions Kalmanson inequalities	see Fig. 16.9	Minimum Contradiction matrices	Problematic

science that corresponds to the search for the shortest path passing once by a number of points and returning back to a starting point. When the distance matrix is Kalmanson, then the solution to the TSP satisfies the Master Tour property. A Master Tour is a solution of the TSP with the property that the optimal tour restricted to a subset of points is also a solution of the reduced TSP. Any restriction of a perfectly ordered distance matrix $Y^n_{i,j}$ to a subset of taxa is therefore perfectly ordered. In contrast to this result, one finds with numerical simulations that, if the minimum contradiction matrix does not fulfill the inequalities for perfect order, the best order is not always preserved when a number of taxa are removed. The order minimizing the contradiction over n taxa does not always minimize the contradiction when restricted to a subset of taxa. It follows that one cannot exclude that the topology of a tree or a split network may change when taxa contradicting perfect order are removed. Deviations from perfect order correspond to problematic regions that have to be interpreted very carefully. For that reason we suggest that minimum contradiction matrices are a useful complement to any distance-based phylogeny.

If Kalmanson inequalities are not fulfilled then the minimum contradiction matrix represents the next level of representation of phylogenetic data. As we will see below, minimum contradiction matrices permit to localize problematic regions in a tree, in which lateral transfers between non-consecutive taxa perturb very significantly the phylogenetic signal. Table 16.1 summarizes the above discussion.

16.5 How to Detect Lateral Transfers

The detection of lateral transfer is quite a difficult task (Boc and Makarenkov 2003; Kunin et al. 2005a; McLeod et al. 2005; Beiko and Hamilton 2006; Zaneveld et al. 2008). A first approach consists in using deviations from the average base composition to detect lateral transfers. In an alternative approach, lateral transfers are identified by comparing the phylogenies on different genes or the difference between species and genes trees. The minimum contradiction method furnishes a different approach to identify and localize some lateral transfers. Its particularity is that the lateral transfer is obtained without having to compare several trees or check the exact composition. Consider an X-tree and a unique lateral transfer between two taxa. If the lateral transfer is between 2 taxa that do not form a cherry and the tree has only 4 leaves, then the existence of a lateral transfer can be identified from the symmetry breaking of the minimum contradiction matrix. The taxa involved in the lateral transfer cannot be detected. In order to determine which taxa may be involved in a lateral transfer, at least 5 taxa are required with a unique lateral transfer between 2 consecutive taxa. In that case too, only partial information on the lateral transfer can be at best recovered. Figure 16.7a gives all the lateral transfers that cannot be detected without prior knowledge of the tree. Figure 16.7b shows an

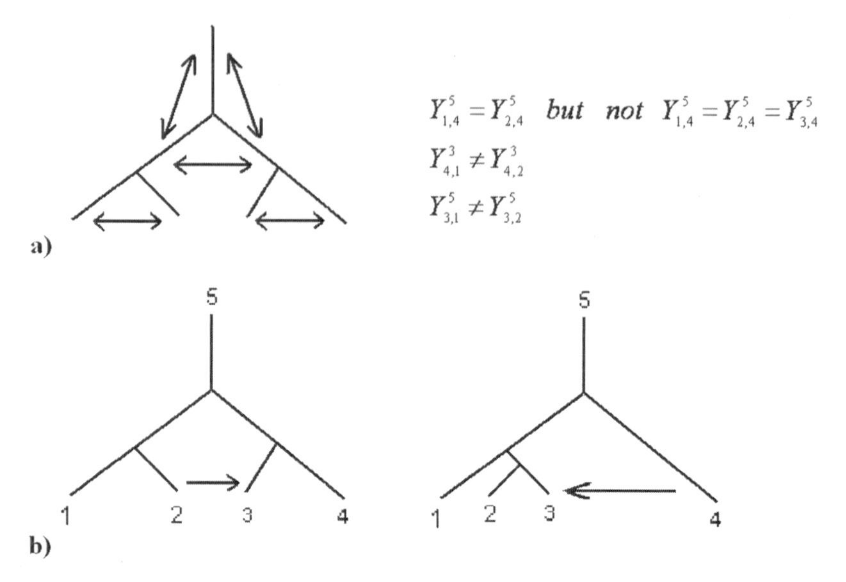

Fig. 16.7 a) A lateral transfer given by any arrow does not change the symmetry of the matrix $Y^n_{i,j}$. b) The symmetries of the distance matrices (Top right) of both trees after lateral transfer are identical

example of two possible topologies for the given symmetries. In that example, the taxon 3 involved in the lateral transfer is identified, while the second taxon is adjacent.

If Kalmanson inequalities are not fulfilled, the identification of the lateral transfers causing the deviations to perfect order can be identified provided there are no two lateral transfers $T(i \leftrightarrow j)$ and $T(k \leftrightarrow l)$ with $i < l < j < k$ (with $T(i \leftrightarrow j)$ representing a transfer between the taxa i and j). If $i < l, k < j$ as in Fig. 16.8, a transfer can be identified by removing all the taxa between i and j until the

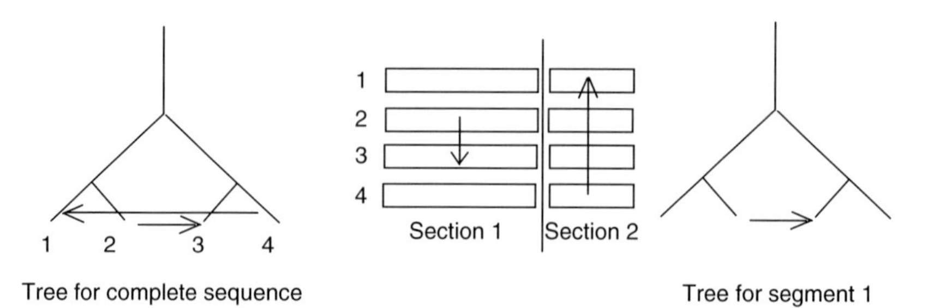

Fig. 16.8 Contradictions resulting from lateral transfers between non-consecutive taxa can be removed either by removing intermediary taxa or by eliminating the segments involved in the lateral transfer

contradiction disappears. It is also possible to recognize which part has been transferred. The removal of characters from the transferred region leads to a decrease of the level of contradiction. It is therefore possible to identify the transferred segment or the ensemble of characters that have been transferred with a stochastic gradient-descent. (The direction of the transfer is unknown. At the single character level this procedure does not work. Several characters have to be taken to obtain a measure of the contradiction).

16.6 Examples with Real Data

16.6.1 Whole Genome Phylogenies

Figure 16.9 shows the minimum contradiction matrix $Y^n_{i,j}$ using *Pirellula*, a bacteria, as a reference taxon. The best order is obtained by minimizing the contradiction using all taxa as a reference vertex at least once. The best order is therefore a kind of "average" best order. The scale on the right of the figure gives the color code used to represent $Y^n_{i,j}$ after rescaling. The minimum values of $Y^n_{i,j}$ correspond to dark blue, while the largest values are coded red. Low values of $Y^n_{i,j}$ are associated to vertices (i,j) having a first common ancestor vertex close to the reference taxa. A cluster of adjacent taxa with large values (red cluster) can be interpreted as a group of close taxa. The distance matrix is computed using the data furnished by the genome phylogeny server (Kunin et al. 2005b) using the genome conservation method. The distance matrix can be given a probabilistic interpretation. The distance matrix $Y^n_{i,j}$ can be approximated by

$$Y^n_{i,j} = 1/2 \cdot (\log_2 (\frac{P(i|n) \cdot P(j|n)}{P(i|j) \cdot m^2})). \tag{16.5}$$

The log term has the form of mutual information and furnishes a measure of the similarity of the genomes i and j in reference to genome n.

One observes that Archaea (Arch.)and Eukaryota (Euka.) are not only adjacent but also form a cluster. All the members of a class or a phylum are neighbors. All proteobacteria are grouped together.

Figure 16.10 shows an example with a clear contradiction when Rickettsiales are taken as reference taxa. Contradictions are identified by looking for regions with $Y^n_{i,j}$ increasing away from the diagonal (i.e., $Y^n_{i,j} < Y^n_{i,k}, i < j < k < n$). One observes a clear contradiction for Eukaryota (dark blue region for Eukaryota compared to light blue for Archaea and Bacteria; region pointed to by an arrow in Fig. 16.10). Considering Fig. 16.9, the values for Eukaryota and Archaea should also be similar in Fig. 16.10. This is clearly not the case. The values of $Y^n_{i,j}$ for Eukaryota are much smaller than for Archaea. The observed contradiction associated to the small values

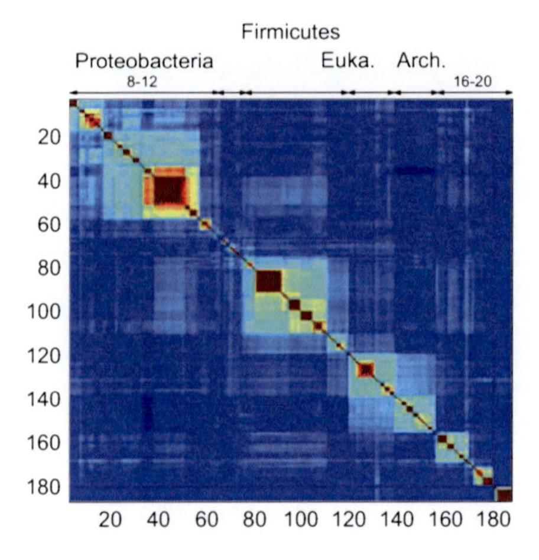

Fig. 16.9 Distance matrix $Y^n_{i,j}$ using the best order and *Pirellula* as reference taxon. For more details see Thuillard (2008)

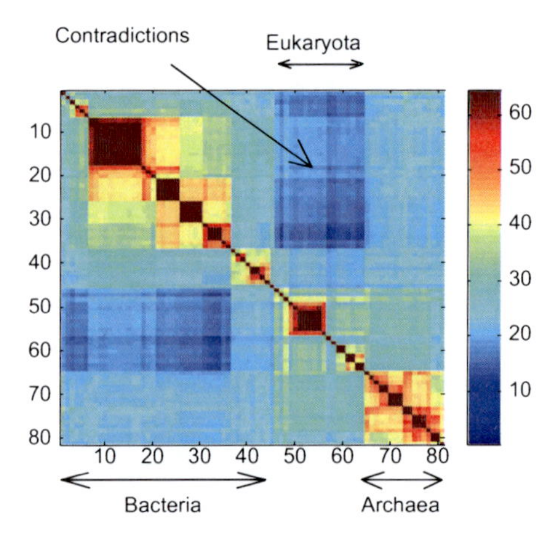

Fig. 16.10 Distance matrix $Y^n_{i,j}$ using Rickettsiales as reference. The region (Eukaryota) with a clear contradiction is pointed to by an arrow. For more details see Thuillard (2008)

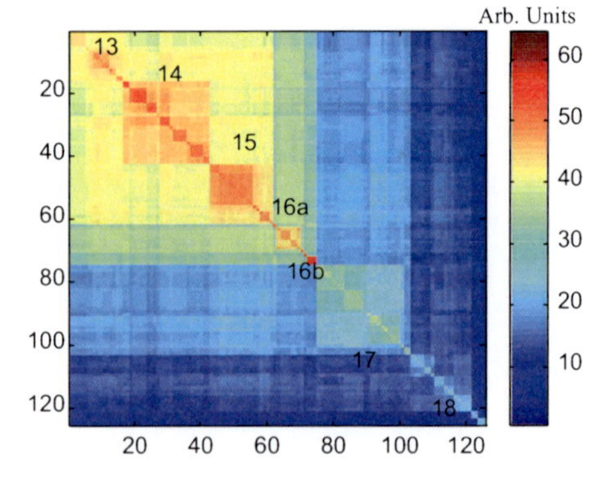

Fig. 16.11 Minimum contradiction matrix corresponding to the transition zone (deep branches) between Archaea and Bacteria. The cluster 16a corresponds to Korarchaeota, while the cluster 16b corresponds to unidentified-Archaeon from some sulfur-emitting vents and deep hydrothermal sources. The sequences corresponding to 16b are the closest to the Aquificales and Thermotogales sequences. For more details see Thuillard (2007)

of $Y_{i,j}^n$ for Eukaryota can be explained by a lateral transfer between the reference taxa (Rickettsiales) and Eukaryota. A lateral transfer between the reference node and Eukaryota leads to smaller values of $Y_{i,j}^n$ for Eukaryota. In the case of an α-lateral transfer between the reference node R and Eukaryota, the expected value $\hat{Y}_{i,j}^n$ is smaller than the value before lateral transfer: $\hat{Y}_{E_1,E_2}^R = (1-\alpha)\cdot Y_{E_1,E_2}^R + \alpha \cdot Y_{R_1,R_2}^R \leq Y_{E_1,E_2}^R$ with R1, R2 the laterally transferred segment after further evolution into the Eukaryota genomes E1, E2. Figure 16.10 is therefore consistent with the hypothesis that mitochondria are the result of an endosymbiotic event involving Rickettsia (Timmis et al. 2004), an event that resulted also into the transfer of some Rickettsia genes into the nucleus of the host.

16.6.2 Deep Branches in SSU rRNA Phylogenies of Archaea

This second example shows that the minimum contradiction approach can be used to explore deep branches. In this example, SSU rRNA sequences for Archaea were extracted from the European ribosomal SSU rRNA database (Wuyts et al. 2004) and completed by the bacterial SSU rRNA for Aquificales and Thermotogales. The distance matrix was computed on the aligned data using the Jukes-Cantor model in the MBEToolbox (Cai et al. 2005). The reference taxon n was chosen among the bacterial rRNA (Thermotogale). The two-dimensional representation of the SSU rRNA data permits to visualize proximity relationships between taxa. Figure 16.11 focuses on the deepest branches corresponding to Korarchaeota and Crenarchaeota. Korarchaeota are closer to bacteria. They show quite a constant distance to Crenarchaeota, an observation that is compatible with Korarchaeota being one of the deepest branches in Archaea. Some unidentified Archaeon found in sulfur-emitting vents and deep hydrothermal sources (Takai and Horikoshi 1999) are the closest neighbors of Aquificales and Thermotogales sequences.

16.7 Conclusions

The minimum contradiction method is a new approach to visualize phylogenetic information as well as to localize lateral transfers. Minimum contradiction matrices can be used to discover problematic regions in phylogenetic trees or networks. The effect of perturbations resulting from lateral gene transfer can be modeled probabilistically. The order of the taxa on a tree is quite robust against lateral transfers. The minimum contradiction approach was applied to the deep branches of Archaea using SSU rRNA and whole genome phylogenies using distances computed with the genome conservation method. In the later case, large deviations from a perfect order were found between Rickettsia and Eukaryota.

References

Bandelt HJ and Dress A (1992) Split decomposition: a new and useful approach to phylogenetic analysis of distance data. Mol Phylogenet Evol 1:242–252

Beiko R and Hamilton N (2006) Phylogenetic identification of lateral transfer events. BMC Evol Biol 6:15

Boc A and Makarenkov V (2003) New efficient algorithm for detection of horizontal gene transfer events, Algorithms in Bioinformatics, Benson G and Page R (eds), Third Workshop on Algorithms in Bioinformatics, Springer-Verlag, New York, pp 190–201

Bryant D, Moulton V (2004) Neighbor-Net: an agglomerative method for the construction of phylogenetic networks. Mol Biol Evol 21: 255–265

Buneman P (1971) The recovery of trees from measures of dissimilarity. In: Hodson FR, Kendall DG, Tautu P (eds). Mathematics in the archaeological and historical sciences. Edinburgh University Press, Edinburgh, pp 387–395

Cai JJ, Smith DK, Xuhua Xia , Kwok-yung Yuen (2005) MBEToolbox: a Matlab toolbox for sequence data analysis in molecular biology and evolution. BMC Bioinfor 6–64

Christopher GE, Farach M, Trick MA (1996) The structure of circular decomposable metrics. In European Symposium on Algorithms (ESA)'96, Lectures Notes in Computer Science 1136: 455–500

Deineko V, Rudolf R and Woeginger G (1995) Sometimes traveling is easy: the master tour problem, Institute of Mathematics, SIAM J Discrete Math 11:81–93

Doolittle WF (2000) Uprooting the tree of life, Scientific American, February issue, 90–95

Dutilh BE, Noort V, Heijden RTJM, Boekhout T, Snel B, Huynen MA. (2007) Assessment of phylogenomic and orthology approaches for phylogenetic inference. Bioinformatics 23:815–824

Fitz-Gibbon ST, House CH. (1999) Whole genome-based phylogenetic analysis of free-living microorganisms. Nucleic Acids Res 27:4718–4222

Fukami-Kobayashi K, Minezaki Y, Tateno Y, Nishikawa K. (2007) A tree of life based on protein domain organizations. Mol Biol Evol 24:1181–1189

Gascuel O, Steel M (2006) Neighbor-joining revealed. Mol Biol Evol 23:1997–2000

Huson D, Bryant D (2006) Application of phylogenetic networks in evolutionary studies. Mol Biol Evol 23:254–267

Kalmanson K. (1975) Edgeconvex circuits and the traveling salesman problem. Can J Math 27:1000–1010

Kunin V, Goldovsky L, Darzentas N, Ouzounis CA. (2005a) The net of life: reconstructing the microbial phylogenetic network. Genome Res 15:954–959

Kunin V, Ahren D, Goldovsky L, Janssen P Ouzounis CA. (2005b) Measuring genome conservation across taxa: divided strains and united kingdoms. Nucleic Acids Res 33(2):616–621

Levy D, Pachter L (2009) The neighbor-net algorithm, Adv Appl Math, in press. http://arxiv.org/abs/math/0702515.

McLeod D, Charlebois R, Doolittle F, Bapteste E (2005) Deduction of probable events of lateral gene transfer through comparison of phylogenetic trees by recursive consolidation and rearrangement. BMC Evol Biol 5:27

Makarenkov V, Leclerc B (1997) Circular orders of tree metrics, and their uses for the reconstruction and fitting of phylogenetic trees. In: Mirkin B, Morris FR, Roberts F, Rzhetsky A. (eds). Mathematical hierarchies and Biology, DIMACS Series in Discrete Mathematics and Theoretical Computer Science. Providence: Am. Math. Soc. pp 183–208

Makarenkov V, Kevorkov D, Legendre P (2006) Phylogenetic network construction approaches. Appl Mycol Biotechnol Intl Elsevier Ser 6. Bioinformatics 61–97

Mihaescu R, Levy D, Pachter L (2006) Why neighbour joining works. arXiv cs.DS/0602041, Accessed 20 May 2007, http://arxiv.org/PS_cache/cs/pdf/0602/0602041v3.pdf.

Saitou N, Nei M (1987) The neighbour-joining method: a new method for reconstructing phylogenetic trees. Mol Biol Evol 4:406–25

Takai K, Horikoshi K (1999) Genetic diversity of Archaea in deep-see hydrothermal vents environments. Genetics 152:1285–1297

Thuillard M (2007) Minimizing contradictions on circular order of phylogenic trees. Evol Bioinformat 3:267–277

Thuillard M (2008) Minimum contradiction matrices in whole genome phylogenies. Evol Bioinformat 4:237–247

Timmis JN, Ayliffe MA, Huang CY Martin W (2004) Endosymbiotic gene transfer: organelle genomes forge eukaryotic chromosomes. Nature Rev Genet 5:123–135

Wuyts J, Perriere G Van de Peer Y (2004) The European ribosomal RNA database. Nucleic Acids Res 32:D101–103

Zaneveld J, Nemergut D, Knight R (2008) Are all horizontal transfers created equal? Prospects for mechanism-based studies of HGT patterns. Microbiology 154:1–15

Part III
Applied Evolutionary Biology

Chapter 17
The Genome Sequence of *Meloidogyne incognita* Unveils Mechanisms of Adaptation to Plant-Parasitism in Metazoa

Etienne G.J. Danchin, and Laetitia Perfus-Barbeoch

Abstract *Meloidogyne incognita* is the most widespread and polyphagous plant-parasitic nematode, a group of nematodes that cause more than $150 billion damage every year. As most control means based upon chemicals are banned from use, new control measures need to be developed. Completion of the genome sequence of this plant-parasitic nematode revealed a set of singularities compared to non-parasitic nematode genomes. One of the most striking features was the presence of a full arsenal of plant cell wall-degrading enzymes probably acquired via horizontal gene transfer. The structure of the genome itself, mostly present as two similar but distinct copies may also account for the parasitic success of this nematode in terms of host spectrum and global distribution. These findings announce mechanisms used for successful establishment of plant-parasitism in metazoan and open the way for the development of more efficient control measures.

17.1 Introduction

Nematoda is an animal clade characterized by diversity and abundance, not only in terms of the number of species and individuals but also in terms of variety of lifestyles and ecological niches. Nematodes comprise over 25,000 described species (with perhaps 10 million undescribed) many of which are parasites of animals or plants (Blaxter 2003). The various possible lifestyles encompass free-living species, feeding on bacteria or fungi, entomopathogens, human and cattle pathogens, endo and ecto parasites (of animals or plants). Nematodes can be found in

---

E.G.J. Danchin and L. Perfus-Barbeoch
INRA, UMR 1301, 400 route des Chappes, F-06903 Sophia-Antipolis, France
CNRS, UMR 6243, 400 route des Chappes, F-06903 Sophia-Antipolis, France
UNSA, UMR 1301, 400 route des Chappes, F-06903 Sophia-Antipolis, France
e-mail: etienne.danchin@sophia.inra.fr

P. Pontarotti (ed.), *Evolutionary Biology: Concept, Modeling, and Application*,
DOI: 10.1007/978-3-642-00952-5_17,

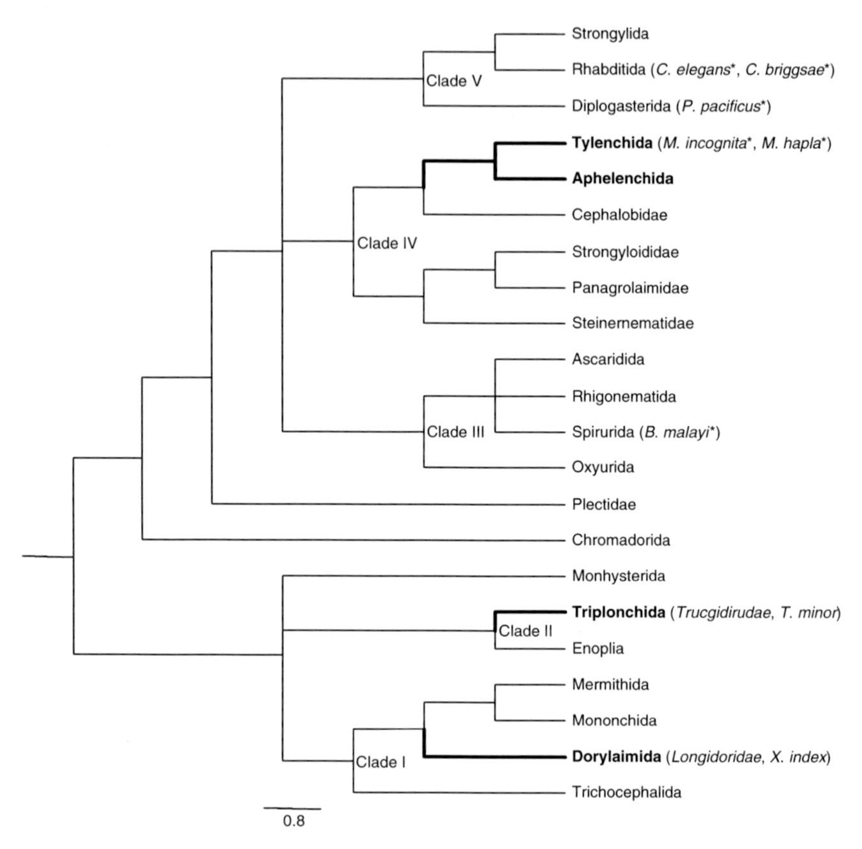

Fig. 17.1 Simplified phylogeny of the nematode phylum. This schematic representation of the phylogeny in the phylum Nematoda is based on Mark Blaxter's classification of nematodes in different clades (Mitreva et al. 2005). Plant-parasitic nematodes appear branched in bold. Example species for each branch of plant-parasites are indicated in parentheses. Species whose genome has been fully sequenced and annotated appear with a star (*) character

environments as diverse as marine, freshwater, soil, or even sand or ice. Plant-parasitism is a lifestyle found in three different clades of the Nematoda tree of life (Fig. 17.1), namely Dorylaimida (clade I), Triplonchida (clade II) and Tylenchida (clade IV). This observation has led nematologists to hypothesize that plant-parasitism appeared at least three times independently during the evolution of nematodes. The almost systematic co-clustering of plant-parasites with fungivorous species also suggested that plant-parasites evolved from fungi-feeding nematodes (Holterman et al. 2006). Plant-parasitic nematodes themselves are responsible for an estimated $150 billion annually in crop damage globally. The southern root-knot nematode *Meloidogyne incognita* is the most widespread and is able to infect the roots of almost all cultivated plants, which possibly renders this species as one of the most damaging crop pathogens in the world (Trudgill and Blok 2001).

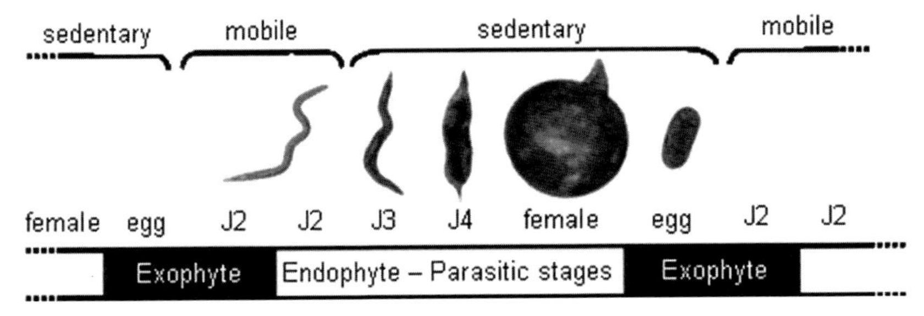

Fig. 17.2 *Meloidogyne incognita* life cycle. *M. incognita* can be found in the soil outside the plant during exophyte developmental stages (black boxes) where mobile infective second juveniles (J2) penetrate the root. Once into the roots, mobile J2 migrate between cells to reach the plant vascular cylinder and induce the formation of a feeding site. Once the feeding site has been established, *M. incognita* remains sedentary. The feeding site provides nutrients to *M. incognita* that go through three molts (J3, J4 and adult female). These three different stages constitute the endophyte phase of *M. incognita* life cycle (white boxes). Asexual reproduction will lead to *egg* production that will be embedded in a protective matrix stuck to the roots, after hatching from the *eggs*, new mobile J2 can repeat the infective cycle

Chemical nematicides have been the most reliable means of controlling root-knot nematodes but their active compounds are notoriously and persistently toxic to humans and the environment. Consequently they are being withdrawn from use. *M. incognita* is an obligatory, sedentary parasite that reproduces strictly by mitotic parthenogenesis. Root-knot nematodes have an intimate interaction with their hosts. Adult females, within the host root, induce the re-differentiation of root cells into specialized "giant" cells, upon which they feed continuously (Fig. 17.2). *M. incognita* can infect the model plant *Arabidopsis thaliana*, making it a key model system for the understanding of adaptations to plant-parasitism (Caillaud et al. 2008a, b).

The model free-living nematodes *Caenorhabditis elegans* and *C. briggsae* have been the subjects of intensive studies (The *C. elegans* Genome Sequencing Consortium 1998; Stein et al. 2003). However, until recently, almost nothing was known at the genomic level about the other members of this diverse phylum and these free-living models will not illuminate the biology of parasitism. The publication of the draft genome sequence of *Brugia malayi*, a human-parasitic species, revealed significant differences from *Caenorhabditis* (Ghedin et al. 2007). Many of these differences were attributed to adaptation to parasitism. In 2008, the publication of the draft genome sequence of *M. incognita* (Abad et al. 2008), highlighted adaptations to plant-parasitism, including the acquisition, probably by lateral gene transfer from bacterial genomes, of a suite of enzymes that likely modify and subvert the host environment to support nematode growth. Later the same year, publication of a paper reporting the genome of another plant-parasite from the same genus, *M. hapla* (Opperman 2008), confirmed many of the findings reported in *M. incognita*. The major difference between the two *Meloidogyne* genus model species is that *M. incognita* reproduces exclusively via mitotic parthenogenesis (i.e., asexually) whereas *M. hapla* can be sexual and is able to do meiosis.

Almost concomitantly, an analysis of the genome sequence of the necromeric beetle-associated nematode, *Pristionchus pacificus*, revealed its own set of singularities probably linked to a versatile lifestyle previously unexplored to date.

In this chapter, we will summarize the data presented during the 12th Evolutionary Biology Meeting about the genome of *M. incognita*, with an updated comparative analysis in light of the nematode genomes newly sequenced. We will focus on singularities that may unveil adaptation of a metazoan genome to plant-parasitism.

17.2 *M. incognita* Genome Organization and Comparison to Other Nematode Genomes

The *M. incognita* genome was sequenced using a whole-genome shotgun strategy. Assembly, performed with Arachne (Jaffe et al. 2003) yielded 2,817 supercontigs, totaling 86 Mb (Table 17.1). This size is almost twice the experimentally-estimated genome size, based on flow cytometry that ranged between 47 and 51 Mb for haploid genome (Leroy et al. 2003). An all-against-all comparison of supercontigs revealed that 648 of the longest (covering 55 Mb) consisted in homologous but diverged segment pairs. These pairs might represent former allelic regions but their origin could also be linked to a hybridization event. About 3.35 Mb of the assembly constitutes a third partial copy aligning with these distinct supercontig pairs. Average sequence divergence between the aligned regions is ~8%. A combination of different processes may explain the observed pattern in *M. incognita*, including auto-polyploidy due to loss of sexual reproduction and hybridization (Triantaphyllou 1985; Castagnone-Sereno 2006); all are frequently associated with polysomy, aneuploidy and asexual reproduction. These observations are consistent with a strictly mitotic parthenogenetic reproductive mode that can permit homologous chromosomes to diverge considerably, as hypothesized for bdelloid rotifers (Mark et al. 2004). In contrast, the facultative parthenogenetic species *M. hapla*, which is able to do meiosis and to reproduce asexually, presents a small genome size (54 Mb, the

Table 17.1 Comparison of genomic feature in Nematode genomes

Feature/ species	*Meloidogyne incognita*	*M. hapla*	*B. malayi*	*Caenorhabditis elegans*	*C. briggsae*	*Pristionchus pacificus*
Assembly size (Mb)	86	*54*	85–86	100	104	**142**
Protein-coding genes	19,212	*14,420*	14,500–17,800	20,060	19,934	**29,201**
% repetitive DNA	**36**	17	12–15	*16.5*	22	17
% GC	31.4	*27.4*	30.5	35.4	37.4	**42**
Gene density "per Mb"	223	**270**	162	235	192	206

Highest values for each row are highlighted in bold.

smallest this far for a nematode) and no evidence for extensive presence of pairs of similar regions (Opperman et al. 2008), Table 17.1. In contrast to what was observed in *B. malayi* (Ghedin et al. 2007), no trace of DNA attributable to bacterial endosymbiont genome(s) was reported in any of the two plant-parasites.

The overall G + C content in *M. hapla* (27.4%) is the lowest found in nematodes this far whereas for *M. incognita* (31.4%); it is similar to that of *B. malayi* (30.5%) but lower than those of *C. elegans* (35.4%), *C. briggsae* (37.4%) and *P. Pacificus* (42%), Table 17.1. The proportion of repetitive non protein-coding DNA represents more than 36% of the *M. incognita* genome (Abad et al. 2008). A proportion substantially higher than those found in other nematodes, as only 17% were reported in *M. hapla* (Opperman et al. 2008); 16.5% and 22%, respectively reported for *C. elegans* and *C. briggsae* (Stein et al. 2003), 17% in *P. pacificus* (Dieterich et al. 2008) and between 12% and 15% in *B. malayi* (Ghedin et al. 2007).

One characteristic feature of nematode genomes is the possibility of trans-splicing of co-transcribed genes that are in close proximity on the same chromosome. The resulting polycistronic pre-mRNAs derived from these operons are resolved by Spliced Leaders (SL) exon at the 5′ end of each gene from the operon. Two main classes of SL exons have been described; SL1 that has been reported in all the nematode genomes considered in this analysis and SL2 which, to date; has only been reported in *C. elegans*, *C. briggsae* and *P. pacificus*. As no SL2-like genes were found so far outside of clade V (Fig. 17.1), it has been hypothesized that they may represent an evolutionary invention of rhabditine nematodes (Guiliano and Blaxter 2006). Using different approaches, operons themselves have been predicted in all the nematode genomes considered here. However, different sets of genes were detected as operonic in each species. Initial analysis of the *M. incognita* genome showed that only one 2-genes operon was found to be strictly conserved between *B. malayi*, *C. elegans* and *M. incognita*. A search against the genomes of *P. pacificus* and *M. hapla* showed that these two genes were also in close proximity in these two other nematodes. Overall the poor conservation supports the idea that operons are a dynamic component of nematode genome architecture that may represent an adaptive component to different lifestyles.

17.3 The Gene Content of Plant-Parasitic Nematodes

Each genome project made use of different strategies to detect protein-coding genes and results are summarized in Table 17.1. Genes were predicted in *M. incognita* by using the integrative gene prediction platform EuGene (Foissac and Schiex 2005) specifically trained on a set of *M. incognita* confirmed genes. A total of 19,212 protein-coding genes were identified. It was shown at the nucleotide level that the majority of the assembly was present as pairs of similar but significantly variable copies. In order to evaluate the outcome at the protein level; predicted proteins were grouped in clusters, of at least 95% identical sequences using the program CD-HIT (Li and Godzik 2006). Clustering showed that more than 69% of protein sequences were less than 95% identical to any other. As this potentially allows functional divergence, all copies were considered to be different genes. The protein-coding genes portion occupies 54.4% of the assembly at an average density of 223 genes per Mb, and 36% are supported by ESTs. In *M. hapla*, a total of 14,420 protein-coding genes were detected with a density of 270 genes per Mb, the highest reported to date for a nematode (Table 17.1).

In *M. incognita*, a comparative analysis of InterPro protein domains relative abundance with *C. elegans*, *B. malayi* and *Drosophila melanogaster* proteomes highlighted several singularities. In particular, it allowed identifying an increased abundance of "pectate lyase," "glycoside hydrolase family GH5" and "peptidase C48 (SUMO)" domains and a decrease in chemoreceptor domains. A similar reduction of chemoreceptor domains was reported in the genome of *M. hapla*. A total of 52 Interpro domains were detected only in *M. incognita* and 27 of these were supported by alignments with ESTs. This *M. incognita*-restricted set included domains involved in plant cell wall degradation and chorismate mutase activity (they will be discussed in a dedicated section "Genes potentially involved in plant-parasitism"). In the *M. incognita* genome paper, the authors performed an OrthoMCL analysis (Li et al. 2003) to predict cluster of orthologous proteins between *M. incognita*, *C. elegans*, *C. brigsae*, *B. malayi*, *D. melanogaster*, *G. zeae*, *M. grisea* and *N. crassa*. Clustering of the proteomes suggested that the core complement of proteins in the phylum Nematoda is relatively small: only 14% of the orthologue groups were shared by *M. incognita*, *C. elegans* and *B. malayi*. Interestingly, as much as 52% of *M. incognita* predicted proteins had no predicted orthologues in the other species. Among them 1,819 proteins were predicted to bear a signal peptide and lack any known domain. A total of 338 of these presumably secreted proteins of unknown function had a supporting EST match. Similarly candidate secreted proteins were searched in the genome of *M. hapla*. The predicted gene models were searched for the presence of putative signal peptides using SignalP (Emanuelsson et al. 2007). The authors additionally filtered out proteins predicted to bear a transmembrane region using TM-HMM (Krogh 1997) leading to a total of 832 candidate secreted proteins without transmembrane region. A similarity search against NCBI's nr protein database and Wormpep (release 185) revealed 434 proteins putatively secreted and without significant similarity in other organisms. Both these sets from *M. incognita* and *M. hapla* constitute an interesting ensemble of genes for the identification of new potential effectors of as-yet-unknown function. Indeed, as stated in the introduction, these plant-parasites are known to secrete into the host via their stylet, a series of proteins and other bio-compounds thought to be key elements of plant-parasitism.

17.4 Genes Potentially Involved in Plant-Parasitism

Compared to their free-living relatives, root-knot nematodes such as *M. incognita* and *M. hapla* have evolved complex biotrophic interactions with their hosts. They penetrate root plant tissues, navigate between cells and induce the re-differentiation of a complex feeding site (Fig. 17.2). The parasite must also elude host defense responses for the several weeks that the feeding site is required to support female development up to *egg* hatching (Jammes et al. 2005). As seen in the previous section, nematode proteins produced in and secreted from, the oesophageal gland cells into the host via its stylet are likely to be important effectors of these processes (Davis et al. 2004; Caillaud et al. 2008). Here we focus on identified gene products that might be involved in the parasitic interaction, particularly those that might degrade or modify plant cell walls.

A few individual Carbohydrate-Active enZymes (CAZymes) involved in plant cell wall degradation had already been identified previously in plant-parasitic nematodes (Davis et al. 2004; Caillaud et al. 2008), including the first ever identified cellulase in an animal species (Smant et al. 1998). However, in the absence of an available assembly and annotation of a plant-parasitic whole genome, it was impossible, until recently, to assess the full spectrum of diversity and abundance of such CAZymes. The recent analysis of two root-knot nematode genomes illuminated this aspect. An unprecedented set of 61 plant cell wall-degrading Carbohydrate-Active enZymes (CAZymes) was identified in the genome of *M. incognita*. This set encompassed 21 cellulases and six xylanases from family GH5, two polygalacturonases from family GH28 and 30 pectate lyases from family PL3. All these enzymes were also reported to be present in the genome of *M. hapla* although at a lower abundance, probably due to the particular structure of the *M. incognita* genome where most regions are present as two diverged copies. Both genome analyses thus converged in highlighting a diverse and abundant repertoire of enzymes for the degradation of plant cell walls. A total of 20 candidate expansins was also reported in *M. incognita*. Expansins are thought to disrupt non-covalent bonds in plant cell walls, making the components more accessible to plant cell wall-degrading enzymes(Qin et al. 2004). These proteins were not reported in the *M. hapla* genome paper. However, using a tBLASTn (Altschul et al. 1997) analysis with *M. incognita* expansins against the *M. hapla* 10X assembly (www.hapla.org); we were able to detect the presence of these expansins in *M. hapla* too. Interestingly, analysis of CAZymes of these two species also allowed identifying families that were not reported before in any metazoan species, including in plant-parasitic nematodes. Two candidate arabinases (family GH43) were detected in *M. incognita* and a tBLASTn analysis confirmed their presence in the genome of *M. hapla*. Their electronically inferred activity remains putative but if they eventually are experimentally confirmed they will be added to the arsenal of plant cell wall-degrading enzymes in root-knot nematodes. Similarly two candidate invertases (also known as b-fructofuranosidase, family GH32) have been reported in the two *Meloidogyne* genomes. Invertases catalyze the conversion of sucrose (the main circulating disaccharide in plants) into glucose and fructose. If their activity is confirmed this would suggest that these enzymes are used to convert sugar circulating in plant hosts into sugars readily appropriate for use as a carbon source by the nematode.

Exploration of the pattern of presence/absence of these CAZyme families shows that except in a few exceptions, none of these enzymes have been reported in any other metazoan species. In particular they are absent from all nematode genomes except plant-parasites. The only exceptions concern families GH5 and GH32. Candidate cellulases from family GH5 have been reported in two phytophagous longicorn beetles, namely *Psacothea hilaris* (Sugimura et al. 2003) and *Apriona germari* (Wei et al. 2006). Recently, the analysis of the genome of *Pristiochus pacificus* (Dieterich et al. 2008) revealed the presence of up to seven enzymes from family GH5. The authors showed through a phylogenetic analysis that these *Pristionchus* enzymes were members of a GH5 subfamily distinct from the one found in plant-parasites. Considering this result; they suggest that the GH5 enzymes have been acquired at least twice independently in nematodes. Concerning the possible activity of GH5s enzymes in *Pristionchus*, a test on carboxy methyl cellulose suggested that these enzymes are cellulases. However, due to possible cross reaction on this artificial substrate, other activities may be considered. A total of 12 different activities have been reported so far inside family GH5 (www.cazy.org) and activities like cellulases and chitosanases are difficult to distinguish. In any case, except in plant-parasitic nematodes, in two phytophagous insects and possibly in *Pristionchus*, cellulases from family GH5 are otherwise not found in any metazoan species so far. The GH32 is the only other enzyme family from this set that possesses representatives in species not restricted to plant-parasitic nematodes. Indeed, enzymes from GH32 family have been identified in two Lepidoptera, *Bombyx mori* (Daimon et al. 2008) and *Helicoverpa armigera* (Pauchet et al. 2008). The invertase activity has been shown in the case of *B. mori*. Interestingly, in both the GH5 and GH32 cases, the only metazoan species in which these enzymes are found are plant-feeding insects.

For all members of the suite of plant cell wall-degrading CAZymes, expansins and associated invertases (including GH5s and GH32s) identified in root-knot nematode genomes, the best blast hits against public databases turned out to be bacterial enzymes. This result appears as a common feature of the two genome papers and suggests that this arsenal of enzymes was probably acquired via horizontal gene transfer (HGT).

Still concerning HGT, another point on which the two plant-parasitic genome analyses converge, was the identification of several secreted chorismate mutases. These enzymes, also most closely resemble bacterial enzymes. Chorismate mutases are central for aromatic amino acids and related products biosynthesis. Root-knot nematodes may use these enzymes to subvert host tyrosine-dependent lignification or defense responses (Lambert et al. 1999).

Overall, these genes suggest a critical role of HGT events in the acquisition of new capabilities linked to the evolution of plant-parasitism in root-knot nematodes.

In parallel to the identification of candidate effectors acquired via HGT, a series of specific gene family expansions was reported in *M. incognita* compared to free-living nematodes. Among the most significant idiosyncrasies in *M. incognita*, was the identification of more than 20 cysteine proteases of the C48 SUMO (small ubiquitin-like modifier) deconjugating enzyme family while *C. elegans* has only 5. Some phytopathogenic bacterial virulence factors are SUMO proteases (Hotson and Mudgett 2004), suggesting that the proteolysis of sumoylated host substrates may be a general strategy used by pathogens to manipulate host plant signal transduction. A profusion of S16 subfamily serine proteases (Lon proteases) was also identified in *M. incognita*. These proteases are known to regulate type III protein secretion in phytopathogenic bacteria (Tang et al. 2006) and may have analogous roles in root-knot nematodes. Though, these expansions need to be confirmed in *M. hapla* for reinforced significance.

Interestingly, a search for previously reported parasitism genes from root-knot and cyst nematodes in the genome of *M. incognita* revealed that while almost all are present in the genome assembly, a majority of them were also present in the animal-parasitic nematode, *B. malayi* and in the free-living species *C. elegans*. Thus, *M. incognita* members putatively involved in parasitism were probably recruited from ancestral nematode genes involved in other biological functions. These included candidate effectors such as venom allergen-like proteins (Gao et al. 2001), glutathione peroxidase (Jones et al. 2004), 14-3-3 (Jaubert et al. 2004), calreticulin (Jaubert et al. 2005), SXP/RAL (Jones et al. 2000; Rao et al. 2000) and cystein proteases (Shingles et al. 2007).

Additionally, twenty-seven previously described *M. incognita*-restricted pioneer genes expressed in oesophageal glands (Huang et al. 2003) were retrieved in the genome assembly or unplaced reads. Eleven additional copies were identified; all remain *Meloidogyne*-specific. These secreted proteins of as-yet-unknown function are likely targets for novel intervention strategies and warrant deeper investigation.

17.5 Other Singularities Potentially Reflecting Adaptation to a Plant-Parasitic Lifestyle

As obligate endo-parasites, root-knot nematodes necessarily have to spend most of their life cycle in the host plant. Life in this particular habitat is expected to have an outcome at the genomic level. Endo-parasites have to elude plant defense and avoid recognition by the host. One aspect of plant-defense responses is the production of cytotoxic oxygen radicals. Surprisingly, the number of genes encoding superoxide dismutases and glutathione peroxidases was lower in *M. incognita* than in *C. elegans*. More striking still was the reduction in glutathione S-transferases (GSTs) and cytochromes P450 (CYPs) enzymes involved in xenobiotic metabolism and protection against peroxidative damage. While *C. elegans* has 44 GSTs from the Omega, Sigma and Zeta classes (Lindblom and Dodd 2006), only five GSTs were detected in *M. incognita* and all from the Sigma class. Sigma class GSTs are involved in protection against oxidants rather than xenobiotics. Interestingly, a comparable reduction in *gst* genes was observed in *B. malayi* (Ghedin et al. 2007). Concerning *cyp* genes, while *C. elegans* has 80 different *cyp* genes from 16 families (Menzel et al. 2001), only 27 full or partial *cyp* genes, from eight families, were identified in *M. incognita*. Family CYP35 and other xenobiotic-metabolizing P450s are lacking in the genome of *M. incognita*. Thus, the arsenal of detoxifying gene products in *M. incognita* is substantially reduced compared to the free-living nematode *C. elegans*. In a similar trend, several immune effectors such as lysozymes, C-type lectins and chitinases were much less abundant in *M. incognita* than in *C. elegans*. Entire classes of immune effectors known from *C. elegans* were also absent from *M. incognita*, including antibacterial genes such as *abf* and *spp* (Alegado and Tan 2008) and antifungal genes of several classes (*nlp*, *cnc*, *fip*, *fipr*) (Ewbank 2006), similar observations were reported in the human-parasite *B. malayi* (Ghedin et al. 2007). Since plant-parasites embedded in root tissues are protected from a variety of biotic and abiotic stresses, we hypothesize that the reduction and specialization of chemical and immune defense repertoires are a result of adaptation to life in this privileged particular environment.

One interesting feature of *C. elegans* is the presence of a broad range of unusual fucosylated *N*-glycan structures (Paschinger et al. 2007) in comparison to other metazoan animals. At the genomic level, this is reflected by a set of candidate fucosyltransferases substantially more abundant than those of other animals. The genome of *M. incognita* encodes almost twice as many candidate fucosyltransferases as *C. elegans*. It has been suggested that peculiar multi-fucosylated structures on the surface of the nematode cuticle may help animal-parasites to evade recognition (Paschinger et al. 2007). A similar role in root-knot nematodes should be considered and investigated in more detail.

17.6 Is the *C. elegans* Genome Representative of Nematode Diversity?

C. elegans is a proven model animal but its genome is not expected to reflect the full spectrum of biodiversity found in the phylum Nematoda. Genes involved in core developmental and regulatory processes (nuclear receptors [NRs], kinases, G-protein coupled receptors [GPCRs], neuropeptides and sex determination) have been annotated in the genome of *M. incognita*. Their abundance and diversity were compared with *C. elegans* and revealed substantial differences. In this section we will focus on differences between the *C. elegans* and *M. incognita* genomes for the core functions described above.

The genome of *C. elegans* encodes a surprisingly large number of NRs but curiously lacks orthologues of many NR types conserved in other animals (Bertrand et al. 2004). Some of these conserved NRs are present in *B. malayi* (Ghedin et al. 2007). Many of the NRs present in *B. malayi* and absent from *C. elegans* were also not found in *M. incognita*. A whole set of *C. elegans* NRs is classified as supplementary NRs (SupNRs), likely derived from a HNF-4-like ancestor (Robinson-Rechavi et al. 2005). Orthologues of SupNRs were found in *M. incognita*, including a 41-member, *M. incognita*-specific expansion. Fourteen SupNRs are one-to-one orthologues between *B. malayi*, *M. incognita* and *C. elegans*, or conserved only between *M. incognita* and *C. elegans*, with secondary losses in *B. malayi*. This suggests that the expansion of SupNRs started before the Clade III (*Brugia*)-Clade IV (*Meloidogyne*)- Clade V (*Caenorhabditis*) separation and has proceeded independently in *C. elegans* and *M. incognita* lineages. It is interesting to note here that a significant reduction of the number of NRs was reported in the genome of *M. hapla* but as no specific analysis of SupNRs was carried out, it is to date impossible to assess whether *M. incognita*-specific expansions of SupNRs are confirmed in *M. hapla*.

A total of 499 kinases were predicted in the genome of *M. incognita*, an abundance comparable to the 411 found in *C. elegans* (Plowman et al. 1999) but higher than the 215 found in *B. malayi* (Ghedin et al. 2007). The kinases from *M. incognita* were grouped into 232 families, 24 of which contained only nematode members, suggesting that they have nematode-specific functions. Four kinase families contained only *M. incognita* and *B. malayi* members, these genes may be involved in parasitism-related functions. Finally, 122 predicted kinase genes appear to be *M. incognita*-specific.

As much as 7% (1,280) of all *C. elegans* genes are predicted to encode GPCR that play crucial roles in chemosensation. These *C. elegans* genes have been divided into three serpentine receptor (SR) superfamilies and five solo families (Robertson and Thomas 2006). *M. incognita* has only 108 GPCR genes and these derive from two of the three superfamilies and one of the solo families. A similar drastic reduction was reported in the genome of *M. hapla* and the authors suggested a gene loss related to life as an internal parasite of plants, an environment homeostatic in comparison with soil. Interestingly, the *M. incognita* chemosensory genes were commonly found as duplicates clustered on the genome, as observed in *C. elegans*.

Given the relative simplicity of the nematodes nervous system their neuropeptide diversity is surprisingly high. In *C. elegans*, 28 FMRFamide-like peptide (*flp*) and 35 neuropeptide-like protein (*nlp*) genes encoding approximately 200 distinct neuropeptides were reported (Marks and Maule 2007). The identified neuropeptide complement of *M. incognita* is much smaller: 19 *flp* genes and 21 *nlp* genes. However, two *flp* (*Mi-flp-30* and *Mi-flp-31*) encode neuropeptides have not been identified in *C. elegans*, suggesting that they could fulfill functions specific to a phytoparasites.

In *C. elegans*, the XX-XO sex determination pathway is intimately linked to the dosage compensation pathway (Zarkower 2006). *M. incognita* reproduces exclusively by mitotic parthenogenesis and males are considered not to contribute genetically to the production of offspring (Castagnone-Sereno 2006). *M. incognita* also displays an environmental influence on sex determination. Indeed, under less favorable environmental conditions far greater numbers of males are produced. These males can also arise through sex reversal (Papadopoulou and Triantaphyllou 1982) and intersexual forms can be observed. *M. incognita* homologues of at least one member of each step of the *C. elegans* sex determination cascade were identified including *sdc-1* from the dosage compensation pathway, *tra-1*, *tra-3* and *fem-2* from the sex determination pathway itself and also downstream genes such as *mag-1* (which represses male promoting genes) and *mab-23* (which controls male differentiation and behavior). A singularity in *M. incognita* was the identification of a large family (~35 genes) of secreted proteins similar to the single *tra-1* gene from *C. elegans* that contains a C2H2 zinc finger motif. Therefore, it is possible that *M. incognita* uses a similar genetic system for sex determination but with the male pathway also modulated by detection of environmental cues.

Taken together, these comparative analyses of genes underpinning important traits suggests that the model *C. elegans* does not fully represent the genomic diversity displayed in the phylum Nematoda on its own. The genomes of *B. malayi* and *P. pacificus* also revealed their own sets of differences and it is expected that every exploration of new branches of the nematode tree of life will highlight new genomic singularities.

17.7 RNAi and Development of New Antiparasitic Drug Targets

RNA interference (RNAi) is a powerful and convenient tool to investigate the effect of gene inactivation in *C. elegans*. This also represents a promising technology for the functional analysis of parasitic nematode genes. In *M. incognita*, RNAi can be induced by feeding, with variable silencing efficiencies depending on the targeted gene (Rosso et al. 2005; Huang et al. 2006). Orthologues to most genes of the *C. elegans* RNAi pathway were found in *M. incognita*, including components of the amplification complex (*ego-1*, *rrf-1*, *rrf-2,* and *rrf-3*). However, no homologues of *sid-1, sid-2, rsd-2* and *rsd-6* genes that are involved in systemic RNAi and dsRNA spreading to surrounding cells, were found. Interestingly, these genes were also not found in *B. malayi* (Ghedin et al. 2007) and *H. contortus* (Zawadzki et al. 2006). Though systemic RNAi can actually spread in *M. incognita*, thus this process may involve novel or poorly conserved factors. A total of 2,958 *C. elegans* genes reported as producing a lethal RNAi phenotype were searched in *M. incognita* based on OrthoMCL clustering of candidate orthologues. Among the 1,083 OrthoMCL clusters that contained a *M. incognita* predicted gene, 148 were only shared with the other nematodes. Because of their lethal RNAi phenotype in *C. elegans* and distinctive sequence properties, these nematode-restricted genes provide an attractive set of new antiparasitic drug targets.

17.8 Conclusion

Comparative analysis of the genome of *M. incognita* with its nematode relatives revealed many traits of interest for studying the fundamentals of plant-parasitism in the phylum Nematoda and more broadly, in Metazoa. Root-knot nematodes can invade and develop within an immunocompetent host plant and transform host tissue into a unique gall-like structure that nourishes the adult female. One striking feature was the identification in *M. incognita* of an abundant and diverse suite of plant cell wall-degrading enzymes, which has no equivalent in any animal studied to date. The presence of this arsenal of enzymes was confirmed in the genome of *M. hapla*. The striking similarity of these enzymes to bacterial homologues strongly suggests that these genes were acquired by multiple HGT events. While interspecies HGT has been a major component of evolution in prokaryotes and some protozoa, HGT into metazoan genomes is relatively rare (Andersson 2005). Similarly to bacterial HGT that involves sets of genes implicated in adaptations to new hosts or food sources, the candidate HGT events into *M. incognita* involved genes with potential roles in interaction with the host. These results suggest that acquisition of new genes via HGT was probably crucial for the establishment of successful plant-parasitism by providing new capabilities to these nematodes. The availability of new genomes including parasites of animals and plants will undoubtedly allow assessing the importance of HGT as a general mechanism for adaptation to parasitism. Both analyses of root-knot nematode genomes allowed identifying gene sets that putatively encode secreted proteins of unknown function. These genes may play roles in host-parasite interaction. Further characterization of this diverse set of putative effectors would be a major step towards understanding how root-knot nematodes manipulate plant cell functions. Transcriptional profiling, proteomic analysis and high throughput RNAi strategies are in progress and will lead to deeper understanding of the processes by which a nematode successfully settles in a plant host and causes plant disease. Combining such knowledge with functional genomic data from the model host plant *A. thaliana* should provide new insights into the intimate molecular dialog governing plant-nematode interactions and allow the further development of target-specific strategies to limit crop damage. In the case of *M. incognita* the majority of the genome assembly was composed of pairs of homologous segments. These pairs of segments are substantially divergent at the nucleotide level and may denote altered former alleles. The origin of these dissimilar copies is still unclear and could reflect a simple loss of sexuality in an ancient sexual ancestor or hybridization between two closely related species. Whatever the hypothesis, this suggests that *M. incognita* is evolving in the absence of sexual reproduction towards effective haploidy through the Meselson effect (Mark Welch and Meselson 2000, 2001; Birky 2004). Functional divergence between ancient alleles of genes involved in the host-parasite interface could explain the extremely wide host range and geographic distribution of this polyphagous nematode. In contrast, *M. hapla*, which is an amphimictic species able to sexually reproduce, presents a substantially narrower host range and global distribution. This observation is counter-intuitive when classically considering sexual reproduction as an adaptive advantage, promoting genetic exchange and variability. This reinforces the idea that currently poorly understood genetic mechanisms such as Meselson effect provide bases for adaptability in *M. incognita*. A detailed comparative genomic analysis will aid understanding genome dynamics and evolutionary processes in asexual *versus* sexual organisms. Looking to the future, the availability of additional free-living, human and plant-parasitic nematode genomes will provide an unparalleled opportunity for comparative genomics to illuminate the evolutionary success of phylum Nematoda.

References

Abad P, Gouzy J, et al (2008) Genome sequence of the metazoan plant-parasitic nematode *Meloidogyne incognita*. Nat Biotechnol 26(8):909–915

Alegado RA, Tan MW (2008) Resistance to antimicrobial peptides contributes to persistence of Salmonella typhimurium in the *C. elegans* intestine. Cell Microbiol 10(6):1259–1273

Altschul SF, Madden TL, et al (1997) Gapped BLAST and PSI-BLAST: a new generation of protein database search programs. Nucleic Acids Res 25(17):3389–3402

Andersson JO. (2005) Lateral gene transfer in eukaryotes. Cell Mol Life Sci 62(11):1182–1197

Bertrand S, Brunet FG, et al (2004) Evolutionary genomics of nuclear receptors: from 25 ancestral genes to derived endocrine systems. Mol Biol Evol 21:1923–1937

Birky CW Jr (2004) Bdelloid rotifers revisited. Proc Natl Acad Sci USA 101(9):2651–2652

Blaxter ML (2003) Nematoda: genes, genomes and the evolution of parasitism. Adv Parasitol 54:101–195

Caillaud MC, Dubreuil G, et al (2008a) Root-knot nematodes manipulate plant cell functions during a compatible interaction. J Plant Physiol 165(1):104–113

Caillaud MC, Lecomte P, et al (2008b) MAP65-3 microtubule-associated protein is essential for nematode-induced giant cell ontogenesis in Arabidopsis. Plant Cell: tpc.107.057422

Castagnone-Sereno P (2006) Genetic variability and adaptive evolution in parthenogenetic root-knot nematodes. Heredity 96(4):282–289

Daimon T, Taguchi T, et al (2008) Beta-fructofuranosidase genes of the silkworm, *Bombyx mori*: insights into enzymatic adaptation of *B. mori* to toxic alkaloids in mulberry latex. J Biol Chem 283(22):15271–15279

Davis EL, Hussey RS, et al (2004) Getting to the roots of parasitism by nematodes. Trends Parasitol 20(3):134–141

Dieterich C, Clifton SW, et al (2008) The *Pristionchus pacificus* genome provides a unique perspective on nematode lifestyle and parasitism. Nat Genet 40(10):1193–1198

Emanuelsson O, Brunak S, et al (2007) Locating proteins in the cell using TargetP, SignalP and related tools. Nat Protoc 2(4):953–971

Ewbank JJ (2006) Signaling in the immune response. WormBook **doi/10.1895/wormbook.1.83.1**: http://www.wormbook.org

Foissac S, Schiex T (2005) Integrating alternative splicing detection into gene prediction. BMC Bioinform 6:25

Gao B, Allen R, et al (2001) Molecular characterisation and expression of two venom allergen-like protein genes in Heterodera glycines. Int J Parasitol 31(14):1617–1625

Ghedin E, Wang S, et al (2007) Draft genome of the filarial nematode parasite *Brugia malayi*. Science 317(5845):1756–1760

Guiliano DB, Blaxter ML (2006) Operon conservation and the evolution of trans-splicing in the phylum Nematoda. PLoS Genet 2(11):e198

Holterman M, van der Wurff A, et al (2006) Phylum-wide analysis of SSU rDNA reveals deep phylogenetic relationships among nematodes and accelerated evolution toward crown Clades. Mol Biol Evol 23(9):1792–1800

Hotson A, Mudgett MB (2004) Cysteine proteases in phytopathogenic bacteria: identification of plant targets and activation of innate immunity. Curr Opin Plant Biol 7(4):384–390

Huang G, Gao B, et al (2003) A profile of putative parasitism genes expressed in the esophageal gland cells of the root-knot nematode *Meloidogyne incognita*. Mol Plant Microbe Interact 16 (5):376–381

Huang GZ, Allen R, et al (2006) Engineering broad root-knot resistance in transgenic plants by RNAi silencing of a conserved and essential root-knot nematode parasitism gene. Proc Natl Acad Sci USA 103(39):14302–14306

Jaffe DB, Butler J, et al (2003) Whole-genome sequence assembly for mammalian genomes: Arachne 2. Genome Res 13(1):91–96

Jammes F, Lecomte P, et al (2005) Genome-wide expression profiling of the host response to root-knot nematode infection in Arabidopsis. Plant J 44(3):447–458

Jaubert S, Laffaire JB, et al (2004) Comparative analysis of two 14-3-3 homologues and their expression pattern in the root-knot nematode *Meloidogyne incognita*. Int J Parasitol 34 (7):873–880

Jaubert S, Milac AL, et al (2005) In planta secretion of a calreticulin by migratory and sedentary stages of root-knot nematode. Mol Plant Microbe Interact 18(12):1277–1284

Jones JT, Reavy B, et al (2004) Glutathione peroxidases of the potato cyst nematode Globodera Rostochiensis. Gene 324:47–54

Jones JT, Smant G, et al (2000) SXP/RAL-2 proteins of the potato cyst nematode Globodera rostochiensis: secreted proteins of the hypodermis and amphids. Nematology 2:887–893

Krogh A (1997) Two methods for improving performance of an HMM and their application for gene finding. Proc Int Conf Intell Syst Mol Biol 5:179–186

Lambert KN, Allen KD, et al (1999) Cloning and characterization of an esophageal-gland-specific chorismate mutase from the phytoparasitic nematode *Meloidogyne javanica*. Mol Plant Microbe Interact 12(4):328–336

Leroy S, Duperray C, et al (2003) Flow cytometry for parasite nematode genome size measurement. Mol Biochem Parasitol 128(1):91–93

Li L, Stoeckert CJ Jr, et al (2003) OrthoMCL: identification of ortholog groups for eukaryotic genomes. Genome Res 13(9):2178–2189

Li W, Godzik A (2006) Cd-hit: a fast program for clustering and comparing large sets of protein or nucleotide sequences. Bioinformatics 22(13):1658–1659

Lindblom TH, Dodd AK (2006) Xenobiotic detoxification in the nematode *Caenorhabditis elegans*. J Exp Zoolog A Comp Exp Biol 305(9):720–730

Mark Welch D, Meselson M (2000) Evidence for the evolution of bdelloid rotifers without sexual reproduction or genetic exchange. Science 288(5469):1211–1215

Mark Welch DB, Meselson MS (2001). Rates of nucleotide substitution in sexual and anciently asexual rotifers. Proc Natl Acad Sci USA 98(12):6720–6724

Mark Welch DB, Cummings MP, et al (2004) Divergent gene copies in the asexual class Bdelloidea (Rotifera) separated before the bdelloid radiation or within bdelloid families. Proc Natl Acad Sci USA 101(6):1622–1625

Marks NJ, Maule AG (2007) Neuropeptides in helminths: occurrence and distribution. Neuropeptide systems as targets for parasite and pest control. In: Geary TG, Maule AG (eds) Neuropeptide systems as targets for parasite and pest control. Landes Bioscience/Eurekah.com, Georgetown, TX

Menzel R, Bogaert T, et al (2001) A systematic gene expression screen of *Caenorhabditis elegans* cytochrome P450 genes reveals CYP35 as strongly xenobiotic inducible. Arch Biochem Biophys 395(2):158–168

Mitreva M, Blaxter ML, et al (2005) Comparative genomics of nematodes. Trends Genet 21 (10):573–581

Opperman CH, Bird DM, et al (2008) Sequence and genetic map of *Meloidogyne hapla*: a compact nematode genome for plant parasitism. Proc Natl Acad Sci USA 105(39):14802–14807

Papadopoulou J, Triantaphyllou AC (1982) Sex-determination in *Meloidogyne incognita* and anatomical evidence of sexual reversal. J Nematol (14):549–566

Paschinger K, Gutternigg M, et al (2007) The N-glycosylation pattern of *Caenorhabditis elegans*. Carbohydr Res 343:2041–2049

Pauchet Y, Muck A, et al (2008) Mapping the larval midgut lumen proteome of *Helicoverpa armigera*, a generalist herbivorous insect. J Proteome Res 7(4):1629–1639

Plowman GD, Sudarsanam S, et al (1999) The protein kinases of *Caenorhabditis elegans*: a model for signal transduction in multicellular organisms. Proc Natl Acad Sci USA 96 (24):13603–13610

Qin L, Kudla U, et al (2004) Plant degradation: a nematode expansin acting on plants. Nature 427 (6969):30

Rao KV, Eswaran M, et al (2000) The Wuchereria bancrofti orthologue of *Brugia malayi* SXP1 and the diagnosis of bancroftian filariasis. Mol Biochem Parasitol 107(1):71–80

Robertson HM, Thomas JH (2006) The putative chemoreceptor families of *C. elegans*. WormBook **doi/10.1895/wormbook.1.66.1**: http://www.wormbook.org

Robinson-Rechavi M, Maina CV, et al (2005) Explosive lineage-specific expansion of the orphan nuclear receptor HNF4 in nematodes. J Mol Evol 60(5):577–586

Rosso MN, Dubrana MP, et al (2005) Application of RNA interference to root-knot nematode genes encoding esophageal gland proteins. Mol Plant Microbe Interact 18(7):615–620

Shingles J, Lilley CJ, et al (2007) *Meloidogyne incognita*: molecular and biochemical characterisation of a cathepsin L cysteine proteinase and the effect on parasitism following RNAi. Exp Parasitol 115(2):114–120

Smant G, Stokkermans JP, et al (1998) Endogenous cellulases in animals: isolation of beta-1, 4-endoglucanase genes from two species of plant-parasitic cyst nematodes. Proc Natl Acad Sci USA 95(9):4906–4911

Stein LD, Bao Z, et al (2003) The genome sequence of *Caenorhabditis briggsae*: a platform for comparative genomics. PLoS Biol 1(2):E45

Sugimura M, Watanabe H, et al (2003) Purification, characterization, cDNA cloning and nucleotide sequencing of a cellulase from the yellow-spotted longicorn beetle, *Psacothea hilaris*. Eur J Biochem 270(16):3455–3460

Tang X, Xiao Y, et al (2006) Regulation of the type III secretion system in phytopathogenic bacteria. Mol Plant Microbe Interact 19(11):1159–1166

The *C. elegans* Genome Sequencing Consortium (1998) Genome sequence of the nematode *C. elegans*: a platform for investigating biology. Science 282(5396):2012–2018

Triantaphyllou AC (1985) Cytogenetics, cytotaxonomy and phylogeny of root-knot nematodes. In: Sasser JN, Carter CC (eds) An advance treatise on *Meloidogyne*, vol 1. North Carolina State University Graphics, Raleigh, NC, pp 113–126

Trudgill DL, Blok VC (2001) Apomictic, polyphagous root-knot nematodes: exceptionally successful and damaging biotrophic root pathogens. Annu Rev Phytopathol 39:53–77

Wei YD, Lee KS, et al (2006) Molecular cloning, expression, and enzymatic activity of a novel endogenous cellulase from the mulberry longicorn beetle, *Apriona germari*. Comp Biochem Physiol B Biochem Mol Biol 145(2):220–229

Zarkower D (2006) Somatic sex determination. WormBook **doi/10.1895/wormbook.1.84.1**: http://www.wormbook.org

Zawadzki JL, Presidente PJ, et al (2006) RNAi in Haemonchus contortus: a potential method for target validation. Trends Parasitol 22(11):495–499

Chapter 18
Ecological Genomics of Nematode Community Interactions: Model and Non-model Approaches

Michael A. Herman, Joseph D. Coolon, Kenneth L. Jones, and Timothy Todd

Abstract The effects of human-induced environmental change are evident at multiple levels of biological organization. To date, most environmental change studies have focused on effects at the ecosystem, community and organismal levels. However, the ultimate controls of biological responses are located in the genome. Thus, genetic and genomic studies of organismal responses to environmental changes are necessary. Recent advances in genome analysis now make such analyses possible. In this chapter we describe a research approach and program that can begin to span this gap by using genome-enabled approaches to characterize organismal changes and then employing a genetically tractable model organism to identify genes involved in the response to environmental perturbations.

Abbreviations

GO Gene ontology
TD_{50} Time to death for 50% of a population

M.A. Herman
Ecological Genomics Institute, Kansas State University, 266 Chalmers Hall, Manhattan, KS, USA;
Division of Biology, Kansas State University, Manhattan, KS, 66506 USA
e-mail: mherman@ksu.edu

J.D. Coolon, K.L. Jones and T. Todd
Ecological Genomics Institute, Kansas State University, Manhattan, KS, USA
J.D. Coolon and K.L. Jones
Division of Biology, Kansas State University, Manhattan, KS, USA

J.D. Coolon
Department of Ecology and Evolutionary Biology, University of Michigan, Ann Arbor, MI, USA

K.L. Jones
Department of Environmental Health Science, University of Georgia, Athens, GA, USA
T. Todd
Department of Plant Pathology, Kansas State University, Manhattan, KS, USA

P. Pontarotti (ed.), *Evolutionary Biology: Concept, Modeling, and Application*,
DOI: 10.1007/978-3-642-00952-5_18,

18.1 Introduction

18.1.1 Global Environmental Change

The world is changing around us at an unprecedented pace (Millennium Assessment, IPCC 2007). The role of human activities in these changes has been understood for some time. In fact, in 2000, the National Science Board in the United States issued a report that stated:

> *Human activities are transforming the planet in new ways and combinations at a faster rate and over broader scales than ever before in the history of humans on Earth. Accelerated efforts to understand Earth's ecosystems and how they interact with the numerous components of human-caused global changes are timely and wise.*

This was a challenge to scientists to study the effects of environmental change. Human-induced changes to the abiotic environment include climatic shifts in temperature and rainfall, effects of pollution and changes in land use, such as conversion of natural landscapes to agriculture (Hannah 1995; Dobson 1997). Of these, the latter appears to be making the greatest impact (Foley 2005). In order to gain the greatest understanding, it is important to study the effects of global environmental change at multiple levels of biological organization.

18.1.2 The Ecological Genomic Approach

The natural environments of organisms present a multitude of biotic and abiotic challenges that require both short-term ecological and long-term evolutionary responses. These responses have long been the subject of biological interest, yet their inherent complexity has made genetic and mechanistic dissection empirically difficult. However, recent technical advances in high-throughput sequencing, genotyping and genome-wide expression profiling, coupled with bioinformatics approaches for handling such data, hold great promise for dissecting these responses with unprecedented resolution. The implementation and application of new techniques requires a multidisciplinary approach, combining organismal analyses with molecular genetics and genomics, laboratory experiments with field studies and all within an ecologically relevant framework. The emerging field of ecological genomics seeks to understand genetic mechanisms underlying the responses of organisms to their natural environment by combining genomic and ecological approaches. These responses include modifications of biochemical, physiological, morphological, or behavioral traits of adaptive significance. Such an integration of fields faces challenges but will revolutionize our understanding of ecological responses at a genetic, genomic and eventually, a mechanistic level (Ungerer et al. 2008).

18.2 Evolutionary Framework for Ecological Genomic Studies

As changing environments are ubiquitous, one of the greatest challenges in biology is understanding and predicting effects of environmental changes on the ecology of the world's biota. Organisms respond to environmental changes on both ecological and evolutionary time scales. The magnitude and extent of human-induced changes to the environment create additional challenges for organisms, including changes to climate (e.g., global temperatures, rainfall patterns and insolation), landscape structure (e.g., urbanization, deforestation, fragmentation of the landscape) and communities (e.g., exotic species in new environments due to agriculture or global commerce/transportation). All of these changes lead to novel interactions among species to which, given the rapidity of human-induced change, organisms must adapt at an unprecedented pace. Recent and growing evidence suggests that organisms may adapt in a microevolutionary sense on decadal time scales to rapid environmental change, a process called contemporary evolution (reviewed in Stockwell et al. 2003; Carroll et al. 2007; Smith and Bernatchez 2008). Contemporary evolution due to human-caused selection is now well documented.

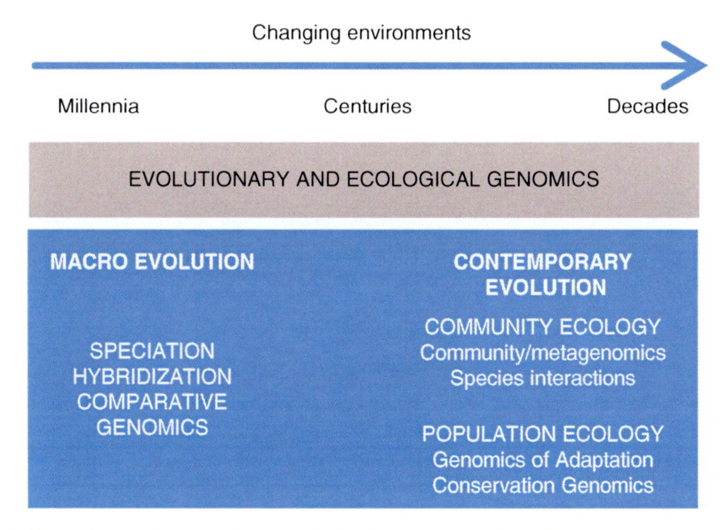

Fig. 18.1 Evolutionary framework for ecological genomics studies. Organisms respond to changing environments through long-term macro-evolutionary and short-term ecological time scales, as depicted by the arrow at the top of the figure. Recent evidence suggests that organisms can adapt to changes in the environment over decadal time scales in a process that has been termed "contemporary evolution." The mechanisms that organisms use to respond to these changes are lodged in the genome, whose discovery requires an evolutionary and ecological genomic approach. Just as organisms respond to environmental change on both ecological and evolutionary time scales, the research addressing these changes must focus on different time scales. Those disciplines and the types of genome-enable approaches they employ are indicated

Rapid adaptive evolution has been shown in flowering time (Franks et al. 2007); photoperiodism (Bradshaw and Holzapfel 2008); the sexual signal of invasive field crickets (Tinghitella 2008); and in response to changes in climate (Reusch and Wood 2007). Change, human-induced or not, elicits organism responses via mechanisms lodged in the genome; whose study requires an evolutionary and ecological genomic approach. Just as organisms respond to environmental change on both ecological and evolutionary time scales, the research addressing these changes must focus on different time scales (Fig. 18.1).

18.3 Nematode Ecological Genomics: Model and Non-model Approaches

18.3.1 Global Environmental Change and the Grassland Ecosystem

Grasslands (Samson and Knopf 1994) perform many essential ecosystem services, such as supplying clean water, recycling essential nutrients and preserving biodiversity (Daily 1997) and are among the most endangered ecosystems on Earth, largely having been replaced by agricultural systems that alter both above- and belowground communities (Baer et al. 2002). In addition, grasslands are among the most sensitive to an array of global change phenomenon (Samson and Knopf 1994; Collins et al. 1998; Field and Chiariello 2000; Buckland et al. 2001; Knapp and Smith 2001; Reich et al. 2001; Briggs et al. 2005). For example, the structure and function of grasslands are determined by patterns of climatic variability and nutrient availability but altered precipitation patterns, enhanced nitrogen deposition and changes in land use (fire and grazing regimes, conversion to agriculture) have the potential to dramatically influence these relationships (Collins et al. 1998; Knapp and Smith 2001; Briggs et al. 2005). Patterns and controls of ecological processes in grasslands and the effects of natural and anthropogenic disturbances have been the focus of long-term research at the Konza Prairie Biological Station (near Manhattan, KS) for more than 25 years (Knapp et al. 1998).

18.3.2 The Importance of Nematode Ecology

We have focused on nematodes because they are among the most abundant invertebrates in soils and are an important component of the microfauna in grasslands (Curry 1994). Nematode species occurring in soils encompass a wide variety of feeding strategies (Freckman 1988), including many free-living species that feed on soil microbes (bacteria or fungi). Microbial-feeding nematodes may be the most important consumers of bacteria and fungi in many soil communities (Blair et al.

2000; Yeates 2003) and their interactions with microbial decomposers affect ecosystem processes including decomposition and nutrient cycling (Freckman 1988; Coleman et al. 1991). Nematodes are also known to be responsive to changing environmental conditions (Freckman and Ettema 1993; Todd 1996; Todd et al. 1999), making them ideal model organisms to assess the potential impacts of global change on soil communities. Several studies have demonstrated that the soil nematode community in tallgrass prairie responds strongly to perturbations, including nutrient enrichment through nitrogen addition, increased soil moisture and different experimental fire regimes (Seastedt et al. 1987; Blair et al. 2000; Todd 1996; Todd et al. 1999; Jones et al. 2006b).

18.3.3 The Nematode Ecological Genomic Approach

The disturbances caused by global environmental change are complex, involving changes in the biotic environment that include microbes, competitors and predators, as well as changes in the abiotic soil environment. To begin to sort out these interactions, we have focused on the responses of microbial-feeding nematodes to the microbial aspects of the grassland biotic environment. We have employed an interdisciplinary approach using high-throughput molecular techniques to first characterize shifts in the nematode community as well as the interacting bacterial community. Next, we have modeled these interactions using the genetic model organism *Caenorhabditis elegans* to begin to understand the interactions of genes with the environment in non-model systems such as the native grassland soil nematode community in grasslands at Konza Prairie. An understanding of the genetic mechanisms underlying ecological interactions should provide a predictive value previously not possible.

18.3.4 C. elegans as a Model Nematode

C. elegans is a free-living nematode found in enriched soils that has been used in genetic research for over 40 years. Its short generation time, small size and ease of maintenance have led to the development of sophisticated genetic tools as evidenced by the thousands of genes that have been isolated and analyzed. In addition, the use of RNA mediated interference (RNAi) induced by treating animals with double-stranded RNA (dsRNA) corresponding to a gene of interest allows one to quickly and easily see the effects of removing, or at least crippling, any gene to determine its function by examining the effects on the phenotype (Fire et al. 1998). Finally, genetic, molecular and sequence data are continually annotated and made available through Wormbase (http://www.wormbase.org). While the high degree of evolutionary conservation allows *C. elegans* to be a good model for the biology of higher organisms, such as humans, it may be an even better model for understanding the responses of soil nematodes.

Model organism systems, such as *C. elegans*, have well developed genetic and genomic tools that allow for powerful analyses. However, they were chosen for characteristics (e.g., small size, small genomes and rapid life cycles) that facilitate genetic analysis but may not be typical of many organisms. While some researchers have chosen to study the ecology of selected model organisms (Roberts and Feder 2000; Weinig et al. 2002) others have chosen to develop genomic capabilities for more ecologically important taxa (Kessler et al. 2004). Both approaches have yielded interesting results. In fact, a combined approach as was done in the use of *Arabidopsis* to discover genes induced by flavinoid release by the invasive species spotted knapweed (Bais et al. 2003), promises to be extremely fruitful. We have chosen this latter approach for our nematode studies.

18.3.5 Non-model Approaches

18.3.5.1 Grassland Nematode Community Responses

To determine the effects on nematode communities, Jones et al. (2006b) used an ongoing long-term experiment at the Konza Prairie Biological Station established in 1986 to address belowground responses to fire, mowing and nutrient enrichment. An understanding of these effects on soil processes, including the soil food web and its invertebrate and microbial components, is integral to predicting the consequences of global change for both natural and managed ecosystems. Nematodes were sampled from four replicates of four treatment combinations (annually burned versus unburned and ammonium nitrate addition versus no addition). We focused specifically on microbial-feeding nematodes and used sequence differences in a 900 base pair (bp) fragment consisting in the 5' 500 bases of the 18S rRNA gene and the entire adjacent internally transcribed spacer region (ITS1) to develop dual-labeled fluorescent probes (e.g., Taqman probes), which were used for detection of 16 different nematode taxa from among 984 individual nematodes samples (Jones et al. 2006a). Sequencing nematodes that were not identified with existing probes identified an additional three taxa. The 19 identified taxa represent three taxonomic families and recent analyses indicate that each of these families belong to different phylogenetic clades (Blaxter 1998; Holterman et al. 2006).

18.3.5.2 Differential Nematode Response

Statistical analyses of relative nematode abundances in each plot revealed that season, nitrogen addition and burning were shown to affect nematode abundance in multiple taxa, with nitrogen addition and season having the most pronounced effects. In addition to these main effects, nematode taxa were differentially affected

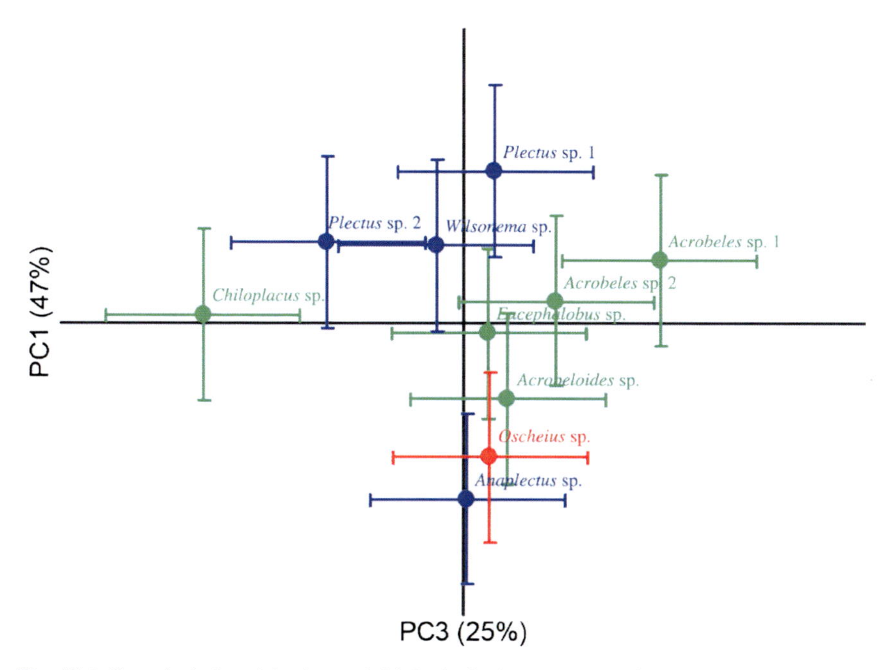

Fig. 18.2 Canonical plot of the first and third principal components of mean adjusted response of the nematode community. Members within taxonomic families are designated by color (Cephalobidae, green; Plectidae, blue; Rhabditidae, red). Data are means ± standard error of the difference (Printed with permission of Molecular Ecology)

by interactions between the burning and nitrogen addition treatments. A principal components analysis illustrating the variation due to burning in the presence of nitrogen (PC1) versus that of the variation due to nitrogen in the presence of burning (PC3) is shown in Fig. 18.2. On the whole, taxon responses were similar within members of a family. However, for each family there was a taxon (*Chiloplacus* sp., *Anaplectus* sp., and *Oscheius* sp.) that responded differently than others within their family. Additionally, although nematodes from different taxonomic groups on average respond differently, similar responses were seen in nematode taxa that span three taxonomic families (e.g., *Acrobeloides* sp., *Oscheius* sp., and *Anaplectus* sp.). What drivers might account for these differential nematode responses? They must involve a combination of indirect and direct effects of the biotic and abiotic environment, respectively. Indirectly, the nematodes may be responding to changes in the community structure (i.e., food resources, parasites/pathogens, competition, or predation). Alternatively, as nematodes live in a film of water and are in direct contact with their environment, sensitivities to soil chemistry may influence the observed responses. As the genetic responses to the biotic and abiotic aspects of the environment are complex, they will need to be dissected separately. However, one must be careful as changes in the abiotic environment may have indirect effects on the nematode's biotic environment. Furthermore, as the biotic interactions affecting the nematode community are highly complex, we first have characterized the nematode response to the bacterial aspects of their biotic interactions.

18.3.5.3 Microbial Community Response to Nitrogen Addition and Burning

One force shaping the bacterial-feeding nematode community could be the response to changes in the microbial community. Thus it might be that nitrogen addition and burning treatments alter the microbial communities, which, in turn, might play a role in structuring nematode communities by altering food resources and pathogens. To demonstrate whether this is possible, we adopted a mass parallel sequencing technique ("454 sequencing") that generated >200,000 short sequences (about 100 bp). To amplify the soil bacteria signal, we used PCR primers that flank the hypervariable V3 region in the 5'-end of the 16S rRNA gene (Baker et al. 2003) on DNA that was extracted directly from the soil. While we were able to derive bacterial sequences for four separate projects (Jones, Coolon, Todd and Herman, unpublished observations), here we only consider the results obtained from the plots in which we previously measured nematode community responses.

These results will be described in detail elsewhere, briefly we developed bioinformatic methods that enabled Operational Taxonomic Unit (OTU) designation across sampled plots. OTUs were generated at each of 18 sequence identity levels (80–98%). At each level of sequence identity, sequences were parsed by plot and used to calculate the frequency of occurrence of all OTUs for each of the plots. The number of OTUs increased as the per cent sequence identity increased from 80% to 98%, following expectations of biological complexity, with OTUs generated at different levels of sequence identity being of different taxonomic resolutions. Using replicated field plots and statistical analysis, we showed reproducible treatment responses within the microbial community. Overall taxonomic richness, dominance and diversity were calculated for each plot and analyzed across treatments by analysis of variance (ANOVA). In order to determine not only whether the community responded but also to infer which level of biological organization (phylum, order, family, etc.) responded, we plotted these community measures at each of the 18 sequence identity levels (80–98%). These analyses demonstrated that richness and diversity increased with sequence identity level while dominance decreased, indicating the levels of biological organization that respond to added nitrogen. For example, treatment elicited differences in richness were consistently significant across all levels of sequence identity, suggesting high order changes in the bacterial community in response to nitrogen addition. These results confirm that bacterial populations, similar to nematodes, are highly responsive, with the magnitude and direction of the changes being different even across taxa of similar taxonomy. Thus, it is plausible that in response to changing environments, such as nitrogen addition, bacterial-feeding nematode communities may be shaped, in part, by responses to changes in the bacterial community.

18.3.6 Model Approaches

18.3.6.1 Use of *C. elegans* to Model Ecological Interactions

So far we have described experiments that documented responses of the soil nematode community to changes in the environment and identifying potential drivers, such as changes in the bacterial community. Next, we modeled these interactions using *C. elegans* in the laboratory to investigate the mechanisms underlying the native nematode responses observed on Konza Prairie. One aim of these studies was to use *C. elegans* as a gene discovery tool to examine gene expression in response to environmental change. Although *C. elegans* has not been found in our experimental plots, other related Rhabditid taxa, specifically *Mesorhabditis* sp., *Oscheius* sp. and *Pellioditis* sp., do occur there. Further, we know from EST databases that *C. elegans* is likely to share 50–80% of gene sequences with most nematode taxa (Parkinson et al. 2004), thus we expect the native Konza taxa more closely related to *C. elegans* (i.e., Rhabdtids) to share more genes than those that are less related. Ultimately we will test the homologs of the candidate genes identified in *C. elegans* for their function in the native soil nematodes.

18.3.6.2 *C. elegans* Genes Involved in Response to Changes in Bacterial Environment

To model naturally occurring nematode–bacterial interactions, as well as to use new environments for gene discovery in the laboratory, we isolated bacteria from grassland prairie soils at the Konza Prairie Biological Station. We isolated three bacteria from Konza soils: *Micrococcus luteus*, *Bacillus megaterium* and *Pseudomonas* sp., of which the latter two were isolated in association with bacterial-feeding nematodes (*Oscheius* sp. and *Pellioditis* sp., respectively). *Pseudomonas fluorescens* was the closest match (98% sequence identity) in the Ribosomal Database Project to the 16S rDNA sequence of the isolated *Pseudomonas* sp.

We used oligonucleotide microarrays to identify *C. elegans* genes that were differentially expressed in response to altered bacterial environments. We compared expression patterns of *wild-type C. elegans*, fed each of these soil bacteria as well as its traditional laboratory food, *Escherichia coli* and all pair-wise comparisons were performed (Coolon et al. 2009). We identified 204 unique genes whose expression was significantly changed in response to bacterial environment. These results indicated that nematode populations express different suites of genes when raised in different bacterial environments.

Within the *C. elegans* genes identified as differentially expressed in response to bacterial environment, metabolism genes were highly represented (9.3%) as expected. Interestingly, genes previously implicated in innate immunity (9.8%) and cuticle biosynthesis or collagens (8.8%) were also found to be highly abundant

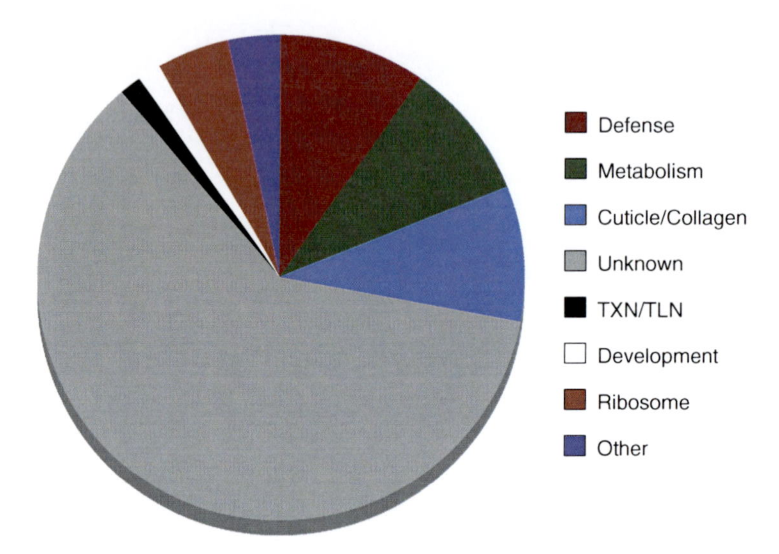

Fig. 18.3 Gene ontology (GO) terms for identified differentially expressed genes. Gene ontology (GO) terms were amended with recently published information and used to categorize the identified differentially expressed genes. Clustering was done manually by grouping GO terms of similar function (Coolon et al. 2009)

within the genes identified. Finally, genes of unknown function made up the largest portion (61% of the total, Fig. 18.3), also as expected since one aim of the work was to determine functions for such genes helping to further characterize the major proportion of the *C. elegans* genome that remains unknown after four decades of genetic dissection. However, ultimately, functional data obtained by interfering with gene function are needed to determine which genes really matter for a particular interaction. To this end, we obtained all available viable non-sterile mutations for the 204 differentially expressed genes in our study (21/204, ~10% of the total genes identified) from the *Caenorhabditis* Genetics Center (CGC) and used them for biological validation of the microarray results (Table 18.1). Functional tests measuring multiple aspects of life history were used to calculate absolute fitness by life table analysis and lifespan was measured with pathogenicity assays in all four bacterial environments. Specifically, age-specific reproduction (m_x) and survival (l_x) were used to calculate intrinsic growth rate ($Ro = \Sigma l_x m_x$), generation time $(\Sigma l_x m_x)/(\Sigma x l_x m_x)$ and Lambda ($\lambda = e^{(\ln Ro/T)}$), which was used as a measure of absolute fitness. Lifespan was measured as time to death for 50% of a population (TD_{50}) (Tan and Ausubel 2000) using survivorship curves and is indicative of the pathogenicity of *C. elegans* food sources. We found that many of the mutations had effects on life history traits that differed significantly from wild type in a given bacterial environment, demonstrating that many of the genes specifically induced in response to different bacteria function to contribute to nematode fitness and longevity in different bacterial environments (Coolon et al. 2009; Table 18.2).

Table 18.1 Genes and alleles used for functional tests

Gene	Allele	Predicted molecular function
acdh-1	ok1489	Acyl-CoA dehydrogenase
C23H5.8	ok651	Unknown function
cey-2	ok902	Cold-shock/Y-box domain containing
cey-4	ok858	Unknown function
cpi-1	ok1213	Homolog of cysteine protease inhibitors (cystatins)
ctl-1	ok1242	Cytosolic catalase
dhs-28	ok450	17-Beta-hydroxysteroid dehydrogenase 4
dpy-14	e188	Type III (alpha 1) collagen
dpy-17	e1295	Cuticle collagen
elo-5	gk182	PUFA elongase
cyp-37A1	ok673	Unknown function
F55F3.3	ok1758	Unknown function
fat-2	ok873	Delta-12 fatty acyl desaturase
gei-7	ok531	Predicted isocitrate lyase/malate synthase
gld-1	op236	Meiotic cell cycle/oogenesis
hsp-12.6	gk156	Predicted heat-shock protein
mtl-2	gk125	Metallothionein
pab-2	ok1851	Polyadenylate-binding protein
rol-6	e187	Cuticle collagen
sqt-2	sc108	Cuticle collagen
Y57A10C.6	ok693	Predicted thiolase

List of 21 mutants used for functional tests, predicted molecular functions are indicated.

18.3.6.3 Specificity of the *C. elegans* Functional Response

In order to compare across bacterial environments we investigated genotype-by-environment interactions (GEI) and examined mutant norms of reaction across bacterial environments (Fig. 18.4). GEI exists when there is re-ranking of the phenotypic responses of genotypes across environments, or genotypes may have more similar phenotypes in one environment than in another, therefore differences in the magnitude of effects exist between different environments (Falconer and Mackay 1996). Reaction norms of fitness (Fig. 18.4a) and lifespan (Fig. 18.4b), revealed differential effects of the bacterial environments on the different mutant genotypes demonstrating the specificity and complexity of mutational effects on these complex traits.

How can we infer whether a particular gene is truly important for a given environmental interaction? A simple assumption that a gene upregulated in an environment positively regulates a particular life history trait predicts that loss of that gene function would cause a reduction in fitness in that environment. One such example is *hsp-12.6* that encodes a heat-shock protein (Hsu et al. 2003) and was found to be upregulated when wild-type *C. elegans* was grown on *E. coli* compared to growth on *B. megaterium*. We found that *hsp-12.6* mutants have a significant reduction in fitness as compared to wild type when the mutant is grown on *E. coli* from that observed on *B. megaterium*. Not only is this difference significant but fitness of *hsp-12.6* mutants was significantly increased relative to wild type when

Table 18.2 Biological validation of identified *Caenorhabditis elegans* genes

Gene	*Escherichia coli* (OP50)		*Micrococcus luteus*		*Pseudomonas* sp.		*Bacillus megaterium*	
	λ	TD_{50}	λ	TD_{50}	λ	TD_{50}	λ	TD_{50}
wt	3.60(0.19)	5.6(0.22)	2.63(0.18)	4.1(0.22)	3.99(0.25)	8.7(0.27)	2.81(0.16)	12.3(0.27)
acdh-1	2.99(0.03)$^{-}$	5.0(0.35)$^{-}$	2.54(0.25)	5.0(0.35)$^{+}$	3.78(0.74)	5.5(0.79)$^{-}$	3.01(0.37)	10.4(0.42)$^{-}$
C23H5.8	2.72(0.03)$^{-}$	7.8(0.57)$^{+}$	2.42(0.04)$^{-}$	3.6(0.42)$^{-}$	3.07(0.02)$^{-}$	6.0(0.79)$^{-}$	3.30(0.04)$^{+}$	8.9(0.74)$^{-}$
cey-2	3.08(0.04)$^{-}$	6.1(0.42)	2.11(0.06)$^{-}$	3.5(0.35)$^{-}$	2.83(0.03)$^{-}$	7.5(0.61)$^{-}$	2.79(0.01)	7.0(0.35)$^{-}$
cey-4	3.51(0.13)	5.6(0.42)	2.84(0.06)$^{+}$	3.6(0.42)$^{-}$	3.57(0.07)$^{-}$	5.9(0.22)$^{-}$	2.95(0.02)	3.7(0.27)$^{-}$
cpi-1	3.25(0.15)$^{-}$	7.6(0.22)$^{+}$	3.01(1.17)	4.4(0.22)	3.65(0.43)	6.6(0.42)$^{-}$	3.19(0.41)	12.4(0.42)
ctl-1	2.91(0.07)$^{-}$	6.2(0.84)	2.53(0.07)	4.8(0.29)$^{+}$	2.77(0.18)$^{-}$	3.9(0.42)	2.29(0.07)$^{-}$	8.5(0.35)$^{-}$
cyp-37A1	3.59(0.08)	8.0(0.50)$^{+}$	2.37(0.06)$^{-}$	4.4(0.42)	3.64(0.03)$^{-}$	8.5(0.35)$^{-}$	2.85(0.04)	9.5(0.50)$^{-}$
dhs-28	2.23(0.18)$^{-}$	6.7(0.27)$^{+}$	2.01(0.21)$^{-}$	3.6(0.22)$^{-}$	2.43(0.14)$^{-}$	7.3(0.27)$^{-}$	1.86(0.27)$^{-}$	10.2(0.76)$^{-}$
dpy-14	1.89(0.44)$^{-}$	2.4(0.22)$^{-}$	1.60(0.07)$^{-}$	2.1(0.22)$^{-}$	1.85(0.17)$^{-}$	3.1(0.42)$^{-}$	0.96(0.02)$^{-}$	4.1(0.42)$^{-}$
dpy-17	2.84(0.52)$^{-}$	4.0(0.35)$^{-}$	2.70(0.34)	3.1(0.42)$^{-}$	3.20(0.45)$^{-}$	3.0(0.35)$^{-}$	2.69(0.80)	12.3(0.57)
elo-5	4.11(0.07)$^{+}$	5.5(0.35)	3.02(0.10)$^{+}$	2.6(0.42)$^{-}$	4.07(0.12)	5.0(0.50)$^{-}$	4.18(0.05)$^{+}$	9.5(0.35)$^{-}$
F55F3.3	3.53(0.15)	3.1(0.55)$^{-}$	2.25(0.14)$^{-}$	2.6(0.55)$^{-}$	2.24(0.07)$^{-}$	5.0(0.35)$^{-}$	2.06(0.07)$^{-}$	5.5(0.35)$^{-}$
fat-2	3.27(0.13)$^{-}$	9.9(0.82)$^{+}$	2.97(0.04)$^{+}$	8.5(0.35)$^{+}$	4.23(0.04)	11.4(0.74)$^{+}$	3.18(0.09)$^{+}$	13.7(1.15)$^{+}$
gei-7	3.52(0.25)	5.7(0.27)	2.73(0.12)	4.5(0.00)$^{+}$	3.77(0.26)	7.6(0.22)$^{-}$	3.27(0.48)	14.3(0.27)$^{+}$
gld-1	3.15(0.13)$^{-}$	5.6(0.22)	2.51(0.28)	3.5(0.35)$^{-}$	3.53(0.06)$^{-}$	4.3(0.57)$^{-}$	2.78(0.04)	5.5(0.35)$^{-}$
hsp-12.6	3.10(0.08)$^{-}$	5.7(0.45)	2.50(0.18)	3.7(0.27)$^{-}$	3.72(0.14)	6.6(1.29)$^{-}$	3.00(0.08)$^{+}$	9.5(1.00)$^{-}$
mtl-2	3.77(0.17)	6.1(0.22)$^{+}$	3.02(0.23)$^{+}$	5.2(0.27)$^{+}$	4.09(0.28)	8.0(0.35)$^{-}$	3.75(0.40)$^{+}$	13.8(0.27)$^{+}$
pab-2	4.14(0.06)$^{+}$	6.6(0.42)$^{+}$	2.72(0.47)	5.4(0.42)$^{+}$	4.29(0.24)	7.7(0.57)$^{-}$	3.20(0.11)$^{+}$	8.9(0.74)$^{-}$
rol-6	2.82(0.22)$^{-}$	3.1(0.82)$^{-}$	2.28(0.11)$^{-}$	2.9(0.22)$^{-}$	3.11(0.09)$^{-}$	7.7(0.45)$^{-}$	2.56(0.04)$^{-}$	10.2(0.76)$^{-}$
sqt-2	2.97(0.01)$^{-}$	6.9(0.42)$^{+}$	2.69(0.06)	3.7(0.57)	3.72(0.06)$^{-}$	4.2(0.57)$^{-}$	3.39(0.47)$^{+}$	7.2(1.35)$^{-}$
Y57A10C.6	3.37(0.18)	6.3(0.45)$^{+}$	2.09(0.09)$^{-}$	4.5(0.00)$^{+}$	3.41(0.33)$^{-}$	8.2(0.57)	2.62(0.23)	15.0(0.35)$^{+}$

Wild-type (N2) and mutant *C. elegans* strains were grown on the four bacterial isolates and absolute fitness (λ) and time to death for 50% of the individuals in a population (TD_{50} in days) were measured. *P*-values are shown for contrasts between environments within strain for fitness and TD_{50}. Standard error (SEM) is given in parenthesis. Additionally, + indicates a significant ($P < 0.05$) increase relative to wild type and – indicates a significant ($P < 0.05$) decrease of the mutant relative to wild type (Coolon et al. 2009).

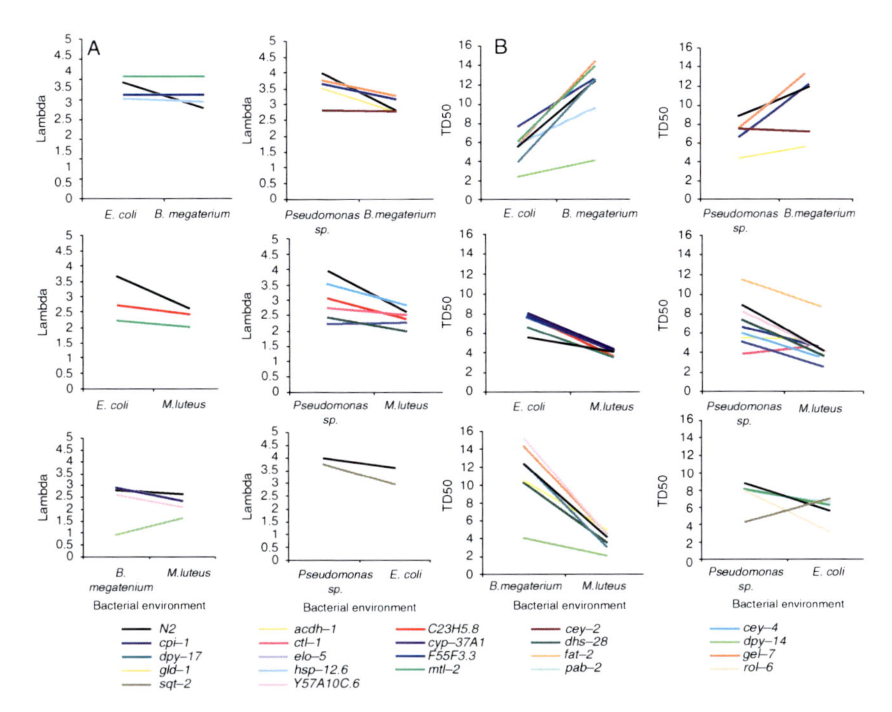

Fig. 18.4 Life history reaction norms with significant gene by environment interactions. Significant gene by environment interactions with Lambda (**a**) and lifespan as measured by TD_{50} (**b**) are illustrated by reaction norms. All pair-wise bacterial comparisons are shown. *B* = *Bacillus megaterium*, *M* = *Micrococcus luteus*, *E* = *Escherichia coli*, *P* = *Pseudomonas* sp.

grown on *B. megaterium* (Fig. 18.4a). This suggests that there was a cost associated with the expression of *hsp-12.6* in an environment in which it was not needed and a detriment to loss of function in an environment in which it was needed. Thus, the *hsp-12.6* allele had an antagonistic pleiotropic effect on fitness in these environments. We observed three other instances of antagonistic pleiotropy (Fig. 18.4a): *cpi-1* that encodes a cysteine protease inhibitor, also in the *E. coli* versus *B. megaterium*, as well as in the *Pseudomonas* sp. versus B. *megatarium* comparisons and *gei-7*, which encodes a isocitrate lyase/malate synthase that has been shown to function in lifespan extension (Tsuboi et al. 2002) also in the *Pseudomonas* sp. versus B. *megatarium* comparison. These observations suggest that these genes are likely under strong stabilizing selection in wild populations, with fitness trade-offs in different environments.

Although we observed examples that met the expectations of the simple prediction that genes positively impact particular life history traits, in many cases the underlying gene regulation may be more complex, involving positive and negative regulation and in some cases in a manner not yet elucidated. Thus in most cases we do not expect to be able to predict the directional effect of a particular mutation on the trait. Instead we predict that we would observe GEI between the environments in which differential expression was found. There were 37 instances of differential expression among the 21 genes tested. ANOVA was used to determine that 49% (18/37) of the contrasts of mutant fitness in the six bacterial comparisons had significant gene by environment interactions (Fig. 18.4a) and that 35/37 (95%) of tests showed significant TD_{50} GEI (Fig. 18.4b). Thus, it appears that the majority of differentially expressed genes are functionally important in the specific environments in which they were regulated illustrating that gene by environment interaction is likely a common feature to genes that are regulated in response to different bacterial environments (Coolon et al. 2009).

18.3.6.4 Do Nematodes "Know" What Is Good for Them?

The *C. elegans* experiments described above were conducted using one environment at a time. However, in the wild, bacterial-feeding nematodes must be faced with many bacterial types as potential food sources, which also may expose them to risks of infection among other interactions. To begin to dissect these more complex interactions, we conducted food preference tests on wild-type *C. elegans* in response to the bacterial isolates and *E. coli*. Using a biased choice assay (Shtonda and Avery 2006) (Fig. 18.5a, upper) we determined food preference for all pair-wise combinations of bacterial isolates (Fig. 18.5a, lower). Comparisons of the pair-wise measures of preferences revealed a hierarchy of food preferences: *Pseudomonas sp* was most preferred, followed by *E. coli*, which were both much more preferred than *B. megaterium*, which was slightly more preferable than *M. luteus*. Interestingly, this hierarchy mirrored the observed trend for fitness in the different bacterial environments (Fig. 18.5b, c), with *C. elegans* preferring *Pseudomonas* sp. on which it was most fit, followed by *E. coli*, *B. megaterium* and *M. luteus*, respectively. Thus *C. elegans* food preference appears to correlate with fitness, with bacterial environments on which worms were most fit being preferred (Coolon et al. 2009).

18.4 Conclusions

One aim of the research program described here was to learn what genetic mechanisms function to allow organisms to respond to the rapid changes to their environment as occurs as a consequence of human activities. This is indeed a great challenge and one biologists are now beginning to tackle using interdisciplinary approaches (Reusch and Wood 2007). What relevance does an understanding of the genetic basis of nematode community responses in the grassland ecosystem have on the larger questions of organismal response to environmental change? We chose to study processes in the grassland ecosystem as it is quite sensitive to global change phenomena (Samson and Knopf 1994; Collins et al. 1998; Field and Chiariello 2000; Buckland et al. 2001; Knapp and Smith 2001; Reich et al. 2001; Briggs et al. 2005). Within that ecosystem, the nematode community has been

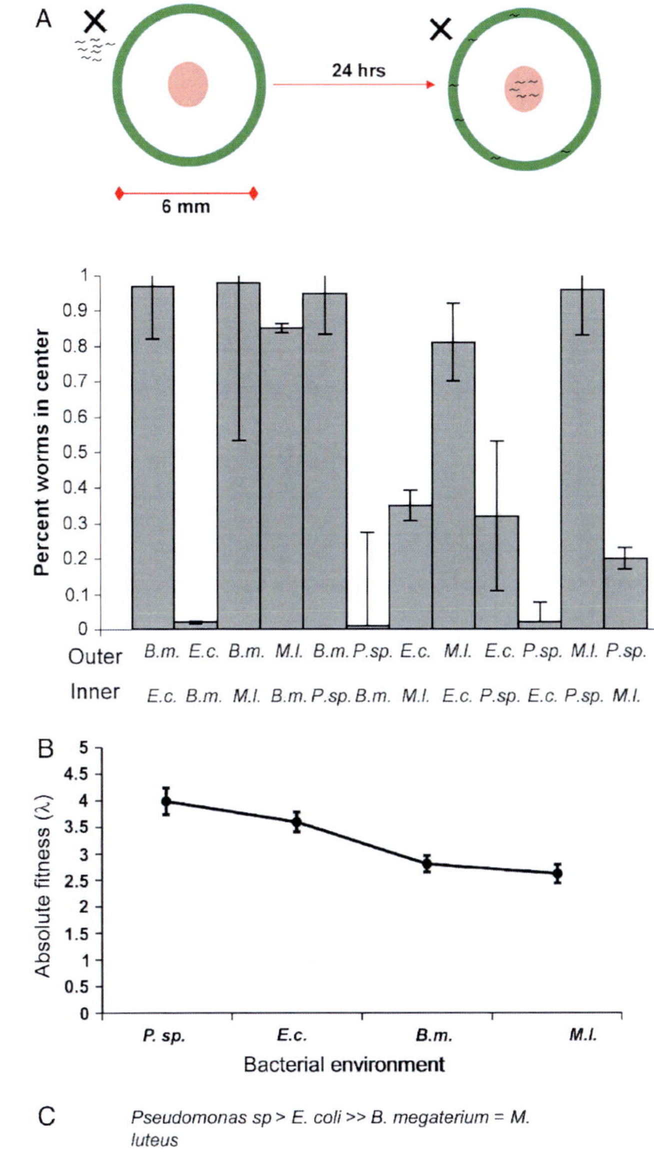

Fig. 18.5 Food preference correlates with fitness. (**a**) Food preferences of wild-type animals were measured in a biased choice assay modified from Shtonda and Avery (2006). (*Upper*) Bacteria were arrayed on an agar plate as shown. Synchronized L1 larvae were placed outside the outer circle (indicated by the X) and the fraction in the center bacterial type was determined after 24 h. (*Lower*) Fraction of nematodes in the center bacterial type is shown for all pair-wise comparisons and reciprocal comparisons were used for *Caenorhabditis elegans* food preference. Standard error for each mean is indicated with error bars. The bacteria listed under each bar were compared and are either outer (outer ring) or inner (inner circle) and B.m. = *Bacillus megaterium*, M.l. = *Micrococcus luteus*, E.c. = *Escherichia coli*, P.sp. = *Pseudomonas* sp. (**b**) Fitness (λ) of *wild-type* animals in the four bacterial environments. Error bars are SEM. (**c**) Hierarchy of food preferences and fitness are correlated

shown to be exquisitely sensitive to the relevant environmental changes and nematodes are good bioindicators of soil health (Bongers and Ferris 1999). Thus, it seems an understanding of the genetic basis of the nematode community response to environmental changes in the grasslands could be important to help us understand and predict the organismal response to global change. Indeed, we have been able to apply high-throughput molecular techniques to document changes in both the nematode and bacterial communities in response to changes in nutrient availability.

The main challenge in identifying the gene functions responsible for these changes in the native nematodes is the lack of available genetic tools. The approach we have taken is to model aspects of changes in the biotic environment using a genetically tractable laboratory nematode, *C. elegans*. To this end, we identified candidate genes that are differentially expressed in response to changes in the bacterial environment and biologically validated our approach by determining gene functions that affect fitness, lifespan and innate immunity. We also found that the hierarchy of food preference for the four bacterial isolates mirrored the trend observed for fitness in the different bacterial environments. This suggests that *C. elegans* prefers the environment in which it will be most fit. It will be interesting to see how *C. elegans* makes this choice and ultimately maximizes fitness. As we have observed that native soil nematodes differ in their susceptibility to the different bacteria in terms of infection/colonization (Coolon and Herman, unpublished data), pathogenicity might also contribute to soil nematode community structure. Taken together we suggest that the expression of metabolism and defense functions may in part drive nematode community dynamics in grassland soil systems.

The next challenge is to determine which gene functions are used in the native soil nematodes to respond to changes in the biotic environment. Since we have discovered several candidate genes in *C. elegans*, one approach is to identify homologs of these genes in the native nematodes and test their functions. While this is feasible, the major impediment to these studies is that the genomes of the relevant nematodes have not been characterized. However, the application of new sequencing methods will allow us to more readily obtain genome sequences for ecological relevant organisms. This promises to begin to close the tractability gap between model versus non-model organisms. An important aspect of this approach will be to be able to test gene function in the native nematodes. While RNA interference (RNAi) works well in *C. elegans* and some other nematode species, it does not work in all and one cannot predict its efficacy based upon phylogenetic relationships (Felix 2008). Thus in cases in which RNAi does not work, other methods will have to be employed.

Another aim of the ecological genomic approach is to better understand genome function in a well-studied genetic organism, which despite decades of research remains largely uncharacterized. The examination of *C. elegans* genome function in new environments uncovered new roles for previously studied genes as well as genes that had not been shown to have a function under standard laboratory conditions. We suggest that only through use of alternate environments does the detailed dissection of genomes become possible. Thus, it is clear that we are already reaping the benefits of the ecological genomic approach by further characterizing genome function of well characterized models. However, work still needs to be done for our ecological genomic approach to identify gene functions that can predict the responses of native organisms to environmental changes. While the challenge is great, we are confident the application of ecological genomic approaches will produce major contributions to understanding organismal responses to global environmental change.

Acknowledgements Thanks to the members of the Ecological Genomics Institute at Kansas State University for discussions and helpful comments. The project was supported by grant number P20RR016475 from the National Center for Research Resources (NCRR), a component of the National Institutes of Health (NIH), NSF grant number 0723862 and KSU Targeted Excellence.

References

Baer SG, Kitchen DJ, Blair JM, Rice CW (2002) Changes in ecosystem structure and function along a chronosequence of restored grasslands. Ecol Appl 12(6):1688–1701

Bais HP, Vepachedu R, Gilroy S, Callaway RM, Vivanco JM (2003) Allelopathy and exotic plant invasion: from molecules and genes to species interactions. Science 301(5638):1377–1380

Baker GC, Smith JJ, Cowan DA (2003) Review and re-analysis of domain-specific 16S primers. J Microbiol Meth 55(3):541–555

Blair JM, Todd TC, Callaham J MA (2000) Responses of grassland soil invertebrates to natural and anthropogenic disturbances. In: Coleman DC, Hendrix PF (eds) Invertebrates as webmasters in ecosystems. CAB International, Wallingford, UK, pp 43–71

Blaxter M (1998) *Caenorhabditis elegans* is a nematode. Science 282(5396):2041–2046

Buckland SM, Thompson K, Hodgson JG, Grime JP (2001) Grassland invasions: effects of manipulations of climate and management. J Appl Ecol 38:301–309

Bradshaw WE, Holzapfel CM (2008) Genetic response to rapid climate change: it's seasonal timing that matters. Mol Ecol 17, 157–166

Briggs JM, Knapp AK, Blair JM, Heisler JL, Hoch GA, Lett MS, McCarron JK (2005) An ecosystem in transition. Causes and consequences of the conversion of mesic grassland to shrubland. Bioscience 55(3):243–254

Bongers T, Ferris H (1999) Nematode community structure as a bioindicator of environmental monitoring. TREE 14:224–228

Carroll SP, Hendry AP, Reznick DN, Fox CW (2007) Evolution on ecological time-scales. Funct Ecol 21:387–393

Collins SL, Knapp AK, Briggs JM, Blair JM, Steinauer EM (1998) Modulation of diversity by grazing and mowing in native tallgrass prairie. Science 280(5364):745–747

Coleman DC, Edwards AL, Belsky AJ, Mwonga S (1991) The Distribution and Abundance of Soil Nematodes in East-African Savannas. Biol Fertil Soils 12(1):67–72

Coolon JD, Jones KL, Todd TC, Carr B, Herman MA (2009) *Caenorhabditis elegans* genomic response to soil bacteria predicts environment-specific genetic effects on life history traits. PLoS Genet 5(6):e1000503. doi:10.1371/journal.pgen.1000503

Curry JP (1994) Grassland invertebrates. Ecology, influences on soil fertility and effects on plant growth. Chapman & Hall, New York

Daily GC (1997) The potential impacts of global warming on managed and natural ecosystem: Implications for human well-being. Abstr Pap Am Chem Soc 213:12-ENVR

Dionisi HM, Layton AC, Harms G, Gregory IR, Robinson KG, Sayler GS (2002) Quantification of *Nitrosomonas oligotropha*-like ammonia-oxidizing bacteria and *Nitrospira* spp. from full-scale wastewater treatment plants by competitive PCR. Appl Environ Microbiol 68:245–253

Dobson AP (1997) Hopes for the future: restoration ecology and conservation biology. Science 277(5325):515–522

Falconer DS, Mackay TF (1996) Quantitative genetics: Longman Harrow, Essex, UK/New York

Felix MA (2008) RNA interference in nematodes and the changes that favored Sydney Brenner. J Biol 7:34

Field CB, Chiariello NR (2000) Global change and the terrestrial carbon cycle: the Jasper Ridge CO2 experiment. In: Ernst WG (ed) Earth systems: processes and issues. Cambridge University Press, Stanford, CA, pp 297–314

Fire A, Xu S, Montgomery MK, Kostas SA, Driver SE, Mello CC (1998) Potent and specific genetic interference by double-stranded RNA in *Caenorhabditis elegans*. Nature 391(6669):806–811

Foley JA (2005) Global consequences of land use. Science 309:570–574

Franks SJ, Sim S, Weis AE (2007) Rapid evolution of flowering time by an annual plant in response to a climate fluctuation. PNAS 104:1278–1282

Freckman DW (1988) Bacterivorous nematodes and organic-matter decomposition. Agric Ecosyst Environ 24:195–217

Freckman DW, Ettema CH (1993) Assessing nematode communities in agroecosystems of varying human intervention. Agric Ecosyst Environ 45:239–261

Garcia F, Rice CW (1994) Microbial biomass dynamics in tallgrass prairie. Soil Sci Soc Am J 58:816–823

Hannah L (1995) Human disturbance and natural habitat—a biome level analysis of a global data set. Biodivers Conserv 4(2):128–155

Hiraishi A, Ueda Y (1994) *Rhodoplanes* gen. nov., a new genus of phototrophic bacteria including *Rhodopseudomonas rosea* as *Rhodoplanes roseus* comb. nov. and *Rhodoplanes elegans* sp. nov. Int J Syst Bacteriol 44:665–673

Holterman M, van der Wurff A, van den Elsen S, van Megen H, Bongers T, Holovachov O et al. (2006) Phylum-wide analysis of SSU rDNA reveals deep phylogenetic relationships among nematodes and accelerated evolution toward crown clades. Mol Biol Evol 23 (9):1792–1800

Hsu AL, Murphy CT, Kenyon C (2003) Regulation of aging and age-related disease by DAF-16 and heat-shock factor. Science 300:1142–1145

IPCC Report (2007) Climate change 2007: physical science basis. Summary for policy makers approved at the 10th session of Working Group I at the IPCC. IPCC, Paris, France, February 2007

Jones KL Todd TC, Herman MA (2006a) Development of taxon-specific markers for high-throughput screening of microbial-feeding nematodes. Mol Ecol Notes 6:712–714

Jones KL, Todd TC, Wall-Beam JL, Coolon JD, Blair JM, Herman MA (2006b) Molecular approach for assessing responses of microbial-feeding nematodes to burning and chronic nitrogen enrichment in a native grassland. Mol Ecol 15(9):2601–2609

Juretschko S, Timmermann G, Schmid M, Schleifer KH, Pommerening-Roser A, Koops HP, Wagner M (1998) Combined molecular and conventional analyses of nitrifying bacterium diversity in activated sludge: Nitrosococcus mobilis and Nitrospira-like bacteria as dominant populations. Appl Environ Microbiol 64:3042–3051

Kaneko T, Nakamura Y, Sato S, et al. (2002) Complete genomic sequence of nitrogen-fixing symbiotic bacterium Bradyrhizobium japonicum USDA110. DNA Res 9:189–197

Kessler A, Halitschke R, Baldwin IT (2004) Silencing the jasmonate cascade: induced plant defenses and insect populations. Science 305(5684):665–668

Knapp AK, et al. (1998) Grassland dynamics: long-term ecological research in tallgrass prairie. Oxford University Press New York

Knapp AK, Smith MD (2001) Variation among biomes in temporal dynamics of aboveground primary production. Science 291(5503):481–484

Konstantinidis KT, Ramette A, Tiedje JM (2006) Toward a more robust assessment of intraspecies diversity, using fewer genetic markers. Appl Environ Microbiol 72:7286–7293

National Science Board (2000) Task Force on the Environment. Report 00–22

Parkinson J, Mitreva M, Whitton C, Thomson M, Daub J, Martin J et al. (2004) A transcriptomic analysis of the phylum Nematoda. Nat Genet 36(12):1259–1267

Purkhold U, Pommerening-Roser A, Juretschko S, Schmid MC, Koops HP, Wagner M (2000) Phylogeny of all recognized species of ammonia oxidizers based on comparative 16S rRNA and amoA sequence analysis: Implications for molecular diversity surveys. Appl Environ Microbiol 66:5368–5382

Regan JM, Harrington GW, Noguera DR (2002) Ammonia- and nitrite-oxidizing bacterial communities in a pilot-scale chloraminated drinking water distribution system. Appl Environ Microbiol 68:73–81

Reich PB, Knops J, Tilman D, Craine J, Ellsworth D, Tjoelker M, et al (2001) Plant diversity enhances ecosystem responses to elevated CO2 and nitrogen deposition. Nature 410 (6830):809–812

Reusch, TBH, Wood TE (2007) Molecular ecology of global change. Mol Ecol 16:3973–3992

Roberts S, Feder M (2000) Changing fitness consequences of hsp70 copy number in transgenic *Drosophila* larvae undergoing natural thermal stress. Funct Ecol 14(3):353–357

Samson F, Knopf F (1994) Prairie conservation in North-America. Bioscience 44(6):418–421

Seastedt TR, Todd TC, James SW (1987) Experimental manipulations of the arthropod, nematode, and earthworm communities in a North American tallgrass prairie. Pedobiologia 30:9–17

Shtonda BB, Avery L (2006) Dietary choice behavior in *Caenorhabditis elegans*. J Exp Biol 209:89–102

Smith, TB, Bernatchez L (2008) Evolutionary change in human-altered environments. Mol Ecol 17:1–8.

Stockwell CA, Hendry AP, Kinnison MT (2003) Contemporary evolution meets conservation biology. TREE 18(2):94–101

Tan MW, Ausubel FM (2000) *Caenorhabditis elegans*: a model genetic host to study *Pseudomonas aeruginosa* pathogenesis. Curr Opin Microbiol 3:29–34

Tinghitella, RM (2008) Rapid evolutionary change in a sexual signal: genetic control of the mutation "flatwing" that renders male field crickets *(Teleogryllus oceanicus)* mute. Heredity 100:261–267

Todd TC (1996) Effects of management practices on nematode community structure in tallgrass prairie. Appl Soil Ecol 3(3):235–246

Todd TC, Blair JM, Milliken GA (1999) Effects of altered soil-water availability on a tallgrass prairie nematode community. Appl Soil Ecol 13(1):45–55

Tsuboi D, Qadota H, Kasuya K, Amano M, Kaibuchi K (2002) Isolation of the interacting molecules with GEX-3 by a novel functional screening. Biochem Biophys Res Commun 292:697–701

Weinig C, Ungerer MC, Dorn LA, Kane NC, Toyonaga Y, Halldorsdottir SS, Makkay TF, Purugganan MD, Schmitt J (2002) Novel loci control variation in reproductive timing in Arabidopsis thaliana in natural environments. Genetics 162(4):1875–1884

Xie C-H, Yokota A (2006) *Sphingomonas azotifigens* sp. nov., a nitrogen-fixing bacterium isolated from the roots of *Oryza sativa*. Int J Syst Evol Microbiol 56:889–893

Yeates GW (2003) Nematodes as soil indicators: functional and biodiversity aspects. Biol Fertil Soils 37(4):199–210

Chapter 19
Comparative Evolutionary Histories of Fungal Chitinases

Magnus Karlsson and Jan Stenlid

Abstract Gene duplication and loss play important roles in the evolution of novel functions and for shaping an organism's gene content. Recently, it was suggested that stress-related genes are frequently exposed to duplications and losses, while growth-related genes show selection against change in copy number. The fungal chitinase gene family constitutes an interesting case study of gene duplication and loss, as their biological roles include growth and development as well as more stress-responsive functions.

We employ a stochastic birth and death model to show that the fungal chitinase gene family evolves non-randomly and we identify eight fungal lineages where larger-than-expected expansions or contractions potentially indicate the action of adaptive natural selection. The result shows that chitinases potentially involved in antagonistic fungal–fungal interactions have expanded in copy numbers in soil borne ascomycetes.

19.1 Introduction

Gene duplications and losses are important evolutionary processes that contribute to phenotypic differentiation between and within species. The rapidly increasing number of whole-genome sequences has made it possible to investigate large-scale genomic differences between species, rather than focusing on comparisons of single orthologous sequences. The fate of a duplicated gene is highly constrained by the functional properties of the protein. Tolerance to duplications and losses is

M. Karlsson and J. Stenlid
Department of Forest Mycology and Pathology, Swedish University of Agricultural Sciences, P.O. 7026, SE-75007, Uppsala, Sweden
email: Magnus.Karlsson@mykopat.slu.se

P. Pontarotti (ed.), *Evolutionary Biology: Concept, Modeling, and Application*,
DOI: 10.1007/978-3-642-00952-5_19,

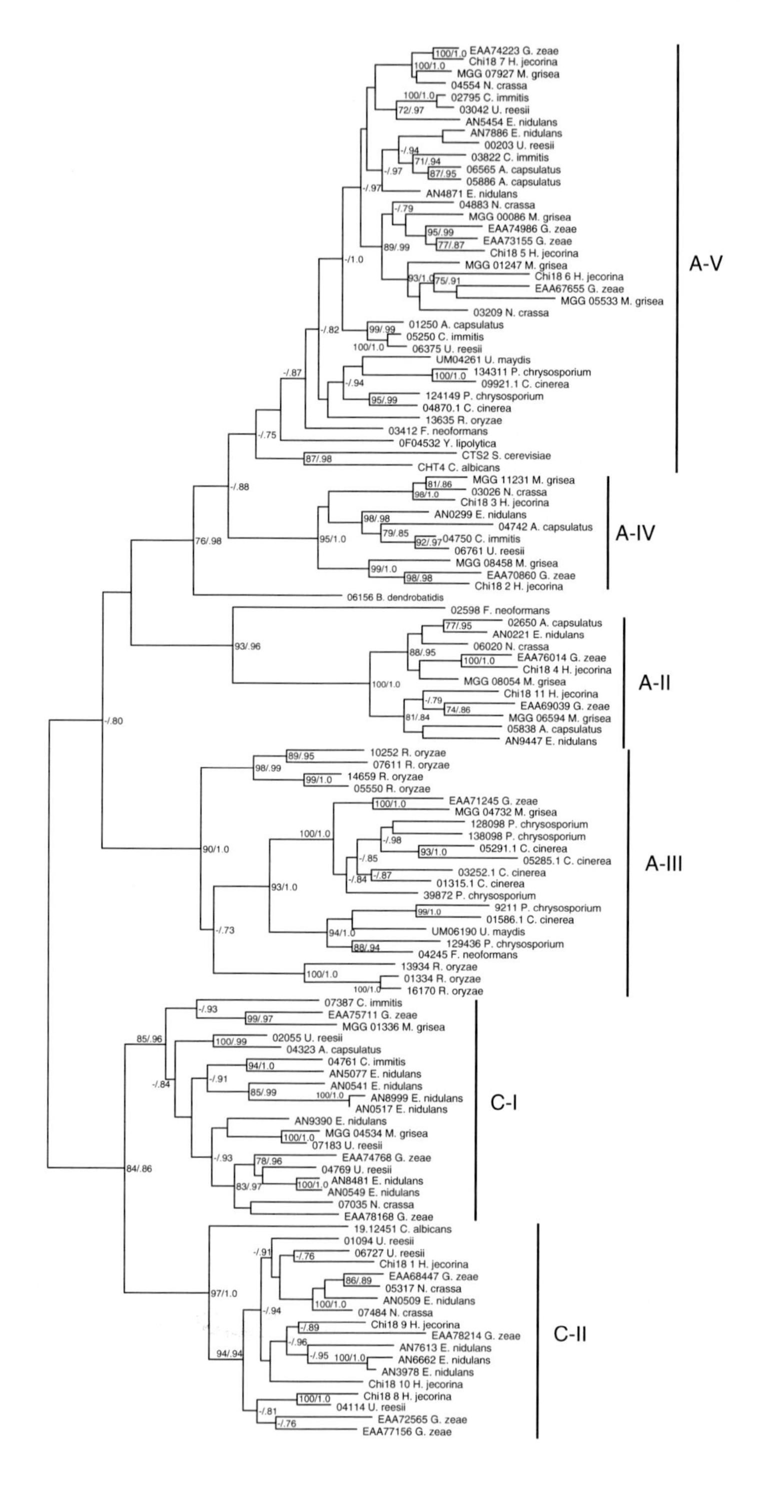
EAA74223 G. zeae
Chi18 7 H. jecorina
MGG 07927 M. grisea
04554 N. crassa
02795 C. immitis
03042 U. reesii
AN5454 E. nidulans
AN7886 E. nidulans
00203 U. reesii
03822 C. immitis
06565 A. capsulatus
05886 A. capsulatus
AN4871 E. nidulans
04883 N. crassa
MGG 00086 M. grisea
EAA74986 G. zeae
EAA73155 G. zeae
Chi18 5 H. jecorina
MGG 01247 M. grisea
Chi18 6 H. jecorina
EAA67655 G. zeae
MGG 05533 M. grisea
03209 N. crassa
01250 A. capsulatus
05250 C. immitis
06375 U. reesii
UM04261 U. maydis
134311 P. chrysosporium
09921.1 C. cinerea
124149 P. chrysosporium
04870.1 C. cinerea
13635 R. oryzae
03412 F. neoformans
0F04532 Y. lipolytica
CTS2 S. cerevisiae
CHT4 C. albicans
A-V
MGG 11231 M. grisea
03026 N. crassa
Chi18 3 H. jecorina
AN0299 E. nidulans
04742 A. capsulatus
04750 C. immitis
06761 U. reesii
MGG 08458 M. grisea
EAA70860 G. zeae
Chi18 2 H. jecorina
A-IV
06156 B. dendrobatidis
02598 F. neoformans
02650 A. capsulatus
AN0221 E. nidulans
06020 N. crassa
EAA76014 G. zeae
Chi18 4 H. jecorina
MGG 08054 M. grisea
Chi18 11 H. jecorina
EAA69039 G. zeae
MGG 06594 M. grisea
05838 A. capsulatus
AN9447 E. nidulans
A-II
10252 R. oryzae
07611 R. oryzae
14659 R. oryzae
05550 R. oryzae
EAA71245 G. zeae
MGG 04732 M. grisea
128098 P. chrysosporium
138098 P. chrysosporium
05291.1 C. cinerea
05285.1 C. cinerea
03252.1 C. cinerea
01315.1 C. cinerea
39872 P. chrysosporium
9211 P. chrysosporium
01586.1 C. cinerea
UM06190 U. maydis
129436 P. chrysosporium
04245 F. neoformans
13934 R. oryzae
01334 R. oryzae
16170 R. oryzae
A-III
07387 C. immitis
EAA75711 G. zeae
MGG 01336 M. grisea
02055 U. reesii
04323 A. capsulatus
04761 C. immitis
AN5077 E. nidulans
AN0541 E. nidulans
AN8999 E. nidulans
AN0517 E. nidulans
AN9390 E. nidulans
MGG 04534 M. grisea
07183 U. reesii
EAA74768 G. zeae
04769 U. reesii
AN8481 E. nidulans
AN0549 E. nidulans
07035 N. crassa
EAA78168 G. zeae
C-I
19.12451 C. albicans
01094 U. reesii
06727 U. reesii
Chi18 1 H. jecorina
EAA68447 G. zeae
05317 N. crassa
AN0509 E. nidulans
07484 N. crassa
Chi18 9 H. jecorina
EAA78214 G. zeae
AN7613 E. nidulans
AN6662 E. nidulans
AN3978 E. nidulans
Chi18 10 H. jecorina
Chi18 8 H. jecorina
04114 U. reesii
EAA72565 G. zeae
EAA77156 G. zeae
C-II

suggested to follow a bipolar principle, where stress-responsive genes can be frequently exposed to copy-number changes throughout evolution while genes that are essential for growth and development show selection against such changes (Wapinski et al. 2007). The underlying idea being that copy-number changes of stress-related genes is beneficial as it allows adaptations to diverse ecological niches, while the high functional constrains on growth-related genes maintain copy numbers. Several methods for studying gene family evolution are based on stochastic birth and death models of gain and loss (Gu and Zhang 2004; Reed and Hughes 2004; Hahn et al. 2005). Studies of comparative gene family evolution are now an important part of genomics and are useful for identifying ecological parameters that contributes to phenotypic differentiation between species, including mammals (Demuth et al. 2006; Niimura and Nei 2007), insects (Hahn et al. 2007) and fungi (Hahn et al. 2005; Wapinski et al. 2007; Martin et al. 2008; Karlsson and Stenlid 2008).

19.2 The Fungal Chitinase Gene Family

19.2.1 Phylogeny and Nomenclature of Fungal Chitinases

Chitin is a biopolymer consisting of *N*-acetylglucosamine monomers (GlcNAc), linked by β-1,4-glucosidic bonds. It is widely distributed in nature and it is a constituent of the exoskeleton of invertebrates, zooplankton and fungal cell walls. Chitinases (EC 3.2.1.14) hydrolyze the bonds between GlcNAc residues releasing oligomeric, dimeric (chitobiose) or monomeric (GlcNAc) products. Chitinases are divided into two different glycoside hydrolase families (18 and 19) based on amino acid sequence similarity (Henrissat 1991; Henrissat and Bairoch 1993). These two families share limited similarity at the amino acid level and have different three-dimensional structures and modes of action (Iseli et al. 1996). These enzymes can display either exo- or endoactivity, depending on the structure of the catalytic site (Terwisscha van Scheltinga et al. 1994; van Aalten et al. 2000, 2001). Fungal chitinases belong to the glycoside hydrolase family 18 and can be divided into two main clusters (A and B) (Figs. 19.1 and 19.2), subdivided into groups

Fig. 19.1 Phylogenetic relationships of fungal cluster A family 18 glycoside hydrolase catalytic domains. Phylogenetic analyses were performed using maximum likelihood methods as implemented in PhyML-aLRT, based on an alignment of family 18 glycoside hydrolase catalytic domain amino acid sequences. Branch support values (bootstrap proportions/approximate likelihood-ratio test probabilities) are associated with nodes, with a dash indicating that the support was <70%/0.70. The bar marker indicates the number of amino acid substitutions. Protein identifiers include protein name, GenBank accession nos. or locus/protein ID from the respective genome projects. Group names are indicated, see text for reference. Cluster B chitinase Chi18–12 from *H. jecorina* was used as an outgroup (not shown) (Adapted from Karlsson and Stenlid 2008)

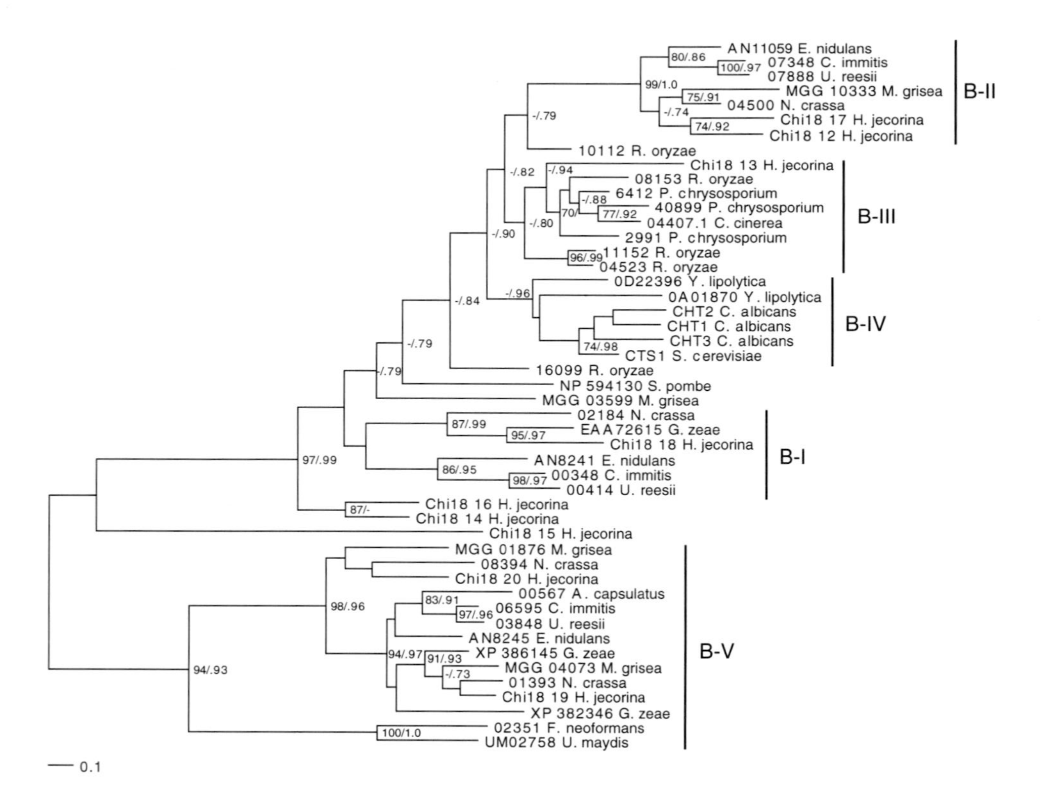

Fig. 19.2 Phylogenetic relationships of fungal cluster B family 18 glycoside hydrolase catalytic domains. Phylogenetic analyses were performed using maximum likelihood methods as implemented in PhyML-aLRT, based on an alignment of family 18 glycoside hydrolase catalytic domain amino acid sequences. Branch support values (bootstrap proportions/approximate likelihood-ratio test probabilities) are associated with nodes, with a dash indicating that the support was <70%/0.70. The bar marker indicates the number of amino acid substitutions. Protein identifiers include protein name, GenBank accession nos. or locus/protein ID from the respective genome projects. Group names are indicated, see text for reference. Cluster A chitinase 06156 from *Batrachochytrium dendrobatidis* was used as an outgroup (not shown) (Adapted from Karlsson and Stenlid 2008)

(Seidl et al. 2005; Karlsson and Stenlid 2008). Fungal chitinases cluster in parallel with bacterial and plant representatives in the main A and B chitinase clusters, hence fungal chitinases do not constitute a monophyletic group (Karlsson and Stenlid 2009). The number of chitinase genes in a single species is highly variable, from 1 in *Schizosaccharomyces pombe* and *Batrachochytrium dendrobatidis* to 34 in *Hypocrea virens*.

19.2.2 Structure of Fungal Chitinases

Chitinases consist in discrete domains, which are variously arranged in different orders in different proteins (Gilkes et al. 1991; Warren 1996; Henrissat and Davies 2000). Besides the catalytic domain there is very often a substrate-binding domain present. These substrate-binding domains are not necessary for chitinolytic activity, although they seem to enhance the efficiency of the enzymes (Suzuki et al. 1999; Limon et al. 2001). These domains can be connected to the catalytic domain by linkers, allowing the domains to function and evolve independently. Evolution of the different domains does not always follow taxonomy, which indicates that domain shuffling is a prominent factor in the evolution of these enzymes (Svitil and Kirchman 1998; Saito et al. 2003). Many chitinases also contain a signal peptide, which targets the protein for secretion or to a subcellular localization (Seidl et al. 2005; Seidl 2008). Subgroups of fungal chitinases are characterized by differences in domain-structure, with chitin-binding type 1 domains present in groups A-II, C-I and C-II, 2-S globulin domains present in group A-IV, peptidoglycan-binding LysM domains present in group C-II, fungal cellulose-binding domains present in group B-II and carbohydrate-binding type V/XII domains present in group B-III (Seidl et al. 2005; Seidl 2008; Karlsson and Stenlid 2008, 2009).

19.2.3 Functions of Fungal Chitinases

Growth and morphological development of fungi makes cell wall remodeling a necessity. Cell expansion and division, spore germination, hyphal branching and septum formation all depend on the activities of hydrolytic enzymes intimately associated with the fungal cell wall, among them chitinases (Adams 2004; Seidl 2008). Chitinases are also implied in autolysis and recycling of older parts of the fungal mycelia (Duo-Chuan 2006). Chitinases also have aggressive roles as fungal pathogenicity factors during infection of other fungi (mycoparasitism), insects and nematodes (Wattanalai et al. 2004; Duo-Chuan 2006; Gan et al. 2007a). Furthermore, chitinases are involved in degradation of chitin for nutritional needs (Duo-Chuan 2006; Lindahl and Finlay 2006). Lysis of the host cell wall and degradation of nematode egg shells are shown to be important steps in the mycoparasitic and nematophagous attack (Howell 2003; Benitez et al. 2004; Gan et al. 2007a) and hence chitinases from various fungi used as biocontrol agents have been cloned and characterized (Felse and Panda 1999; Hoell et al. 2005; Klemsdal et al. 2006; Gan et al. 2007a, b; Dong et al. 2007). In summary, accumulating data from chitinase phylogeny, domain-structure and expression patterns supports the view that these diverse proteins are involved in multiple biological roles during different parts of the fungal life cycle (Seidl 2008).

19.3 Chitinase Gene Family Evolution

The fungal chitinase gene family provides an interesting case study of gene duplication and loss, as the biological roles of the enzymes include growth and development as well as more stress-responsive functions. Hence it is possible to test the hypothesis that growth-related genes display selection against duplications and losses while stress-related genes tolerate more changes in copy number, within a single gene family.

19.3.1 Analysis of Gene Gain and Loss

In order to statistically test whether the size of the fungal chitinase gene family is compatible with a stochastic birth and death model we used the program CAFE (Computational Analysis of gene Family Evolution) (De Bie et al. 2006), which is based on the probabilistic framework developed by Hahn et al. (2005). From a specified phylogenetic tree and the gene family size in extant species, we inferred the most likely gene family size at internal nodes, tested for accelerated rates of gene family expansions or contractions and identified the branches that are responsible for the non-random evolution.

The fungal chitinase gene family data in extant species that were used in the current analysis are found in Table 19.1. Fungal chitinases can be divided into two major phylogenetic clusters, A and B and further subdivisions within these clusters are made (Seidl et al. 2005; Karlsson and Stenlid 2008). Therefore we analyzed the data in three ways; cluster A chitinases separately, cluster B chitinases separately and all chitinases merged. CAFE assumes that the gene family under study is present in the most recent common ancestor of all taxa included in the analysis. Therefore *B. dendrobatidis* was excluded from the analysis of cluster B chitinases. The phylogenetic relationships between the species that were included in the analysis are shown in Fig. 19.3, with branch lengths in millions of years. Phylogenetic relationships and estimations of divergence times were taken from previous publications (Bowman et al. 1996; Kasuga et al. 2002; Padovan et al. 2005; Taylor and Berbee 2006), assuming that the Devonian ascomycete *Paleopyrenomycites*

Table 19.1 Number of chitinase genes in different fungal species (Adapted from Karlsson and Stenlid 2008)

Species	Class	Cluster A chitinase genes	Cluster B chitinase genes	Total no. of chitinase genes
Batrachochytrium dendrobatidis	Chytridiomycetes	1	0	1
Rhizopus oryzae	Mucormycotina	9	6	15
Schizosaccharomyces pombe	Schizosaccharomycetes	0	1	1
Yarrowia lipolytica	Saccharomycetes	1	2	3
Candida albicans	Saccharomycetes	2	3	5
Saccharomyces cerevisiae	Saccharomycetes	1	1	2
Emericella nidulans	Eurotiomycetes	17	3	20
Ajellomyces capsulatus	Eurotiomycetes	7	2	9
Uncinocarpus reesii	Eurotiomycetes	11	3	14
Coccidioides immitis	Eurotiomycetes	6	3	9
Gibberella zeae	Sordariomycetes	16	3	19
Hypocrea jecorina	Sordariomycetes	11	9	20
Magnaporthe grisea	Sordariomycetes	11	4	15
Neurospora crassa	Sordariomycetes	8	4	12
Ustilago maydis	Ustilaginomycetes	2	1	3
Filobasidiella neoformans	Tremellomycetes	3	1	4
Coprinopsis cinerea	Agaricomycetes	7	1	8
Phanerochaete chrysosporium	Agaricomycetes	7	3	10

devonicus (Taylor et al. 2005) represents Pezizomycotina (Taylor and Berbee 2006), which gives an estimated age of 923 millions of years for the fungal phylum.

Alternative estimates of divergence times can be made by assuming that *P. devonicus* represents Sordariomycetes (estimated age of the fungal phylum at 1,630 millions of years) or Ascomycota (estimated age of the fungal phylum at 495 millions of years) as outlined in Taylor and Berbee (2006), although these alternative estimates resulted in more improbable age estimates when compared with age estimates in other phyla. However, these alternative estimates were included in the analysis although *Coccidioides immitis*, *Uncinocarpus reesii*, *Ajellomyces capsulatus* and *B. dendrobatidis* were excluded because of incompatibility of divergence estimates (Bowman et al. 1996; Kasuga et al. 2002; Padovan et al. 2005; Taylor and Berbee 2006).

The birth and death parameter (λ) was estimated from the data (De Bie et al. 2006) and was 0.001 for all data sets. *P*-values were computed using 1,000 re-samplings and identification of the branch, which was the most likely cause of deviations from a random model, was determined by Viterbi, branch-cutting and likelihood-ratio test procedures (De Bie et al. 2006). We considered *P*-values ≤ 0.05 or likelihood ratios above 50 to be significant for branch identification.

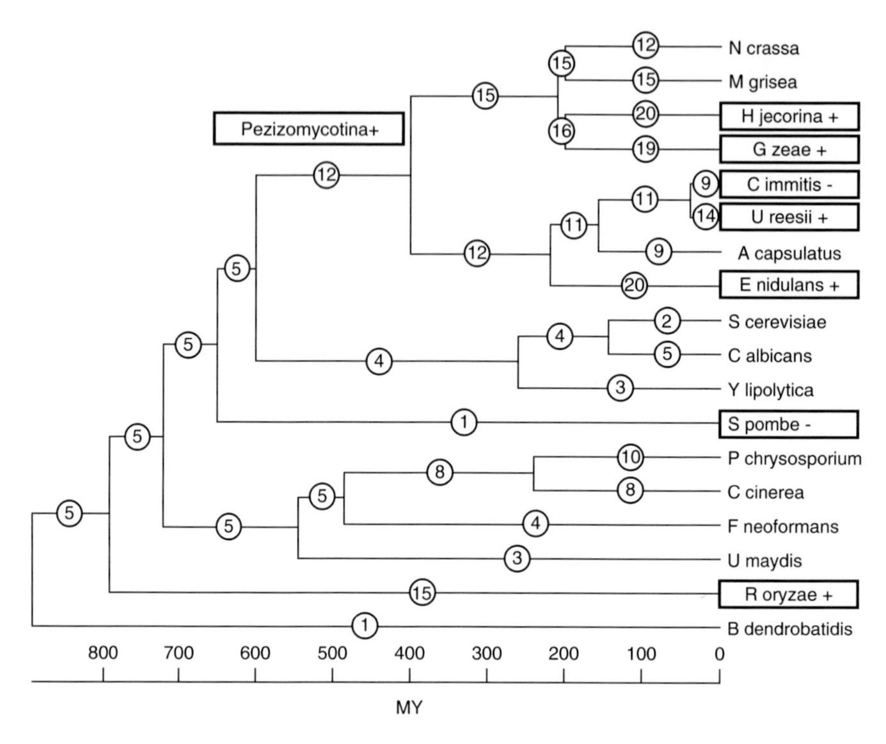

Fig. 19.3 Distribution of chitinase gain and loss among fungal lineages. Phylogenetic relationships among the fungal species used in the current study are shown, including divergence dates in millions of years (Taylor and Berbee 2006). Circled numbers represent total number of chitinase genes in extant species and estimates of total number of chitinase genes for ancestral species. Boxed taxon names indicate a significant (P-values ≤ 0.05 or likelihood ratios > 50) expansion (+), or a significant contraction (−) of the chitinase gene family size (Adapted from Karlsson and Stenlid 2008)

19.3.2 Expansions and Contractions of Chitinases

To investigate the evolutionary change of the number of chitinase genes in fungi, we estimated the number of genes in ancestral species and the number of gene gains and losses for each branch of the phylogenetic tree of the fungal species (Fig. 19.3). The analysis showed that the fungal chitinase gene family, analyzing both cluster A and B together, as well as cluster A chitinases alone have evolved non-randomly ($P < 0.001$, Table 19.2). Six branches were identified as contributing to this non-random pattern, including five expansions and one contraction (Table 19.2). These branches included the ancestor to the Pezizomycotina clade, as well as extant species *C. immitis*, *U. reesii*, *E. nidulans*, *Gibberella zeae* and *Rhizopus oryzae*. Analysis of gene phylogenies of chitinase subgroups, identified subgroups C-I and C-II as the likely targets responsible for the non-random expansion seen in *E. nidulans* and *U. reesii* (Fig. 19.1). The contraction in *C. immitis* probably took

Table 19.2 Non-randomly evolving branches in the fungal chitinase gene family (Adapted from Karlsson and Stenlid 2008)

Data set	Branch ID	*P*-value[a]	Likelihood ratio[a]	Change[b]
All chitinase genes		<0.001		
	Pezizomycotina	0.005	36	7
	Coccidioides	0.032	48	−2
	Uncinocarpus	0.009	31	3
	Emericella	0.002	14	8
	Rhizopus	0.010	1	10
Cluster A chitinase genes		<0.001		
	Pezizomycotina	0.002	34	6
	Gibberella	0.026	5	5
	Coccidioides	0.002	161	−3
	Uncinocarpus	0.053	57	2
	Emericella	0.004	24	8
	Rhizopus	0.043	1	6
Cluster B chitinase genes		0.568		
	Hypocrea	0.002	14	5
	Rhizopus	0.018	1	5

[a]See De Bie et al. (2006) for reference.
[b]Gene family size change as compared with the most recent ancestor.

place in subgroup C-II (Fig. 19.1). It was not possible to identify the target subgroup in *G. zeae*, although C-I and C-II appeared to have expanded compared to other Sordariomycetes (Fig. 19.1, Seidl et al. 2005). It should be noted though, that an additional gene was also present in subgroups A-III and A-V, as compared with the closely related *H. jecorina* (Fig. 19.1, Seidl et al. 2005). The expansion seen in cluster A chitinases in *R. oryzae* took place in subgroup A-III (Fig. 19.1). Although cluster B chitinases did not display a non-random pattern of evolution as a whole ($P = 0.568$), two branches were still identified where significant expansions took place (Table 19.2). For both *H. jecorina* and *R. oryzae* this expansion took place in the large B-I/II/III/IV cluster (Fig. 19.2). Analyzing all chitinases together using *R. oryzae* as the most recent common ancestor of all taxa instead of *B. dendrobatidis* resulted in a significant contraction in *S. pombe* ($P = 0.023$).

Taylor and Berbee (2006) also published two alternative estimates of divergence times of fungal taxa, although these alternative estimates resulted in more improbable age estimates when compared with age estimates in other phyla. Analyzing the evolution of the chitinase gene family using these alternative estimates was performed to assess the robustness of the analysis to differences in divergence dates. The two more recent estimates of divergence dates (estimated age of the fungal phylum at 495 or 923 millions of years) both showed that the chitinase family have evolved non-randomly ($P < 0.001$) and identified the same branches as contributing to this non-random pattern. The oldest estimate (estimated age of the fungal phylum at 1,630 millions of years) gave no significant changes in size of the chitinase gene family.

The involvement of chitinase subgroups C-I and C-II in both expansions and contractions in ascomycetes qualified them for further study. Based on genome sequences, C-I members are characterized by Chitin-binding type 1 domains (InterPro acc. no. IPR001002), while C-II members contain LysM peptidoglycan-binding domains (InterPro acc. no. IPR002482) in addition to Chitin-binding type 1 domains. These proteins show considerable similarity with the α-subunit of the yeast killer toxin from *Kluyveromyces lactis*. The A-III members in *R. oryzae* were short and contained no other domains than the chitinase catalytic domain. The domain-structure of cluster B chitinases in *H. jecorina* has been described before (Seidl et al. 2005) and was characterized by fungal cellulose-binding domains (InterPro acc. no. IPR000254). The domain-structure of *R. oryzae* cluster B members included both carbohydrate-binding family V/XII domains (InterPro acc. no. IPR003610) and GH18 carbohydrate-binding domains (InterPro acc. no. IPR005089).

19.4 Correlations Between Gene Family Evolution and Fungal Lifestyles

An expansion in size of a particular gene family or subgroup within a gene family, such as chitinases, suggests that this gene family or subgroup has been important for the fitness of the species during evolution. The observed variation among species could possibly be attributed to differences in morphology, growth patterns, nutrient acquisition or antagonistic ability. This approach can be used to establish links between phylogeny of the chitinase gene family with the ecological role of the species and to identify specific subgroups as important evolutionary targets in specific fungal lineages.

19.4.1 Chitinase Expansions in Filamentous Ascomycetes

Filamentous ascomycetes generally possess a larger number of chitinase genes as compared with other fungal groups. This larger chitinase gene family size can possibly be attributed to a larger gene copy number in certain subgroups but more importantly to the presence of several chitinase subgroups that appear to be unique for filamentous ascomycetes (A-II, C-I, and C-II). Subgroups C-I and C-II are identified as the most likely target for the observed expansion in *E. nidulans*, *U. reesii*, *G. zeae* and the ancestor to Pezizomycotina. These chitinase genes share extensive homology with the α -subunit of the yeast killer toxin from *K. lactis*. This yeast killer toxin, zymocin, consists in three subunits (α, β and γ) where toxicity relies solely on the γ-subunit and the α - and β-subunits function in the delivery of γ inside the cell by permeabilization of the yeast cell wall and membrane (Stark et al. 1990; Magliani et al. 1997). This has led to the suggestion that the C-I and C-II chitinases are involved in a similar mechanism in aggressive fungal–fungal interactions, by permeabilization of the cell wall and membrane to enable penetration of antifungal molecules into the antagonist (Seidl et al. 2005). This role is supported by expression data; the C-II member *chi18–10* from *H. atroviridis* is only expressed during growth on fungal cell walls and during plate confrontation assays but not by carbon starvation or chitin exposure (Seidl et al. 2005). The expansion of C-I and C-II chitinases suggests that interspecific interactions are important processes for soil borne ascomycetes. It also supports the idea that genes involved in stress-related functions can tolerate, or are even under selection for, increases in copy number (Wapinski et al. 2007).

Another intriguing result is the different evolutionary trajectories of the chitinase gene family between the human pathogen *C. immitis* and the closely related *U. reesii*. An expansion of chitinase subgroup C-II in saprotrophic *U. reesii* is in contrast to a contraction in the same subgroup in the pathogenic *C. immitis*, although these species are very closely related (Bowman et al. 1996; Kasuga et al. 2002). This difference should be related to the different life-styles of the two fungi and we hypothesize that the expansion of C-II seen in *U. reesii* is a consequence of the need for antagonistic interactions with other soil-dwelling fungi. The contraction in *C. immitis* may be reflecting adaptation to a pathogenic lifestyle, where antagonistic interactions with other fungi are minimized. Another human pathogen, *A. capsulatus*, also contains the same number of chitinase genes (9) as *C. immitis* and also lacks subgroup C-II members completely, which indicate that the function performed by C-II members of the chitinase protein family is dispensable for the human pathogenic lifestyle.

The significant expansion of the fungal chitinase gene family in the ancestor of the Pezizomycotina probably reflects the emergence of the unique C-I and C-II chitinases in filamentous ascomycete fungi, although the ecological factors driving the selection for these subgroups remain obscure. Although very speculative, the emergence of C-I and C-II chitinases coincide with the estimated time of colonization of terrestrial environments by plants (Sanderson 2003), which suggests the possibility that the emergence of terrestrial plants created new ecological niches where filamentous ascomycetes could expand into to compete for space and nutrients.

Another expansion took place in cluster B chitinases in *H. jecorina*, which is closely related to mycoparasitic fungi such as *H. lixii*, *H. virens* and *H. atroviridis*. Seidl et al. (2005) reported that certain cluster B chitinase genes from *H. jecorina* have high similarity to chitinases from entomopathogens, such as *Metarhizium anisopliae*, which suggests an aggressive role of these proteins in chitin degradation. Again, there is expression data that support this role; *chi18–13* from *H. atroviridis* is up-regulated during growth on fungal cell walls and during plate confrontation assays (Seidl et al. 2005). The fact that two chitinase subgroups that are implied in aggressive fungal–fungal interactions have expanded significantly during fungal evolution suggests that interspecific antagonistic interactions are important determinants of fungal evolution, community development and functioning. This result is in line with the idea that genes involved in stress-related functions can tolerate, or are even under selection for, increases in copy number (Wapinski et al. 2007).

19.4.2 *Methodological Assumptions with Implications for Data Interpretation*

This analysis shows the usefulness of the combination of a stochastic birth and death model and phylogenetic information in a probabilistic framework for identification of lineages with unusually evolving gene families. However, the birth and death model assumes independence among individual genes. This means that any large-scale chromosome duplication, deletion, or polyploidization that acts on several gene family members at once violates the assumption of the model (Hahn et al. 2005). Interpretation of gene family size differences between taxa that are separated by genome duplications should be made with caution. There are indications of a recent polyploidization in *R. oryzae* (Taylor and Berbee 2006) that suggests the possibility that the observed non-random chitinase family size in this species may not be entirely related to adaptive selection. On the other hand, the two chitinase genes in the duplicated *S. cerevisiae* genome (Kellis et al. 2004) belong to the fundamentally different A and B clusters. This indicates that the loss of one of the duplicated *CTS1* and *CTS2* paralogues have been selected during evolution after whole-genome duplication. Cts1 and cts2 are involved in cell separation during budding and in sporulation (Kuranda and Robbins 1991; Giaever et al. 2002), which is in line with the idea that genes involved in growth-related functions are under selection against changes in copy number (Wapinski et al. 2007).

Another violation against the assumption of independence among individual genes may be seen in *Neurospora crassa*. This species has the lowest number of chitinase genes (12) among the Sordariomycetes, which may be attributed to the presence of a wide array of genome defense mechanisms, including repeat-induced point mutations, greatly slowing down the creation of new genes (Galagan et al. 2003). The low number of chitinase genes in this species is not significantly violating a random process ($P = 0.256$) but interpretation of data for other gene families in *N. crassa* should be done with caution.

19.5 Conclusions

Here we use fungal genome data in comparative work to infer phylogenetic relationships in the fungal chitinase gene family and to detect non-random expansions and contractions. This approach can be used to establish links between phylogeny of the chitinase gene family with the ecological role of the species and identify specific chitinase subgroups as important evolutionary targets in specific fungal lineages. Within the fungal chitinase gene family we observe selection against changes in copy number in chitinases involved in growth and development as well as selection for increased copy number in chitinases involved in stress-related functions, supporting the idea of a bipolar principle that governs tolerance to duplications and losses (Wapinski et al. 2007). The results also indicate that antagonistic fungal–fungal interactions constitute an important evolutionary force in soil borne ascomycetes.

References

Adams DJ (2004) Fungal cell wall chitinases and glucanases. Microbiology 150:2029–2035

Benitez T, Rincon AM, Limon MC, Codon AC (2004) Biocontrol mechanisms of *Trichoderma* strains. Int Microbiol 7:249–260

Bowman BH, White TJ, Taylor JW (1996) Human pathogeneic fungi and their close nonpathogenic relatives. Mol Phyl Evol 6:89–96

De Bie T, Cristianini N, Demuth JP, Hahn MW (2006) CAFE: a computational tool for the study of gene family evolution. Bioinformatics 22:1269–1271

Demuth JP, De Bie T, Stajich JE, Cristianini N, Hahn MW (2006) The evolution of mammalian gene families. PLoS One 1:e85

Dong LQ, Yang JK, Zhang KQ (2007) Cloning and phylogenetic analysis of the chitinase gene from the facultative pathogen *Paecilomyces lilacinus*. J Appl Microbiol 103:2476–2488

Duo-Chuan L (2006) Review of fungal chitinases. Mycopathologia 161:345–360

Felse PA, Panda T (1999) Regulation and cloning of microbial chitinase genes. App Microbiol Biotechnol 51:141–151

Galagan JE, Calvo SE, Borkovich KA, Selker EU, Read ND, et al. (2003) The genome sequence of the filamentous fungus *Neurospora crassa*. Nature 422:859–868

Gan Z, Yang J, Tao N, Liang L, Mi Q, Zhang K-Q (2007a) Cloning of the gene *Lecanicillium psalliotae* chitinase *Lpchi1* and identification of its potential role in the biocontrol of root-knot nematode *Meloidogyne incognita*. Appl Microbiol Biotechnol 76:1309–1317

Gan Z, Yang J, Tao N, Yu Z, Zhang K-Q (2007b) Cloning and expression analysis of a chitinase gene *Crchi1* from the mycoparasitic fungus *Clonostachys rosea* (syn. *Gliocladium roseum*). J Microbiol 45:422–430

Giaever G, Chu AM, Ni L, Connelly C, Riles L, et al. (2002) Functional profiling of the *Saccharomyces cerevisiae* genome. Nature 418:387–391

Gilkes NR, Henrissat B, Kilburn DG, Miller RC, Warren RAJ (1991) Domains in microbial beta-1,4-glycanases—sequence conservation, function, and enzyme families. Microbiol Rev 55:303–315

Gu X, Zhang H (2004) Genome phylogeny inference based on gene contents. Mol Biol Evol 21:1401–1408

Hahn MW, De Bie T, Stajich JE, Nguyen C, Cristianini N (2005) Estimating the tempo and mode of gene family evolution from comparative genomic data. Genome Res 15:1153–1160

Hahn MW, Han MV, Han S-G (2007) Gene family evolution across 12 *Drosophila* genomes. PLoS Genet 3:e197

Henrissat B (1991) A classification of glycosyl hydrolases based on amino acid sequence similarities. Biochem J 280:309–316

Henrissat B, Bairoch A (1993) New families in the classification of glycosyl hydrolases based on amino acid sequence similarities. Biochem J 293:781–788

Henrissat B, Davies GJ (2000) Glycoside hydrolases and glycosyltransferases. Families, modules, and implications for genomics. Plant Physiol 124:1515–1519

Hoell IA, Klemsdal SS, Vaaje-Kolstad G, Horn SJ, Eijsink VGH (2005) Overexpression and characterization of a novel chitinase from *Trichoderma atroviride* strain P1. Biochim Biophys Acta Prot Proteom 1748:180–190

Howell CR (2003) Mechanisms employed by *Trichoderma* species in the biological control of plant diseases: the history and evolution of current concepts. Plant Dis 87:4–10

Iseli B, Armand S, Boller T, Neuhaus J-M, Henrissat B (1996) Plant chitinases use two different hydrolytic mechanisms. FEBS Lett 382:186–188

Karlsson M, Stenlid J (2008) Comparative evolutionary histories of the fungal chitinase gene family reveal non-random size expansions and contractions due to adaptive natural selection. Evol Bioinform 4:47–60

Karlsson M, Stenlid J (2009) Evolution of microbial family 18 glycoside hydrolases: Diversity, domain-structures and phylogenetic relationships. J Mol Microbiol Biotechnol 16:208–223

Kasuga T, White TJ, Taylor JW (2002) Estimation of nucleotide substitution rates in eurotiomycete fungi. Mol Biol Evol 19:2318–2324

Kellis M, Birren BW, Lander ES (2004) Proof and evolutionary analysis of ancient genome duplication in the yeast *Saccharomyces cerevisiae*. Nature 428:617–624

Klemsdal SS, Clarke JHL, Hoell IA, Eijsink VGH, Brurberg MB (2006) Molecular cloning, characterization, and expression studies of a novel chitinase gene (ech30) from the mycoparasite *Trichoderma atroviride* strain P1. FEMS Microbiol Lett 256:282–289

Kuranda MJ, Robbins PW (1991) Chitinase is required for cell-separation during growth of *Saccharomyces cerevisiae*. J Biol Chem 266:19758–19767

Limon MC, Margolles-Clark E, Benitez T, Penttilä M (2001) Addition of substrate-binding domains increases substrate-binding capacity and specific activity of a chitinase from *Trichoderma harzianum*. FEMS Microbiol Lett 198:57–63

Lindahl B, Finlay R (2006) Activities of chitinolytic enzymes during primary and secondary colonization of wood by basidiomycetous fungi. New Phytol 169:389–397

Magliani W, Conti S, Gerloni M, Bertolotti D, Polonelli L (1997) Yeast killer systems. Clin Microbiol Rev 10:369–400

Martin F, Aerts A, Ahrén D, Brun A, Danchin EGJ, et al. (2008) The genome of Laccaria bicolor provides insights into mycorrhizal symbiosis. Nature 452:88–92

Niimura Y, Nei M (2007) Extensive gains and losses of olfactory receptor genes in mammalian evolution. PLoS One 8:e708

Padovan ACB, Sanson GFO, Brunstein A, Briones MRS (2005) Fungi evolution revisited: Application of the penalized likelihood method to a Bayesian fungal phylogeny provides a new perspective on phylogenetic relationships and divergence dates of ascomycota groups. J Mol Evol 60:726–735

Reed WJ, Hughes BD (2004) A model explaining the size distribution of gene and protein families. Math Biosci 189:97–102

Saito A, Fujii T, Miyashita K (2003) Distribution and evolution of chitinase genes in *Streptomyces* species: involvement of gene-duplication and domain-deletion. Antonie Van Leeuwenhoek 84:7–15

Sanderson MJ (2003) Molecular data from 27 proteins do not support a Precambrian origin of land plants. Am J Bot 90:954–956

Seidl V (2008) Chitinases of filamentous fungi: a large group of diverse proteins with multiple physiological functions. Fungal Biol Rev 22:36–42

Seidl V, Huemer B, Seiboth B, Kubicek CP (2005) A complete survey of *Trichoderma* chitinases reveals three distinct subgroups of family 18 chitinases. FEBS J 272:5923–5939

Stark MJR, Boyd A, Mileham AJ, Romanos MA (1990) The plasmid-encoded killer system of *Kluyveromyces lactis*—a review. Yeast 6:1–29

Suzuki K, Taiyoji M, Sugawara N, Nikaidou N, Henrissat B, Watanabe T (1999) The third chitinase gene (chiC) of *Serratia marcescens* 2170 and the relationship of its product to other bacterial chitinases. Biochem J 343:587–596

Svitil AL, Kirchman DL (1998) A chitin-binding domain in a marine bacterial chitinase and other microbial chitinases: implications for the ecology and evolution of 1,4-beta-glycanases. Microbiology 144:1299–1308

Taylor JW, Berbee ML (2006) Dating divergences in the fungal tree of life: review and new analyses. Mycologia 98:838–849

Taylor TN, Hass H, Kerp H, Krings M, Hanlin RT (2005) Perithecial ascomycetes from the 400 million year old Rhynie chert: an example of ancestral polymorphism (vol 96, pg 1403, 2004). Mycologia 97:269–285

Terwisscha van Scheltinga ACT, Kalk KH, Beintema JJ, Dijkstra BW (1994) Crystal-structures of hevamine, a plant defense protein with chitinase and lysozyme activity, and its complex with an inhibitor. Structure 2:1181–1189

van Aalten DMF, Komander D, Synstad B, Gåseidnes S, Peter MG, Eijsink VGH (2001) Structural insights into the catalytic mechanism of a family 18 exo-chitinase. Proc Natl Acad Sci USA 98:8979–8984

van Aalten DMF, Synstad B, Brurberg MB, Hough E, Riise BW, Eijsink VGH, Wierenga RK (2000) Structure of a two-domain chitotriosidase from *Serratia marcescens* at 1.9-angstrom resolution. Proc Natl Acad Sci USA 97:5842–5847

Wapinski I, Pfeffer A, Friedman N, Regev A (2007) Natural history and evolutionary principles of gene duplication in fungi. Nature 449:54–61

Warren RAJ (1996) Microbial hydrolysis of polysaccharides. Ann Rev Microbiol 50:183–212

Wattanalai R, Boucias DG, Tartar A (2004) Chitinase gene of the dimorphic mycopathogen, *Nomuraea rileyi*. J Invertebr Pathol 85:54–57

Chapter 20
Aging: Evolutionary Theory Meets Genomic Approaches

George L. Sutphin and Brian K. Kennedy

Abstract Modern evolutionary theory describes aging as the result of an accumulation of late-acting, deleterious genes caused by reduced force of natural selection late in life, combined with selection for genes that are beneficial early in life but damaging late in life. Theories based on this logic predict that organisms will be optimized for overall fitness as opposed to maximum longevity. Recent advances in genomics combined with large-scale methods for single gene knockout in several common aging models have allowed the first genome-wide studies of life span. These studies provide insight into several aspects of the biology of aging that relate to evolution, including the scope of cellular processes that influence longevity and the conservation of longevity determinants between organisms. Here we review the evolution of the aging field over the past several years and the implications of the move toward genomics. We also highlight key results and discuss their importance and relation to evolutionary theories of aging.

20.1 Introduction

"When one or more individuals have provided a sufficient number of successors they themselves, as consumers of nourishment in a constantly increasing degree, are an injury to those successors. Natural selection therefore weeds them out, and in many cases favors such races as die almost immediately after they have left successors" (Alfred Russel Wallace, 1865–1870).

G.L. Sutphin
Departments of Pathology, University of Washington, Seattle, WA 98195, USA; The Molecular and Cellular Biology Program, University of Washington, Seattle, WA 98195, USA
e-mail: lothos@u.washington.edu

B.K. Kennedy
Department of Biochemistry, University of Washington, Seattle, WA 98195, USA
e-mail: bkenn@u.washington.edu

P. Pontarotti (ed.), *Evolutionary Biology: Concept, Modeling, and Application*,
DOI: 10.1007/978-3-642-00952-5_20,

Aging is commonly defined as a degenerative process characterized by a progressive decline in fitness resulting in mortality. How does a process that ultimately results in the death of each individual in a species evolve by natural selection? The first attempts by evolutionary biologists to answer this question followed closely on the heels of the theory of natural selection itself. The earliest written argument was made by Alfred Russel Wallace in an informal note, the essence of which is captured in the above quote (Wallace 1889). August Weismann, a German biologist and evolutionary theorist, expanded Wallace's ideas into a theory describing aging as a programmed process to end an organism's life in the absence of accident or predation in order to make way for succeeding generations (Weismann 1889). In modern evolutionary biology, aging is viewed not as an advantageous trait that is selected for directly but as a side-effect of a decline in the force of natural selection with increasing age combined with selection for other traits. Although the early "programmed aging" theories have lost favor, they illustrate the magnitude of the problem aging posed to evolutionary theory.

Invertebrate organisms have emerged as the preeminent model systems in aging research. The most prominent are the budding yeast *Saccharomyces cerevisiae*, the nematode *Caenorhabditis elegans* and the fruit fly *Drosophila melanogaster*, which share a number of characteristics that make them ideal for aging research: relatively short life span, rapid production of large numbers of offspring, ease of maintenance and manipulation in the laboratory environment, well characterized biology, fully sequenced genomes and the availability of powerful genetic tools. A great deal of effort has gone into the identification and characterization of interventions and genetic pathways that influence longevity in these systems. Much of this work has been accomplished by looking at secondary age-associated phenotypes, such as stress resistance and fecundity, or by looking for genes associated with pathways already known to influence aging. This approach has yielded valuable insight and an understanding of specific pathways and processes that influence life span but does not inform with respect to the total number of genes and pathways that affect aging. Are there only a few aging genes or many? The past decade has seen the creation of an open reading frame (ORF) deletion collection in yeast and RNAi libraries in nematodes and fruit flies, allowing for the first time the development of unbiased methods for looking at life span on a genomic-scale.

The desire to apply findings from studies in diverse organisms to human aging raises another central question: to what degree are the molecular mechanisms involved in the determination of life span conserved between evolutionarily divergent organisms? A number of interventions have been identified that extend life span in divergent organisms, including dietary restriction (Chapman and Partridge 1996; Fabrizio et al. 2004; Good and Tatar 2001; Jiang et al. 2000; Lakowski and Hekimi 1998; Lin et al. 2000; McCay et al. 1935), reduced insulin/IGF-1-like signaling (IIS) (Bluher et al. 2003; Holzenberger et al. 2003; Kenyon et al. 1993; Tatar et al. 2001), increased sirtuin activity (Kaeberlein et al. 1999; Rogina and Helfand 2004; Tissenbaum and Guarente 2001) and reduced target of rapamycin (TOR) signaling (Kaeberlein et al. 2005; Kapahi et al. 2004; Powers et al. 2006;

Vellai et al. 2003). Results from the first genome-wide longevity studies indicate that a large number of genes are likely to play a role in longevity and provide the first quantitative evidence for evolutionary conservation of longevity determinants. Here we discuss the transition of the aging field into genome-scale research and the implications of recent findings in the context of evolutionary aging theory.

20.2 The Evolution of Aging: Why Not Immortality?

The models proposed by Wallace and Weismann describe aging as a programmed process that evolved through direct selection on senescence as a beneficial trait (Weisman 1889). These models suffer from a reliance on group theory and do not provide an explanation for why an individual with a mutation that confers increased life span—and thus an increased opportunity to produce offspring—would not be selected over individuals without such a mutation (Kirkwood 2005). This view of aging was challenged in 1952, when Peter Medawar proposed the "mutation accumulation" theory, revolutionizing the way most biologists think about the evolution of aging (Medawar 1952). The underlying reasoning is based on the observation that, even in the absence of aging or other intrinsic decline, most species experience a substantial rate of mortality from external forces such as accident, disease, or predation. For a given species, fewer individuals live to progressively older ages, diminishing the force of natural selection in an age-dependent manner and resulting in stronger selection against genes that are deleterious early in life relative to genes that are deleterious late in life (Fisher 1930; Haldane 1941). The mutation accumulation theory states that genes with late-acting deleterious effects will accumulate in the germline, resulting in an increase in mortality with age (Medawar 1946, 1952).

George C. Williams refined Medawar's reasoning by incorporating the concept of pleiotropy. In the "antagonistic pleiotropy" model of aging, age-dependent increase in mortality is caused by an accumulation of genes that function to the benefit of the organism early in life but become deleterious with advanced age, thus providing a means by which senescence can be selected for indirectly (Williams 1957). A third related theory, proposed by Thomas Kirkwood and termed "disposable soma," states that natural selection will favor genes that promote redirection of resources from maintenance of soma to reproduction, resulting in an accumulation of damage that increases with age (Kirkwood 1977).

The theories of mutation accumulation, antagonistic pleiotropy and disposable soma all represent aging as a result of negligible natural selection with advanced age rather than a programmed process. An important extension of these arguments is that natural selection is concerned with overall fitness, which does not necessarily correlate with enhanced longevity (Kirkwood and Holliday 1979). Organisms should therefore possess genes that optimize fitness and not maximize life span.

20.3 Measuring and Interpreting Life Span Phenotypes

Overall fitness may provide an explanation for why we age, but to develop an understanding of how we age, longevity is the most relevant characteristic. However, life span is a complex phenotype and interpretation can be complicated. From a technical standpoint, one issue arises from variation in the definition of life span in different organisms. At what point do we say something has died? When dealing with macroscopic animals the answer to this question is fairly straight forward. An animal is considered dead at the point of failure of the majority of its macro-scale systems (e.g. the respiratory system or the circulatory system). Individual cells that are technically still alive will only remain so for a period of time that is short relative to the life span of the animal. To an extent this is also true for smaller animals that are on the verge between the micro- and macroscopic worlds, such as worms and flies. While the failure of major systems is more difficult to judge, by general acceptance these animals are considered dead at the point when they fail to respond to external stimuli.

The path is less clear for single cell systems. *S. cerevisiae* is one of the most prominent organisms in aging research, but at what point can you say a yeast cell is dead? Two models of aging have been developed in yeast, one mitotic and one non-mitotic (Steinkraus et al. 2008) (Fig. 20.1). The first, termed replicative life span (RLS), measures the number of cell divisions an individual cell completes before undergoing replicative senescence (Mortimer and Johnston 1959). The second, termed chronological life span (CLS), measures the duration of time that a cell remains viable in a growth arrested state (Fabrizio and Longo 2003). RLS might be considered analogous to aging in mitotic tissue, such as skin and blood and CLS to aging in non-mitotic tissue, such as heart or brain, although it is unclear if this analogy, based on the proliferative potential of mammalian tissues, truly holds.

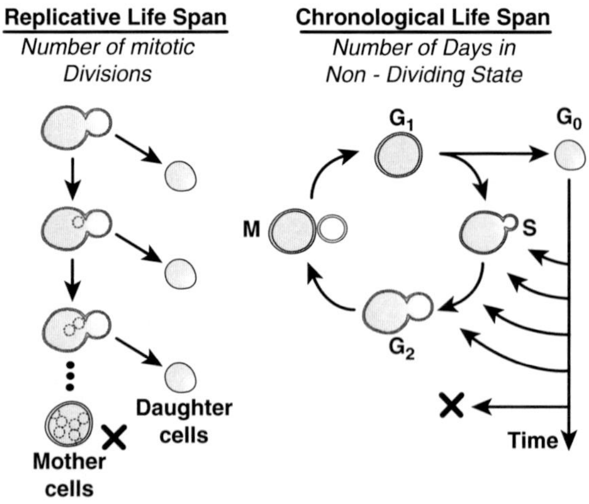

Fig. 20.1 Yeast aging models. Two models of aging are used in yeast. Replicative life span (*left*) is a measure of the number of cell divisions a cell undergoes before undergoing replicative senescence. Chronological life span (*right*) is a measure of the number of days a cell remains viable in a post-replicative state

Notably, accumulating evidence suggests a considerable degree of conservation in factors that influence aging, even between mitotic models, such as yeast replicative aging, and primarily non-mitotic models, such as *C. elegans*, in which only the germline is mitotically active during adulthood. Inter-organism conservation of longevity determinants is discussed in detail below.

A second complexity in interpreting life span relates to the inherent interdependence of the genetic pathways that converge on longevity. At least three pathways are known to modulate life span in diverse organisms: IIS, sirtuins and TOR signaling. While these pathways are at least partially independent, they interact both upstream, by responding to similar environmental queues and downstream, by influencing overlapping sets of downstream targets to regulate complex processes such as metabolism or growth. Modifying the action of even a single gene that only directly plays a role in a single pathway can potentially alter the contributions to the mortality of many other pathways (Kennedy 2008). By measuring life span, we are effectively looking at the integrated contributions from each pathway. Because of these complexities, simple analyses, such as traditional epistasis, may not always be straight forward to interpret. Take the controversial role of the yeast histone deacetylase Sir2 in DR for example. Sir2 was originally proposed as a mediator of DR based on the observations that overexpression of *SIR2* extends RLS (Kaeberlein et al. 1999) and that Sir2 is activated in a NAD-dependent manner (Lin et al. 2000). Indeed, DR was shown to be ineffective at extending RLS in yeast lacking *SIR2* (Lin et al. 2000). By classical interpretation, this result led to the conclusion that DR requires Sir2 in order to extend life span. However, it was subsequently shown that DR robustly extends RLS in strains where both *SIR2* and *FOB1* are deleted (Kaeberlein et al. 2004; Lamming et al. 2005) and in long-lived strains that either lack *FOB1* or overexpress *SIR2* (Kaeberlein et al. 2004). Thus Sir2 is not necessary for extension of life span by DR and likely controls life span via a separate mechanism. Importantly, while *SIR2* and DR provide an illustrative counterexample, classical epistasis analysis can work with respect to life span and probably does in most cases. In *C. elegans* for example, *daf-2*, the gene encoding the IIS receptor, negatively regulates the activity of *daf-16*, the gene encoding the IIS FOXO-family transcription factor. Mutations in *daf-2* dramatically increase life span and mutations in *daf-16* shorten life span relative to wild type. Epistasis analysis correctly predicts that life span extension via reduced activity of DAF-2 is blocked by mutations in DAF-16 (Kenyon et al. 1993).

A final caution applies to interpreting life span phenotypes in the context of natural selection. As discussed in Section 20.2, longevity should not be equated to fitness. In fact, in a system optimized for fitness, mutations that increase longevity will generally incur a fitness cost. However, the cost/benefit analysis is not always obvious or straightforward. For example, wild-type guppies are short-lived, grow quickly and reproduce early in the wild where predation is high. Place the same guppies in a predator free environment and they both live longer and produce more progeny (Reznick et al. 2004). The combined benefit associated with longer life and increased total reproduction is outweighed by the ability to grow and reproduce rapidly in a high mortality environment. In *C. elegans*, *age-1* (PI-3 kinase) mutants are outcompeted by wild type worms when subject to periods of starvation (Walker et al. 2000) despite increased survival when exposed to increased temperature and longer life span (Lithgow et al. 1995). Some long-lived fruit fly mutants, such as those lacking *indy*, an amino acid transporter, actually have more progeny than wild type (Rogina et al. 2000). Long-lived mutants commonly studied in aging research are not found in nature, which implies that there will be some cost associated with the beneficial phenotypes. These examples demonstrate that longevity phenotypes are often context dependent and the strictly-defined, low-variability environment of the laboratory can sometimes obscure what is really going on.

Our goal in this section is to highlight a few of the practical challenges faced when studying aging and working with phenotypes with complex origins, such as longevity. Life span remains the most relevant phenotype to aging and epistasis in the context of longevity will remain at the forefront of aging research. However, we encourage vigilance when analyzing and interpreting results.

20.4 Conservation of Longevity Control

One of the primary goals of aging research is the development of interventions to impede the aging process in an effort to fight age-associated pathologies in humans. It would be theoretically optimal to investigate interventions in mammalian systems. Unfortunately, studying longevity directly is difficult in mammals due to their long life spans. For example, life span experiments require approximately 3 years in mice, 25 years in rhesus monkeys and are ethically and functionally impractical in humans.

Invertebrate models offer many powerful advantages in the context of aging research (discussed in Section 20.1) and a substantial portion of the knowledge we possess about how and why organisms age has been generated using these models. An important consideration when interpreting evidence from invertebrate systems is relevance to human aging. Several environmental interventions, such as DR and transient heat shock, are known to influence longevity in multiple evolutionarily divergent organisms suggesting that they may do so in humans as well (Table 20.1). DR in particular has been shown to extend life span in yeast, worms, flies, mice, spiders, rats, dogs and hamsters (Kennedy et al. 2007; Masoro 2005; Weindruch and Walford 1988). There are likely a set of key environmental conditions that induce a similar set of responses—increased longevity, enhanced resistance to stress, reduced fecundity—in a wide range of organisms. However, this does not necessarily guarantee that the molecular mechanisms that mediate these responses are the same in different organisms and there are cases where it appears that certain age-associated responses, including increased life span, are mechanistically implemented in different ways in different organisms. Thus we have another key question in aging research: are the genetic pathways involved in the determination of life span evolutionarily conserved among diverse organisms? In the past several decades three pathways have emerged as conserved regulators of longevity: IIS,

Table 20.1 Conserved environmental and genetic interventions known to influence life span. Arrows indicate whether the intervention increases (↑), decreases (↓), is not applicable to (n/a) or has an unknown effect on (?) life span in each aging model

		Yeast		Worms	Flies	Mice
		Replicative	Chronological			
Environmental interventions	Dietary restriction	↑	↑	↑	↑	↑
	Transient exposure to stress	↑	?	↑	↑	↑
	Antioxidants	?	↑	?	↑	↑
Genetic interventions	Reduced IIS	n/a	n/a	↑	↑	↑
	Increased sirtuin activity	↑	↓	↑	↑	?
	Reduced TOR signaling	↑	↑	↑	↑	?

sirtuins and TOR signaling (Table 20.1). Each pathway is at least partially distinct, though there is evidence for some interaction between pathways through both influence from environmental conditions and action on downstream targets. Each pathway has been independently proposed as a potential mediator of the beneficial effects of DR.

20.4.1 IIS Promotes Aging

IIS pathways are conserved among multicellular eukaryotes and share a set of core components, including insulin-like proteins, insulin/IGF-1-like receptors, a phosphatidylinositol-3 (PI-3) kinase, an Akt kinase and a FOXO-family transcription factor. IIS inhibits activity of the FOXO-family transcription factors by preventing their nuclear localization, controlling expression of downstream target genes in response to environmental queues. Worms and flies each possess a single insulin/IGF-1-like receptor and numerous insulin-like signaling molecules (Bartke 2008; Toivonen and Partridge 2008). Reducing IIS extends life span and increases stress resistance in both species (Clancy et al. 2001; Dorman et al. 1995; Hercus et al. 2003; Kenyon et al. 1993; Martin et al. 1996; Murakami and Johnson 1996; Tatar et al. 2001; Tu et al. 2002). Unlike invertebrates, mammals possess only three insulin-like ligands—insulin, IGF-1 and IGF-2—and five dimeric insulin/IGF-1-like receptors, including different receptors for insulin and IGF-1, which form from different combinations of one insulin receptor and two IGF-1-receptor monomer subtypes (Taguchi and White 2008). Increased longevity has been linked to reduced activity of both the insulin receptor (Bluher et al. 2003) and the IGF-1 receptor (Holzenberger et al. 2003), as well as the related growth hormone signaling pathway (Brown-Borg et al. 1996; Coschigano et al. 2003; Flurkey et al. 2002).

20.4.2 Sirtuins: Playing Both Sides?

Sir2-orthologs (sirtuins) are NAD-dependent protein deacetylases that have been identified in eukaryotic species from yeast to humans (Imai et al. 2000; Landry et al. 2000; Smith et al. 2000; Tanner et al. 2000). The first evidence for a role in aging for sirtuins came with the discovery that overexpression of Sir2 is sufficient to extend replicative life span in yeast (Kaeberlein et al. 1999) and life span extension has subsequently been demonstrated for overexpression of *sir-2.1* in worms (Tissenbaum and Guarente 2001) and dSir2 in flies (Rogina and Helfand 2004). A role for sirtuins in mammalian longevity has yet to be definitely demonstrated, though transgenic mice overexpressing SIRT1 have several phenotypes that are associated with increased life span, including improved metabolic profiles (Banks et al. 2008; Bordone et al. 2007), delayed disease progression in neurodegenerative disease models (Kim et al. 2007) and reduced incidence of colon cancer (Firestein et al. 2008).

The role of sirtuins in aging has an intriguing feature. While sirtuins appear to modulate longevity in a diverse set of eukaryotic organisms, current data suggests that they do so through at least partially disparate mechanisms. In yeast, Sir2 deacetylase activity primarily targets histones and promotes silencing specifically at the ribosomal DNA (rDNA), the silent mating (HM) loci and regions near the telomeres (Aparicio et al. 1991; Bryk et al. 1997; Gottschling et al. 1990; Ivy et al. 1986; Rine and Herskowitz 1987; Smith and Boeke 1997). One cause of replicative senescence in yeast is the accumulation of extrachromosomal rDNA circles (ERCs) with age (Sinclair and Guarente 1997). Sir2 is thought to regulate replicative life span primarily by repressing ERC formation via promotion of rDNA genomic stability (Kaeberlein et al. 1999). Sirt1, the mammalian counterpart of yeast Sir2, may also affect chromatin stability through chromatin interactions (Oberdoerffer et al. 2008). However, they have a variety of other targets as well, including stress response factors and FOXO-family transcription factors, among others (Brunet et al. 2004; Gerhart-Hines et al. 2007; Luo et al. 2001; Motta et al. 2004; Rodgers et al. 2008; van der Horst et al. 2004; Vaziri et al. 2001; Viswanathan et al. 2005). There is no evidence that ERCs contribute to aging, interact with sirtuins, or even accumulate with age in multicellular eukaryotes. In worms, *sir-2.1* may modulate life span by interacting with *daf-16*, the IIS FOXO-family transcription factor, in a 14-3-3 dependent manner (Berdichevsky et al. 2006; Wang et al. 2006), while in flies there is evidence suggesting a longevity-related interaction between dSir2 and Rpd3 histone deacetylase (Rogina and Helfand 2004).

Recent evidence suggests that sirtuins may both promote and antagonize the aging process. For example, in contrast to the protective roll Sir2 plays in yeast replicative aging overexpression of Sir2 limits yeast chronological life span (Fabrizio et al. 2005; Kennedy et al. 2005) and SIRT1 deficiency in mouse embryonic fibroblasts (MEFs) promotes resistance to replicative senescence and increases replicative potential under chronic oxidative stress (Chua et al. 2005). Furthermore, studies in mice and flies suggest that both increasing and decreasing sirtuin activity may be neuroprotective (Kim et al. 2007; Li et al. 2008; Rogina and Helfand 2004). Thus, while sirtuins are known to play a role in determining longevity, further work will be necessary to understand the full range of sirtuin interactions and how they influence mortality.

20.4.3 Reduced TOR Signaling Provides Consistent Life Span Extension

TOR is a nutrient-responsive kinase with high evolutionary conservation among eukaryotes. TOR acts as part of two complexes, TOR complex 1 (TORC1) and TOR complex 2 (TORC2) (De Virgilio and Loewith 2006; Martin and Hall 2005), though only TORC1 is thought to be involved in aging. TORC1 is a central regulator of response to nutrients, growth signals and energy status (Wullschleger et al. 2006) and both TORC1 and TORC2 are required for viability (Guertin et al. 2006; Helliwell et al. 1998). Aside from DR, reducing TOR signaling is the only intervention known to extend life span in worms (Jia et al. 2004; Vellai et al. 2003), flies (Kapahi et al. 2004) and both yeast paradigms (Kaeberlein et al. 2005; Powers et al. 2006). The role of TOR signaling in mammalian aging is unknown, though a longevity study of mice fed a diet supplemented with rapamycin, a pharmacological inhibitor of TORC1, is currently underway at the National Institute on Aging Interventions Testing Program (Miller et al. 2007). As a counterexample to sirtuins, the TOR signaling pathways are conserved both upstream and downstream of TORC1, including an S6 kinase and processes regulated downstream such as mRNA translation, autophagy, stress response and mitochondria metabolism.

20.4.4 DR and the Search for a Mechanism

DR is defined as a decrease in dietary intake without malnutrition and is the most widely effective and intensely studied intervention known to extend life span. DR has been shown to enhance longevity and increase stress resistance in eukaryotic species from yeast to mice and intense effort has gone into uncovering the underlying molecular mechanisms. Each of the pathways discussed above have independently been proposed as a key mediator of life span extension via DR. The best and most consistent evidence is for TOR signaling. Life span extension from reduced TOR signaling and DR are non-additive in yeast replicative aging, worms and flies (Hansen et al. 2007; Juhasz et al. 2007; Kaeberlein et al. 2005). A recent study in yeast found that the starvation-responsive *GCN4* transcription factor attenuates life span extension by deletion of *TOR1* (Steffen et al. 2008). Interestingly, *GCN4* and TOR signaling influence many of the same cellular processes and *GCN4* expression is primarily regulated by translation, suggesting a model in which TOR signaling or DR influences *GCN4* target genes by translationally regulating *GCN4* expression (Steffen et al. 2008; Valenzuela et al. 2001; Yang et al. 2000). TOR has yet to be definitively linked to DR with respect to aging in mice or the yeast chronological paradigm.

IIS appears to influence life span by mechanisms at least partially distinct from DR in worms, flies and mice (Bartke et al. 2001; Giannakou et al. 2008; Kaeberlein et al. 2006), though there are potentially conflicting findings in flies (Clancy et al. 2002) and evidence for interaction with respect to other age-associated phenotypes in all three organisms (Bluher et al. 2003; Gershman et al. 2007; Greer et al. 2007; Iser and Wolkow 2007; Libert et al. 2007).

The interaction between sirtuins and DR is unresolved. The role of sirtuins in mediating life extension by DR is a source of ongoing controversy in yeast aging (Guarente and Picard 2005; Kaeberlein 2006; Kaeberlein et al. 2007; Kennedy et al. 2005; Lamming et al. 2005) though the majority of evidence suggests that they influence life span via independent mechanisms (Kaeberlein and Powers 2007). DR and *sir-2.1* appear to act in parallel in worms to extend life span (Lee et al. 2006; Tsuchiya et al. 2006), while the opposite is supported by evidence in flies (Rogina and Helfand 2004; Rogina et al. 2002). The relationship between DR and SIRT1 has not been determined in mice. Thus, while the available evidence supports a hypothesis placing TOR signaling downstream of DR with respect to longevity, it is not known for certain what other pathways are involved. The quest for a mechanism continues.

20.5 Aging Genomics

The ongoing research surrounding DR, IIS, sirtuins and TOR signaling demonstrates that conserved aging factors exist but does not provide any indication as to how common genes involved in determination of longevity are in the genome. To investigate the number of longevity genes in a given organism or the degree of conservation of longevity control between evolutionarily diverse organisms we must follow the way of the modern geneticist: genomics.

Aging has often been studied indirectly by looking at secondary phenotypes associated with longevity, such as enhanced stress resistance or decreased fecundity, or by looking at genes that interact with specific pathways known to play a role in aging. While these types of studies are valuable and help to improve our understanding of the biological systems that influence aging, they cannot provide information about global age-associated changes, an unbiased estimate of the number of genes in a given organism that act to modulate life span, or an estimate as to what degree these genes are conserved among divergent organisms. The past decade has seen a response to these issues in the form of aging genomics. To date two approaches have been used to look at aging on the genome scale: microarrays and large-scale genetic screens for longevity phenotypes (Kaeberlein 2004; Steinkraus et al. 2008).

20.5.1 Microarrays Uncover Age-Associated Gene Expression Patterns

Two strategies have been used in the application of microarrays to aging. The most common is comparison of gene expression patterns between young and old individuals to find changes that correlate with age. Studies have found that factors involved in oxidative stress response are upregulated with age in flies (Landis et al. 2004; Pletcher et al. 2002; Zou et al. 2000), mice (Weindruch et al. 2001) and monkeys (Kayo et al. 2001), which is consistent with an observed increase in expression of oxidative stress genes in young individuals from long-lived *C. elegans* strains (McElwee et al. 2003; Murphy et al. 2003). One group used microarray analysis to compare age-associated gene expression changes between *C. elegans* and *D. melanogaster* (McCarroll et al. 2004) and found a similar age-related gene expression program involving mitochondrial metabolism and DNA repair, among others. Further studies of this type will be of interest, particularly involving comparison of gene expression patterns between invertebrates and mammals.

The second microarray strategy compares gene expression patterns between young, age-matched individuals with different longevity phenotypes (resulting from differences in environmental exposure, genotype, or both) with the goal of identifying genetic programs that contribute to increased life span when activated early in life. For example, this strategy was used to demonstrate that gene expression for Ames dwarf mice is different from wild-type mice subject to DR and that changes in gene expression in response to DR are different for wild-type and Ames dwarf mice (Masternak et al. 2004). This is in agreement with the independent action of DR and IIS on life span in *C. elegans* (Houthoofd et al. 2003; Kaeberlein et al. 2006; Lakowski and Hekimi 1998; Lee et al. 2006). Similar studies have linked DR to osmotic stress and increased respiration in yeast (Kaeberlein et al. 2002; Lin et al. 2002) and to growth hormone signaling in mice (Miller et al. 2002). Microarrays have also been used to demonstrate that gene expression changes associated with DR occur quickly relative to life span in mice (Dhahbi et al. 2004), which is consistent with a rapid decrease in mortality in response to DR in flies (Mair et al. 2003) and mice (Dhahbi et al. 2004) and the observation that DR extends worm life span even when initiated late in life (Smith et al. 2008). The ability to identify short-term changes in gene expression with potential long-term consequences on life span opens the possibility of screening for pharmacological agents that mimic the beneficial effects of DR (Kaeberlein 2004).

A study combining the two microarray strategies compared the age-associated changes between mice fed a control diet and mice subject to DR and found that DR reversed a subset of the age-related gene expression changes (Weindruch et al. 2001). The findings from aging microarray studies to date demonstrate their potential for identifying global changes associated with advanced age or enhanced longevity. Notably, early attempts have suffered from a myriad of technical and analytical problems, such as limited sample size or lack of rigorous statistical analysis.

Nevertheless, the field remains optimistic that microarrays will be prevalent in the future of aging research and addressing the technical challenges and discussion of novel approaches to using microarrays in aging has been the topic of many reviews (Becker 2002; Golden et al. 2006; Han et al. 2004; Melov and Hubbard 2004; Nair et al. 2003; Werner 2007). Technical issues aside, microarrays are inherently limited in that they are observational in nature. Another method is needed to identify genes mechanistically involved in aging.

20.5.2 Genome-Scale Life Span Screens Identify a Large Number of Longevity Genes

The methods and findings discussed so far demonstrate that life span is under genetic control and that the mechanisms of control are conserved across divergent species, at least to a degree. The next task is to determine whether the majority of genes involved in life span control are already known, or whether there are still a substantial number of longevity genes yet to be discovered. This requires the ability to look at a large fraction of the genes in a particular genome. To reiterate an earlier point, looking at life span phenotypes is the only way to directly identify genes involved in longevity control. The search for genes that influence life span went genome-wide with the creation of large-scale genetic libraries for *S. cerevisiae* and *C. elegans*. These libraries are being used to screen for mutations that extend life span. Looking specifically for increased life span is particularly important when screening at the genomic level, where the potential for false positives from mutations that shorten life span independent of aging is vast.

In *C. elegans*, RNA interference (RNAi) can be used to knock down the expression of any given open reading frame (ORF) by feeding animals bacteria expressing double-stranded RNA corresponding to that ORF (Timmons and Fire 1998). Two RNAi libraries were created using *Escherichia coli* that together cover more than 90% of the ORFs in the *C. elegans* genome (Kamath et al. 2003; Rual et al. 2004). Two large-scale longevity screens (Hamilton et al. 2005; Hansen et al. 2005) and several screens targeting specific subsets of genes (Chen et al. 2007a; Curran and Ruvkun 2007; Dillin et al. 2002; Kim and Sun 2007; Lee et al. 2003) were performed using these libraries, resulting in the identification of 276 genes that extend life span when knocked down. The two genome-wide screens represent the first unbiased approach to the discovery of novel aging genes.

The implication of the discovery of a large number of genes that influence life span in independent studies is that a substantial fraction of the genes in the *C. elegans* genome are likely to play a role in aging. The question, "why so many?," brings us back to evolutionary theory. One implication of post-Medawar aging theory is that organisms will be selected for overall fitness and not for maximum longevity. Under this model for selection you might expect to find a large number of genes that increase life span when their expression is altered. A further extension of the relationship between fitness and longevity is that mutations that increase longevity should also have a detrimental effect on fitness. Indeed, long-lived *C. elegans* mutants were found to have reduced fitness relative to wild type in both a demographic survival analysis (Chen et al. 2007b) and in direct competition assays (Jenkins et al. 2004; Walker et al. 2000). Mutations resulting in enhanced longevity are also associated with reduced performance in other areas reproduction in particular in worms (Apfeld and Kenyon 1999; Van Voorhies and Ward 1999) and flies (Buck et al. 2000; Burger et al. 2007; Marden et al. 2003; Mockett and Sohal 2006).

The *S. cerevisiae* library consists of approximately 4,800 yeast strains in a common strain background, each with a single non-essential gene deletion (Winzeler et al. 1999). The strategy of completely knocking out a gene removes the problems associated with variability in the efficiency of gene knock down by RNAi experienced in the *C. elegans* RNAi screens but has the disadvantage of excluding all essential genes. The yeast deletion collection was used to screen for long-lived mutants in both the replicative and chronological paradigms. Measurement of replicative life span is manual labor intensive and the screen is still ongoing. In an initial report for 564 of the single gene mutants 13 (2.3%) were found to be long-lived, five of which are known to function in the TOR signaling pathway (Kaeberlein et al. 2005). The chronological life span screen used a high-throughput method to measure life span for all ~4,800 genes in the deletion collection (Powers et al. 2006). Of the 90 longest-lived strains, 16 have been implicated in TOR signaling and nutrient uptake (Powers et al. 2006). Thus the first unbiased longevity screens in yeast strongly implicate TOR signaling as a central regulator of aging and longevity.

20.6 The Search for Conserved Longevity Determinants: The Genome-Wide Multi-organism Approach

Early applications of genomics to longevity have provided the first glimpse of the true scope of the aging landscape and identified potential key players in the aging process. Large-scale, unbiased screens for life span in different organisms also provide the first opportunity to address in a quantitative manner the degree of conservation of aging determinants between these organisms. This directly impacts the question of relevance to human aging. On the evolutionary timeline, yeast and nematodes are separated by approximately 1.5 billion years, while nematodes and humans are separated by only approximately 1 billion years (Wang et al. 1999). Thus, if we can identify genes that play a conserved role in modulating life span between yeast and nematodes, a subset is likely to play a similar role in mammalian aging as well (Fig. 20.2).

A recent study provided the first quantitative evidence for conservation of longevity control between *S. cerevisiae* and *C. elegans* (Smith et al. 2008).

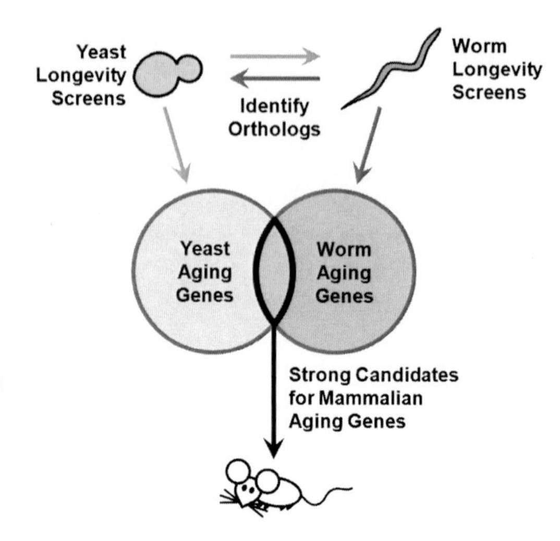

Fig. 20.2 A genome-wide, multi-organism approach to studying genes involved in aging. Cross-examining orthologs of longevity genes between evolutionarily divergent organisms, such as yeast and worms, allows the identification of genes that play a conserved role in longevity determination. These genes are strong candidates for aging genes in diverse organisms, such as mammals

Smith et al. (2008) measured replicative life span for single gene deletion strains corresponding to each of the 276 worm genes identified in the RNAi longevity screens (see Section 20.5.2). First, yeast genes were selected based on protein homology using a two-tiered approach. Yeast orthologs for each worm gene were identified using a high-stringency modified BLASTp reciprocal best match criterion, allowing selection of two yeast genes for a single worm gene when two yeast paralogs had BLASTp scores within 10% of each other. Up to 6 yeast homologs were then selected for each worm gene, requiring at least 20% protein sequence identity and at least 10% amino acid alignment. In total, 264 yeast homologs were selected for analysis, of which 76 met the high-stringency requirements.

Replicative life span analysis identified 25 (9.5%) long-lived mutants from the set of 264 analyzed (Smith et al. 2008). Of the 76 orthologs that met the high-stringency requirement, 11 (14.5%) were long-lived. This is a substantial enrichment for longevity determinants as compared to the expected 2.3% (3.4% if only yeast genes with worm potential orthologs are considered) as estimated by the unbiased screen of 564 genes (Kaeberlein et al. 2005), demonstrating conservation of genes that control aging between yeast and worms. Notably, 15 of the 25 genes identified have clear human orthologs (Smith et al. 2008).

20.7 Uncovering the Mechanisms Behind Conserved Longevity Factors: A Central Role for Translation?

If aging is not under direct evolutionary selection, why should genes influencing longevity be conserved among disparate species? A clue came from the long-lived high-stringency ortholog pairs identified by Smith et al. (2008), which were significantly enriched (6 out of 11) for genes involved in TOR signaling and more specifically, mRNA translation. Apart from TOR itself, these factors include S6 kinase, translation initiation factors and ribosomal subunits. Combining these findings with the evidence for TOR signaling as a conserved pro-aging pathway and strong candidate for mediation of the beneficial effects of DR (see Section 20.4) suggests mRNA translation as a central process in the modulation of life span. Unlike aging, the organismal response to environmental nutrient levels is critical to organismal fitness. We propose that longevity is conserved primarily because altered responses to nutrient levels profoundly impact longevity. In low nutrient environments (DR or mutations that impair nutrient signaling), organisms devote energy to somatic maintenance and live longer (Kirkwood 1977). Therefore, the conserved effects of certain pathways on longevity in disparate species reflect a longstanding impact of the proper response to environmental nutrient levels on fitness and longevity is tightly coupled to this response.

Several mechanisms have been suggested for how reduced translation might enhance longevity, including altered resource allocation (potentially away from investment in reproduction, in agreement with the disposable soma theory of aging), differential translation of subsets of mRNA and improved protein homeostasis (Kaeberlein and Kennedy 2008). We expect translation to be a central topic in aging research over the next few years and anticipate further clarification of its mechanistic role in longevity determination.

20.8 Conclusion

From the evolutionary insights by Medawar in the 1950s, our understanding of the aging process progressed steadily for roughly the next 30 years through the study of interventions that influence life span, such as DR and secondary age-associated phenotypes (e.g. stress resistance). The advances in genetics in the 1990s led to rapid identification of several key pathways and processes involved in the determination of life span and the demonstration that these factors play a conserved role in aging across multiple species. The genomics-revolution of the past decade and the development of methodology for studying aging and life span determinants on the genomic-scale have provided the first glimpse of the large number of factors involved in controlling aging and the degree of conservation of these factors between evolutionarily divergent organisms.

Our understanding of the aging process and its interaction with disease has also started to yield its first clinically application in humans. Pharmacological agents targeting sirtuins and TOR signaling are now in clinical trials for treatment of age-associated pathologies such as cancer and metabolic disease. The discovery of new longevity genes and novel aspects of known aging pathways has accelerated at a break-neck pace and we expect that this knowledge will rapidly be translated to clinical practice as well.

References

Aparicio OM, Billington BL, Gottschling DEe (1991) Modifiers of position effect are shared between telomeric and silent mating-type loci in *Saccharomyces cerevisiae*. Cell 66:1279–1287

Apfeld J, Kenyon C (1999) Regulation of lifespan by sensory perception in *Caenorhabditis elegans*. Nature 402:804–809

Banks AS, Kon N, Knight C, Matsumoto M, Gutierrez-Juarez R, Rossetti L, Gu W, Accili D (2008) SirT1 gain of function increases energy efficiency and prevents diabetes in mice. Cell Metab 8:333–341

Bartke A (2008) Impact of reduced insulin-like growth factor-1/insulin signaling on aging in mammals: novel findings. Aging Cell 7:285–290

Bartke A, Wright JC, Mattison JA, Ingram DK, Miller RA, Roth GS (2001) Extending the lifespan of long-lived mice. Nature 414:412

Becker KG (2002) Deciphering the gene expression profile of long-lived snell mice. Sci Aging Knowledge Environ 2002:pe4

Berdichevsky A, Viswanathan M, Horvitz HR, Guarente L (2006) C. elegans SIR-2.1 interacts with 14-3-3 proteins to activate DAF-16 and extend life span. Cell 125:1165–1177

Bluher M, Kahn BB, Kahn RC (2003) Extended longevity in mice lacking the insulin receptor in adipose tissue. Science 299:572–574

Bordone L, Cohen D, Robinson A, Motta MC, van Veen E, Czopik A, Steele AD, Crowe H, Marmor S, Luo J, Gu W, Guarente L (2007) SIRT1 transgenic mice show phenotypes resembling calorie restriction. Aging Cell 6:759–767

Brown-Borg HM, Borg KE, Meliska CJ, Bartke A (1996) Dwarf mice and the ageing process. Nature 384:33

Brunet A, Sweeney LB, Sturgill JF, Chua KF, Greer PL, Lin Y, Tran H, Ross SE, Mostoslavsky R, Cohen HY, Hu LS, Cheng HL, Jedrychowski MP, Gygi SP, Sinclair DA, Alt FW, Greenberg ME (2004) Stress-dependent regulation of FOXO transcription factors by the SIRT1 deacetylase. Science 303:2011–2015

Bryk M, Banerjee M, Murphy M, Knudsen KE, Garfinkel DJ, Curcio MJ (1997) Transcriptional silencing of Ty1 elements in the *RDN1* locus of yeast. Genes Dev. 11:255–269

Buck S, Vettraino J, Force AG, Arking R (2000) Extended longevity in Drosophila is consistently associated with a decrease in developmental viability. J Gerontol A Biol Sci Med Sci 55: B292–301

Burger JM, Hwangbo DS, Corby-Harris V, Promislow DE (2007) The functional costs and benefits of dietary restriction in Drosophila. Aging Cell 6:63–71

Chapman T, Partridge L (1996) Female fitness in *Drosophila melanogaster*: an interaction between the effect of nutrition and of encounter rate with males. Proc Biol Sci 263:755–759

Chen D, Pan KZ, Palter JE, Kapahi P (2007a) Longevity determined by developmental arrest genes in *Caenorhabditis elegans*. Aging Cell 6:525–533

Chen J, Senturk D, Wang JL, Muller HG, Carey JR, Caswell H, Caswell-Chen EP (2007b) A demographic analysis of the fitness cost of extended longevity in *Caenorhabditis elegans*. J Gerontol A Biol Sci Med Sci 62:126–135

Chua KF, Mostoslavsky R, Lombard DB, Pang WW, Saito S, Franco S, Kaushal D, Cheng HL, Fischer MR, Stokes N, Murphy MM, Appella E, Alt FW (2005) Mammalian SIRT1 limits replicative life span in response to chronic genotoxic stress. Cell Metab 2:67–76

Clancy DJ, Gems D, Harshman LG, Oldham S, Stocker H, Hafen E, Leevers SJ, Partridge L (2001) Extension of life-span by loss of CHICO, a Drosophila insulin receptor substrate protein. Science 292:104–106

Clancy DJ, Gems D, Hafen E, Leevers SJ, Partridge L (2002) Dietary restriction in long-lived dwarf flies. Science 296:319

Coschigano KT, Holland AN, Riders ME, List EO, Flyvbjerg A, Kopchick JJ (2003) Deletion, but not antagonism, of the mouse growth hormone receptor results in severely decreased body weights, insulin, and insulin-like growth factor I levels and increased life span. Endocrinology 144:3799–3810

Curran SP, Ruvkun G (2007) Lifespan regulation by evolutionarily conserved genes essential for viability. PLoS Genet 3:e56

De Virgilio C, Loewith R (2006) The TOR signalling network from yeast to man. Int J Biochem Cell Biol 38:1476–1481

Dhahbi JM, Kim HJ, Mote PL, Beaver RJ, Spindler SR (2004) Temporal linkage between the phenotypic and genomic responses to caloric restriction. Proc Natl Acad Sci USA 101:5524–5529

Dillin A, Hsu AL, Arantes-Oliveira N, Lehrer-Graiwer J, Hsin H, Fraser AG, Kamath RS, Ahringer J, Kenyon C (2002) Rates of behavior and aging specified by mitochondrial function during development. Science 298:2398–2401

Dorman JB, Albinder B, Shroyer T, Kenyon C (1995) The age-1 and daf-2 genes function in a common pathway to control the lifespan of *Caenorhabditis elegans*. Genetics 141:1399–1406

Fabrizio P, Longo VD (2003) The chronological life span of *Saccharomyces cerevisiae*. Aging Cell 2:73–81

Fabrizio P, Battistella L, Vardavas R, Gattazzo C, Liou LL, Diaspro A, Dossen JW, Gralla EB, Longo VD (2004) Superoxide is a mediator of an altruistic aging program in *Saccharomyces cerevisiae*. J. Cell Biol 166:1055–1067

Fabrizio P, Gattazzo C, Battistella L, Wei M, Cheng C, McGrew K, Longo VD (2005) Sir2 blocks extreme life-span extension. Cell 123:655–667

Firestein R, Blander G, Michan S, Oberdoerffer P, Ogino S, Campbell J, Bhimavarapu A, Luikenhuis S, de Cabo R, Fuchs C, Hahn WC, Guarente LP, Sinclair DA (2008) The SIRT1 deacetylase suppresses intestinal tumorigenesis and colon cancer growth. PLoS ONE 3:e2020

Fisher RA (1930) The genetical theory of natural selection. Oxford Press, Oxford

Flurkey K, Papaconstantinou J,Harrison DE (2002) The Snell dwarf mutation Pit1(dw) can increase life span in mice. Mech Ageing Dev 123:121–130

Gerhart-Hines Z, Rodgers JT, Bare O, Lerin C, Kim SH, Mostoslavsky R, Alt FW, Wu Z, Puigserver P (2007) Metabolic control of muscle mitochondrial function and fatty acid oxidation through SIRT1/PGC-1alpha. Embo J 26:1913–1923

Gershman B, Puig O, Hang L, Peitzsch RM, Tatar M, Garofalo RS (2007) High-resolution dynamics of the transcriptional response to nutrition in Drosophila: a key role for dFOXO. Physiol Genomics 29:24–34

Giannakou ME, Goss M, Partridge L (2008) Role of dFOXO in lifespan extension by dietary restriction in *Drosophila melanogaster*: not required, but its activity modulates the response. Aging Cell 7:187–198

Golden TR, Hubbard A, Melov S (2006) Microarray analysis of variation in individual aging *C. elegans*: approaches and challenges. Exp Gerontol 41:1040–1045

Good TP, Tatar M (2001) Age-specific mortality and reproduction respond to adult dietary restriction in *Drosophila melanogaster*. J Insect Physiol 47:1467–1473

Gottschling DE, Aparicio OM, Billington BL, Zakian VA (1990) Position effect at *Saccharomyces cerevisiae* telomeres: reversible repression of Pol II transcription. Cell 63:751–762

Greer EL, Dowlatshahi D, Banko MR, Villen J, Hoang K, Blanchard D, Gygi SP, Brunet A (2007) An AMPK-FOXO pathway mediates longevity induced by a novel method of dietary restriction in *C. elegans*. Curr Biol 17:1646–56

Guarente L, Picard F (2005) Calorie restriction—the SIR2 connection. Cell 120:473–482

Guertin DA, Guntur KV, Bell GW, Thoreen CC, Sabatini DM (2006) Functional genomics identifies TOR-regulated genes that control growth and division. Curr Biol 16:958–970

Haldane JBS (1941) New paths in genetics. Allen & Unwin, London

Hamilton B, Dong Y, Shindo M, Liu W, Odell I, Ruvkun G, Lee SS (2005) A systematic RNAi screen for longevity genes in *C. elegans*. Genes Dev 19:1544–1555

Han ES, Wu Y, McCarter R, Nelson JF, Richardson A, Hilsenbeck SG (2004) Reproducibility, sources of variability, pooling, and sample size: important considerations for the design of high-density oligonucleotide array experiments. J Gerontol A Biol Sci Med Sci 59:306–315

Hansen M, Hsu AL, Dillin A, Kenyon C (2005) New genes tied to endocrine, metabolic, and dietary regulation of lifespan from a *Caenorhabditis elegans* genomic RNAi screen. PLoS Genet 1:119–128

Hansen M, Taubert S, Crawford D, Libina N, Lee SJ, Kenyon C (2007) Lifespan extension by conditions that inhibit translation in *Caenorhabditis elegans*. Aging Cell 6:95–110

Helliwell SB, Howald I, Barbet N, Hall MN (1998) TOR2 is part of two related signaling pathways coordinating cell growth in *Saccharomyces cerevisiae*. Genetics 148:99–112

Hercus MJ, Loeschcke V, Rattan SI (2003) Lifespan extension of *Drosophila melanogaster* through hormesis by repeated mild heat stress. Biogerontology 4:149–156

Holzenberger M, Dupont J, Ducos B, Leneuve P, Geloen A, Even PC, Cervera P, Le Bouc Y (2003) IGF-1 receptor regulates lifespan and resistance to oxidative stress in mice. Nature 421:182–187

Houthoofd K, Braeckman BP, Johnson TE, Vanfleteren JR (2003) Life extension via dietary restriction is independent of the Ins/IGF-1 signalling pathway in *Caenorhabditis elegans*. Exp Gerontol 38:947–954

Imai S, Armstrong CM, Kaeberlein M, Guarente L (2000) Transcriptional silencing and longevity protein Sir2 is an NAD- dependent histone deacetylase. Nature 403:795–800

Iser WB, Wolkow CA (2007) DAF-2/insulin-like signaling in *C. elegans* modifies effects of dietary restriction and nutrient stress on aging, stress and growth. PLoS ONE 2:e1240

Ivy JM, Klar AJ, Hicks JB (1986) Cloning and characterization of four SIR genes of *Saccharomyces cerevisiae*. Mol Cell Biol 6:688–702

Jenkins NL, McColl G, Lithgow GJ (2004) Fitness cost of extended lifespan in *Caenorhabditis elegans*. Proc Biol Sci 271:2523–2526

Jia K, Chen D, Riddle DL (2004) The TOR pathway interacts with the insulin signaling pathway to regulate *C. elegans* larval development, metabolism and life span. Development 131:3897–3906

Jiang JC, Jaruga E, Repnevskaya MV, Jazwinski SM (2000) An intervention resembling caloric restriction prolongs life span and retards aging in yeast. FASEB J 14:2135–2137

Juhasz G, Erdi B, Sass M, Neufeld TP (2007) Atg7-dependent autophagy promotes neuronal health, stress tolerance, and longevity but is dispensable for metamorphosis in Drosophila. Genes Dev 21:3061–3066

Kaeberlein M (2004) Aging-related research in the "-omics" age. Sci Aging Knowledge Environ pe39

Kaeberlein M (2006) Longevity and aging in the budding yeast. In: Conn PM (ed) Handbook of models for human aging. Elsevier, Boston, MA

Kaeberlein M, Kennedy BK (2008) Protein translation, 2008. Aging Cell 7:777–782

Kaeberlein M, Powers RW, 3rd (2007) Sir2 and calorie restriction in yeast: a skeptical perspective. Ageing Res Rev 6:128–140

Kaeberlein M, McVey M, Guarente L (1999) The *SIR2/3/4* complex and *SIR2* alone promote longevity in *Saccharomyces cerevisiae* by two different mechanisms. Genes Dev 13:2570–2580

Kaeberlein M, Andalis AA, Fink GR, Guarente L (2002) High osmolarity extends life span in *Saccharomyces cerevisiae* by a mechanism related to calorie restriction. Mol Cell Biol 22:8056–8066

Kaeberlein M, Kirkland KT, Fields S, Kennedy BK (2004) Sir2-independent life span extension by calorie restriction in yeast. PLOS Biol 2:1381–1387

Kaeberlein M, Powers III RW, Steffen KK, Westman EA, Hu D, Dang N, Kerr EO, Kirkland KT, Fields S, Kennedy BK (2005) Regulation of yeast replicative life-span by TOR and Sch9 in response to nutrients. Science 310:1193–1196

Kaeberlein M, Burtner CR, Kennedy BK (2007) Recent developments in yeast aging. PLoS Genet. 3:655–660

Kaeberlein TL, Smith ED, Tsuchiya M, Welton KL, Thomas JH, Fields S, Kennedy BK, Kaeberlein M (2006) Lifespan extension in *Caenorhabditis elegans* by complete removal of food. Aging Cell 5:487–494

Kamath RS, Fraser AG, Dong Y, Poulin G, Durbin R, Gotta M, Kanapin A, Le Bot N, Moreno S, Sohrmann M, Welchman DP, Zipperlen P, Ahringer J (2003) Systematic functional analysis of the *Caenorhabditis elegans* genome using RNAi. Nature 421:231–237

Kapahi P, Zid BM, Harper T, Koslover D, Sapin V, Benzer S (2004) Regulation of lifespan in *Drosophila* by modulation of genes in the TOR signaling pathway. Curr Biol 14:885–890

Kayo T, Allison DB, Weindruch R, Prolla TA (2001) Influences of aging and caloric restriction on the transcriptional profile of skeletal muscle from rhesus monkeys. Proc Natl Acad Sci USA 98:5093–5098

Kennedy BK (2008) The genetics of ageing: insight from genome-wide approaches in invertebrate model organisms. J Intern Med 263:142–152

Kennedy BK, Smith ED, Kaeberlein M (2005) The enigmatic role of Sir2 in aging. Cell 123:548–550

Kennedy BK, Steffen KK, Kaeberlein M (2007) Ruminations on dietary restriction and aging. Cell Mol Life Sci 64:1323–1328

Kenyon C, Chang J, Gensch E, Rudner A, Tabtiang R (1993) A *C. elegans* mutant that lives twice as long as wild type. Nature 366:461–464

Kim D, Nguyen MD, Dobbin MM, Fischer A, Sananbenesi F, Rodgers JT, Delalle I, Baur JA, Sui G, Armour SM, Puigserver P, Sinclair DA, Tsai LH (2007) SIRT1 deacetylase protects against neurodegeneration in models for Alzheimer's disease and amyotrophic lateral sclerosis. Embo J 26:3169–3179

Kim Y, Sun H (2007) Functional genomic approach to identify novel genes involved in the regulation of oxidative stress resistance and animal lifespan. Aging Cell 6:489–503

Kirkwood TB (1977) Evolution of ageing. Nature 270:301–304

Kirkwood TB (2005) Understanding the odd science of aging. Cell 120:437–447

Kirkwood TB, Holliday R (1979) The evolution of ageing and longevity. Proc R Soc Lond B Biol Sci 205:531–546

Lakowski B, Hekimi S (1998) The genetics of caloric restriction in *Caenorhabditis elegans*. Proc Natl Acad Sci USA 95:13091–13096

Lamming DW, Latorre-Esteves M, Medvedik O, Wong SN, Tsang FA, Wang C, Lin SJ, Sinclair DA (2005) *HST2* mediates *SIR2*-independent life-span extension by calorie restriction. Science 309:1861–1864

Landis GN, Abdueva D, Skvortsov D, Yang J, Rabin BE, Carrick J, Tavare S, Tower J (2004) Similar gene expression patterns characterize aging and oxidative stress in *Drosophila melanogaster*. Proc Natl Acad Sci USA 101:7663–7668

Landry J, Sutton A, Tafrov ST, Heller RC, Stebbins J, Pillus L, Sternglanz R (2000) The silencing protein SIR2 and its homologs are NAD-dependent protein deacetylases. Proc Natl Acad Sci USA 97:5807–5811

Lee GD, Wilson MA, Zhu M, Wolkow CA, de Cabo R, Ingram DK, Zou S (2006) Dietary deprivation extends lifespan in *Caenorhabditis elegans*. Aging Cell 5:515–524

Lee SS, Lee RY, Fraser AG, Kamath RS, Ahringer J, Ruvkun G (2003) A systematic RNAi screen identifies a critical role for mitochondria in *C. elegans* longevity. Nat Genet 33:40–48

Li Y, Xu W, McBurney MW, Longo VD (2008) SirT1 inhibition reduces IGF-I/IRS-2/Ras/ERK1/2 signaling and protects neurons. Cell Metab 8:38–48

Libert S, Zwiener J, Chu X, Van Voorhies W, Roman G, Pletcher SD (2007) Regulation of Drosophila life span by olfaction and food-derived odors. Science 315:1133–1137

Lin SJ, Defossez PA, Guarente L (2000) Requirement of NAD and SIR2 for life-span extension by calorie restriction in *Saccharomyces cerevisiae*. Science 289:2126–2128

Lin SJ, Kaeberlein M, Andalis AA, Sturtz LA, Defossez PA, Culotta VC, Fink GR, Guarente L (2002) Calorie restriction extends *Saccharomyces cerevisiae* lifespan by increasing respiration. Nature 418:344–348

Lithgow GJ, White TM, Melov S, Johnson TE (1995) Thermotolerance and extended life-span conferred by single-gene mutations and induced by thermal stress. Proc Natl Acad Sci USA 92:7540–7544

Luo J, Nikolaev AY, Imai S, Chen D, Su F, Shiloh A, Guarente L, Gu W (2001) Negative control of p53 by Sir2alpha promotes cell survival under stress. Cell 107:137–148

Mair W, Goymer P, Pletcher SD, Partridge L (2003) Demography of dietary restriction and death in Drosophila. Science 301:1731–1733

Marden JH, Rogina B, Montooth KL, Helfand SL (2003) Conditional tradeoffs between aging and organismal performance of Indy long-lived mutant flies. Proc Natl Acad Sci USA 100:3369–3373

Martin DE, Hall MN (2005) The expanding TOR signaling network. Curr Opin Cell Biol 17:158–166

Martin GM, Austad SN, Johnson TE (1996) Genetic analysis of ageing: role of oxidative damage and environmental stresses. Nat Genet 13:25–34

Masoro EJ (2005) Overview of caloric restriction and ageing. Mech Ageing Dev 126:913–922

Masternak MM, Al-Regaiey K, Bonkowski MS, Panici J, Sun L, Wang J, Przybylski GK, Bartke A (2004) Divergent effects of caloric restriction on gene expression in normal and long-lived mice. J Gerontol A Biol Sci Med Sci 59:784–788

McCarroll SA, Murphy CT, Zou S, Pletcher SD, Chin CS, Jan YN, Kenyon C, Bargmann CI, Li H (2004) Comparing genomic expression patterns across species identifies shared transcriptional profile in aging. Nat Genet 36:197–204

McCay CM, Crowell MF, Maynard LA (1935) The effect of retarded growth upon the length of life and upon ultimate size. J Nutr 10:63–79

McElwee J, Bubb K, Thomas JH (2003) Transcriptional outputs of the *Caenorhabditis elegans* forkhead protein DAF-16. Aging Cell 2:111–121

Medawar P (1952) An unsolved problem in biology. H.K. Lewis, London

Medawar PB (1946) Old age and natural death. Mod 1:30–56

Melov S, Hubbard A (2004) Microarrays as a tool to investigate the biology of aging: a retrospective and a look to the future. Sci Aging Knowledge Environ re7

Miller RA, Chang Y, Galecki AT, Al-Regaiey K, Kopchick JJ, Bartke A (2002) Gene expression patterns in calorically restricted mice: partial overlap with long-lived mutant mice. Mol Endocrinol 16:2657–2666

Miller RA, Harrison DE, Astle CM, Floyd RA, Flurkey K, Hensley KL, Javors MA, Leeuwenburgh C, Nelson JF, Ongini E, Nadon NL, Warner HR, Strong R (2007) An aging interventions testing program: study design and interim report. Aging Cell 6:565–575

Mockett RJ, Sohal RS (2006) Temperature-dependent trade-offs between longevity and fertility in the Drosophila mutant, methuselah. Exp Gerontol 41:566–573

Mortimer RK, Johnston JR (1959) Life span of individual yeast cells. Nature 183:1751–1752

Motta MC, Divecha N, Lemieux M, Kamel C, Chen D, Gu W, Bultsma Y, McBurney M, Guarente L (2004) Mammalian SIRT1 represses forkhead transcription factors. Cell 116:551–563

Murakami S, Johnson TE (1996) A genetic pathway conferring life extension and resistance to UV stress in *Caenorhabditis elegans*. Genetics 143:1207–1218

Murphy CT, McCarroll SA, Bargmann CI, Fraser A, Kamath RS, Ahringer J, Li H, Kenyon C (2003) Genes that act downstream of DAF-16 to influence the lifespan of *Caenorhabditis elegans*. Nature 424:277–283

Nair PN, Golden T, Melov S (2003) Microarray workshop on aging. Mech Ageing Dev 124:133–138

Oberdoerffer P, Michan S, McVay M, Mostoslavsky R, Vann J, Park SK, Hartlerode A, Stegmuller J, Hafner A, Loerch P, Wright SM, Mills KD, Bonni A, Yankner BA, Scully R,

Prolla TA, Alt FW, Sinclair DA (2008) SIRT1 redistribution on chromatin promotes genomic stability but alters gene expression during aging. Cell 135:907–918

Pletcher SD, Macdonald SJ, Marguerie R, Certa U, Stearns SC, Goldstein DB, Partridge L (2002) Genome-wide transcript profiles in aging and calorically restricted *Drosophila melanogaster*. Curr Biol 12:712–723

Powers RW, 3rd, Kaeberlein M, Caldwell SD, Kennedy BK, Fields S (2006) Extension of chronological life span in yeast by decreased TOR pathway signaling. Genes Dev 20:174–184

Reznick DN, Bryant MJ, Roff D, Ghalambor CK, Ghalambor DE (2004) Effect of extrinsic mortality on the evolution of senescence in guppies. Nature 431:1095–1099

Rine J, Herskowitz I (1987) Four genes responsible for a position effect on expression from HML and HMR in *Saccharomyces cerevisiae*. Genetics 116:9–22

Rodgers JT, Lerin C, Gerhart-Hines Z, Puigserver P (2008) Metabolic adaptations through the PGC-1 alpha and SIRT1 pathways. FEBS Lett 582:46–53

Rogina B, Helfand SL (2004) Sir2 mediates longevity in the fly through a pathway related to calorie restriction. Proc Natl Acad Sci USA 101:15998–16003

Rogina B, Reenan RA, Nilsen SP, Helfand SL (2000) Extended life-span conferred by cotransporter gene mutations in Drosophila. Science 290:2137–2140

Rogina B, Helfand SL, Frankel S (2002) Longevity regulation by Drosophila Rpd3 deacetylase and caloric restriction. Science 298:1745

Rual JF, Ceron J, Koreth J, Hao T, Nicot AS, Hirozane-Kishikawa T, Vandenhaute J, Orkin SH, Hill DE, van den Heuvel S, Vidal M (2004) Toward improving *Caenorhabditis elegans* phenome mapping with an ORFeome-based RNAi library. Genome Res 14:2162–2168

Sinclair DA, Guarente L (1997) Extrachromosomal rDNA circles-a cause of aging in yeast. Cell 91:1033–1042

Smith ED, Tsuchiya M, Fox LA, Dang N, Hu D, Kerr EO, Johnston ED, Tchao BN, Pak DN, Welton KL, Promislow DEL, Thomas JH, Kaeberlein M, Kennedy BK (2008) Quantitative evidence for conserved longevity pathways between divergent eukaryotic species. Genome Res 18:564–570

Smith JS, Boeke JD (1997) An unusual form of transcriptional silencing in yeast ribosomal DNA. Genes Dev 11:241–254

Smith JS, Brachmann CB, Celic I, Kenna MA, Muhammad S, Starai VJ, Avalos JL, Escalante-Semerena JC, Grubmeyer C, Wolberger C, Boeke JD (2000) A phylogenetically conserved NAD^+-dependent protein deacetylase activity in the Sir2 protein family. Proc Natl Acad Sci USA 97:6658–6663

Steffen KK, MacKay VL, Kerr EO, Tsuchiya M, Hu D, Fox LA, Dang N, Johnston ED, Oakes JA, Tchao BN, Pak DN, Fields S, Kennedy BK, Kaeberlein M (2008) Yeast lifespan extension by depletion of 60S ribosomal subunits is mediated by Gcn4. Cell 133:292–302

Steinkraus KA, Kaeberlein M, Kennedy BK (2008) Replicative aging in yeast: the means to the end. Annu Rev Cell Dev Biol 24:29–54

Taguchi A, White MF (2008) Insulin-like signaling, nutrient homeostasis, and life span. Annu Rev Physiol 70:191–212

Tanner KG, Landry J, Sternglanz R, Denu JM (2000) Silent information regulator 2 family of NAD-dependent histone/protein deacetylases generates a unique product, 1-O-acetyl-ADP-ribose. Proc Natl Acad Sci USA 97:14178–14182

Tatar M, Kopelman A, Epstein D, Tu MP, Yin CM, Garofalo RS (2001) A mutant Drosophila insulin receptor homolog that extends life-span and impairs neuroendocrine function. Science 292:107–110

Timmons L, Fire A (1998) Specific interference by ingested dsRNA. Nature 395:854

Tissenbaum HA, Guarente L (2001) Increased dosage of a sir-2 gene extends lifespan in *Caenorhabditis elegans*. Nature 410:227–230

Toivonen JM, Partridge L (2008) Endocrine regulation of ageing and reproduction in Drosophila. Mol Cell Endocrinol

Tsuchiya M, Dang N, Kerr EO, Hu D, Steffen KK, Oakes JA, Kennedy BK, Kaeberlein M (2006) Sirtuin-independent effects of nicotinamide on lifespan extension from calorie restriction in yeast. Aging Cell 5:505–514

Tu MP, Epstein D, Tatar M (2002) The demography of slow aging in male and female Drosophila mutant for the insulin-receptor substrate homologue chico. Aging Cell 1:75–80

Valenzuela L, Aranda C, Gonzalez A (2001) TOR modulates GCN4-dependent expression of genes turned on by nitrogen limitation. J Bacteriol 183:2331–2334

van der Horst A, Tertoolen LG, de Vries-Smits LM, Frye RA, Medema RH, Burgering BM (2004) FOXO4 is acetylated upon peroxide stress and deacetylated by the longevity protein hSir2 (SIRT1). J Biol Chem 279:28873–28879

Van Voorhies WA, Ward S (1999) Genetic and environmental conditions that increase longevity in *Caenorhabditis elegans* decrease metabolic rate. Proc Natl Acad Sci USA 96:11399–11403

Vaziri H, Dessain SK, Eaton EN, Imai SI, Frye RA, Pandita TK, Guarente L, Weinberg RA (2001) hSIR2(SIRT1) functions as an NAD-dependent p53 deacetylase. Cell 107:149–159

Vellai T, Takacs-Vellai K, Zhang Y, Kovacs AL, Orosz L, Muller F (2003) Genetics: influence of TOR kinase on lifespan in *C. elegans*. Nature 426:620

Viswanathan M, Kim SK, Berdichevsky A, Guarente L (2005) A role for SIR-2.1 regulation of ER stress response genes in determining *C. elegans* life span. Dev Cell 9:605–615

Walker DW, McColl G, Jenkins NL, Harris J, Lithgow GJ (2000) Evolution of lifespan in *C. elegans*. Nature 405:296–297

Wallace AR (1889) The action of natural selection in producing old age, decay and death [a note by Wallace written "some time between 1865 and 1870"]. In: Weismann A (ed) Essays on hereditary and kindred biological problems. Clarendon, Oxford

Wang DY, Kumar S, Hedges SB (1999) Divergence time estimates for the early history of animal phyla and the origin of plants, animals and fungi. Proc Biol Sci 266:163–171

Wang Y, Wook Oh S, Deplancke B, Luo J, Walhout AJM, Tissenbaum HA (2006) *C. elegans* 14-3-3 proteins regulate life span and interact with SIR-2.1 and DAF-16/FOXO. Mech Ageing Dev 127:741–747

Weindruch R, Walford RL (1988) The retardation of aging and disease by dietary restriction. Charles C. Thomas, Springfield, IL

Weindruch R, Kayo T, Lee CK, Prolla TA (2001) Microarray profiling of gene expression in aging and its alteration by caloric restriction in mice. J Nutr 131:918S–923S

Weismann A (1889) Essays upon hereditary and kindred biological problems. Clarendon, Oxford

Werner T (2007) Regulatory networks: linking microarray data to systems biology. Mech Ageing Dev 128:168–172

Williams GC (1957) Pleiotropy, natural selection and the evolution of senescence. Evolution 11:398–411

Winzeler EA, Shoemaker DD, Astromoff A, Liang H, Anderson K, Andre B, Bangham R, Benito R, Boeke JD, Bussey H, Chu M, Connelly C, Davis K, Dietrich F, Dow SW, El Bakkoury M, Foury F, Friend SH, Gentalen E, Giaever G, Hegemann JH, Jones T, Laub M, Liao H, Liebundguth N, Lockhart DJ, Lucau-Danila A, Lussier M, M'Rabet N, Menard P, Mittmann M, Pai C, Rebischung C, Revuelta JL, Riles L, Roberts CJ, Ross-MacDonald P, Scherens B, Snyder M, Sookhai-Mahadeo S, Storms RK, Veronneau S, Voet M, Volckaert G, Ward TR, Wysocki R, Yen GS, Yu K, Zimmerman K, Philippsen P, Johnston M, Davis RW (1999) Functional characterization of the *S. cerevisiae* genome by gene deletion and parallel analysis. Science 285:901–906

Wullschleger S, Loewith R, Hall MN (2006) TOR signaling in growth and metabolism. Cell 124:471–484

Yang R, Wek SA, Wek RC (2000) Glucose limitation induces GCN4 translation by activation of Gcn2 protein kinase. Mol Cell Biol 20:2706–2717

Zou S, Meadows S, Sharp L, Jan LY, Jan YN (2000) Genome-wide study of aging and oxidative stress response in *Drosophila melanogaster*. Proc Natl Acad Sci USA 97:13726–13731

Part IV
Applications in Other Fields

Chapter 21
Galaxies and Cladistics

Didier Fraix-Burnet

Abstract The Hubble tuning fork diagram, based on morphology and established in the 1930s, has always been the preferred scheme for classification of galaxies. However, the current large amount of multiwavelength data, most often spectra, for objects up to very high distances, asks for more sophisticated statistical approaches. Interpreting formation and evolution of galaxies as a 'transmission with modification' process, we have shown that the concepts and tools of phylogenetic systematics can be heuristically transposed to the case of galaxies. This approach, which we call 'astrocladistics', has successfully been applied on several samples. Many difficulties still remain, some of them being specific to the nature of both galaxies and their diversification processes, some others being classical in cladistics, like the pertinence of the descriptors in conveying any useful evolutionary information.

21.1 Introduction

Galaxies have been discovered quite recently with respect to the History of Humanity, since it was in 1922 by Edwin Hubble. Our own galaxy, called the Milky Way, was rather well-known at this epoch. Hubble found that there are indeed many other similar objects situated at very far distances from us, much farther away than the dimension of the Milky Way. Galaxies are now known to be fundamental entities of the Universe and can be defined as self-gravitating ensemble of stars, gas and/or « dust » (dust being mainly grains for astronomers).

Typical big galaxies like our own are about 100,000 light years across. Shapes are basically those defined by Hubble, which are spiral, elliptical and irregular. This reflects the orbits of the stars that can be distributed in a disk (spirals), in a

D. Fraix-Burnet
Laboratoire d'Astrophysique de Grenoble, UMR5571 CNRS/Université Joseph Fourier, BP 53, F-38041 Grenoble cedex 9, France
e-mail: didier.fraix-burnet@obs.ujf-grenoble.fr

P. Pontarotti (ed.), *Evolutionary Biology: Concept, Modeling, and Application*, DOI: 10.1007/978-3-642-00952-5_21,

regular three dimension structure (ellipticals), or in a more disturbed fashion (irregulars). In spiral galaxies, density waves create spiral arms, sometimes with a bar-like structure around the centre of the galaxy. These density waves are not necessarily long-lasting but they tend to concentrate gas and dust so that star formation is mostly associated with them.

Self-gravitating objects probably appeared early in the Universe (the age is now established to be $13.7 \cdot 10^9$ years), in association with the very slightly inhomogeneous distribution of gravitational matter. The structure of the Universe is shaped by dark energy (70% of the Universe total energy), which causes the expansion, and the gravitational field that is fashioned essentially by the dark matter (26%). Galaxies are made up of baryonic matter that represents only 4% of the composition of the Universe. The primeval inhomogeneities of the gravitational potential are observed at the recombination epoch (time when the Universe became transparent at an age of about 400,000 years), thanks to their thermal radiation the expansion of the Universe has now 'cooled' down to 3 K (e.g., Hinshaw et al. 2009). These early overdensities have grown up by condensation to form bigger and bigger structures called halos and filaments. The baryonic matter is gravitationally affected by the evolution of the dark matter distribution and collapses of small-scale entities created the first light emitting objects (first stars at about 400 millions years) whose distribution can be observed now in the infrared (Kashlinsky et al. 2005). Currently, most distant galaxies can be seen at redshifts about 6 or 7, equivalent to distances of about 12 Gyr or 90% of the age of the Universe in lookback time. Since the Universe is in expansion, it was denser at these earlier epochs, so that interactions and collisions between self-gravitating objects were frequent. This certainly explains why galaxies observed at large distances seem rather disturbed and do not fit the Hubble morphological classification (van den Bergh 1998).

There are typically about 10^8–10^{12} stars in a galaxy. Once formed from the collapse of a cloud of gas and dust, stars evolve by themselves depending on their initial mass and chemical composition. The environment has very little effect on them, they essentially never collide, die, nor disappear. Only the most massive ones explode as supernovae, injecting into the interstellar medium some of their gas that is enriched in heavy atomic elements ('metallic', i.e., heavier than beryllium and lithium) and leaving a dense remnant that can be a neutron star or a small black hole.

Galaxies generally have a lot of gas and dust that evolve because of stellar radiation and density perturbations. The chemistry of atoms, molecules and grains is local and very complex but the orbital motions rather rapidly homogenise somehow the chemical properties within a galaxy. In particular, the collapse of molecular clouds that leads to the formation of stars generally happens more or less simultaneously over a significant fraction of the volume of the galaxy. This phenomenon relates global characteristics of galaxies with small-scale processes but it also smears out traces of past localised events.

The properties of the three fundamental constituents (stars, gas and dust) fully describe a given galaxy, provided we have access to them at all places in the three dimensions. This implies that the full description of a galaxy is already complex.

Moreover, any gravitational perturbation, either caused by external passing-by galaxies or by internal interactions, modifies the shape of the galaxy and possibly properties of gas and dust. Even internal events, like the explosion of many supernovae at nearly the same time or density waves, may have a strong influence. Since there are so many components in a single galaxy, their mutual interactions through radiation, shock waves, gravitational perturbations, chemical and physical processes and so on, are essentially non-linear and chaotic, the evolution of a galaxy truly belongs to the complex sciences.

Nowadays, observations are becoming very detailed and numerous. Systematic surveys feed databases comprising millions of galaxies for which we have images and spectra. Big telescopes and very sensitive detectors allow us to observe objects from the distant past (galaxies at high redshifts). This is somewhat reminiscent of paleontology and like evolutionary biologists, astronomers want to understand the relationships between distant and nearby galaxies, like our own. Strangely enough, the Hubble classification, born from the Hubble diagram, is still frequently used as a support to describe galaxy evolution, even though it ignores all observables except morphology (i.e., Hernandez and Cervantes-Sodi 2006; Cecil and Rose 2007). Indeed, the Hubble classification is very successful not only because it is simple but also because the global shape of a galaxy grossly summarises several physical properties. In particular, star formation is more efficient in disks and it seems that spiral galaxies have also more gas than ellipticals. Obviously, the distribution of the stellar orbits is related to the history of a galaxy. Numerical simulations are easier to make with stars and gravitation only, the gas and dust components requiring more complicated equations and much more computer power. Thus, many simulations have shown precisely how the structures of galaxies transform themselves during interactions or merging (Bournaud et al. 2005). For instance, ellipticals have been shown to result from the merging of two galaxies of similar masses. However, the inclusion of gas in the simulations now leads to other possible formation scenarii (Bournaud et al. 2007; Ocvirk et al. 2008).

Other classifications of galaxies are somehow inherited from this morphological Hubble classification and are most often dictated by the instrument used to make the observations of the sample. Correlation plots are used to make crude classification in a very few number of categories (often two like blue and red, high and low intensity for a given wavelength, more or less metallic, ...). This is, most of the time, sufficient for some physical modelling but it cannot describe the huge diversity of galaxies across the Universe and their now recognised complex history in a very objective and synthetic way.

The understanding of galaxy formation and evolution, which we prefer to call galaxy diversification, is a major challenge of contemporary astrophysics. Abandoning the one-parameter classification approach and using all available descriptors means taking a methodological step equivalent to the one biologists took after Adanson and Jussieu in the eighteenth century. Today the tools do exist and ordination methods, essentially the one of Principal Component Analysis, are being used more often, mainly to automatically separate stars and quasars from galaxies on large images of the sky (e.g., Cabanac et al. 2002). A few attempts to apply clustering methods have been made recently, still with little success in identifying new classes that convince the astronomical community (e.g., Chattopadhyay and Chattopadhyay 2006). From our point of view, there are two difficulties here. The first one is that the PCA components are non-physical, hence very difficult to interpret and model in the way astrophysicists are used to. The second one, more important from our point of view, is that evolution, an unavoidable fact, is not at all taken into account. By mixing together objects at different stages of evolution, any physical significance of a classification is undoubtedly lost. This was precisely our motivation for developing astrocladistics in 2001 (Fraix-Burnet et al. 2006a, b, c).

21.2 The Astrocladistics Project

21.2.1 A Phylogenetic Framework for the Galaxies

As we have seen in Section 21.1, the baryonic matter is immersed in the gravitational field that is shaped by the dominating dark matter. Even though its nature is totally unknown, the dark matter is supposedly affected only by gravitation. This is relatively simple physics and it enabled the first numerical simulations of the cosmological evolution of the gravitational field in the expanding Universe from the observed primordial fluctuations up to the present time. Since matter attracts matter, tiny overdensities, called halos, progressively grow in mass and size by merging together. This is the hierarchical model of formation of the large structures in the Universe.

The evolution of dark matter halos is generally represented by a ‘merger tree’, a typical one being shown on Fig. 21.1a (Stewart et al. 2008). Many small halos at the top (redshift of about 7 equivalent to about 13 Gyr ago) of the tree merge while time goes downward to yield large dark matter halos observed in galaxy clusters. These merger trees indeed represent the genealogy of a single halo. A schematic representation of these trees is often used but they tend to suggest that small halos disappear with time. This is not true and the tree on Fig. 21.1a clearly shows that some of them ‘survive’. Hence, at any epoch in the Universe, halos of different sizes with different merging histories coexist.

From the astrocladistics point of view, the hierarchical evolution of the dark matter halos is better represented on a phylogenetic tree, where mass (or size) is the only criterion for diversity (Fig. 21.1b). This tree describes the evolution of the environment in which galaxies form and evolve.

Galaxies are made up of baryonic matter that is sensitive to gravitation but also to electromagnetic, weak and strong interactions, as well as radiation and thermodynamics. Its physics is thus very complex and there are many processes that can strongly affect its gravitational behaviour. Probably because of the first cosmological numerical simulations that were able to take the sole gravitation into account,

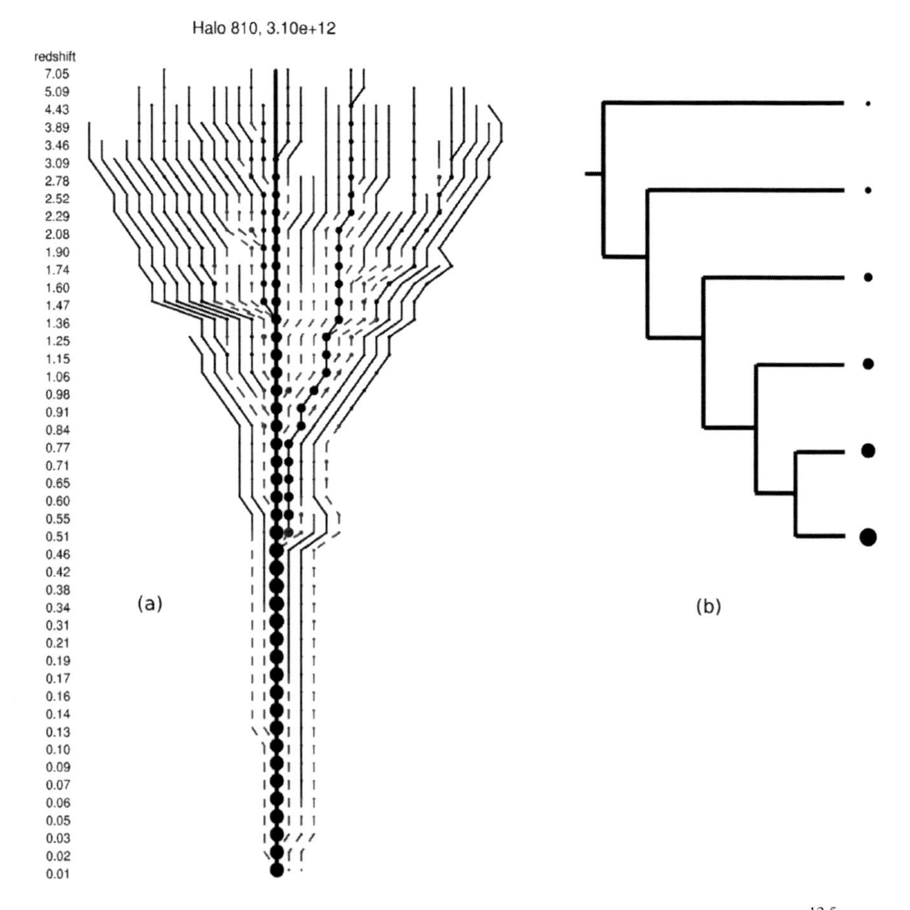

Fig. 21.1 (**a**) A 'typical' merger tree for an individual dark matter halo with mass $\simeq 10^{12.5}$ solar masses at present time (redshift = 0). Time progresses downward, with the redshift printed on the left-hand side. The bold, vertical line at the centre corresponds to the main progenitor, with filled circles proportional to the radius of each halo. The minimum mass halo shown in this diagram has $m = 10^{9.9}$ solar masses. Solid and dashed lines and circles correspond to isolated field halos, or subhalos, respectively. The dashed lines that do not merge with the main progenitor represent surviving subhalos at $z = 0$. (Taken from Stewart et al. 2008 and reproduced by permission of the AAS.) Such trees are indeed genealogical trees for individual halos. (**b**) The corresponding phylogenetic tree as seen by astrocladistics. There is no time scale in this representation. Mass or size is the sole diversification criterion and increases downward. Each bifurcating node corresponds to a merger event and indicates the first occurrence in the Universe of a halo of a given mass. The chronology thus goes from left to right

galaxies are generally thought of as being at the centre of dark matter halos. As a consequence, the hierarchical formation scenario seems to apply naturally to galaxies and has been very popular up to now (e.g., Baugh 2006; Avila-Reese 2007).

However, the old formation scenario for galaxies, called the monolithic model, gains new consideration thanks to new numerical simulations and observations.

This model indeed describes the collapse of a cloud of gas and includes all the necessary physics that is not included in the pure hierarchical model. Merging of galaxies does exist but it cannot be of the same nature as for the dark matter since the gravitational energy can be transformed into heat, shocks or radiation. Consequently, the diversification of galaxies cannot follow the same simple hierarchical scenario as that of the dark matter represented in Fig. 21.1.

Astrocladistics makes it clear that the formation and the evolution of galaxies, governed by a complex physics, occur in an environment whose evolution, distinct and largely independent, is governed by sole gravitation. Galaxies interact between each other while they move in the gravitational field shaped by the dark matter. To a first approximation, galaxies can be good tracers of the dark matter but as recent observations show, the two can sometimes be well disconnected.

It is still difficult to tell what the very first objects were. Since the Universe was essentially homogeneous at the recombination epoch, the primordial gas was certainly not very clumpy. With time, with the growth of the tiny initial fluctuations in dark matter halos, gas density began to increase in some places. When these seeds became gravitationally bound, we could call them galaxies. It is probable that the first stars formed very rapidly during this process. These very first galaxies were probably already quite diverse in mass content (gas/stars) but very similar otherwise. They also subsequently merged rapidly, interacted strongly with their neighbours, forming new and still more diverse objects. The expansion of the Universe definitively separated large regions that make what we call superclusters and clusters of galaxies. These can be seen as islands in which the corresponding populations of galaxies have evolved on their own, possibly creating some evolutionary divergence.

However, physics is the same everywhere (cosmological principle), the environment is purely gravitational and differs from place to place only by its shape in the curved space–time. We thus expect a lot of similarities between different 'populations' of galaxies due to convergent and parallel evolutions. Unfortunately, our quest to understanding diversification is somewhat complicated by the finite velocity of light. We do not have access to the entire Universe, hence to the entire galaxy diversity, which is contemporary to us because when galaxies are a bit far away, we see them as they where when the light was emitted. We are like paleontologists, except that we do not have access to the entire present diversity. Nonetheless recent observations seem to suggest that galaxy evolution has been very gentle in the last 1 Gyr or so.

21.2.2 Transmission with Modification Among Galaxies

Galaxy formation is now acknowledged to occur continuously during the Hubble time (Hoopes et al. 2007). In extragalactic astrophysics, evolution and formation are generally not clearly distinguished. Indeed, galaxy evolution is more like galaxy transformation, which leads to the formation of new kinds of objects. Formation often seems to be employed as if big spiral and elliptical galaxies are at the end of evolution. From our point of view, 'galaxy formation and evolution' can be more precisely called galaxy diversification, in which evolution means transformation and formation refers to diversity.

There are five transformation processes for galaxies: assembling, secular evolution, interaction, accretion/merging, ejection/sweeping. In all these cases, the fundamental material of a galaxy, stars, gas and dust, is transmitted to the new object, somewhat modified. These modifications are mainly kinematical because the gravitational perturbations change the orbits, hence the distribution of the objects. The stars are otherwise unaffected, except that new ones can be formed from the perturbed gas and dust. These latter two constituents can also be modified in density, temperature and composition.

Schematically, galaxy diversification can be depicted like in Fig. 21.2. At a given time, a galaxy undergoes a transformation event to form a new object made with the same constituents but modified. Between two such events, stars get older and essentially cooler (redder). This is part of the secular evolution, which is the evolution of a totally isolated galaxy. Secular evolution is a transformation process driven by internal phenomena (stellar ageing, interaction between the three constituents, density waves, shock waves due to supernovae, instabilities, ...), which is generally slow and non-violent. However, after a certain time, the galaxy can be so different that it can be described as a new object and even belong to a different class of objects. Like for living species, the modifications are gradual and at some level of details, the splitting in different species is somewhat arbitrary. Thus, secular evolution can also be a transformation event participating to the diversification of galaxies.

Figure 21.2 shows that it is possible to see the transformation of galaxies as a transmission with modification process. It is not exactly a Darwinian process in the sense that there is no natural selection (even though big galaxies survive more easily and tend to swallow small ones!). Also, there is no duplication but each transformation process is a replication mechanism, because the new object has inherited all constituents and most of their properties. Since the transformation events are always random, this is equivalent to a replication with 'spontaneous mutations'. In addition, the environment, which is gravitational, has a strong influence on the occurrence and properties of all the galaxy transformation processes.

Galaxy diversification is thus characterised by replication, randomness and environment. It follows that a branching pattern can be expected. This is why

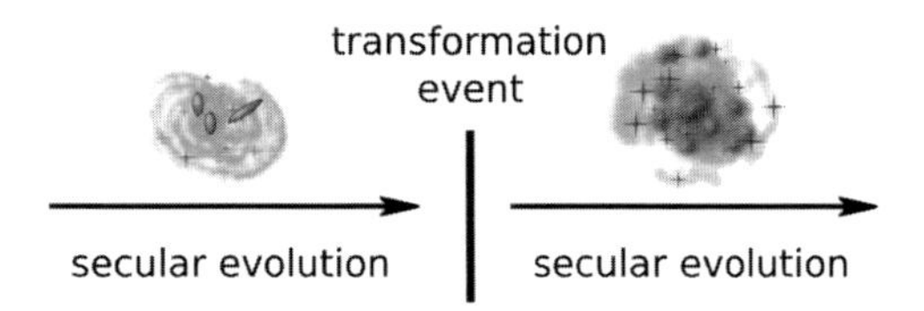

Fig. 21.2 A schematic view depicting the transmission with modification process in galaxy diversification. For any transformation event, gas, dust and stars are transmitted to the new object generally with some modification of their properties. Secular evolution is defined as the evolution of a totally isolated galaxy. It thus occurs all the time and even if it is not violent in general, it can sometimes lead to a significant transformation

cladistics should be applicable to this problem. However, it might not be so simple since mergers are probably frequent at least near the beginning of the Universe, so that reticulation ('hybridation' or horizontal transfer) could be present. Cladistic analyses of several samples of galaxies can investigate this. As shown in Fraix-Burnet et al. (2006a, b) and confirmed by all the robust cladograms we have obtained so far, reticulation does not seem to be dominant. Yet, this conclusion must be seen as preliminary and more investigation should be done, particularly with very distant objects.

21.3 Applying Cladistics to Galaxies

Astrophysics is an observational science, not an experimental one. It is not possible to weigh a galaxy or to return it to see what is behind. We simply collect photons, at all wavelengths (from gamma-rays and x-rays to the radio domain, with the ultraviolet, optical, infrared, far-infrared and submillimetre in- between), which are emitted by every component at every place in a galaxy. The level of details is limited by the faintness and the small apparent projected size of the objects, particularly those at very high distances from us. Both improve with the diameter of the photon collector (the mirror of the telescope) and with the sensitivity of the detectors. Requirements for the astronomical observations are a strong thrust for technology research and development and recent progresses in this domain have changed the scale in the amount of data concerning galaxies. But still, each picture element ('pixel', i.e., the smallest detail) is a mixture of the light emission from many stars and gas/dust clouds situated along the line of sight and across a region whose size depends on the distance of the galaxy.

Basically, observations result in the recording of spectra. Apart from the structure, given by imagery or relative spatial positioning, spectra contain all the physical and chemical information that is possible to obtain from remote objects. Systematic surveys of the sky now provide spectra for millions of galaxies. However, in addition to the limitations mentioned above, there are limits coming from the spectral resolution of the detectors and from Doppler effects due to the differential motions of the galaxy components that enlarge and blur absorption and emission lines. As a consequence, physical and chemical conditions at each part of a galaxy are always partial and averaged.

Measuring characteristics from spectra is also not sufficient to derive physical or chemical quantities, some combinations must be used like ratios of several emission lines. Moreover, models must most often be used to disentangle the complexity of the emitting region and to translate spectral information into physical quantities. Unfortunately, these models necessarily introduce some subjectivity. Thus, truly intrinsic descriptors of galaxies are not easy to obtain and this is certainly a specific difficulty of astrocladistics.

Anyhow, descriptors in extragalactic astrophysics are always continuous variables since, apart from spectral information, the other information we can measure is the dimension (size and position of individual components within the galaxy). The only exception is the morphology. It is traditionally based on the Hubble classification represented by the Hubble diagram, most often with the use of a discrete scale distinguishing different kinds of elliptical and spiral galaxies. However, in our astrocladistic project, we have chosen not to use this parameter for several reasons. The first one is that this is a very crude way of discretising a shape parameter because it is essentially made by eye. The second reason is that this discretisation is not homogeneous in the sense that differences between spirals are given the same weight as between ellipticals, while they are not of the same nature (shape and number of the spiral arms that are density waves, not physical structures, for the first ones and estimation of roundness for the latter ones; de Vaucouleurs 1994). The third reason is that it is not clear at what level this descriptor can be informative regarding evolution. Finally, we know that the Hubble classification is correlated with quite a few global properties of galaxies (kinematics, amount of gas and star formation most notably), so it is certainly redundant with other descriptors, quantitative and objective, coming from the spectrum.

For such an exploratory project like astrocladistics, choices have to be made right at the beginning to focalise the research. Up to now, we have chosen to discretise all observables into 10 or 30 bins, depending on the error bars estimated by the observers and on the distribution of values among the sample. This allows us to use parsimony as the optimisation criterion, which seems to be the simplest strategy to implement for a cladistic analysis. These first choices already enabled us to obtain very positive results. Other paths can now be explored.

Last but not least, we do not have a multivariate classification of galaxies at our disposal. We have to assume that each galaxy represents a 'species' that will have to be defined later on. It is consequently difficult to define an outgroup that is necessary to root the tree and help the interpretation of the evolutionary scenario. Most generally, we root the trees with the object that is the less metallic, that is, in which the relative abundance of heavy atomic elements is the lowest (heavy elements are only produced in stars and cannot be primordial). But we are aware that this choice is made on one character only.

21.4 The First Extragalactic Trees

The first application of astrocladistics was performed on a sample of galaxies issued from a cosmology simulation. This work used 50 objects and 50 characters (observables), corresponding to five 'lineages' sampled at 10 epochs. In such simulations, each entity, called a galaxy, which appears, is given a number. Only accretion or merging is considered and a new number is attributed each time such an event occurs. It is then easy to follow the tree of transformations of an initial galaxy into its different descendants. The simulations calculate the radiation from galaxies and transpose it into the same observables as obtained from real telescopes. This work was intended to illustrate the concepts and practical methodology of astrocladistics and to show how the correct 'genealogy' can be reconstructed (Fraix-Burnet et al. 2006a, b) from usual observables.

More importantly, a small sample of real galaxies (Dwarf Galaxies of the Local Group, 36 objects, 24 characters) have also successfully been analysed, providing the first evolutionary galaxy tree ever established (Fraix-Burnet et al. 2006c). These dwarf galaxies are very numerous throughout the Universe. Most of them are satellites of big galaxies like our own. They supposedly have been the first galaxies at the beginning of the Universe and are thus believed to be the building blocks that merged to make bigger and bigger galaxies. A fully resolved tree obtained on a subset is represented in Fig. 21.3. Bootstrap and decay values indicated at the nodes show that it is very robust. The groupings are consistent with our current knowledge of these particular objects, with a more refined classification. This tree clearly

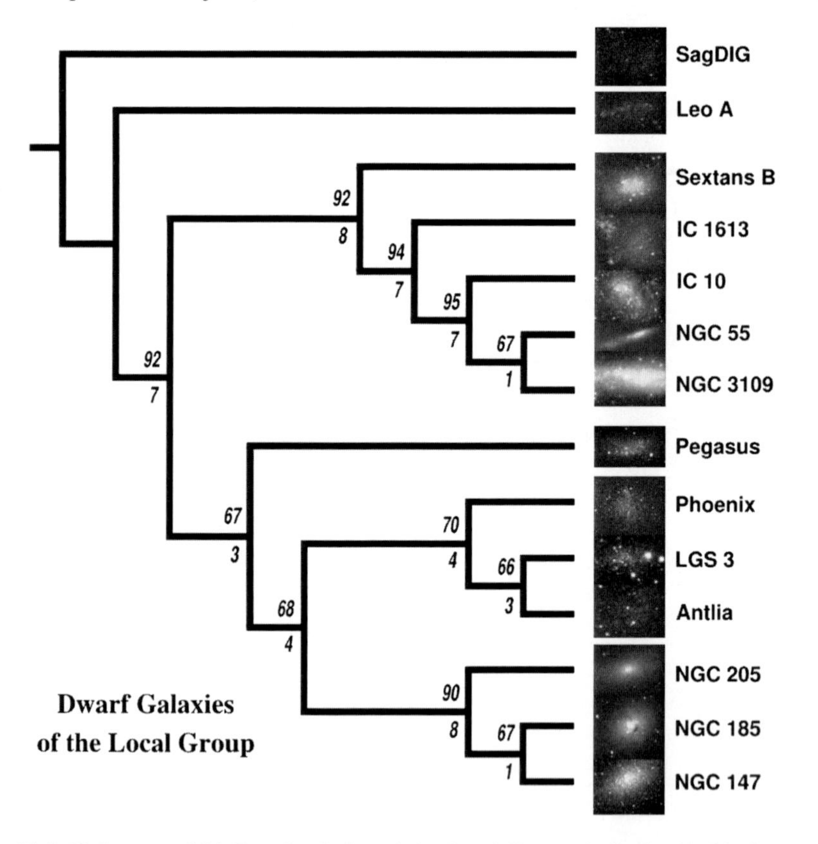

Fig. 21.3 Cladogram of 14 Dwarf galaxies of the Local Group obtained with 24 characters (observables and derived quantities). Bootstrap values (*above*) and decay indices (*below*) are indicated for each node. The outgroup (SagDig) has been chosen because it contains the lowest amount of metallic material, suggesting that it is made up of more primordial material (see Fraix-Burnet et al. 2006c for details on this work). Image credits: NGC~55 : David Malin; LGS~3 and Pegasus~DIG: Deidre A. Hunter; Antlia: Mike Irwin; NGC~185: David M. Delgado; NGC~147: Walter Nowotny; Sag~Dig: Hubble Heritage Team (AURA / STScI), Y. Momany (U. Padua) et al., ESA, NASA; Leo~A, Sextans~B, IC~1613 and IC~10: Corradi, R.L.M. et al., 2003, ING Newsletter No.~7, p. 11; NGC~3109: NASA/ STScI; Phoenix: Knut Olsen (CTIO) \& Phillip Massey (Lowell Observatory), (NOAO / CTIO / KPNO); NGC~205: Atlas Image [or Atlas Image mosaic] courtesy of 2MASS/UMass/IPAC-Caltech/NASA/NSF

implies that the nature of the ancestors is of irregular shape. The other type, the spheroidals, is consequently more evolved. This question is debated in the astrophysical community and is naturally addressed by astrocladistics in a multivariate way. It is important to note that the characters used in this analysis are a mixture of observables (like colours, fluxes in different spectral lines and composition ratios) and derived quantities (like masses, maximum rotational velocities, velocity dispersion, star formation rate). Such kind of information is more difficult to obtain with other more distant galaxies.

We subsequently performed a cladistic study of 222 galaxies of the Virgo cluster using 41 characters that were all observables. Because of too many unknown parameters, we were able to build a robust tree for 123 objects only and used it to construct a supertree. Figure 21.4 shows the cladogram with 123 galaxies. It took us some time to understand the astrophysical consequence of the regularity of the (largely unbalanced) tree that is also found on the supertree. In fact, this is due to the sole stellar evolution that is hidden in nearly all of the observables. It is consequently highly redundant and dominant in this multivariate analysis. The astrophysical outcome is thus slightly disappointing. We obtained similar results on other samples: poor gas galaxies of the field (comprising 227 objects and 37 characters), galaxies from a big survey (500 objects with 60 characters analysed so far).

From the astrocladistic point of view, all these studies show that our approach is successful in constructing cladograms of galaxies in several different types of samples. Nevertheless, the regular trees found when using solely observables reveal that stellar evolution is universally present and acts as a parallel evolution. It does not bring much information on galaxy diversity and disturbs the multivariate analyses. We call it 'cosmic evolution' because it is universal and does not lead to a sufficient transformation of the galaxy, like secular evolution in general or the other transformation processes. Even in the most violent cases, a given stellar population continues to age unperturbed and its radiation properties are unaffected unless the population itself is changed for instance with the addition of newborn stars. Somehow, we could think that cosmic evolution is a kind of ontogeny for galaxies. We looked for a way to remove it from the observables because it obviously hampers the search for 'speciation'. We used various synthetic stellar populations, modelling artificial lineages with specific properties. We restricted ourselves to small samples (30 objects). Our idea was to scrutinise each observable in a very well controlled sample. By trial and error, we found some combinations of several observables that make the corrected characters to remain essentially constant between transformation events (Fraix-Burnet et al. in preparation). The physical interpretation of these corrected characters is still unclear but the important result is that they lead to physically consistent clades for these simple synthetic stellar populations. Application to real objects is under progress.

Globular clusters are self-gravitating ensembles of stars without dust or gas. In a sense, they are simple galaxies, somehow like our synthetic stellar populations. Taking advantage of our experience gained so far, we performed a cladistic analysis of 54 globular clusters of our Galaxy by choosing four properties. Three of them are intrinsic and come from their conditions of formation. They are not affected by

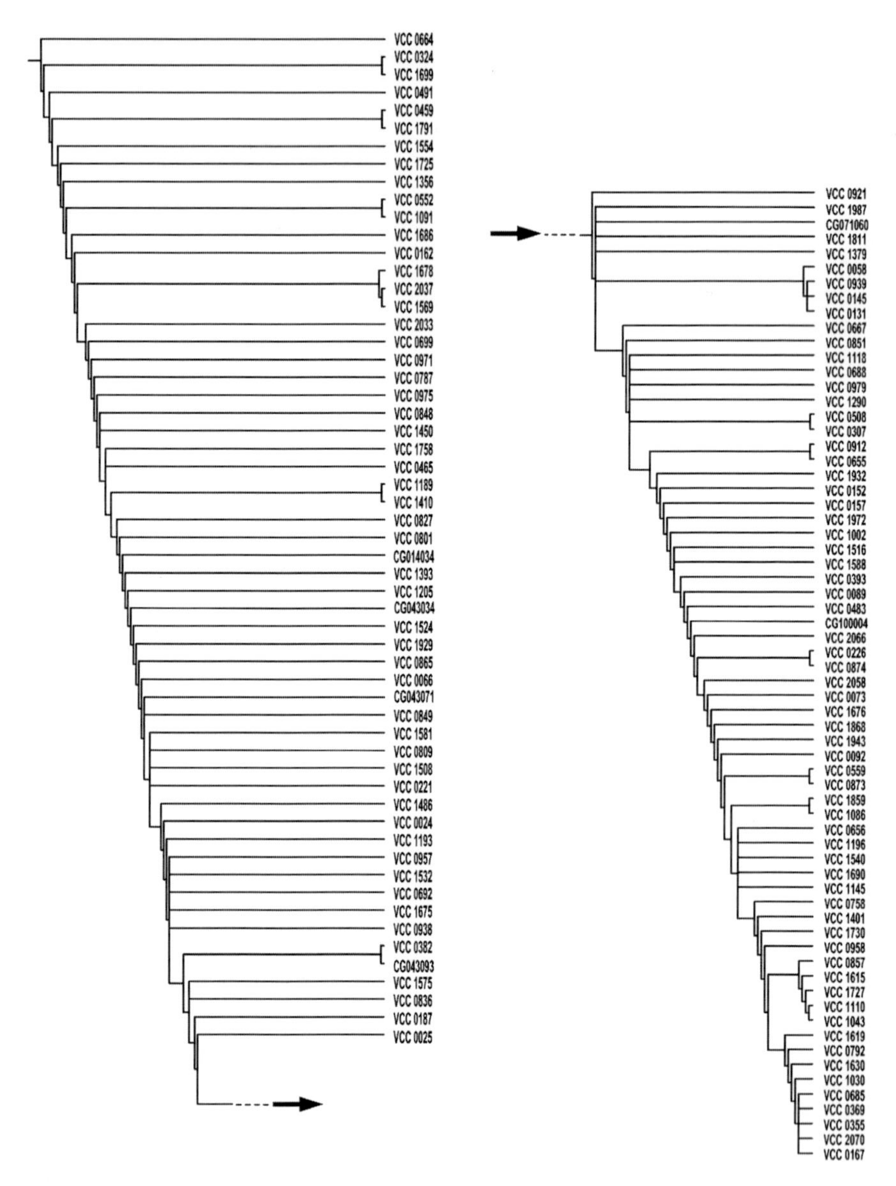

Fig. 21.4 Cladogram for 123 galaxies of the Virgo cluster with 41 observables. Note the regularity of the tree indicating the cosmic evolution (star ageing), which is highly redundant

cosmic evolution. The other one is the colour and measures the age of the stars, it is thus a direct indicator of the cosmic evolution. We gave it a weight half of the other ones and effectively rank the clusters chronologically within the lineages or groupings that are defined by the other three characters. Our result tree gathers the objects in a new classification that brings a very convincing scenario for the formation of these objects and consequently for some episodes in the building up of our own galaxy. This result is the first spectacular outcome of astrocladistics and will hopefully convince more astronomers to invest in this novel approach (Fraix-Burnet et al. submitted).

21.5 Some Open Questions

It is now clear that cladistics can be applied and be useful to the study of galaxy diversification. Many difficulties, conceptual and practical, have been solved but many remain in the exploration of this large research field. To begin with, it is possible that probabilistic methods, instead of parsimony, might be better suited for the quantitative nature of the variables and the evolution of galaxies. In this respect, the use of continuous data without discretisation will have to be investigated seriously, even with parsimony (Wiens 2001; MacLeod 2002; Goloboff et al. 2006; Gonzàlez-José et al. 2008).

There are difficulties that seem to be intrinsic to astrophysics. Most notably, we have millions of objects but a few tens of descriptors. Of course the situation will improve with time, in particular with integral-field spectroscopy (spectra of detailed regions in the galaxy, e.g., Emsellem et al. 2007). Spectra might not be currently employed at their full capacity of description. But it is not clear whether they would lead to the discrimination of hundreds of classes. Perhaps this is an erroneous target, perhaps galaxies cannot be classified with such a refinement. But this is already a matter of using multivariate clustering methods and interpreting their results usefully. We are convinced that improvements can be made here. In any case, cladistics is supposed to identify clades that are evolutionary groups, whereas the concept of 'species' is not defined at all in astrophysics and we have even not converged toward groupings based on multivariate analyses.

This is probably more than a conceptual question because clustering with continuous data and large intraspecific variance is a very complicated problem in itself. Sophisticated statistical tools must be used but the question of characterising the groups in this context is not very clear to us at this time. This probably requires a different culture that is not yet present in astrophysics. Incidentally, this also points to the relevance of discretisation we have used so far, particularly when measurement uncertainties are sometimes quite large. Our results definitively show that we are not wrong but it might be possible to do better.

The possible predominance of reticulation is a concern for cladistics in general and we foresee to investigate this question very soon. In some ways, galaxy diversification has similarities with bacteria diversification for which reticulation methods have been considerably developed recently. We intend to take advantage of these advances. It would already be interesting to detect some reticulation in the samples we have studied so far. The equivalent of hybridation or horizontal transfer certainly comes through the merging of two galaxies. Since the Universe is in expansion, object density was higher at earlier times and interactions and merging were more common. As a consequence, reticulation could be a problem for distant galaxies, which is when the Universe was young.

A very important step for astrocladistics is our assessment of the influence of the cosmic evolution and our ability to correct for it or avoid it in the character selection. But there are probably other kinds of parallel or convergent evolutions and reversals as well. Hence, we are now concentrating our efforts on finding the most pertinent descriptors. For this we must detail all the possible quantities we can obtain from the observations, to see how they behave during the different transformation processes and whether they can be reliable tracers of past events from a cladistic point of view. Biologists pretty well know that this is crucial work, apparently without an end.

21.6 Conclusion

The astrocladistic project has now reached a solid base as far as concepts and tools are concerned. First results have shown the usefulness of this approach and unavoidable difficulties are being solved one after one. This triggers the hope of being able to map galaxy evolution from the very first objects of the Universe, some 14 Gyr ago but there is still a long way to go. Astrocladistics opens a new way to analyse galaxy evolution and a path towards a new systematics of galaxies.

Extragalactic astrophysics is currently undergoing a scaling revolution. The number of objects and the amount of data it has to cope with already impose new statistical tools to be used and this will certainly get worse with still bigger telescopes and more sensitive detectors. Less than a century after their discoveries, galaxies come into our databases by the millions. This is a cultural revolution, in particular because the complexity of these objects and their evolution become more obvious.

Astrocladistics is fascinating in combining a multivariate statistical philosophy with evolutionary concepts. It is fascinating as an interdisciplinary research field. But it is also quite a difficult approach, not intuitive at all and does not fit well in the usual methods of physicists. This is a huge research field, very exploratory, with many possible paths to follow. We still lack spectacular astrophysical outcomes but most astrophysicists are very enthusiastic, because they agree that there is an obvious need for multivariate and evolutionary methods to study galaxy diversification.

References

Avila-Reese V (2007) Understanding galaxy formation and evolution. Ap&SS Proceedings, 'Solar, stellar and galactic connections between particle physics and astrophysics', Carramiñana A, Guzmán FS, Matos T (eds), Springer, pp 115–165. http://arxiv.org/abs/astro-ph/0605212

Baugh CM. (2006) A primer on hierarchical galaxy formation: the semi-analytical approach. Rep Prog Phys 69(12):3101–3156. http://arxiv.org/abs/astro-ph/0610031

Bournaud F, Jog CJ, Combes F (2005) Galaxy mergers with various mass ratios: Properties of remnants. Astron Astrophys 437:69–85. http://arxiv.org/abs/astro-ph/0503189

Bournaud F, Jog CJ, Combes F. (2007) Multiple minor mergers: formation of elliptical galaxies and constraints for the growth of spiral disks. Astron Astrophys 476:1179–1190. http://arxiv.org/abs/0709.3439

Cabanac RA, de Lapparent V, Hickson P (2002) Classification and redshift estimation by principal component analysis. Astron Astrophys 389:1090–1116. http://arxiv.org/abs/astro-ph/0206062

Cecil G, Rose JA (2007) Constraints on galaxy structure and evolution from the light of nearby systems. Rep Prog Phys 70:1177–1258. http://arxiv.org/abs/0706.1332v2

Chattopadhyay T, Chattopadhyay AK (2006) Objective classification of spiral galaxies having extended rotation curves beyond the optical radius. Astron J 131:2452–2468

de Vaucouleurs G (1994) Global physical parameters of galaxies. In: Quantifying galaxy morphology at high redshift. Space Telescope Science Institute, Baltimore, MD, April 27–29 1994 http://www.stsci.edu/institute/conference/galaxy-morphology/program2.html

Emsellem E, Cappellari M, Krajnović D, van de Ven G, Bacon R, Bureau M, Davies RL, de Zeeuw PT, Falcón-Barroso J, Kuntschner H, McDermid R, Peletier RF, Sarzi M (2007) The SAURON project – IX. A kinematic classification for early-type galaxies. Mon Not R Astron Soc 379:401–417. http://arxiv.org/abs/astro-ph/0703531v3

Fraix-Burnet D, Choler P, Douzery E, Verhamme A (2006a) Astrocladistics: a phylogenetic analysis of galaxy evolution. I. Character evolutions and galaxy histories. J Classification 23:31–56. http://arxiv.org/abs/astro-ph/0602581

Fraix-Burnet D, Douzery E, Choler P, Verhamme A (2006b) Astrocladistics: a phylogenetic analysis of galaxy evolution. II. Formation and diversification of galaxies. J Classification 23:57–78. http://arxiv.org/abs/astro-ph/0602580

Fraix-Burnet D, Choler P, Douzery E (2006c) Towards a phylogenetic analysis of galaxy evolution: a case study with the dwarf galaxies of the local group. Astron Astrophys 455:845–851. http://arxiv.org/abs/astro-ph/0605221

Goloboff, PA, Mattoni CI, Quinteros AS (2006) Continuous characters analyzed as such. Cladistics 22:589–601

Gonzàlez-José R, Escapa I, Neves WA, Cúneo R, Pucciarelli HM (2008) Cladistic analysis of continuous modularized traits provides phylogenetic signals in Homo evolution. Nature 453:775–779

Hernandez X, Cervantes-Sodi B (2006) A dimensional study of disk galaxies. 11th Latin-American Regional IAU Meeting 2005. RevMexAA (Serie de Conferencias), 26, 97–100 (2006) http://arxiv.org/abs/astro-ph/060225

Hinshaw G, Weiland JL, Hill RS, Odegard N, Larson D, Bennett CL, Dunkley J, Gold B, Greason MR, Jarosik N, Komatsu E, Nolta MR, Page L, Spergel DN, Wollack E, Halpern M, Kogut A, Limon M, Meyer SS, Tucker GS, Wright EL (2009) Five-year Wilkinson Microwave Anisotropy Probe (WMAP) observations: data processing, sky maps, and basic results. Astrophys J Suppl 180:225–245. http://arxiv.org/abs/0803.0732

Hoopes CG, Heckman TM, Salim S, et al (2007) The diverse properties of the most ultraviolet-luminous galaxies discovered by GALEX. Astrophys J Suppl Series 173:441–456. http://arxiv.org/abs/astro-ph/0609415

Kashlinsky A, Arendt RG, Mather J, Moseley SH (2005) Tracing the first stars with fluctuations of the cosmic infrared background. Nature 438:45–50. doi:10.1038/nature04143

MacLeod N (2002). Phylogenetic signals in morphometric data. In: MacLeod N, Forey PL (eds) Morphology, shape, and phylogeny. Taylor & Francis, London, pp 100–138

Ocvirk P, Pichon C, Teyssier R (2008) Bimodal gas accretion in the Horizon-MareNostrum galaxy formation simulation. Mon Not R Astron Soc 390:1326–1338

Stewart KR, Bullock JS, Wechsler RH, Maller AH, Zentner AR (2008) Merger histories of galaxy halos and implications for disk survival. Astrophys J 683:597–610. http://arxiv.org/abs/0711.5027

van den Bergh S (1998) Galaxy morphology and classification. Cambridge University Press, Cambridge

Wiens JJ (2001) Character analysis in morphological phylogenetics: problems and solutions. Syst Biol 50:689–699

Chapter 22
Economics Pursuing the Mold of Evolutionary Biology: "Accident" and "Necessity" in the Quest to make Economics Scientific

Michael H. Turk

Abstract Efforts to ground economics as a science have long been based upon an alignment of it with the physical or natural sciences. Leon Walras championed a model drawn from physics; Alfred Marshall saw a closer relationship with biology. These models tend to emphasize the centrality of foundational concepts like maximization and equilibrium. As an alternative basis for such an alignment, one might look to methodology rather than content. In particular, the distinction between "accident and necessity," advanced by Jacques Monod in evolutionary biology, bears a striking resemblance to the tension between the idiographic and the nomothetic raised by economists. Ultimately, notwithstanding the recent efforts by Paul Krugman, Paul David and Brian Arthur to bridge the gap between the "accidents" of history and the "necessity" of economic rule-making, economics continues to fall short, limiting the discipline's ability to form what Thorstein Veblen called an "evolutionary science."

22.1 Introduction

What makes economics scientific? For that matter, should economics, a discipline in the social sciences, be comprehended as a science akin to astronomy, biology, chemistry, or physics? This is a matter of some contention, as economics has long held out claims to having, or at least of aspiring to have, the form and rigor of the natural or physical sciences. To some degree a change in nomenclature captures this tension, as the study of "political economy," as the discipline was originally called, became the "science of economics" in the latter part of the nineteenth century. It was during this period that various groups of economists in England, France, Switzerland

M.H. Turk
Professor of Economics, Department of Economics, Fitchburg State College, History, and Political Science, Fitchburg, 160 Pearl Street, 01420 MA, USA
e-mail: mturk@fsc.edu

P. Pontarotti (ed.), *Evolutionary Biology: Concept, Modeling, and Application*,
DOI: 10.1007/978-3-642-00952-5_22,

and Austria known as the marginalists engaged in a rethinking of economics that led to the emergence of what came to be known as neoclassical economics and, over time, its increasingly mathematized and often highly abstracted treatment.

This quest to be scientific did not come without complications. The countervailing view of economics as a fundamentally social science, emphasizing the role of institutions, human agency, historical experience, or the interactive nature of human behavior in economic affairs, where the interpretation of events may itself influence the actions that follow, did not vanish. More often than not, advocates of this approach have been dissenters from the mainstream of economic thought over the last century and they have spanned the range of ideological perspectives, bracketing the neoclassical mainstream from the left and the right.

Nor was the appropriate form, or even the measure of rigor, for scientific economics clearly settled—or readily comprehended. For example, did the mathematization of economics offer up only the appearance of rigor, without necessarily establishing the substance or structure of a suitably scientific discipline? Moreover, if economics was to be understood as bearing some kinship with or resemblance to a "hard" natural science, which such science would best serve as the model?

It is this last question that brings into focus attempts that have been made to fit economics into the mold of evolutionary biology. At the outset there is a certain irony here, since if economics has borrowed from evolutionary biology, the reverse is also the case. One need look no farther than Charles Darwin's *Autobiography* to note the source cited as the origin of his idea of selective evolution arising from a struggle for survival that was won by those best adapted to do so, namely Thomas Malthus' *Essay on Population*, an immensely popular and influential, if also draconian and chilling work of political economy, first published in 1797:

> In October 1838, that is, fifteen months after I had begun my systematic enquiry, I happened to read for amusement Malthus on *Population*, and being well prepared to appreciate the struggle for existence which everywhere goes on from long-continued observation of the habits of animals and plants, it at once struck me that under those circumstances favourable variations would tend to be preserved, and unfavourable ones to be destroyed. The result of this would be the formation of new species. Here, then, I had at last got a theory by which to work. (Darwin 2005: 98–99)

Nor should any facile association of economic considerations with evolution be deemed sufficient to establish a solid, substantive link between economics and evolutionary biology. As the economist Kenneth Arrow cautioned twenty years ago, while contemplating the possibility of better results in economic theorizing: "The analogy between evolution and technological progress has been almost a commonplace since the time of Darwin... but has in fact not led to much" (Anderson et al. 1988: 280).

22.2 The "Nomothetic Paradox" in Economics

An exploration of the basis for a possible fit of economics with evolutionary biology, as well as the limitations attendant upon that effort, requires a preliminary inquiry into the ways that economics might be seen as conforming to a structure, framework, or set of expectations generally associated with scientific disciplines, or as falling short in this regard.

The place to begin is in the realm of methodology. To a considerable extent the scientific claims of economics rest upon what might be termed its "nomothetic paradox." The tension between the nomothetic and the idiographic was derived from late nineteenth century German idealist philosophy, primarily the work of neo-Kantians like Hans Rickert and Wilhelm Windelband, though the use of these two descriptives to sort out the scientific claims of economics was promoted only decades later by the Austrian-American economist Fritz Machlup, in articles with titles like, "The Inferiority Complex of the Social Sciences" and "Are the Social Sciences Really Inferior?" (Machlup 1978a, Machlup 1978b)

Rickert and Windelband distinguished among intellectual fields of inquiry by positing an essential difference between universal rule-making or law-giving, characteristic of the nomothetic sciences, and individual fact gathering, the mark of idiographic disciplines. In Windelband's terms it was a distinction between sciences based upon laws, generally speaking the natural or physical sciences and those built out of events, the human sciences. As a philosophical inquiry it had been prompted by concerns about the solidity of the foundations for all the "human sciences," informed as they are by human activity, interaction and construction. For most of those engaged in this inquiry, economics, like history and political science, most likely belonged together as human sciences, though there was no unanimity on the subject.

However, a much sharper distinction between the generalizing quality of abstract principles and the more individuated amassing of facts figured centrally in the *Methodenstreit*, or "quarrel over method," waged between the Austrian economist Carl Menger and the German economist Gustav Stolper in the early 1880s. By and large, Menger, one of the leading marginalists noted above, prevailed: economics as a science was conceived of as built upon deduced universal principles and the power to abstract was embraced as essential to the method of economics (Menger 1963 [1882]). But this presented a paradox, for it meant that to be scientific economics would have to be decoupled from historical narrative and the daily economic affairs of humankind, from which it originally sprang, whether it be the accounts of British commerce or French agriculture that both formed and informed the first major writings in political economy.

How satisfactory was this result? And if economics was to be regarded as essentially nomothetic, how were its laws to be determined? In particular, what place did individual facts and societal events have in establishing or confirming these economic laws? Further, what kind of laws were these? Were they inevitable and inexorable, like the law of gravitational attraction, or more on the order of universally recurring patterns, likely tendencies that might be discerned?

The assertion of the nomothetic character of economics led to an alignment of economics with classical physics, which was taken as the paradigmatic science of law-making. For economists so inclined this meant consciously emulating the achievement of Isaac Newton in classical physics in reducing the multiplicity of the actions and movements of all objects, both at close range as well as at astronomical distances, to the simplicity and great explanatory power of a few laws of motion with universal application. And while the lure of Newton had attracted economists from the earliest days of political economy, drawing in such diverse figures as Adam Smith and Karl Marx, it was especially strong in the period when economics came to be conceived of as a science. Perhaps no one was a more ardent champion of a linkage between economics and physics than the French-Swiss economist, Leon Walras, who envisioned an economics as structured by analogy with classical mechanics but with its own—corresponding—fundamental principles and elementary elements. He espoused the use of mathematics in economics, which would make possible the resolution of markets by solving equations where the only critical variables were prices for and quantities of goods produced. While Walras' attempt to establish a general equilibrium across an economy on that basis has been widely critiqued, the turn to classical physics as a model for economics entered the mainstream of economic thought and has remained there (Turk 2006).

22.3 Possible Model in Biology

Even as this alignment of economics with physics took center stage in economic thought, though, there were alternatives posed by economists who still sought a suitable model to enhance the scientific legitimacy of economics through a correspondence of some sort with a broadly accepted scientific discipline but had doubts about the aptness of the analogy of economics to physics. In particular, these economists tended to look toward biology as a model. One might in fact note that a biological model, specifically the flow of blood through the body, inspired the French Physiocrat, Francois Quesnay, a political economist, administrator and court physician to Louis XV, in his adumbration of the circular flow of goods in the national economy, found in his *Tableau Economique*. Quesnay's construction is now widely regarded as the first substantial effort at system-building in economics.

But it is with the emergence of evolutionary biology that the countervailing approach to align economics with biology took hold. The first step in this regard took place with the recognition of a kinship between economics and biology as both being subjects of complexity. The British economist Alfred Marshall, who was active in the late nineteenth and early twentieth century, is often seen as the first major figure in the era of the science of economics who favored a correspondence between economics and biology. Marshall is credited with adapting marginalist economics into the broader synthesis of neoclassical economics and his textbook, *Principles of Economics,* shaped the general contours of microeconomics, as well as the graphical and mathematical techniques used in its analysis. These techniques clearly contributed to the comprehending of economics as akin to classical mechanics, a point acknowledged by Marshall himself.

Yet at the same time Marshall stated plainly in the *Principles* that economics more nearly resembled biology than mechanics, on at least two counts. First of all, economics, like biology, was concerned with living things; this immediately set it apart from physics. Second, the multiplicity of human activities could not ultimately be reduced to a few universal principles. Instead, economics was a subject whose phenomena and variables reflected the diversity of actions of living beings.

As Marshall noted in the *Principles of Economics*:

> [T]here are many [economic laws] which may rank with the secondary laws of those natural sciences, which resemble economics in dealing with the complex action of many heterogeneous and uncertain causes. The laws of biology, for instance.... (Marshall, 1898: 104)

Moreover, even if not stated, this also implied that economic phenomena might well be better comprehended as acting within a system.

Over the course of the twentieth century philosophical disquisitions about the nature of biology, whether by philosophers of science or evolutionary biologists, emphasized the central place of complexity in the framework of biology as a science. Thus herein lay a possible linkage between economics and biology.

A second strand in the thread tying economics to evolutionary biology can be found in the work of Marshall's contemporary and critic, the American economist Thorstein Veblen. Veblen faulted neoclassical economics for failing to take up the scientific challenge posed by Darwin's theory of evolution. Veblen saw himself and was seen by others, as a "Darwinian" among economists. He felt that the universal principles enunciated by economists, classical, or neoclassical, produced a self-contained and self-fulfilling system out of touch with the experience of everyday life and effectively prevented economics from claiming the mantle of science. Instead, Veblen envisioned a far more dynamic approach to economics, where social and cultural phenomena were integrated into the depiction and analysis of economic life and where, crucially, changes in economic life took place as adaptations of current mores and practices, producing an evolution of economic experience over time. Veblen's critique of the failure of economics to comprehend evolutionary change and development was presented forcefully in a 1919 paper entitled, "Why is economics not an evolutionary science?" (Veblen 1919)

Veblen, and similarly-minded economists, constituted a school of economic thought known as Institutionalism, which flourished in the United States in the first decades of the twentieth century. They dissented from the mainstream of economic thought in that period, often challenging the complacency of neoclassical economists in accepting the economic *status quo* and evoked a general theme: changes in economies occur on a cumulative basis, as economies evolve over time.

As a school Institutionalism helped spawn an interest in and the development of an "evolutionary economics" for which dynamic processes, institutional changes and cumulative effects were seen as inextricably linked. In recent decades this approach has to some degree taken on a more conservative cast, in line with a "new Institutionalism" largely reconciled to the virtues of the *status quo*.

A further variant upon the notion of an "evolutionary economics" took hold in the latter part of the twentieth century, influenced by the thinking of the heterodox economist Kenneth Boulding. In this instance the tack taken was fundamentally ecological, as economies were viewed as ecosystems, where, according to Boulding, "[e]ach economy is then seen as a segment of the larger evolutionary process of the universe in space and time . . . ." (Boulding 1991: 9). At this stage it is apparent that a more general idea of evolution, with only a tangential connection to evolutionary biology, informed this infusing of economics into the totality of the environment.

Overall, the notion of "cumulative change" constituted another possible point of contact between economics and evolutionary biology. For both these economists and evolutionary biologists the concerns were similar: how might one explain how changes, whether economic or biological, take place over time and then how might one demonstrate, establish, or posit the connectedness of such changes through a process of adaptation? For economics this reintroduced the problem of reintegrating history into economic analysis.

In general terms, the effort to align economics more closely with evolutionary biology has drawn strength from the notion that both fields of inquiry might be viewed as relying more upon concepts than laws, Marshall's ready and expansive use of the term "law" across disciplines notwithstanding. The physicist and science popularizer Alan Lightman has provided a neat formulation for distinguishing how biologists think about science as opposed to physicists: while the former emphasizes concepts and complexity, the latter is built upon laws and simplicity.

> Physicists make simplifications and idealizations and abstractions until the final problem is so simple that it can be solved by a mathematical law. . . . Biologists think differently. . . . [B]iology deals with living things, [a]nd life requires the interaction between elements in a system. Thus, biology usually deals with systems. (Lightman 2006: xii)

Lightman goes on to say:

> Where physics might ponder the electrical force between two electrons, biology would be concerned with how the electrical charges on both sides of a cell membrane regulate the passage of substances across the membrane and thus connect the cell to the rest of the organism. Roughly speaking, physics has laws, while biology has concepts. (Lightman 2006: xiii)

Hence, to place economics more directly in league with biology requires a rethinking, even a reconstruction, of economic laws by replacing them with economic concepts but nonetheless derived from the same experiences and constructions as those economic laws.

This process would entail treating "equilibrium" and "maximization" as central concepts in economics, rather than as mathematically-reached physical states, say as the result of constrained optimization, that reflected the resolution of economic forces obtained through the most efficient use of resources, as one might balance a lever. Equilibrium might then be regarded as a temporal state, of unstated duration, while maximization would be taken to be the most efficacious form of adaptation. In that form both equilibrium and maximization could readily be understood as corresponding to similar notions in evolutionary biology. One might, for example, see a parallel between the "punctuated equilibrium" of evolutionary biology (Mayr 2001), where a seeming stability in biological states holds for a considerable period of time and the apparently smooth and uninterrupted growth of an economy, only for both to suddenly undergo rapid and transforming changes, the bases for which lay in more gradual, barely perceived events.

22.4 The Parallel Provided by "Accident" and "Necessity"

There is another possible source of alignment of economics with evolutionary biology that can be found in their methodologies, which I believe is a more fundamental way of grasping the formative elements linking the two disciplines. This approach opens the way to establishing a more specific correspondence between the construction of the two disciplines rooted in the parallel between "accident" (or "chance") and "necessity" in evolutionary biology and the idiographic and the nomothetic in economics. Thus this correspondence would be formed from their natures as disciplines and requires a discerning of the evolutionary mechanism in both.

In *Chance and Necessity*, first published in 1971, Jacques Monod recounted the broad implications of his research into the nature of the regulation of gene expression and through it the basis for evolutionary change in biology. Monod identified an evolutionary interplay between accidents, individual and effectively random events, which might produce genetic variations in biological development and a more generally determined and adaptive means of selection that would lead to the permanent inclusion of those variations that would advance the species. Hence, the randomness of the particular was linked to an adaptive mechanism that acted in accordance with general principles.

Monod described the interplay of chance and necessity as the essence of the process of evolution:

> The initial elementary events which open the way to evolution in the intensely conservative systems called living beings are microscopic, fortuitous, and utterly without relation to whatever may be their effects upon teleonomic functioning. (Monod 1971: 118)

He then noted the "translation" of accident into necessity:

> But once incorporated in the DNA structure, the accident—essentially unpredictable because always singular—will be mechanically and faithfully replicated and translated . . . . Drawn out of the realm of pure chance, the accident enters into that of necessity, of the most implacable certainties. For natural selection operates at the macroscopic level, the level of organisms. (Monod 1971: 118)

In the last several decades of the twentieth century, nearly contemporaneous with the publication of Monod's work, a number of younger economists, like Brian Arthur, Paul David and Paul Krugman, were grappling with the significance of scale effects, loosely speaking the gains achieved by the concentration of economic activity and their impact upon the way that economies changed over time.

The stakes involved in these inquiries were high. If scale or other concentrating effects, sometimes known as agglomeration effects, proved to be essential to the actual changing configuration of economic activity by region or country, over time, then the dominant model of purely competitive economic arrangements, abstracted from all historical experience and contexts, faced a serious challenge. It would mean that economics in time, seen as occurring in an historical evolution, would turn out to be largely a story of imperfect competition, with less-than-fully competitive markets playing the larger role. I have treated their work as economists in rethinking this subject more extensively in a paper entitled, "The Arrow of Time in Economics," which is set for publication by the *European Journal of the History of Economic Thought* in September 2010.

Each of these economists drew upon the notion of a critical interplay between chance events and the general principles embodied in the laws governing economic activity, hence presenting at least implicitly a parallel with Monod's vision of an evolutionary interplay between chance and necessity in biology.

Indeed, one might see the appeal of this interplay of accident and necessity to both evolutionary biologists and historically-minded economists presaged in Windelband's characterization of an ambiguous status for biology, posed somewhere between the nomothetic and the idographic, in light of evolution, which introduced events and historical time into the study of a natural science (Windelband 1913: 50). Moreover, Windelband's depiction was effectively echoed a century later by Ernst Mayr:

> When a biologist tries to answer a question about a unique occurrence . . . he cannot rely on universal laws . . . .[Instead] he constructs a historical narrative. (Mayr 1997: 64)

It is a similar matter ultimately driving the inquiry of these economists: how to set an economics based upon a science of general rules in real or historical time.

Nonetheless, there are differences in the focus of the three economists. In works like *The Spatial Economy* Paul Krugman was concerned with the nature of the spatial or geographic specialization of economic activity, the resultant clustering of industry by region and patterns of international trade. Krugman identified the likely bifurcation of economic activity, say between more heavily concentrated manufacturing regions and largely agricultural areas, on the basis of tipping points reached because of locational differentials in resources, price levels, or wages that had existed at an earlier time, or because of a set of initiating conditions or events that reinforced over time the advantages accruing to a certain region. Krugman referred to these events as the "accidents of history." (Krugman 1991a, Krugman 1991b)

Arthur and David explored the connectedness of previous technological choices as events establishing the actual subsequent pathway of technological change and economic development. These pathways might inform a narrative of economic history, making sense of the past economic experience of the United States or various European economies, or be presented as a theoretical construct. For his part, Arthur emphasized the concept of technological "lock-in," where earlier events and decisions set a fairly rigid path for future development. David famously proffered the QWERTY thesis, whereby a combination of economic and historical circumstances led to the adoption, standard use and long-term dominance of a particular typewriter keyboard, even in the face of technologically superior versions (David 1985).

There are also some significant differences in conceptualization among these three economists, even as they all sought to create an economics that was both historically-minded and scientific in character or operation. Krugman stressed the importance of initial conditions as the accidents of history upon which economic forces, cast by analogy in physical terms as either centripetal or centrifugal, then act. This leads to a cumulative process, which may also be understood as an evolution that is ultimately a matter of distribution and configuration: to what extent does economic activity become concentrated or broadly dispersed? It would be a mistake to conceive of this approach as laying out a path, as Krugman has consciously constructed his spatial model on the basis of thresholds or tipping points.

By contrast, both David and Arthur do establish path dependence as central to their constructs of economic change occurring in time. For both a critical distinction exists between states of equilibrium and disequilibria and it is with regard to the latter that positive externalities, in David's terms, or agglomeration effects, in Arthur's terms, are associated with economic changes unfolding in time. That is because path dependence reflects the uni-directedness and lack of any necessary stability of changes produced by events at each stage or step.

Such path dependence was analogized with physical states as described and categorized in thermodynamics. David and Arthur applied the notion of ergodicity, derived from Boltzmann's hypothesis about an independent pathway of motion of gases for which past states were of no causal consequence, to two different sets of economic affairs. In the one that was ergodic, time—and certainly historical time—played no part; these produced a static equilibrium, consistent with neoclassical rules regarding the efficient allocation of resources. In the other, which was nonergodic, historical time held an essential part, as time itself was irreversible and dynamic processes were at work (David 1997, Arthur 1994a).

22.5 Arthur and Monod

In many ways Brian Arthur shows the closest—and deepest—affinity with Monod and this is not altogether coincidental. While Arthur's work can be understood in one sense as a blending of economics with engineering in his pursuit of a mathematically-based and scientifically-sound model, his thinking on the subject of economics has been very much shaped by his reading of disparate and diverse writers from different fields, making for a substantially more heterodox outlook about the workings of economics. In the late 1970s Arthur was struck in particular by new ideas emerging from biology and their possible application to economic theorizing. His introduction to these new ideas came first from Horace Judson Freeland's *The Eighth Day of Creation*, which was published in 1979. This led

Arthur to Jacques Monod's *Chance and Necessity*, which he himself described as formative to his thinking (Delorme and Hodgson 2005: 17ff).

Economists like Arthur favored comprehending economics as a subject of "complexity," an approach advanced significantly by a multi-disciplinary and cross-disciplinary institute, the Santa Fe Institute, among whose leading thinkers included the economist Kenneth Arrow and Arthur himself. They have tended therefore to look at least as much toward biology as physics as the source of a scientific model or set of correspondences. Thus, I think it is fair to say that Arthur was attracted to the pair of "accident" or "chance" with "necessity" at least in part because it came from the natural but not the physical sciences.

Drawing upon the dichotomous pairing of chance and necessity, Arthur clearly set as foundational the distinction between the 'accidental' as historical and the "necessary" as economic. Arthur called the accidents of history "small events," by which he was referring to choices among technologies. The necessity of economic laws refers to the rules governing economic agglomeration, like increasing returns to scale, where the magnification of inputs into production leads to an even greater magnification of output produced.

The process Arthur described was essentially evolutionary. In a seminal analysis of "competing technologies, increasing returns and lock-in," Arthur wrote:

> Under increasing returns however, static analysis is no longer enough. Multiple outcomes are possible and to understand how one outcome is selected we need to follow step by step the process by which small events cumulate to cause the system to gravitate toward that outcome rather than the others. (Arthur 1994b, p. 28)

From Arthur's perspective increasing returns have certain salient characteristics:

> To the list of already known increasing-returns properties like potential inefficiency and nonpredictability, a dynamic approach adds two new ones: inflexibility, in that allocations gradually rigidify, or lock-in, in structure; and nonergodicity, in that small events early on may decide the larger course of structural change. The dynamics thus take on an evolutionary flavor, with a 'founder effect' akin to that of genetics. (Arthur 1994b: 28)

There are two fundamentally complementary options: one in which the static analysis of equilibria holds; the other in which dynamic processes, shaped by agglomeration effects or other concentrating factors, like increasing returns, hold sway. Here "accidental historical order of choice," "geographical attractiveness" (Arthur 1994c: 51) and economic rule combine to establish the direction, or what Arthur calls the "course of structural change," owing to technology, which is seen as the economic counterpart of speciation.

Arthur acknowledges that the dichotomy presented by static situations, which are set in isolation and dynamic processes, which evolve over time and – critically, from the perspective of economic theorizing – in time, that is, within the framework of historical experience and events, leads to two quite different conclusions:

> Our analysis of the validity of the historical-dependence viewpoint does not imply that the conventional viewpoint, the unique-equilibrium state one, is wrong. The two Weltanschauungen are complementary. The validity of each depends on the degree to which agglomeration economies are present or absent. (Arthur 1994c" 28)

22.6 The Flaw in the Analogy

But does the correspondence with evolutionary biology, as sought by Arthur, actually work? Monod's necessity is that of a certain purposefulness of evolution itself: accidents that matter are those that advance the species, while those that do not remain inconsequential. Monod saw the mechanism of natural selection as progressive and purposeful by dint of its teleology, embodied in the "teleonomic apparatus." According to Monod "[i]t is the teleonomic apparatus. . . that lays down the essential *initial conditions* for the admission, temporary or permanent, or rejection of the chance-bred innovative attempt" (Monod 1971: 119–120).

Further, "[i]t is teleonomic performance, the aggregate expression of the properties of the network of constructive and regulatory interactions, that is judged by selection . . . " (Monod 1971:120).

Thus, in the language of evolutionary biology advanced by Jacques Monod and his colleague Francois Jacob in the 1950s and 1960s, it was the "teleonomic apparatus" that served as the mechanism linking the accidental event to the necessary and general rule. From a more contemporary perspective one might think of it in terms of establishing the "fitness" of any possible variation. As Jacob noted in his own popular account of the nature and implications of evolutionary biology, *The Logic of Life,* this introduces a teleology into evolutionary biology (Jacob 1973: 8–9). Without such a teleology it would be necessary to adopt an instrumental hypothesis and assume that the results themselves demonstrate the fitness of the variations that were incorporated into the species.

Is there, then, a similar sorting out in economics, whether conceived of as a "teleonomic apparatus" or not, whereby historical, that is, "small-event" accidents, will either initiate a new economic pathway, or not? One might think of the possibility of an overarching economic efficiency as a grand rule, operating thereby as an evolutionary principle within economics itself. Yet economic rules do not express an inner, driving logic of their own; even when touted as universal, that is, without any limitation of time or place, they are applied to the situation at hand, typically on an *ad hoc* basis.

Elias Khalil, who has taken a strong interest in the philosophical underpinnings of the idea of economic complexity, has argued that the "[economic process] is about the production of goods by purposeful activity" (Khalil 1990: 164); in effect, contending that a teleology along the lines suggested by Monod does obtain in economics. But Khalil then goes on to limit the scope of such a teleology by distinguishing the purposefulness of economic organization from economic structures, like the business cycle. These latter "are unintended results of acting efficiently' as 'response[s] to or. . .anticipation[s] of " (Khalil 1994: 191) whatever such productive activity, or even the possibilities of such activities, might offer. Moreover, the persistence of the business cycle itself, with its turns toward vicious circles of decline through panics and recessions, as well as toward virtuous circles of growth and prosperity, is a signal of the inability of economies to proceed along an economic path in which the better outcome is always pursued and successfully attained.

What Arthur has done is graft Monod's dichotomy into the longstanding distinction between the nomological and the idiographic, substituting "necessity" for the former and "accident" for the latter; while at the same time seeking to establish the interrelation between the two. This grafting follows from a flawed correspondence, as it lacks the critical element, akin to the "teleonomic apparatus," which would bridge the idiographic and the nomological by translating one into the other. Nonetheless, Arthur's central insights, like those of David and Krugman, are not insignificant, advancing the notion that economics must be understood as moving forward in time and emphasizing the importance of geographic specialization and imperfect competition.

This borrowing from evolutionary biology, based upon a methodological parallelism, should be distinguished from simply relying upon the power of the idea of complexity. In general, the advocates for treating economics as a subject of complexity have met with mixed results in grafting this notion from evolutionary biology and this includes many of those who have participated with Arthur in workshops at the Santa Fe Institute. Writing in *The Quark and the Jaguar*, first published in 1994, the physicist Murray Gell-Mann, who was also one of the collaborators in the Santa Fe Institute, offered up at best a tentative assessment of its success (Gell-Mann 1994: 320–324). Gell-Mann's description of the scope of the economics initiative at the institute is telling:

> [A] number of scholars. . .have directed their efforts toward studying economies as evolving complex adaptive systems composed of adaptive economic agents endowed only with bounded rationality, possessing imperfect information, and acting on the basis of chance as well as perceived self-interest. (Gell-Mann 1994: 322)

The framework Gell-Mann sets forth is highly abstracted: if it is intended to allow for the depiction of, or an explanation for, economies evolving, it appears to be divorced from any historical experience, or history itself. It would produce a paradox of a new sort in economics: a nomothetic system that describes evolutionary processes out of historical time. Nor does it address the central concerns raised in the work of Krugman, David, or Arthur in trying to mesh economic rules with historical experience.

Moreover, as a final note, oftentimes it has been matters of finance, especially in the estimation of stock options or the workings of financial markets that have been deemed most suitable for treatment as subjects of complexity. Here the notion of a random walk is often invoked. In fact, there is a long history of such exploration, dating back more than a century, to the use of a model of dissipative motion like Brownian motion to track the movement of stock prices, preceding even the pathbreaking examination of Brownian motion in physics at the beginning of the twentieth century.

But, as the reference to the business cycle, noted above, shows, this approach has not extended to broader models capturing the pertinent elements in the workings of the economy as a whole. It thus leaves open to further question the significance and consequences of the fault line in economics between microeconomics, with its modeling of the behavior of individual economic agents through a set of basic postulates and macroeconomics, where abstract models about the overall state of the economy share the stage with statistical characterizations and economic history. This presents yet another gap in the analogy drawn between economics and evolutionary biology, for Monod saw the link between chance and necessity as also bridging a parallel pairing in biology of the microscopic and the macroscopic. In the world of economics, though, notwithstanding all the efforts undertaken to establish the microfoundations of macroeconomics, no neat or seamless interweaving of the two has been effected.

References

Anderson PW, Arrow KJ, Pines D (eds) (1988) The economy as an evolving complex system. Addison-Wesley, Redwood City, CA

Arthur WB (1994a) Increasing returns and path dependence in the economy. The University of Michigan Press, Ann Arbor

Arthur WB (1994b) Competing technologies, increasing returns and lock-in by historical events. In: Arthur WB, Increasing returns and path dependence in the economy. The University of Michigan Press, Ann Arbor

Arthur WB (1994c) Industry location patterns and the importance of history. In: Arthur WB, Increasing returns and path dependence in the economy. The University of Michigan Press, Ann Arbor, pp. 49–67

Boulding K (1991) What is evolutionary economics? J Evol Econom 1(1):9–17

Darwin C (2005) [1958] The autobiography of Charles Darwin, 1809–1882, In: Nora Barlow (ed). W.W. Norton, London

David P (1985) Clio and the economics of QWERTY. Am Econom Rev 75(2):332–337

David P (1997) Path dependence and the quest for historical economics. Discussion Papers in Economic and Social History, No. 20. University of Oxford, Oxford

Delorme R, Hodgson Gey (2005) Complexity and the economy: an interview with W. Brian Arthur. In: John F, Magali O (eds), Complexity and the Economy. Edward Elgar, Cheltenham, UK, pp. 17–32

Fujita M, Krugman P, Venables AJ (1999) The spatial economy. MIT Press, Cambridge, MA

Gell-Mann M (1994) The Quark and the Jaguar. Holt Paperbacks, New York

Jacob F (1973) The logic of life, trans. by Betty E. Spillman. Pantheon Books, New York

Khalil E (1990) Entropy law and exhaustion of natural resources: is Nicholas Georgescu-Roegen's Paradigm Defensible? Ecol Econom 2(2): 163–178

Khalil E (1994) Entropy and economics. In: Geoffrey MH, Warren JS, Marc R (eds) The elgar companion to institutional and evolutionary economics A-K. Edward Elgar Publishing, Cheltenham, UK, pp. 186–193

Krugman P (1991a) Increasing returns and economic geography. J Pol Econ 99(3): 483–499

Krugman P (1991b) Geography and trade. MIT Press, Cambridge, MA

Lightman A (2006) The discoveries: great breakthroughs in 20th-century science. Vintage Books, New York

Machlup F (1978a) The inferiority complex of the social sciences. In: Methodology of economics and other social sciences. Academic Press, New York, pp. 333–344

Machlup F (1978b) Are the social sciences really inferior?. In: Methodology of economics and other social sciences. Academic Press, New York, pp. 345–367

Marshall A (1898) [1891] Principles of economics, 4th edn. Macmillan, London

Mayr E (1997) This is biology: the science of the living world. Harvard University Press, Cambridge, MA

Mayr E (2001) What evolution is. Basic Books, New York

Menger C (1963) [1882] Problems of economics and sociology, trans. by Francis J. Nock. University of Illinois Press, Urbana, IL

Monod J (1971) Chance and necessity, 1st American ed.; trans. by Austryn Wainhouse. Knopf, New York

Rickert H (1921) Kulturwissenschaft und Naturwissenschaft, 4th ed, J.C.B. Mohr: Tubingen

Turk M (2006) The fault line of axiomatization: Walras' linkage of physics with economics. Eur J History Econom Thought 13(2):195–212

Veblen T (1919) Why is economics not an evolutionary science? In: The place of science in modern civilization and other essays. B.W. Huebsch, New York

Windelband W (1913) The principles of logic. In: Arnold Ruge (ed.) The encyclopedia of philosophy, vol. 1: Logic, trans. by B. Ethel Meyer. Macmillan, London

Index